Quality and Reliability Engineering
Recent Trends and Future Directions

Quality and Reliability Engineering
Recent Trends and Future Directions

EDITORS

Boby John and U.H. Acharya
SQC & OR Unit
Indian Statistical Institute, Bangalore

Ashis K. Chakraborty
SQC & OR Division
Indian Statistical Institute, Kolkata

ALLIED PUBLISHERS PVT. LTD.

New Delhi • Mumbai • Kolkata • Lucknow • Chennai
Nagpur • Bangalore • Hyderabad • Ahmedabad

ALLIED PUBLISHERS PRIVATE LIMITED

1/13-14 Asaf Ali Road, **New Delhi**–110002
Ph.: 011-23239001 • E-mail: delhi.books@alliedpublishers.com

47/9 Prag Narain Road, Near Kalyan Bhawan, **Lucknow**–226001
Ph.: 0522-2209942 • E-mail: lko.books@alliedpublishers.com

17 Chittaranjan Avenue, **Kolkata**–700072
Ph.: 033-22129618 • E-mail: cal.books@alliedpublishers.com

15 J.N. Heredia Marg, Ballard Estate, **Mumbai**–400001
Ph.: 022-42126969 • E-mail: mumbai.books@alliedpublishers.com

60 Shiv Sunder Apartments (Ground Floor), Central Bazar Road,
Bajaj Nagar, **Nagpur**–440010
Ph.: 0712-2234210 • E-mail: ngp.books@alliedpublishers.com

F-1 Sun House (First Floor), C.G. Road, Navrangpura,
Ellisbridge P.O., **Ahmedabad**–380006
Ph.: 079-26465916 • E-mail: ahmbd.books@alliedpublishers.com

751 Anna Salai, **Chennai**–600002
Ph.: 044-28523938 • E-mail: chennai.books@alliedpublishers.com

5th Main Road, Gandhinagar, **Bangalore**–560009
Ph.: 080-22262081 • E-mail: bngl.books@alliedpublishers.com

3-2-844/6 & 7 Kachiguda Station Road, **Hyderabad**–500027
Ph.: 040-24619079 • E-mail: hyd.books@alliedpublishers.com

Website: www.alliedpublishers.com

© 2013, Quality and Reliability Engineering: Recent Trends and Future Directions

ISBN: 978-81-8424-831-9

Acknowledgements

William Q. Meeker
Iowa State University, USA

T.S. Arthanari
University of Auckland, New Zealand

Raghu Kacker
National Institute of Standards and Technology, USA

Jeff (Yu) Lei
The University of Texas at Arlington, Texas, USA

Lesley Walls
University of Strathclyde, Scotland

Peihua Qiu
University of Minnesota, USA

Ram Ramanathan
University of Bedfordshire, UK

Shigeru Yamada
Tottori University, Japan

Gopinath Chattopadhyay
CQUniversity, Australia

Asokan Mulayath Variyath
Memorial University of Newfoundland, Canada

A.K. Chaudhuri
ADAAP Process Solutions, Bangalore, India

B.C. Sheshadri
Infosys, Bangalore, India

To the Loving Memory

of

Varghese John Paramban

Preface

The International Conference on Quality and Reliability Engineering was organized during 20–22 December, 2011 at Hotel Atria, Bangalore. This conference was organized to provide a common platform between universities, research institutions and industry for sharing knowledge and experiences on recent trends and future directions of quality, reliability and related fields. The objectives of the conference were to facilitate sharing of:

- Research based knowledge related to quality control, quality management, process performance modeling and reliability engineering among academicians and practitioners.
- Case studies and novel applications of statistical tools and techniques in quality and reliability engineering.

To disseminate the state of the art knowledge on quality and reliability engineering, we had invited twelve world class scholars to share their ideas on recent trends and future directions in the aforementioned disciplines. Also included in the conference were ninety two papers selected from over one hundred, submitted from India, Asia-Pacific, United States, Middle East, Europe and Africa in response to the call of papers. To preserve the information and ideas presented in this platform and make it accessible to knowledge seekers beyond the conference delegates, it was decided to publish this book with selected papers from the conference. All the contributed papers were subjected to double fold blind review process and based on the recommendations and comments of the reviewers, thirty seven papers are selected for publication.

We hope that the readers will find this book useful. Apart from the theoretical chapters, it also includes chapters on applications and case studies in quality and reliability engineering and six sigma. This book demonstrates the great returns of providing a common platform between researchers from academia and practitioners from industry!!

The engineering students can use this book as a supplementary text book for quality and reliability engineering courses. The researchers can use this book to update their knowledge on the recent developments in reliability and quality engineering. The industrial professionals can gain insight on solving industrial problems using six sigma, quality and reliability engineering methodologies. In addition, the readers can also learn from the many successful case studies presented in the book.

The conference as well as the publication of this book is supported by Indian Statistical Institute, especially we are grateful to Sankar K. Pal (Director, Indian Statistical Institute), K.K. Chowdhury (Head, SQC and OR Division, Indian Statistical Institute) and U.H. Acharya (Head, SQC and OR unit of Indian Statistical Institute, Bangalore) for endorsing and supporting the conference idea when it was proposed in 2009. In addition to Indian

Statistical Institute, we are indebted to Council of Scientifc and Industrial Research, Hinduja Global Solutions, Minitab and QualMind Global for providing financial support for organizing the conference and publication of this book.

Many have contributed their time and effort for the success of this endeavour. In particular, Ashis K. Chakraborty and I. Islam of Indian Statistical Institute deserve special mention for their participation in every detail from the beginning.

We are grateful to the following experts for their comprehensive review and recommendations for selection of papers: U.H. Acharya, Ashis K. Chakraborty, K.K. Chowdhury, I. Islam and E.V. Gijo. On behave of the editorial board, we thank all the contributing authors, whose works published in this book will widen the frontiers of knowledge of anyone who reads this book.

We are indebted to S. Manjula and team from Allied Publishers for carefully formatting, preparing the camera ready copy of the papers and publication of this book.

From the beginning, it is the team work that made this endeavour a great success. We extend our great appreciation to the current team leaders: Bimal Roy (Director, Indian Statistical institute), N.S. Narasimha Sastri (Head, Indian Statistical Institute, Bangalore). Amitava Banerjee (Head, SQC and OR Division, Indian Statistical Institute) and Somnath Ray (Head, SQC and OR Unit, Indian Statistical Institute, Bangalore) and the team members especially our colleagues at Indian Statistical Institute, Bangalore. We are greatly indebted to those who worked behind the scenes in particular K.S. Prasanna Kumari, J. Devadas, A. Krishnaiah, and Gnanananda.

Editorial Team

Contents

Acknowledgements .... *v*

Preface .... *vii*

1. Quality and Reliability Engineering: An Introduction 1
   Boby John and U.H. Acharya

PART–I: Quality Engineering

2. Fuzzy Inference Technique for Assessment of Software Quality Index 5
   for Mission Critical Software
   Lalit Kumar and B. Valsa

3. Integrated Balanced Risk Assessment in Manufacturing Planning 15
   Gerhard Schoepf

4. Role of Binary Logit Regression in Survey Analysis 31
   Neena Mital and P.R. Lakshmikanthan

5. Fuzzy Multinomial Cumulative Sum and Exponentially Weighted 52
   Moving Average Control Charts
   Hassen Taleb and Ali Achouri

6. Application of Taguchi's Beta Correction (Partial Correction) 62
   Technique for Process Control
   Vidya Sagar Akkinapalli and U.H. Acharya

7. Statistical Economic Design of $\bar{X}$ Control Chart with Pareto In-Control Times 70
   Neelufur

8. A Hybrid Software Quality Forecasting Model 80
   Arun Duraibalan

9. A VIKOR Multiple—Attribute Decision Method for Improving 90
   Supply Chain Leanness
   P. Parthiban, H. Abdul Zubar, Richa Tiwari and Surapaneni Bhavana

10. Tuning Sprint Goal Achievement in Different Project Scenarios 97
    Vijay Wade

11. A Bouquet of Unusual Statistical Applications for Consumer Insights 105
    Ranjan Samanta, Gireesh Sabari and Jones Joseph

12. A Study on TPM Implementation: Factors and Performance 117
    P.K. Suresh and Mary Joseph

13. ISO 9001: A Platform to Bring Cultural Changes within 130
    the Organization: A Case Study
    Sudipta Gouri

14. Re-Engineering Business Processes— A Winning Approach to Transform 142
    Businesses Using the BVEM
    Senthil Anantharaman

15. Role of Quality in Meeting Global Manufacturing Challenges 151
    K. Balasubramanyam

16. Service Quality of Public Sector Banks in India 161
    Sitaram Vikram Sujir

17. Vendor-Buyer Model Considering Imperfect Items, Trade Credit 174
    and Volume Agility under Inflation
    V. Gupta and S.R. Singh

18. Process Monitoring through Application of Principal Component 187
    Analysis in a Process Industry
    S.M. Subhani

19. Some Properties of CG (u, v) 203
    Moutushi Chatterjee and Ashis Kumar Chakraborty

PART–II: Reliability Engineering

20. Goodness-of-Fit Comparisons of Change-Point Models 213
    for Software Reliability Assessment
    Shinji Inoue and Shigeru Yamada

21. Impact of Complexity of Object-Oriented Design on Software Reliability 222
    Shrihari A. Hudli and Anand V. Hudli

22. Bayesian Accelerated Life Testing Under Competing Exponential 229
    Causes of Failure
    Soumya Roy and Chiranjit Mukhopadhyay

23. Reliability in Medical Devices: An Experience 246
    Saraswathi Deora

24. Stochastic Analysis of a Complex System with inspection 252
    in Different Weather Conditions
    Beena Nailwal and S.B. Singh

25. Safety Critical Software of Software Reliability Growth Model 288
    Considering Log-Logistic Testing-Effort and Imperfect Debugging
    S.P.V.N.D. Suneetha and O. NagaRaju

26. Determining Optimum Software Release Time with Euler Distribution 302
    as a Prior for the Number of Undiscovered Bugs
    Satadal Ghosh, Soumya Roy and Ashis K. Chakraborty

27. Physics of Failure Based Reliability Prediction Method for a Two-Wheeler 313
    C. Sasun and Sushant Mohan Dewal

28. An Extension of General Class of Change Point and Change Curve Modeling 328
    for Life Time Data Considering More than One Change
    Point Present in the Data
    Ishita Basak and Ashis K. Chakraborty

29. Performance Analysis of Industrial System under Corrective and Preventive 342
    Maintenance
    *Manwinder Kaur, Arvind Kumar Lal, Satvinder Singh Bhatia and Akepati Sivarami
    Reddy*

PART–III: Six Sigma

30. Six Sigma, Quality Function Deployment and TRIZ—An Amalgamation 355
    M. Shanmugaraja, M. Nataraj and N. Gunasekaran
31. The Use of Design of Experiments in Design for Six Sigma 366
    for Improving Cable Toughness
    Arup Ranjan Mukhopadhyay
32. Pend Volumes Reduction: A Case Study 377
    Joyson Peter Coelho
33. Application of DFSS Framework in Adaptive Radiation Therapy 388
    Planning Proof of Concept
    Prashant Kumar and Yogish Mallya
34. Error Reduction: A Case Study 394
    A.M. Romesh Kumar Corera
35. Improved Up Selling Using FMEA 403
    Gurupreet Singh Khanuja
36. Reduction of Rework Percentage for Traditional Claims Process: A Case Study 417
    Shereena Mody
37. Application of DMAIC Approach for Improving the Accuracy 430
    of Output from the Current Level of 89% to 98%
    Moses Davala

Author Index 443

Quality and Reliability Engineering: An Introduction

Boby John[1] and U.H. Acharya[2]

Indian Statistical Institute, Bangalore
E-mail: [1]boby@isibang.ac.in; [2]uha@isibang.ac.in

1. INTRODUCTION

In globally competitive business environment, quality and reliability are the two key drivers for maximizing market penetration and customer loyalty. The application of quality and reliability engineering has enabled industries to produce high quality products and deliver high quality services at low cost. The development in quality and reliability engineering disciplines are phenomenal in the recent past and will continue to inspire the imaginations of researchers as well as practitioners.

This book consists of a collection of papers on recent trends and future directions in the field of Quality and Reliability engineering authored by a group of academic researchers and industrial practitioners. The purpose of this book is to encourage sharing of knowledge between academia and industry. The book discusses the theoretical developments and applications of the quality and reliability engineering methodologies in industrial problem solving. It also features a handful of successful six sigma case studies from industry. The book is organized in three major parts as follows.

2. QUALITY ENGINEERING

Quality has become an important factor influencing the customer's decision on product selection. The effective implementation of quality improvement methodology has resulted in increased productivity and reduced cost. Since as quality improves, the cost of rework and rejections decreases and the first pass yield increases.

Today quality engineering is a key factor determining the organizational growth and market penetration across industries. Quality engineering is a set of operational, managerial and engineering activities that a company uses to ensure that the product meets the customer expectations (Montgomery, 1997). This can be achieved through minimizing the variation in the processes. Since variability can be measured only in statistical terms, the statistical methods play an important role in quality improvement. This session has ten chapters discussing the recent developments in the application of statistical methods. The methodologies discussed in these chapters are fuzzy inference for quality assessment, binary logistic regression for survey data analysis, CUSUM and EWMA control charts based on fuzzy theory, β correction technique, Pareto in control time based control charts, software quality estimation using support vector machines, multiple attribute decision making using lean enablers, an integrated inventory model for a volume agility manufacturing system for imperfect items, principal component analysis application in process industry and a study on multivariate process capability indices.

Eight of the ten chapters of quality engineering are dedicated to the discussions on various aspects of quality management. This include risk assessment in manufacturing planning, sprint goal achievement by agile scrum, conjoint analysis, study on TPM implementation, ISO 9001 platform for cultural changes, business transformations using business value enhancement methodology, role of quality in research and development, service quality of public sector banks in India.

3. RELIABILITY ENGINEERING

The recent developments in technology have resulted in inventing highly sophisticated and complex products and systems. Any failure of these systems can have significant effect on society as a whole. For example, the breakdown of a nuclear power plant or failure of air traffic control systems can have catastrophic effects (Kuo and Zuo, 2003). Today the customer expectation is not only the products and systems should be free from defects but also required to perform the intended function for a specified period of time (Birolini, 2004). The nine chapters in reliability engineering session discuss the various aspects of reliability engineering discipline including software reliability, Bayesian approach for accelerated life testing, reliability of medical devices, k-out-of-n-systems, automobile reliability, change point hazard rate models and maintainability.

4. SIX SIGMA

Today any book on quality will not be complete without a discussion on Six Sigma. Six sigma is a methodology to significantly improve customer satisfaction (Stamatis, 2002). It aims at reducing the defects and minimizing the variation in every aspect of the business. Today six sigma is the most popular quality initiative in the corporate world. In fact, six sigma has provided a common language which all people in industry can understand and speak (Harry and Schroder, 2000).

This book has eight chapters on six sigma. The focus of these chapters are on less known facets of six sigma like DFSS, TRIZ, case studies service and software industries.

5. CONCLUSION

In short, a person reading this book will not only get familiarized with the recent theoretical developments but also learn the practical difficulties encountered by practitioners while applying quality, reliability and six sigma methodologies to industrial problems and how they could be overcome.

REFERENCES

[1] Birolini, A. (2004), *Reliability Engineering: Theory and Practice*, 4th edition, Springer, Germany.
[2] Harry, M. and Schroder, R. (2000), *Six Sigma*, Currency, USA.
[3] Kuo, W. and Zuo, M.J. (2003), *Optimal Reliability Modeling*, John Wiley and Sons Inc., USA.
[4] Montgomery, D.C. (1997), *Introduction to Statistical Quality Control,* 3rd edition, John Wiley and Sons, Inc. USA.
[5] Stamatis, D.H. (2002), *Six Sigma and Beyond: Foundations of Excellent Performance*, Vol. 1, St. Lucie Press, USA.

PART–I
Quality Engineering

Fuzzy Inference Technique for Assessment of Software Quality Index for Mission Critical Software

Lalit Kumar[1] and B. Valsa[2]

Vikram Sarabhai Space Centre, Thiruvananthapuram
E-mail: [1]lalit_kumar@vssc.gov.in; [2]b_valsa@vssc.gov.in

ABSTRACT: The Flight Software residing in On-Board computers of Launch Vehicle performs the mission critical function of computation and directing the vehicle to the target based on its calculations. From time to time, a number of methods have evolved for predicting quality and reliability of software. Due to constraints on availability of data during the initial developmental phases of software, it is extremely difficult to use literature proposed models for the purpose of predicting the quality index of flight software. For using the available data, it needs to be categorized in the general context of development and then can be used, based on the Fuzzy inference logic. In this paper, Fuzzy Inference Technique is used for assessing the Quality Index of Flight software. Inference mechanism which is used here is mainly based on the experiences from previous missions and categorization of available data and its impact on Mission Critical software. Based on the above classification, a model has been developed for calculating the quality index of software in crisp values. This model has been validated with available data and results are found to be in accord. This method enables prediction of complete software quality index based on the information provided to it. This helps in avoiding the errors along with the gradual learning of system for predicting the result for future if used along with Neural Network.

Keywords: Fuzzy Inference, Quality Index, Mission Critical Software, Quality Assessment, Fuzzy Rules, Defuzzification.

1. INTRODUCTION

Software is a vital and integral part of aerospace systems for their successful operation. Realization of zero defect/high reliability software for Mission critical application is a challenging and complex task. Flight software residing in onboard computer, performs the critical task of guiding the launch vehicle to its required orbit and precisely injecting the satellite. This requires very high quality and reliable software. The target of building such software that works right for the first time and every time, is a challenging task for both designers and quality assurance engineers, as the final validation of software is in actual flight of vehicle. Any error in software results in mission failure, which is very expensive in terms of both money and re-work involved for its correction (Valsa *et al.*, 2005).

2. DEVELOPMENT ENVIRONMENT

Software development environment for satellite launch vehicle is highly dynamic, where the mission requirements are evolving, design parameters are very frequently changing and mission constraints are revised from time to time. These problems often accompany any project whose development spans a few years.

In flight-software different software components are defined to do different functions. These software developed requires interfacing with different hardware as well as software. Software components like Navigation, Guidance, Control, Sequencing, etc. are designed to perform specific set of tasks (Valsa *et al.,* 2005). These software components are developed by different design teams and are verified by respective quality assurance team and are finally integrated to realize complete launch vehicle software. In order to complete these activities effectively and correctly, Iterative Waterfall Software development life cycle model is adopted. As per the life cycle model, various plans are generated and quality assurance agencies ensure that all the steps are being followed to realize quality product. A number of versions emerge as the software undergoes verification and validation.

3. VERIFICATION AND VALIDATION CYCLE

Verification and Validation (V&V) of launch vehicle software is entirely different from those of conventional software and plays a very prominent role in the success of a mission. Unlike conventional software, V&V activities for mission critical software start from the very beginning of project. A number of documents are generated for streamlining the development process.

For mission critical software, major V&V activities carried out are—review of documents namely Functional Requirements, Software Requirements Specification and Software Design Document. Functional Requirements Document (FRD) is reviewed by the mission expert team including Quality Assurance (QA) expert followed by the review of Software Requirements Specification (SRS). In SRS review, it is ensured that all requirements in FRD is directly mapped to SRS and is correct and consistent. After rigorous review of SRS, it is translated into Software Design Document (SDD) where software level mapping to code is described. All of the above stated documents also undergo document assessment in order to ensure that all review recommendations are correctly implemented.

Code Inspection and testing of the evolved software is carried out after the completion of the above stated activities. Modified Fagan's Code Inspection method is adopted for the purpose of code inspection (Fagan, 1986). After code inspection is completed, the final version of code is subjected to white-box testing, black-box testing, profile testing and target code testing. The observations of all phases are brought out after the respective stage in the form of formal reports, which classify the categories of observation and are reviewed and resolved.

4. ERRORS

All the observations from the V&V phases of software are generally classified as errors.

- *SRS Review:* Output of this review is different types of errors in the SRS, which gives a insight into its quality. Here, we are able to verify that all the requirements are spelt out unambiguously, no functional requirements are missed, requirements mentioned are consistent throughout and functional requirements listed are not different from the actual requirements.

- *SDD Review:* this review is carried out in order to verify software requirements are mapped correctly into software design. Here, detailed design is verified and ensured that it is free from errors like design/requirement mismatch, traceability mismatch, document corrections and flow graph errors. All interfaces are also verified for its correctness.
- *Code Inspection:* brings out guideline violations like complexity, nesting level, fan-in, fan-out etc., increasing beyond recommended limits. Translation of software design to code is verified for logical and implementation correctness.
- *Testing:* during testing, different aspects of software are evaluated using different sets of input. It brings out interfacing errors, overflow/underflow conditions, precision difference error, implementation errors, infeasible paths, dead codes, etc.

5. SOFTWARE QUALITY ASSESSMENT

Based on the errors, an assessment about the quality of the software has to be made. While working with hard real time systems like satellite launch vehicle, all the above errors play a vital role in the quality of software. In order to get fault free and reliable software, the impact of errors found in different phases of the quality assurance of the software has to be mapped. These errors are categorized according to the phases at which they are detected.

If the quality of software can be assessed prior to its use in actual environment, it would enable management to assess the risk associated with using that software. Based on the risk factor a decision on the usage of the particular software can be made.

Numerous methods have evolved over a period of time and are suggested in literature. These methods mainly focus on statistics based software quality inference techniques and derive their output from errors detected. Since in the area of mission critical software development, especially in mission critical software domain very huge databases of errors do not exist, application of statistical techniques are not very effective. Again these kinds of software are not very frequently reused because of the changing mission requirements from project to project. Methods employing execution time or calendar time for the purpose are not very relevant. Fuzzy logic on the other hand tries to deduce the software quality index using non-numerical method. In this paper, a different approach is proposed for predicting the quality index for mission critical software.

6. FUZZY INFERENCE TECHNIQUE

This technique is basically derived from the principle of human brain anticipation method. Fuzzy Inference is a technique which accepts noisy data over a range instead of crisp value as input. Based on these input values, it applies de-fuzzyfication rules over the data and infers the output in the form of crisp and absolute output.

This technique is especially useful in deriving the quality index of mission critical software, as inputs which are expected from developmental cycle may not be having very accurate and crisp values. Fuzzy Inference technique is very efficient and accurate in terms of usage of input data from sparsely related process and derives the required output. Because of this property, the outputs derived here do not deviate grossly from the expected output.

7. ADVANTAGES

Fuzzy Inference technique comes with a bucket of advantages:

- It circumvents the need for rigorous mathematical modeling.
- Unlike reasoning based on classical logic, fuzzy reasoning aims at the modeling of reasoning scheme based on uncertain or imprecise information.
- It uses human expertise developed over years, in order to frame its rules.
- Its effectiveness of inferring results out of certain set of input is not bounded by very stringent mathematical constraints.
- It can a use set of rules from many heterogeneous problem domains.
- Fuzzy rules, once formulated can be used again and again.
- Fuzzy rules can be easily modified to suit different domain and set of problems.

8. MODEL USED

Mamdani rule base is used to model the system as it is intuitive, having widespread acceptance as well as very well suited to human input.

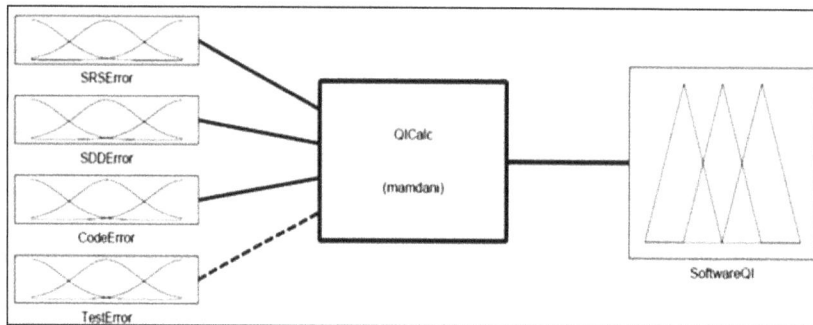

Fig. 1: Model

Mamdani rule base is a crisp model of a system, i.e. it takes crisp inputs and gives crisp outputs. It works with the use of user-defined fuzzy rules on user-defined fuzzy variables. The idea behind using a Mamdani rule base to model crisp system behavior is that the rules for many systems can be easily described by humans in terms of fuzzy variables. Thus we can model a complex non-linear system, with common-sense rules on fuzzy variables.

Designing a Mamdani rule base requires three steps:

(a) Determining appropriate fuzzy set over the input domain and output range.
(b) Determining a set of rules between the fuzzy inputs and fuzzy outputs that model system behavior.
(c) Create a frame work that maps crisp input to crisp output with help of point 'a' and 'b'.

9. MAMDANI RULE BASE

Operation of the Mamdani rule base can be broken down into four parts (Davis, *http://plaza.ufl.edu/badavis/CIS6930_Project1.doc*) and (MATLAB documentation-fuzzy logic toolbox):

(a) Mapping each of the crisp inputs into a fuzzy variable (fuzzification).
(b) Determining the output of each rule given its fuzzy attendance.
(c) Determining the aggregate output(s) of all the fuzzy rules.
(d) Mapping a fuzzy output(s) to crisp output(s) (defuzzification).

10. FUZZIFICATION

Since the Mamdani rule base models a crisp system, it has crisp inputs and outputs. The rules are given in terms of fuzzy variables. The membership of each fuzzy input variable is evaluated for the given crisp input based on experience, and the resulting value is used in evaluating the rules. In the process of fuzzification of membership function, their range is classified as "small", "moderate", "large" and "very large".

Table 1: Error Range for Fuzzifization of Membership Function

	SRS	*SDD*	*Code*	*Testing*
Small	0–5	0–9	0–12	0–15
Moderate	3–10	7–20	9–25	10–24
Large	7–20	16–30	17–40	20–42
Very Large	16 & above	25 & above	36 & above	38 & above

- *Functional and Software Requirements Phase (SRS):* Figure 2 provides the fuzzified membership function for SRS phase using Gaussian distribution.

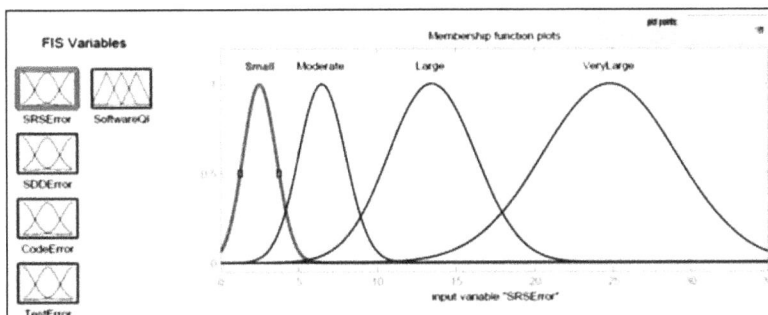

Fig. 2: SRS Membership Function - Gaussian

- *Software Design Phase (SDD):* Fuzzified membership function for SDD phase is provided in Figure 3, using Gaussian distribution.

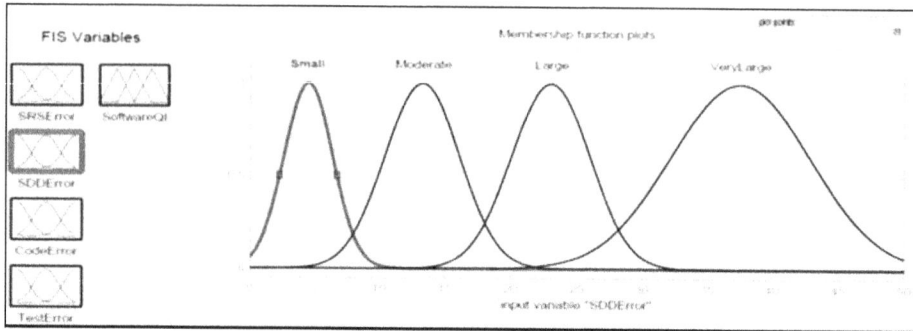

Fig. 3: SDD Membership Function-Gaussian

- *Code Inspection Phase:* Figure 4 gives fuzzification of membership function for code inspection phase using Gaussian distribution.

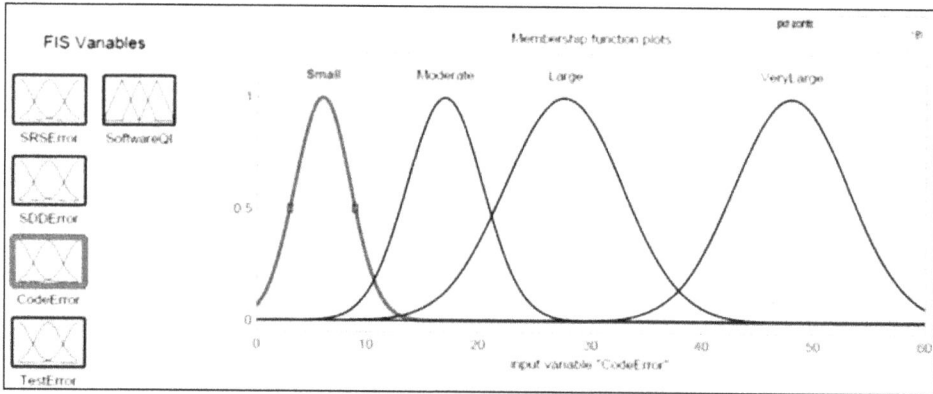

Fig. 4: Code Membership Function – Gaussian

- *Testing Phase:* As stated in code inspection phase, Figure 5 gives the fuzzified membership function.

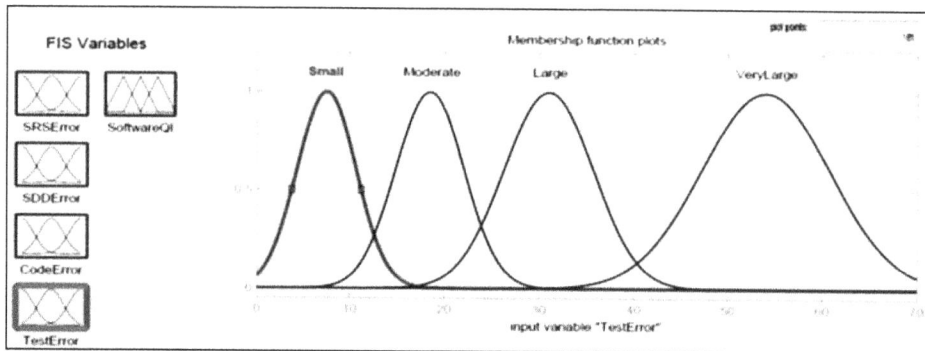

Fig. 5: Testing Membership Function-Gaussian

11. EVALUATING THE RULES

Using the membership values of determined during fuzzification, the rules are evaluated according to compositional rule of inference. The result is an output fuzzy set that is some clipped version on the user-specified output fuzzy set. Range of this output fuzzy set depends on the combination of membership functions.

An alternative to evaluating the fuzzy rules could be assignment of weight to the fuzzy membership functions. Once weight is added, different combination of fuzzy rule shall yield different sets of output. Along with weight assigned to membership function, degree of fuzzification of membership function also plays important role in determination of output. If degree of fuzzification is high, its effect on the output will be less and vice versa.

12. AGGREGATING THE RULES

After evaluating the rules, we have fuzzy output defined for each of the rules in the rule base. We then need to combine these fuzzy outputs into a single fuzzy output. Mamdani defines that the output of the rule base should be the maximum of the outputs of each rule (Davis, *http://plaza.ufl.edu/badavis/CIS6930_Project1.doc*).

Another thing to consider is that some of the rules might be more important than the other rules in determining the system behaviour. But here while modeling this system we have not assigned any type of weight to any rule.

13. FUZZY RULES

Fuzzy rules (about 120) have been given for the fuzzification of input membership function. Input membership functions are combined together using "AND" operator.

Some typical rules are tabulated Table 2.

Table 2: Fuzzy Rules

Number of Errors				S/w QI
SRS	SDD	Code	Test	
Small	Small	Small	Small	Excellent
Small	Small	Small	Mod*	Excellent
Small	Small	Mod*	Large	Good
Small	Small	Mod*	Mod*	Excellent
Small	Small	Large	Large	Good
Small	Mod*	Small	Mod*	Good
Small	Mod*	Small	VL**	Fair
Small	Mod*	Mod*	Large	Fair
Mod*	Small	Small	VL**	Fair
Small	Large	Large	VL**	Poor
VL**	VL**	Large	Small	Poor

*Mod is Moderate, **VL is Very Large.

14. DE-FUZZIFICATION

Fuzzy logic is a rule-based system written in the form of horn clauses (i.e., if-then rules). These rules are stored in the knowledge base of the system. The input to the fuzzy system is a scalar value that is fuzzified. The set of rules is applied to the fuzzified input. The output of each rule is fuzzy. These fuzzy outputs need to be converted into a scalar output quantity so that the nature of the action to be performed can be determined by the system. The process of converting the fuzzy output is called defuzzification. Here for the purpose of defuzzification, "centroid" method is used. This method is also known as center of gravity or center of area defuzzification. This technique was developed by Sugeno in 1985. This is the most commonly used technique and is very accurate. The centroid defuzzification technique (Pant and Holbert, 2004) can be expressed as:

$$x^* = \{\textstyle\int u_i\,(x)\,x\,dx\}/\{\textstyle\int u_i\,(x)\,dx\,\}$$

where x^* is the defuzzified output, $u_i\,(x)$ is the aggregated membership function and x is the output variable.

In Figure 6 de-fuzzification of output is given using Gaussian distributions.

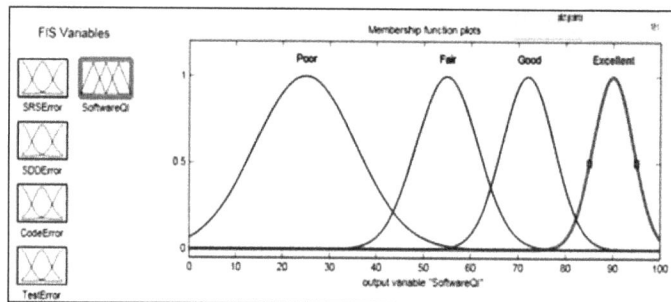

Fig. 6: S/W QI Membership Function-Gaussian

15. DEFUZZIFIED QUALITY INDEX

Following are the defuzzified output of software quality index. It is given in the form of 3D graphs as well as the some typical crisp values are also tabulated in order to get the better understanding.

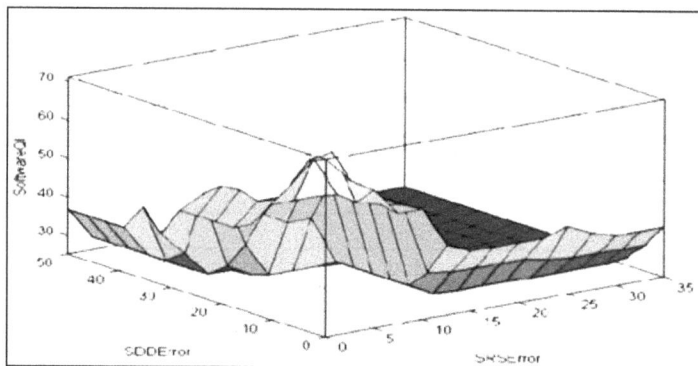

Fig. 7: Graph of Defuzzified Output (SRS & SDD Vs S/W QI)

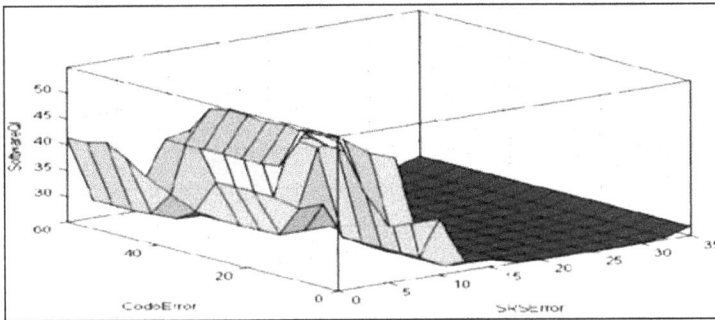

Fig. 8: Graph of Defuzzified Output (SRS and Code Vs S/W QI)

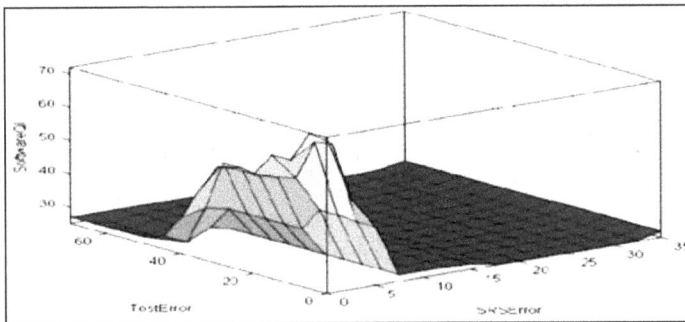

Fig. 9: Graph of Defuzzified Output (SRS and Test Vs S/W QI)

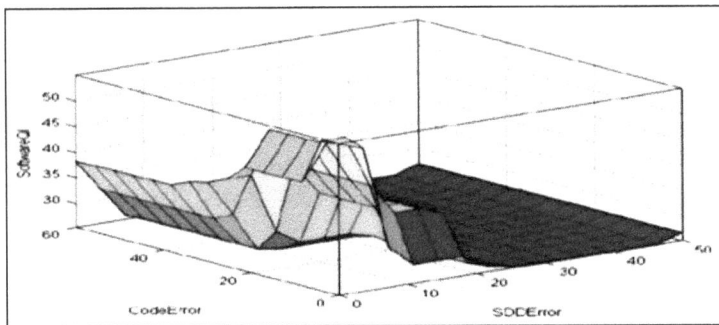

Fig. 10: Graph of Defuzzified Output (SDD and Code Vs S/W QI)

Table 2 provides the defuzzified crisp output, when input is fuzzified using Gaussian distribution.

Based on the crisp software quality index obtained, software can be classified under following category:

(a) 0.80 < Quality Index < 1.0; Quality is "Excellent". Using the software has no risk.

(b) 0.60 < Quality Index < 0.85; Quality is "Good". Using the software has low risk.

(c) 0.40 < Quality Index < 0.70; Quality is "Fair". Using the software has moderate risk.

(d) 0.00 < Quality Index < 0.50; Quality is "Poor". Using the software has high risk.

Table 3: Crisp Output Using Gaussian Distribution

Errors in SRS	Errors in SDD	Errors in Code	Errors in Test	S/W QI
0	1	1	1	89.4
2	2	4	7	89.3
4	6	9	14	80.9
4	7	14	17	73.2
7	7	12	22	68.4
9	11	18	25	61.9
12	11	10	10	65.5
12	11	22	20	53.3
15	9	22	10	55.6
15	15	25	30	34.7

16. CONCLUSION

This model was developed to assess whether fuzzy inference technique can be used for assessing quality of mission critical software. Results obtained from the model are encouraging. It matches in terms of performance of the software observed in actual environment. Future studies can be done by categorizing the errors in each phase and giving weight depending on the impact of the error on the software. Currently all errors are assumed to have same impact. Also if a Neural Network is integrated, a neuro-fuzzy model which can learn and predict errors as well as give the quality index can be developed.

REFERENCES

[1] Davis, B., "Evaluation of Implication on Fuzzy Sets in Describing a Given System", University of Florida http://plaza.ufl.edu/badavis/CIS6930_Project1.doc

[2] Fagan, M.E., "Advances in Software Inspections", *IEEE Transactions on Software Engineering*, V. SE-12, N. 7, July 1986, pp. 744–751.

[3] Fuzzy control programming. Technical report, International Electrotechnical Commission, 1997.

[4] MATLAB documentation, Fuzzy Logic Toolbox, MathWorks.

[5] Pant, S.N. and Holbert, K.E., "Fuzzy Logic in Decision Making and Signal Processing", Arizona State University, 2004. http://enpub.fulton.asu.edu/PowerZone/ FuzzyLogic/chapter%206/frame6.htm

[6] Valsa, B., Vikraman Nair, R. and Kaimal, M.R., A Neural Network Model for Assessing Software Quality for Mission Critical Applications, *International Aeronautical Conference 2005 (IAC-03-U.2.a.05).*

Integrated Balanced Risk Assessment in Manufacturing Planning

Gerhard Schoepf

Hana Pačaiová, Juraj Sinay, TU Kosice
E-mail: gerhard.schoepf@gmx.de

ABSTRACT: Risk assessment tools *i.e.* Failure Mode and Effects Analysis can be applied for all phases of a project starting from the early beginnings until end of a production and dismantling/recycling phase. Integrated means that risk assessment should be integrated in all phases of a project. Firstly, a rough FMEA should be carried out to support feasibility studies and decision making in early concept phases. Furthermore, a project risk evaluation should be carried out for complementing the feasibility and profitability study. Based on a risk factor determined for the project and results of the concept FMEA, suitable actions for project planning and control can be derived to reduce risks as low as reasonably practicable.

In the following planning and realization phase, a "classical" Process FMEA can be applied in an improved manner to identify failure modes and causes in detail. In order to carry out a successful project, all of the mentioned tools *i.e.* Concept FMEA, Project Risk Evaluation and Process FMEA should be combined and risks should be reduced in a balanced way.

Keywords: Risk Assessment, Integrated, Feasibility, Balanced.

1. INTRODUCTION

Every company should strive to reduce quality costs to a minimum. Quality costs include "failure costs" (e.g. rework, scrap and external costs), preventive costs (e.g. FMEA) and costs for testing (for equipment and personnel). It is important to apply established management and quality tools e.g. Risk Management and Six Sigma and to set up an excellent organization to reach the optimum on quality costs.

Pfeiffer has proven that early risk assessment and failure prevention is necessary in order to reduce failure costs. The occurrence of failure should be prevented by suitable actions *i.e.* by removing causes for failure or by actions to mitigate effects of failure.[15]

The graph in Figure 1 shows the connection between failure prevention and failure detection. Pfeiffer describes the "rule of 10" *i.e.* the costs for failure increase for the factor 10 by each project step/phase.

Ways and procedures of risk management are described amongst others in ISO 31000ff, ONR49000 and IEC 62198.[18, 19, 20]

Different types of risk assessment methods e.g. creative techniques, scenario analysis, indicator analysis, statistical methods and functional analysis can be applied. Useful methods for functional analysis are FMEA, HAZOP, HACCP and hazard analysis.

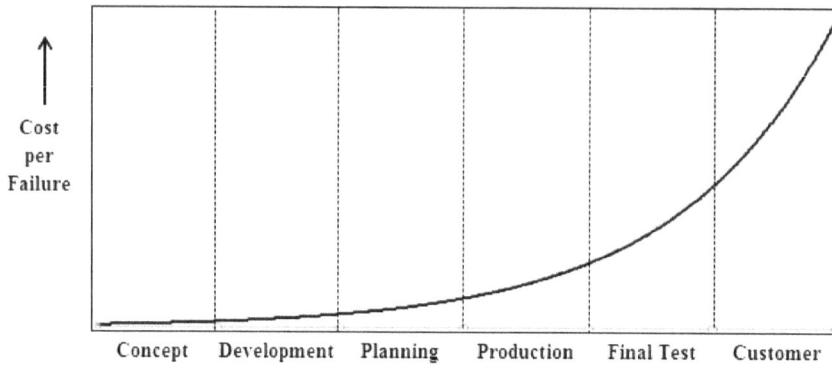

Fig. 1: Rule of Ten[15–17]

The following Figure 2 shows in which phases of the life cycle certain types of FMEA can be applied.

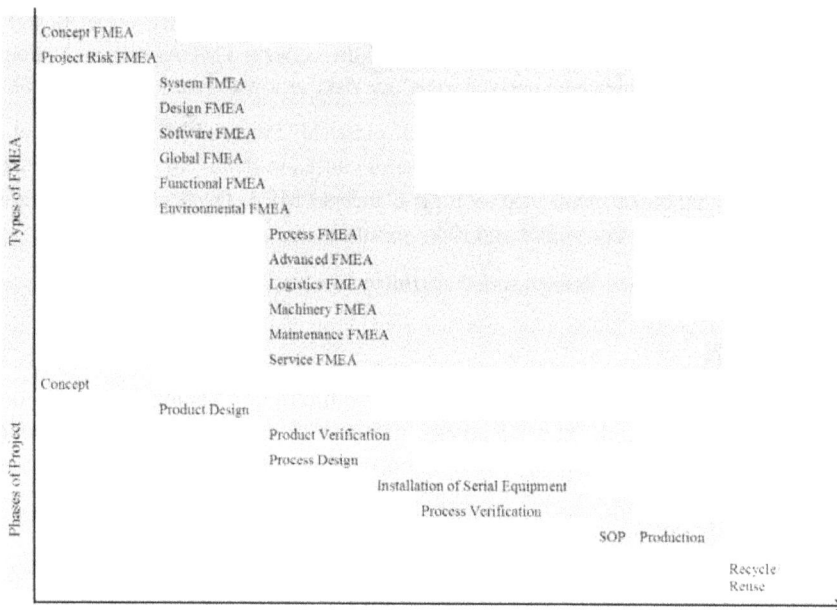

Fig. 2: Application of FMEA Types in Different Phases of the Life Cycle

For different project phases and applications, various types of FMEAs have been developed.

The focus in this paper will be on a modified type of Concept and Process FMEA supported by a simplified project risk analysis.

Concept FMEA and project risk assessment are effective if these are applied at a very early stage of a project. Important impacts on a project and its combination must be analyzed. Figure 3 shows examples of major impacts on a project.

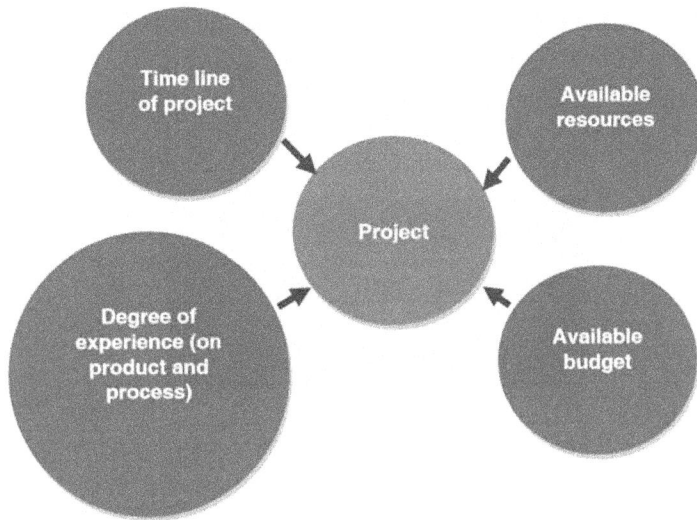

Fig. 3: Important Impacts on a Project

Actions for project planning must be derived from concept FMEA and project risk assessment.

Inputs from the concept FMEA as well as Design-Inputs are necessary to go deeper into process FMEA. Especially issues with significant and critical characteristics need to be transferred and considered in the process FMEA.[9, 22]

Many important decisions are taken at an early stage of a project. Concepts are developed, decisions about technologies, production and test concepts are taken. Financial evaluation of a project is carried out *i.e.* the budget and available resources for the project are determined at an early stage. One can say major decisions are done before starting with a system, design or process FMEA.

Table 1 shows original definitions for three types of FMEAs. Specific modification/optimization of these types of FMEAs will be discussed in this paper.

Table 1: Original Definition for Concept, Project Risk and Process FMEA

CFMEA	Concept FMEA	Is a Design or Process FMEA carried out at an early stage of development. "Hardware has not been defined yet". Goal is to identify all failure modes caused by interactions. Important failures might be eliminated by concept changes.[9]
RFMEA	Project Risk FMEA	Is applied to quantify and analyze project risks. It is a modification of process, design and service FMEA technique. Goals are amongst others focusing on most imminent risks, prioritizing risk contingency planning, developing improved risk controls.[23]
PFMEA	Process FMEA	Is focusing on processes e.g. supporting manufacturing process planning/development by analyzing possible failures and causes and developing suitable preventive and detection actions in detail in order to avoid or reduce risks.[6]

The following flow chart shows typical steps in a project and possible application of Concept FMEA, Project Risk Evaluation and Process FMEA.

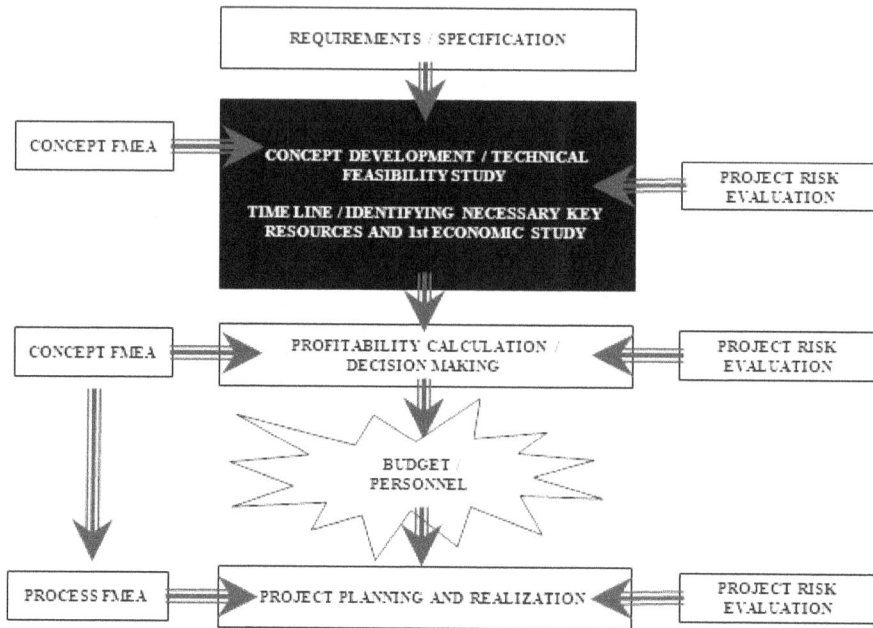

Fig. 4: Applying Concept FMEA, Project Risk Evaluation and Process FMEA

1.1 Concept FMEA

For concept development/planning of manufacturing processes, basic tasks need to be carried out *i.e.* a flow chart and an equipment list must be developed. Furthermore, quality planning *i.e.* an estimation of quality performance/data must be carried out for every project.

These basic tasks can be integrated and combined with a rough quantitative process FMEA. In the definition/context of this chapter this fundamental FMEA is named concept FMEA. The concept FMEA should include relevant process steps and all major production and test equipment.

Quantitative risk assessment is more reliable and controllable than the most common qualitative risk assessment which has got many weaknesses.[3] Prior to showing the new proposed way of carrying out the risk assessment, different weaknesses of qualitative risk assessment will be explained.

The qualitative RPN usually is calculated as follows:

$$RPN = S \times O \times D \qquad \qquad \ldots (1)$$

Some weaknesses in qualitative risk assessment *i.e.* in RPN calculation are shown in ICE 60812. These are amongst others, gaps in the ranges of RPN, duplicate RPNs, sensitivity to small changes, inadequate scaling of occurrence table as well as inadequate scale of RPN. Furthermore, Schoepf has described that the value of risk can be significantly influenced by assigning actions either to preventive or detection actions due to unclear definitions in standards. Schoepf made a proposal of a clear definition for distinguishing actions in order to harmonize FMEAs and to improve the reliability of RPN calculation/risk determination.[2]

In addition, the result of the qualitative RPN calculation does not distinguish between the efficiency of actions *i.e.* both combinations in Table 2 have got the same value of risk. The risk can be significantly different for the same RPN as shown in Table 2 due to the fact that the failure is prevented in the first case, and there is an absolutely intolerable rate of rejected parts in the second case.

Table 2: Example of Same Values of Risk with Actual Significantly Different Risks

Example	RPN	S	O	D
1.	100	10	1	10
2.	100	10	10	1

Standard qualitative risk assessment neglects internal risks mostly due to the way of calculating a RPN. One failure mode may have external and internal consequences. Internal consequences are neglected in many cases due to the fact that the value for the external severity is mostly higher than the value for the internal severity. However, the actual internal risk may be significantly higher.

The following equation shows that the maximum value for the severity of the external and internal risk is only of relevance for calculating the RPN. *i.e.* in Pareto analysis usually only external issues can be found in the top ranking.

$$\text{RPN} = \max S_{(\text{external, internal})} \times O_{(\text{of failure mode})} \times D_{(\text{of failure mode or failure cause})} \qquad \dots (2)$$

Where:

$\quad$ RPN $\quad$: Risk Priority Number
$\quad$ S $\qquad$: Severity of Effect (with rating 1–10)
$\quad$ O $\qquad$: Occurrence of Failure (with rating 1–10)
$\quad$ D $\qquad$: Detection of Failure (with rating 1–10).

A more reliable quantification of the risk can be achieved by a modified way of calculating a value of risk. Instead of applying qualitative values the severity can be quantified monetarily. This financial approach is shown amongst others in ONR 49002–2:2010. The ONR standard includes an example of a table with different risk criteria and an assignment of costs.[19] Furthermore, J. Braband applied a monetary value for the severity to verify his semi-quantitative procedure for calculating a value of risk. [12, 13]

$$\text{RV}_{\text{ipm}} = S_{(\text{external})} \times \alpha \times \gamma + S_{(\text{internal})} \times \alpha \times (1-\gamma) \qquad \dots (3)$$

Where:

$\quad$ RV_{ipm} $\quad$: Risk Value improved monetarily
$\quad$ α $\qquad$: Failure mode ratio
$\quad$ γ $\qquad$: Conditional probability of detecting a failure mode or its failure cause
$\quad$ $S_{(\text{external})}$: Average of external costs for hazards, field defects, customer rejects
$\quad$ $S_{(\text{internal})}$: Average of internal costs for touch up, rework, repair, scrap for the failure mode.

Available quality data from installed processes/already running products should be basis for selecting values for α and γ. For completely new manufacturing/test equipment. An estimation may be carried out together with experienced institutions/possible supplier of equipment. This way of calculating a monetary value of risk can support decision making very easily. The impact of concept changes on the business result can be easily shown.

Table 3: Example of a Quantitative Concept FMEA with Financial Evaluation

Process Step	Potential Failure Effect (external)	Severity External, Average Costs [$]	Potential Failure Effect (internal)	Severity Internal, Average Costs [$]	Failure Mode	Occurrence [ppm]	Preventive Actions	Detection [%]	Detection Actions	Number of Parts Produced over Life Time	Risk [internal] over Life Time [$]	Risk [external] over Life Time [$]	Risk [total] over Life Time [$]
Welding	Customer reject (99%), Accident (1%)	1.000.000	Rework (90%), Scrap (10%)	80	Welding Failure	30	Capable approved machine (type)	99,995	Automated X-Ray (type)	2300000	2069,9	3450	5519,9

Similar to the As Low As Reasonably Practicable (ALARP) principle boundaries for broadly acceptable region, ALARP or tolerability region and the unacceptable region can be defined.[24] In the original application of the ALARP principle the unit for defining the boundaries has been the fatal accident rate.[27, 24] Principles for Cost Benefit Analysis (CBA) in support of ALARP decisions have been described by HSE.[28] In the context of this paper, ALARP boundaries are defined monetarily only. Figure 5 shows several process steps with different risks. Project steps with unacceptable risks must be improved by further actions e.g. Process step 1 in Figure 5. On the other hand, project steps in the broadly acceptable region

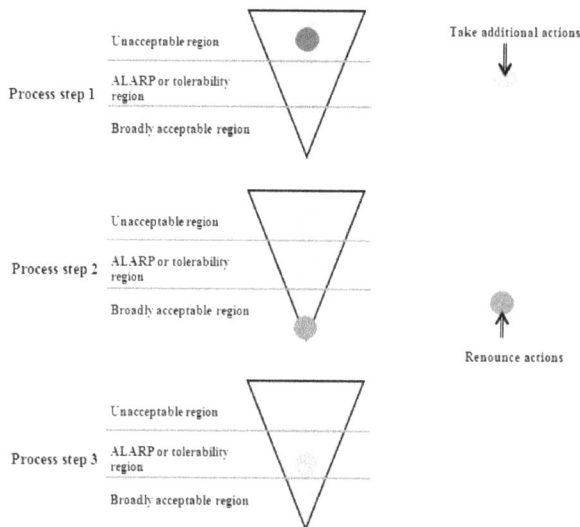

Fig. 5: ALARP Principle[27]

may be modified as well e.g. process step 2 may be over engineered, redundant or nice to have features may have been planned. Budget may be shifted from a process step with very low risk to another process step with negligible negative impact on risks. If this is not reasonably possible, improvements and additional budget are necessary. In overall (general) a balanced distribution of the risk should be achieved.

Criteria for determining the boundaries could be turnover of a company, turnover made with a product over life time, IRR, total costs of internal/external failure. However, fix boundaries are not recommended for FMEA. This is described amongst others in the VDA and in the AIAG FMEA manual.[4,6] Boundaries may be specified industry, process or project specific amongst others due to significant technical differences in processes and their capabilities. A basic example for defining boundaries is shown below. This simple example considers the total number of process steps *i.e.* does not distinguish between single process steps.

Boundary

for tolerable region
$$\frac{\text{Total costs of poor quality}}{\text{Turnover (for product over life time)}} \quad < 1\% \text{ tolerable} \quad \dots (4)$$

for broadly acceptable region
$$\frac{\text{Total costs of poor quality}}{\text{Turnover (for product over life time)}} \quad < 0, 2\% \text{ broadly acceptable} \quad \dots (5)$$

Boundaries for different process steps can be defined separately/company specific by individual criteria.

2. PROJECT RISK EVALUATION

In parallel to the concept FMEA a project risk assessment should be carried out. Decisions and actions should be derived based on this assessment.

2.1 Classical Project Risk Evaluation

The original project risk analysis, amongst others applied by NASA in the past, works as follows:

Besides technical consequences, the impact of costs and schedule on a program/project is considered. Risks are ranked from minimal to high.[30]

$$RF = PF + CF - PF \times CF \qquad \dots (6)$$

$$CF = \frac{CFC + CFS + CFT}{3} \qquad \dots (7)$$

Where,
RF   : Risk Factor.
PF   : Probability of Failure.
CF   : Consequence of Failure.
CFS : Consequence of Failure due to changes in schedule (with rating 0.1–0.9).

CFC : Consequence of Failure due to changes in costs (with rating 0.1–0.9).
CFT : Consequence of Failure due to technical factors (with rating 0.1–0.9).

The risk evaluation process *i.e.* a flow chart showing how the risk assessment process works is included in NASA REPORT. Three categories of risks are distinguished depending on RF: Low risk with RF < 0.3, medium risk with 0.3 < RF < 0.7 and high risk with RF > 0.7. For RF with medium and high risk, risk report and risk abatement actions are mandatory. Furthermore, a special review team must take care of issues with high risk.

3.2 Alternative Simplified Supporting Project Risk Assessment

In project risk assessment, the macro system can be examined besides the micro system. B. Bruehwiler describes 9 topics in a hazard list with several sub-items which may be taken into account for risk analysis. Relevant/generic for project realization could be amongst others the topic innovation, products and services with the sub-items, new products/life cycle, technological development, innovation change management and project management. Furthermore, operative performance processes are listed.[26] Risk may be of different nature. In Six Sigma CT trees and CT matrixes are applied to identify/analyze customer requirements as well as impacts on a company's business. As a generic term, criticality to satisfaction (CTSA) has been established. Furthermore, satisfaction is divided/expressed by criticality to quality, delivery and costs *i.e.* CTQ, CTD and CTC.

Another aspect for project risk assessment which should be considered separately is the criticality to launch a product on time CTL. *I.e.* the delivery to the customer can not start on time or necessary actions to avoid risks if production equipment is not completed on time. In addition, criticality to safety CTS and criticality to environment CTE can be considered similar to the RCM procedure.

The following table shows some concrete aspects which can be considered for project risk evaluation. Different issues may be of relevance for concept development and project planning. The example shows different type of criticalities which could be weighted from 1 to 10 and assigned to various issues. The impact of the issue on the different types of criticality can be rated from 1 to 10. The total criticality can be determined by adding all products of weighting and rating of different criticalities for each issue.

Table 4: Project Risk Assessment Table

Issues to be Considered	Critical to	Weighting	Rating	Total Criticality Per Issue	Risk Treatment
Micro System					
Product Specific					
New technology in product design (level of innovation)	CTL	10	10	100	External support by experienced institute
New (critical) materials/components/ auxiliaries applied in design	CTL	10	8	80	

Issues to be Considered	Critical to	Weighting	Rating	Total Criticality Per Issue	Risk Treatment
Uncertainties	CTC	8	7	56	
Safety relevance of product	CTS	10	9	90	
Degree of complexity of product					
Many different components (low number of identical components with other projects)					
Many variants					
Process specific					
New technology in production process					
New type of machine					
Dangerous production technology (i.e. explosion)					
Manufacturability of product					
Number of Interfaces/initial expense					
Appropriate responsibility for interfaces					
Number of process steps/operations					
Quality requirements					
General project issues					
Time line	CTL	10	10	100	Provide additional resources/fast approval of budget
New customer					
New supplier for important equipment	CTL	10	10	100	Approval process for new supplier
Ability of supplier	CTS	10			
Resources of supplier					
Turnover of supplier in relationship to equipment costs					
Maintenance strategy					
Clear organization (responsibilities)					
Project organization					
Plant organization					
Sufficient internal resources					
Capability of resources (training)					
Approval processes					
Macro System					
Infrastructure					
Environment					

Issues to be Considered	Critical to	Weighting	Rating	Total Criticality Per Issue	Risk Treatment
Education					
Political					
Culture					
Communication system					
Energy supply					
Waste disposal					
Procurement market					
Geological					
Earthquake					
Floodings					
Typhoons					

Even more simple could be a verbal documentation of risk without weighting and rating of factors for different criteria/the different issues. Similar to the RCM procedure, there could be a simple verbal description yes/no risk based on experience. Rules for risk treatment may be implemented company specific. Important is that identified risks are treated to avoid intolerable risks. Some examples have been shown in Table 4.

3. PROCESS FMEA

Process FMEA is mandatory for projects in various industries. PFMEA is described in many standards.[3–6,8,9] PFMEA usually includes the elements failure mode and effects analysis and a criticality analysis. Criticality analysis can be done by different methods e.g. by calculating a RPN, by risk matrixes or by top down sorting of severity, occurrence and detection. A very valuable tool is the severity/occurrence risk matrix. The risk matrix includes 3 areas. Risks from items in the red zone must be reduced by suitable actions. It is recommended to reduce risks in the yellow zone as well if reasonably practicable. The green zone is equivalent to the broadly acceptable region.

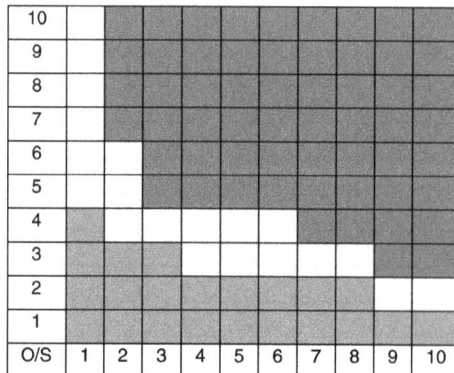

Fig. 6: Risk Matrix[4]

Besides the severity, only the occurrence of the failure mode or the failure cause is considered in this matrix. The detection of a failure is completely neglected. This could have the consequence that detection actions are assigned in the preventive actions column in the form sheet. Reason for this can be the goal of reducing the value of the occurrence in order to achieve an acceptable result of the risk matrix. However, this does not reflect the actual process. The yield should be the criteria to assign preventive and detection actions.[2] The development of an additional S/O^{TC} matrix is described in Chapter 4.1. This matrix can be applied in addition to the common S/O Matrix and can be used to show that the risk to the customer is low despite red items in the S/O Matrix.

3.1 Severity—"Occurrence to Customer" Risk Matrix

In order to determine the Risk to the customer, the detection must be considered e.g. in an additional risk matrix. The Occurrence to the customer O^{TC} has been determined by multiplying a quantitative value for occurrence and detection. For certain combinations of O and D, values for O^{TC} have been determined. The qualitative value for the Occurrence to the customer has been derived from the quantitative value *i.e.* based on the rating table for O shown in Figure 8 qualitative values have been selected/determined. The following 2 diagrams show a possible assignment of quantitative values to rating scales 1–10 for occurrence and detection. Figure 7 shows for decreasing quantitative values of detection increasing qualitative values for D. Furthermore, the conditional probability of detecting a failure or its failure cause γ applied in equation 3 has been displayed in the Table.

D (%)	γ	D
100	0	1
99,9	0,001	2
99	0,01	3
96	0,04	4
90	0,1	5
80	0,2	6
66	0,34	7
50	0,5	8
30	0,7	9
0	1	10

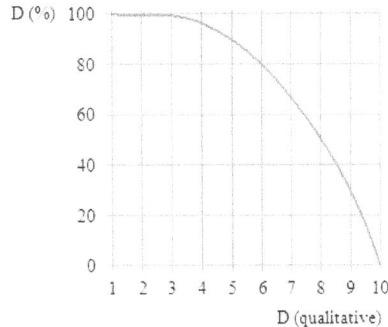

Fig. 7: Detection Qualitative-Quantitative

ppm	p / 1000	O
0	0,0000	1
14	0,0143	2
45	0,0452	3
143	0,1429	4
452	0,4518	5
1429	1,4286	6
4518	4,5175	7
14286	14,2857	8
45175	45,1754	9
142857	142,8571	10

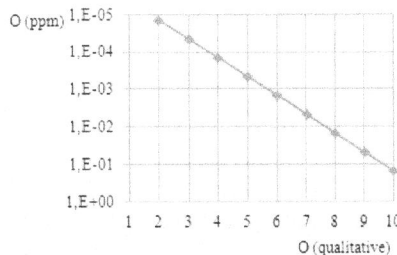

Fig. 8: Occurrence Qualitative-Quantitative

The graph in Figure 8 for O qualitative/quantitative considers logarithmic scaling for O = 2 to 10 as applied by J. Braband for his semi quantitative RPN.[12, 13] However, for O = 1 a failure rate of 0 has been chosen similar to the latest AIAG FMEA manual.

The following Table 5 is based on the ratings of the graphs shown before. It shows quantitative O^{TC} values for different combinations of quantitative values of occurrence and detection.

Table 5: Occurrence to Customer (Quantitatively)

Detection of Failure — Bad outgoing quality (ppm)

D	D (%)										
10	0	142857	45175	14286	4518	1429	452	143	45	14	0
9	30	100000	31623	10000	3162	1000	316	100	32	10	0
8	50	71429	22588	7143	2259	714	226	71	23	7	0
7	66	48571	15360	4857	1536	486	154	49	15	5	0
6	80	28571	9035	2857	904	286	90	29	9	3	0
5	90	14286	4518	1429	452	143	45	14	5	1	0
4	96	5714	1807	571	181	57	18	6	2	1	0
3	99	1429	452	143	45	14	5	1	0	0	0
2	100	143	45	14	5	1	0	0	0	0	0
1	100	0	0	0	0	0	0	0	0	0	0
		142857	45175	14286	4518	1429	452	143	45	14	0	ppm
		10	9	8	7	6	5	4	3	2	1	O

Occurrence of Failure

Qualitative values for O^{TC} have been derived from this quantitative matrix applying ratings according Figure 8.

Table 6: Occurrence to Customer (Qualitatively)

Detection of Failure — Occurrence to customer

D	D (%)										
10	0	10	9	8	7	6	5	4	3	2	1
9	30	10	9	8	7	6	5	4	3	2	1
8	50	10	9	8	7	6	5	4	3	2	1
7	66	10	9	8	7	6	5	3	2	2	1
6	80	9	8	7	6	5	4	3	2	2	1
5	90	8	7	6	5	4	3	2	2	1	1
4	96	8	7	6	5	4	2	2	1	1	1
3	99	6	5	4	3	2	2	1	1	1	1
2	100	4	3	2	2	1	1	1	1	1	1
1	100	1	1	1	1	1	1	1	1	1	1
		142857	45175	14286	4518	1429	452	143	45	14	0	ppm
		10	9	8	7	6	5	4	3	2	1	O

Occurrence of Failure

Both S/O Matrix and S/O^{TC} Matrix should be evaluated within risk assessment. The following graph shows an example for applying the information from both risk matrixes in a FMEA form sheet.

Effect	Severity	Failure Mode	Failure Cause	Preventive Action	Occurrence	Detection Action	Detection	O^{TC}	S/O area	S/O^{TC} area	Decision Making
leak pressure loss	10	Welding joint too thin	Welding speed to high	Manual process, Instruction of worker	4	Manual visual inspection	8	4			
					4	Sample Size (random control last part release): Check of shape/ dimensions via gauge blocks, digital height gauge, dial test indicator according test spec.	4	2			Further Actions - Yes/ No (e.g. based on RPN$_{ip}$ and costs) Documentation of deciders result
				Automated welding by controlled roboter, constant speed acc. spec. ensured First part release: Check of shape/ dimensions via gauge blocks, digital height gauge, dial test indicator according test spec.	1		4	1			

Fig. 9: Example of Using Input from S/O Matrix and S/O^{TC} Matrix in FMEA form Sheet

The following table includes a recommendation for risk treatment for all possible combinations of results from S/O Matrix and S/O^{TC} Matrix. Highest focus should be on following combinations: Items in red area of S/O Matrix + items in red area of S/O^{TC} Matrix and also on items in red area of S/O Matrix + items in yellow area of S/O^{TC} Matrix.

Table 7: Risk Treatment Based on Result of Risk Matrixes

en	S/O^{TC} Matrix Red	S/O^{TC} Matrix Yellow	S/O^{TC} Matrix Green
S/O Matrix red	Team: prepare Decision Making, e.g. analysis by attributes, applying RV$_{ipm2}$/Review by Management	Team: prepare Decision Making/Review by Management	Team: discuss costs for rework/ scrap, consider further actions if reasonably practicable
S/O Matrix yellow	n/a	Team: prepare Decision Making/Review by Management	Team: consider further actions if reasonably practicable
S/O Matrix green	n/a	n/a	no further actions

Of prime importance, items in the red area should be investigated in detail e.g. an extended risk assessment of critical items can be carried out by applying quantitative methods. RV$_{ipm2}$ *i.e.* RV$_{ipm}$ extended by considering the conditional probability of the failure mode effect β can be applied to support decision making.

$$RV_{ipm2} = S_{(internal)} \times \alpha \times \gamma + S_{(external)} \times \alpha \times (1 - \gamma) \times \beta \qquad \dots (8)$$

Where:

RV$_{ipm2}$: Risk Value improved monetarily considering β

β : Conditional probability of the current failure modes failure effect.

The result of the risk assessment/decision making for critical items should be documented in the FMEA form sheet e.g. names of decider, date.

4. CONCLUSIONS

Failure mode and effects analysis and risk assessment can be carried out in all phases of a project. Risk Assessment for Manufacturing should not be limited to qualitative Process FMEA. Projects should be supported by a Concept FMEA and Process Risk assessment. The greatest benefit can be achieved if failures are prevented at an early stage of a project. There are different possibilities/methods which can be applied. Depending on the stage of a project, on the information available and the necessary efforts, suitable methods can be chosen and applied. Qualitative FMEA can be improved by applying both, S/O and S/O^{TC} Matrix. Quantitative Risk Assessment can be applied to support the assessment of critical items from qualitative FMEA.

NOMENCLATURE

Greek symbols

α	Failure mode ratio
γ	Conditional probability of detecting a failure mode or its failure cause
β	Conditional probability of the current failure modes failure effect

Subscripts

ALARP	As low as Reasonably Practicable
FMEA	Failure Mode and Effects Analysis
CFMEA	Concept FMEA
PFMEA	Process FMEA
RFMEA	Project Risk FMEA
S	Severity
O	Occurrence
O^{TC}	Occurrence to Customer
D	Detection
RPN	Risk Priority Number
RV_{ipm}	Risk Value improved monetarily
RV_{ipm2}	Risk Value improved monetarily considering β
CTSA	Critical to Satisfaction
CTC	Critical to Costs
CTD	Critical to Delivery
CTE	Critical to Environment
CTL	Critical to Launch
CTQ	Critical to Quality
CTS	Critical to Safety
HACCP	Hazard Analysis and Critical Control Points

HAZOP   Hazard and Operability Study
IRR   Internal Rate of Return
RCM   Reliability Centered Maintenance
SOP   Start of Production

REFERENCES

[1] Schöpf, G. (2010), Qualitative und Quantitative Risikoanalysen mittels FMEA, XVI APIS Benutzertreffen.

[2] Schöpf, G. (2011), Kategorisierung von Vermeidungs- und Entdeckungsmaßnahmen für Prozess-FMEA's zur Steigerung der Vergleichbarkeit und Zuverlässigkeit bei der Risikobewertung, 6. Osnabrücker FMEA Forum.

[3] IEC 60812, VDE Verlag, 2006.

[4] VDA 4.4 Quality Management in the Automobile Industry, Quality Assurance Prior to Serial Production, Product and Process FMEA, 2006.

[5] SAE J1739/SAE ARP 5580.

[6] AIAG FMEA reference manual 4th Edition.

[7] MIL-STD-1629A.

[8] TM 5-698-4 (2006), Failure Modes, Effects and Criticality Analysis (FMECA) for Command, Control, Communications, Computer, Intelligence, Surveillance and Reconnaissance (C4ISR) facilities, Headquarters, Department of the Army.

[9] Ford Handbook Failure Mode and Effect Analysis (with Robustness Linkages) 4.1.

[10] DRBFM (Toyota FMEA).

[11] Becker G. (2009), Design Review Based on Failure Mode, bfk-ingenieure, www.bfk-ingenieure.de.

[12] Braband J., (2007), FQS Tagung 2007 und Walter-Massing Preis, In Zukunft beherrschte Risiken?; QZ Qualität und Zuverlässigkeit, pp. 18–19.

[13] Braband, J. (2008), Mit neuem Ansatz zu verlässlichen Risikoprioritätszahlen, Beschränktes Risiko; QZ Qualität und Zuverlässigkeit, pp. 28–33.

[14] Vetter, R., http://home.t-online.de/home/ralf.vetter/kosten.htm

[15] Timischl, W., Qualitätssicherung, Statistische Methoden, 3. Auflage, p. 6.

[16] RW TH Aachen FVW I – Vorlesung "Prüfung, Qualitätsmanagement", Ausmaß der Fehlerentstehung und Behebung (Jahn).

[17] Jahn, H., Erzeugnisqualität, die logische Folge von Arbeitsqualität; VDI-Z 130 (1988).

[18] ISO 31000: 2009, Risk Management—Principles and guidelines, International Organization for Standardization.

[19] ONR 49000ff.: 2010, Risikomanagement für Organisationen und Systeme, Austian Standard plus.

[20] IEC 62198, VDE Verlag, 2001.

[21] ISO/DIS 26262-1 to 10, Road vehicles—Functional safety, Edition: 1, Current stage date: 2009-12–10.

[22] VDA Produktentstehung, Prozessbeschreibung besondere Merkmale (BM), 2010.

[23] Carbone, T. and Tippet, D. (2004), Project Risk Management Using the Project Risk FMEA, Engineering Management Journal Vol. 16 No. 4.

[24] Rausand, M. (2004), Risk Analysis an introduction, System reliability theory (2nd ed), Wiley.

[25] Laser Atmospheric Wind Sounder, DR-20, Vol. II, Final Report, LMSC-HSV TR F320789-11. George C. Marshall Space Flight Center, 1992.

[26] Brühwiller, B., Risikomanagement nach ISO 31000 und ONR 49000, 2009.

[27] Maag, T. (2004), Risikobasierte Beurteilung der Personensicherheit von Wohnbauten im Brandfall unter Verwendung von Bayes´schen Netzen, ETH Zürich

[28] HSE, HSE principles for Cost Benefit Analysis (CBA) in support of ALARP decisions, http://www.hse.gov.uk/risk/theory/alarpcba.htm

[29] HSE, Reducing Risks, Protecting People, HSE's decision-making process, ISBN 0 7176 2151 0, 2001.

[30] Laser Atmospheric Wind Sounder, DR-20, Vol. II, Final Report, LMSC-HSV TR F320789-11. George C. Marshall Space Flight Center, 1992.

Role of Binary Logit Regression in Survey Analysis

Neena Mital[1] and P.R. Lakshmikanthan[2]

[1]Ram Lal Anand College, University of Delhi, India
[2]SQC and OR Unit, Indian Statistical Institute, Delhi, India
E-mail: n_mital@yahoo.com

ABSTRACT: In clinical/industrial problems many effects are studied as yes/no type, for example: a person having a cardiac attack or not, a person is using a foot-over bridge or not, whether to change a tool or not in an industrial process, whether to send a machine for maintenance or not are some examples. In such cases, the response variable becomes dichotomous. Binary Logistic Regression is a suitable model for studying these kinds of situations which will identify the key factors affecting the response. In this paper, results of an industrial survey is presented which is primarily using this modeling approach for accessing the Quality Status of the industries and thereby finding important causal variables.

Keywords: Scoring Method, p-Value, Kruskal-Wallis Test, Wilcoxon-Signed Rank Test, Binary Logistic Regression.

1. INTRODUCTION

Regression analysis tries to find out causal relationship with certain indicators. Statistical Regression Analysis establishes a very well formed mathematical relationship in this regard. This relationship can be used for prediction along with an associated error of predictions with certain degree of confidence. General empirical relationship is of the form:

$$z = \beta_0 + \beta_1 x_1 + \beta_2 x_2 + \beta_3 x_3 + \ldots + \beta_k x_k + error \qquad \ldots (1)$$

In conventional regression analysis both the indicator as well as causal variables must be continuous. Indicator variables are supposed to be independent of each other. In practical situations response variable may not be continuous. In many survey data response may be in YES/NO type only. Examples are:

- Consumer chooses particular brand (1) or not (0)
- A quality defect occurs (1) or not (0)
- A person is hired (1) or not (0)
- Evacuate home during hurricane (1) or not (0)
- Credibility assessment in the loan sanction (1) or not (0).

Here, decision may depend on number of factors which may again be binary, categorical or numerical and may or may not be independent of each other. In such situations, Binary logistic regression can give appropriate models. This model gives the intensity of severity in terms of probability of occurrence of the response. Logit model is used for prediction of the probability of occurrence, of an event, by fitting data to a logit function (logistic curve). Logistic regression has been extensively used in the medical and social sciences fields, as well as marketing applications such as prediction of a customer's propensity to purchase a product or cease a subscription. For example, the probability that a person has a heart attack within a specified time period might be predicted from knowledge of the person's age, sex and body mass index. Logistic function or the logit response function, alike probabilities, always takes values between zero and one. A typical logistic curve form is shown below in figure 1 and logistic model is given by:

$$f(z) = \frac{e^z}{e^z + 1} = \frac{1}{1 + e^{-z}}$$

$$= \frac{\exp(\beta_0 + \beta_1 x)}{1 + \exp(\beta_0 + \beta_1 x)}$$

... (2)

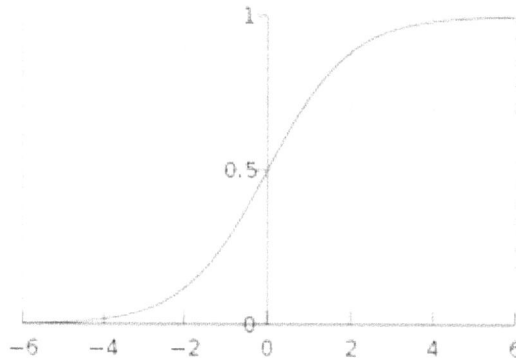

Fig. 1: Logistic Curve

The variable z represents the exposure to some set of independent variables, while $f(z)$ represents the probability of a particular outcome, given that set of explanatory variables. The variable z is a measure of the total contribution of all the independent variables used in the model and is known as the logit.

The method of estimation in conventional regression analysis uses the principle of least squares and error follows normal distribution with 0 and variance σ^2 whereas in the binary logistic regression estimation procedure is maximum likelihood and error distribution is binomial with parameters n and π. The analysis and the inferences drawn using Binary logistic regression on survey data is presented in this paper.

2. SURVEY OBJECTIVE

Survey was conducted to find level of Quality control and application of statistical techniques in Indian industries. The survey was carried out in Delhi predominantly for automobile industries. Survey was conducted on the following six aspects and relation between various quality aspects and responsible factors is studied.

- Quality Status
- Research and Development
- Data Analysis
- Use of Statistical Techniques
- Manufacturing:
    - R&D Regarding Factors
    - Levels of Factors in Production Process
    - Noise Factors in Production Process
- Market Customer Complaint.

On each aspect questions were asked in retrospect of the aspect it covers. On Quality Status aspect the following four questions were asked by each respondent.

Q1. *What quality means to you?*

1. As per standard requirement.
2. Customer's Satisfaction.
3. Product Satisfies Customer's Minimum Satisfaction.
4. Should Lie between the Specification Limits.
5. Robust, Reliable and Easy to use.

Q2. *Do you have quality control department?*

1. No department.
2. Small department but no expert headed by an engineer.
3. Medium department with an engineer trained in QC.
4. Big department with an engineer trained in statistics.
5. Full fledged department with a statistician.

Q3. *Who heads your quality control team?*

1. No head workers look for standards.
2. An experienced engineer.
3. An engineer trained for QC tools.
4. Statistician.

Q4. *Does your company have any certificate of quality viz. ISO 9000, QS 9000, IS 14000, etc.?*

1. Our own quality standards.
2. ISO 9000–2000, ISO 9000–2001, ISO 9000–2008, etc.
3. IS 14000 plus above.
4. TS 16949/any specific for your process plus above.

The Scenario of the Companies' surveyed is shown in Figure 2.

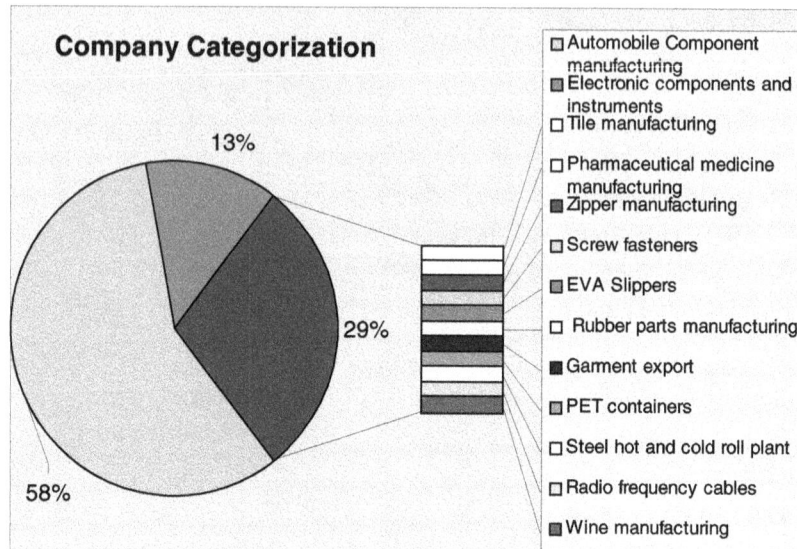

Fig. 2: Company Categorization

3. BACKGROUND OF THE SURVEY

Survey is conducted to know the status of quality maintained in industries. Data is collected through the well designed and validated questionnaire. Data is then converted into valid scores on the scale of 0–10. Suitable variance test is applied to see differentiation among questions within each aspect. Using test values appropriate weights are given to each question. This gives importance of each facet in respect to each other within an aspect, which is used to give average score to each company on each aspect. This statistical test is used for every main aspect to find comprehensive scores. This concept is based on a conjecture that all the sub questions are given equal weight (Null hypothesis). If the conjecture is disproved it indicates that sub questions are different from each other. In this concept of differentiation, a statistical concept called p-value (It indicates the risk of differentiation between two propositions status quo vs. changes) is used. Since, the data is perception based, this P-value is fixed at 10% indicating thereby the results can be guaranteed to a level of 90%. In other words, we can say maximum pre-calculated risk of 10% is taken on the judgment done under this project. Therefore, test is done at $\alpha = 0.10$ level of significance. Statistical test Kruskal-Wallis and p-value is used for testing. The weights are given to questions in each category wherever required, using Z-score (It is a normalized score obtained by dividing the difference of referral value and average value by the standard deviation). The details of the analysis on the aspect of Quality status are given below in section 4. The methodology for the analysis is given by the flow chart in Figure 3.

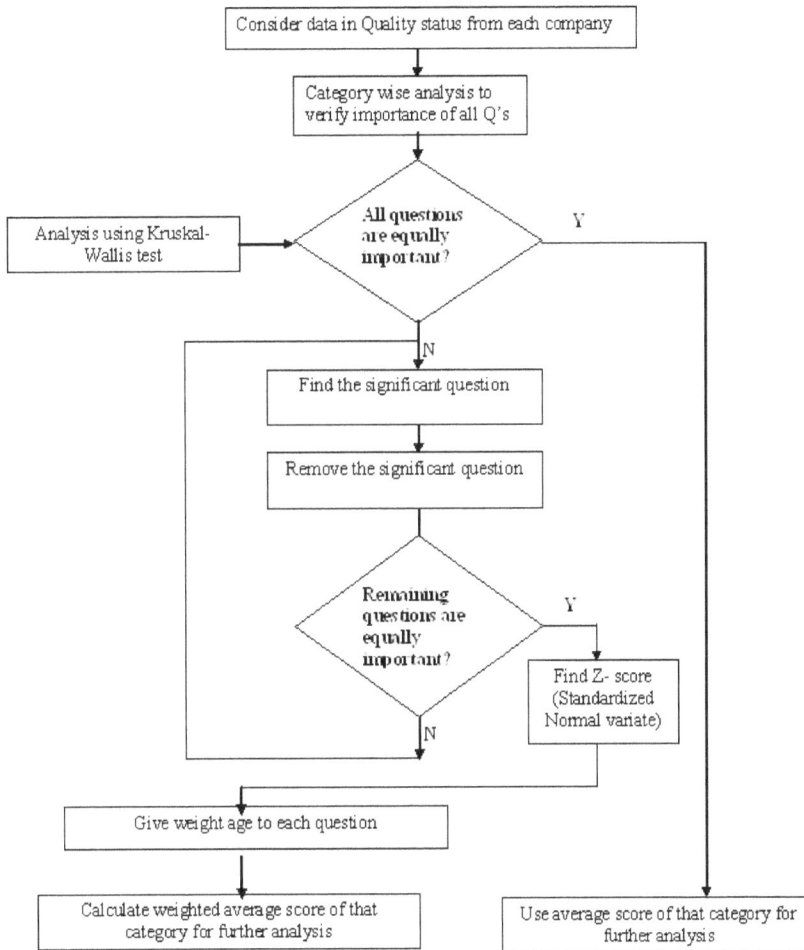

Fig. 3: Analysis Flow-Chart

4. ANALYSIS OF QUALITY STATUS ASPECT

Since the survey is aimed at finding the quality standard maintained by the company, so, first aspect deals with finding the status maintained as well as understood by the quality personals of the company. Statistical test is used to find comprehensive scores on this aspect (Refer to Section 2). The analysis on this aspect is discussed in detail in this section. The hypothesis defined in this respect is:

H_0: All the sub questions (facets) are given equal importance

H_1: All the sub questions (facets) are given differential importance.

The data does not follow normality so non-parametric-Kruskal-Wallis test is used to test the above conjecture and Table 1 below gives the results.

Table 1: Kruskal Wallis Test: A: Quality Status versus Facets

Facets	N	Median	Ave Rank	Z
1	38	4	58.6	−2.9
2	38	6	69.2	−1.18
3	38	8	87.6	1.79
4	38	8	90.7	2.29
Overall		152	76.5	

$H = 13.68 \; DF = 3 \; P = 0.003$
$H = 14.63 \; DF = 3 \; P = 0.002$ (adjusted for ties)

Since, P = 0.002 is fairly smaller than α = 0.10 our test level of significance, we have strong evidence that disproves our conjecture that sub-questions can be viewed on the same wavelength by all the companies The test reveals that as per pre-calculated risk in quality status the companies could not view all questions in harmony. There are evidences that companies do not rate all the facets on similar levels and some seems to be more relevant compared to others. The test reveals there are three categories in the facets 1, 2 and 3 and 4. Accordingly, their average ranks are coming out to be 58.6, 69.2 and 89.15 (average of ranks of facet 3 and 4). In order to give an accurate effect of each facet the small differential between facet 3 and 4 is also taken into consideration while giving the weights for each facet as shown below in Table 2.

Table 2: Average Ranks and Weights

Facet	1	2	3	4
Avg. Rank	58.6	69.2	87.6	90.7
Weights	58.6/58.6 = 1	69.2/58.6 = 1.18	87.6/58.6 = 1.49	90.7/58.6 = 1.55

The average score in each aspect based on overall performance of the companies is calculated. Using this approach, average score of each company is calculated with respect to the aspect of quality status and is shown in Table 3 below:

Table 3: Scores on Quality Status Aspect

Company Code	C1	B1	C2	B2	B3	C3	C4	C5	C6	B4	A1	A2	A3	A4	A5
Score	5.98	5.14	2.57	7.38	4.1	4.45	6.11	1.93	2.52	6.54	6.92	7.14	7.38	8.52	7.31
Company Code	A6	B5	C7	A7	A8	A9	A10	A11	A12	A13	C8	C9	A14	A15	C10
Score	6.47	6.34	6.54	8.52	6.92	8.52	6.33	7.69	7.38	8.28	4.55	4	5.59	7.38	6.33
Company Code	A16	A17	A18	C11	A19	A20	A21	A22							
Score	7.38	8.07	8.28	3.41	7.38	6.34	6.78	6.92							

The survey predominantly is carried out on different types of companies situated in Delhi to get a broader spectrum of company profile. Through the analysis an attempt was made to get a comprehensive score representing the aspect of quality status in the companies situated in and around Delhi. The overall average of all the companies on Quality status is 6.3.

Anderson-Darling normality test reveals that data does not follow normality. Thus, instead of mean, median score is ideal for this representation of data. The median score of Quality status is 6.6. The median score says that 50% of companies score is less than 6.6. Looking at the scale of 1–10 this score is not very healthy. Lot of improvement is required in this aspect to become internationally competitive. We first study the significance of the median score with respect to overall companies. Accordingly non-parametric-Wilcoxon-signed rank test is applied to see whether all industries may be ranked equally on quality status or they are significantly different. The results of the test are shown below in Table 4.

Table 4: Wilcoxon Signed Rank CI: Weighted Average

		Confidence Interval			
	N	Est. Median	Achieved Confidence	Lower	Upper
Wt. Avg	38	6.54	95	5.74	7.03

The test gives the estimated median of 6.54 with 95% C.I. (5.74, 7.03). A sensitivity test is required to be carried out to find the significant value of the overall average quality status. A Sensitivity test is normally carried on data to find out various types of sampling and testing errors and their impact on ultimate decisions. For this purpose, a sensitivity analysis for the 95% C.I. region is done on median of quality status. The result of the test in this region is given below in Table 5.

Table 5: Wilcoxon Signed Rank Test: Weighted Average

		Test of est. median = 6.540 versus median < 6.540			
	N	N for test	Wilcoxon Statistic	P	Est. Median
Wt. Avg	38	369	0.494	6.536	

The test shows that median at level of 6.54 is not significant. However, as it can be seen there are quite a few companies whose comprehensive score is very much more than average median score of 6.54 (marked as * in Table 1) showing distinctively superior kind of performance with respect to quality status aspect. These companies may be termed as bench marked companies as quality is always considered to be an important parameter for any type of company to be in market. To verify whether these companies are really a benchmark companies a sensitivity analysis was carried out and test median of 7.0 showed significance and test results are shown below in Table 6.

Table 6: Wilcoxon Signed Rank Test: Weighted Average

		Test of median = 7.0 versus median < 7.0			
	N	N for Test	Wilcoxon Statistic	P	Est. Median
Wt. Avg	38	38	243	0.033	6.536

The test shows that median at level of 7.0 is highly significant, which signifies higher median that is higher perception towards quality aspect is a significant phenomena. Incidentally, in general the companies scoring above 7.0 are automobile companies. It means that Automobile and Non-automobile industries may differ in their perception towards Quality Status, maybe because of strong competitive environment for auto-companies and hence this necessitates a break up of the companies into Automobile and Non-automobile sector. The indices were prepared for auto and non-auto separately also and is given in Table 7.

Table 7: Average Scores for Different Types of Companies

Type of Company	Average Score
Automobile	7.58
Non-automobile	4.84

It can be seen clearly that automobile companies in general are giving more importance to Quality Status as compared to non-automobile. This could be due to heavy competition or influence of QS-9000 and other certifications. The driving force primarily is to get quality products at economic costs to satisfy fairly highly enlightened customers. As we go to non-automobile sector, low score could be because of practically no competition and target of consumers may not be demanding much.

Even though quality consciousness and need to satisfy customer's requirement are the two important paradigms for any company, the reasons exhibited by the companies who are having score of 7 and above on quality status were studied and it is found that these reasons could be attributed to:

1. Type of company (Nature of product)
2. Age of the company
3. Turnover of the company
4. Employees strength

Developing an appropriate mathematical model for this, companies' good performance maybe attributed to some factors for improvement in quality status.

4.1 Model

$$Y = f (X_1, X_2, X_3, X_4) \qquad \qquad \dots (2)$$

where Y is a response on quality status which is a function of four factors namely type of company (X_1), age of the company (X_2), turnover of the company (X_3) and employees strength (X_4). We want to develop this model to understand relative importance of each factor in achieving certain degree of quality status.

4.1.1 *Type of Company (X₁)*

In general, the automobile companies behaved differently from others as shown by score of quality status also as they have to maintain the standards as per international market and from the remaining also electronics component manufacturing industries also faired better due to technicality and heavy competition. Considering this the scaling is done in Table 8 on the first factor of type of industry as:

Table 8: Company Categorization

Type of Company (X_1)	Scale Factor
Automobile companies and its ancillaries	3
Electronics component manufacturing	2
Other general companies	1

4.1.2 *Age of Company (X_2)*

Age of the company is the period in years since the company is in existence. Since, age is supposed to be measurable data, it is expected to follow normal distribution. In real life, apparently, age does not come out to be a true data and hence it does not follow test for normality and same is shown by Anderson-Darling test. In analytical approach data transformation such as square-root, $\log_e$ are tried to find out whether transformed data follows normal distribution. Accordingly, square root transformation was found to be appropriate as shown by Anderson-Darling test result in Figure 4.

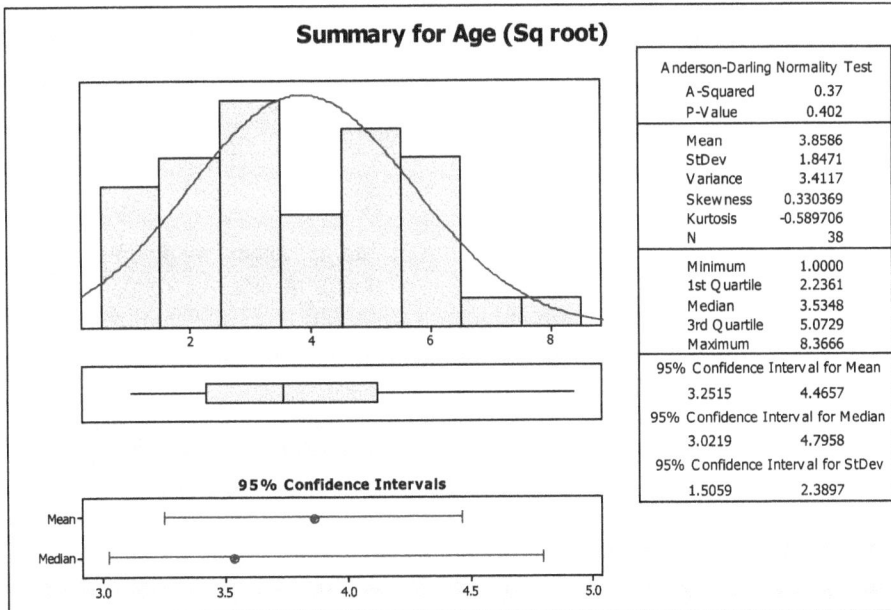

Fig. 4: Normality Check for Transformed (Sq-root) Data on Age of Company

Since the transformed data is following normal the properties of normal distribution were used to get decisive cut-offs. Accordingly standardized cut-offs were developed and are given below in Table 9.

Table 9: Age Distribution Cut-Offs in Line to Normal Distribution

▪	$\mu-3\sigma = -1.68$	▪	$\mu+\sigma = 5.71$
▪	$\mu-2\sigma = 0.16$	▪	$\mu+2\sigma = 7.56$
▪	$\mu-\sigma = 2.01$	▪	$\mu+3\sigma = 9.4$
▪	$\mu = 3.86$		

As a cut-off of $\mu-3\sigma$ (= −1.68) comes out to be negative which is not a feasible range as age can only be a positive factor, quartile distribution of age was taken into consideration. The cut-offs on the quartiles are given below:

$Q_1 = 2.24$

$Q_2 = 3.53$

$Q_3 = 5.1$

$Q_4 = 8.4$

These quartile ranges were re-transformed into the original data viz. age (in years). Accordingly the scales were developed as shown in Table 10. The Table 10 gives the frequency distribution of the companies in various age groups.

Table 10: Age Categorization

Quartiles Cut-Offs	Age Interval (in years)	Scale	Freq.
0–25%	0–5	1	8
25%–50%	5–12	2	8
50%–75%	26 Dec.	3	13
75%–100%	26–70	4	9

4.1.3 *Turnover of Company (X₃)*

Turnover is expressed in terms of annual business (in Indian rupees) by a company. Turnover of a company may depend on volume of sales as well as unit price of a product. On similar lines the cut-offs on the quartiles and frequency distribution of the companies in various turnover ranges is shown in Table 11.

Table 11: Turnover Categorization

Quartiles Cut-offs	Turnover Interval (in Crores)	Scale	Freq.
0–25%	0–32	1	09
25%–50%	32–108	2	09
50%–75%	108–276	3	10
75%–100%	276–812	4	09
Extreme data point	>812	5	01

4.1.4 *Employees Strength of Company (X₄)*

Employees' strength is defined as total number of employees working in a company right from top executive to workers. The distribution of employees in different areas of a company plays a vital role. On similar lines the cutoffs on the quartiles frequency distribution of the companies in various employees' strength ranges is shown below in Table 12.

Table 12: Employees Strength Categorization

Quartiles Cut-offs	Employees strength	Scale	Freq.
0–25%	0–92	1	9
25%–50%	92– 475	2	10
50%–75%	475–762	3	10
75%–100%	762–3505	4	9

A mathematical model developed to find out factor propensity to improve w.r.t aspect viz. quality status. The response parameter of this model is dichotomous in nature with 0 as the companies whose quality score is less than tested median score and equal and above as level 1. The same model is used in other aspects also to see dominancy of factors in attaining the corresponding response level.

5. BINARY LOGISTIC REGRESSION ANALYSIS: QUALITY STATUS ASPECT

The purpose of the analysis is to assess the effects of multiple explanatory variables, which can be numeric or categorical, on the outcome variable using model in equation 2. The results of binary logistic regression done w.r.t all the four factors X_1, X_2, X_3 and X_4 is shown below:

5.1 Binary Logistic Regression on Quality Status Aspect at the Estimated Median Level

Response = 6.54 (estimated median score) versus factors X_1, X_2, X_3 and X_4.
Link Function: Logit

Response Information

Variable	Value	Count
Response binary	1	21 (Event)
	0	17
	Total	38

Logistic Regression Table

Predictor	Coef	SE Coef	Z	P
Constant	−6.94711	2.45761	−2.83	0.005
X_1	1.77415	0.629098	2.82	0.005
X_2	−0.296184	0.619660	−0.48	0.633
X_3	0.586137	0.527795	1.11	0.267
X_4	0.912944	0.684283	1.33	0.182

Log-Likelihood = −13.514
Test that all slopes are zero: G = 25.229, DF = 4, P-Value = 0.000

Measures of Association: (Between the Response Variable and Predicted Probabilities)

Pairs	Number	Percent
Concordant	327	91.6
Discordant	26	7.3
Ties	4	1.1
Total	357	100.0

The above analysis done on binary logistic regression shows decision to predict is in accordance with actual data having higher concordance of 91.6%. The overall p-value = 0.0 for the model predicts high significance of the model. It can be seen that all the regressor coefficients are having positive sign except for X_2 (age). Age has got a negative sign implying that as age increases the propensity to improve in quality status decreases, since it is not a significant parameter as indicated by p = 0.633, the variable X_2 was dropped. With the remaining factors rerun of the analysis with different combination was made with highly significant factor X_1 having p = 0.005. Binary logistic regression performed is shown below.

5.1.1 *Binary Logistic Regression on Quality Status*

Response = 6.54 (estimated median score) versus factors X_1 and X_4.

Logistic Regression Table

Predictor	Coef	SE Coef	Z	P
Constant	−6.98045	2.28024	−3.06	0.002
X_1	1.95860	0.616915	3.17	0.001
X_4	1.08616	0.500320	2.17	0.030

Log-Likelihood = −14.182
Test that all slopes are zero: G = 23.894, DF = 2, P-Value = 0.000

Measures of Association: (Between the Response Variable and Predicted Probabilities)

Pairs	Number	Percent
Concordant	311	87.1
Discordant	26	7.3
Ties	20	5.6
Total	357	100.0

In the above analysis, concordance is still high at 87.1%. The overall p-value = 0.0 signifies that model is highly significant. It can be seen that all the regressor coefficients are having positive sign implying that propensity to improve in quality status is directly proportional to factors X_1 and X_4, as well as they are significant parameter as indicated by p = 0.001 and 0.030 respectively. Taking the coefficients of these factors the predicted response model may be given by:

5.1.2 *Response Model: Quality Status Aspect at the Estimated Median Level*

$$Y = -6.98 + 1.96 \times X_1 + 1.1 \times X_4 \qquad \qquad \dots (3)$$

Through the model we are trying to develop a policy oriented approach for General, Electronic and Automobile Company to move up in a quality status scale which is going to be primary target for all the companies for survival. Using the developed predictive response model given by equation 3 the probabilities are worked out for different combinations, using exponential distribution and are shown below in Table 13.

$$\text{Predictive Prob.} = e^Y/(1 + e^Y)$$

Table 13: Predictive Probabilities

X_1 \ X_4	1	2	3	4
1	0.019265	0.055201	0.148047	0.340740
2	0.122389	0.293178	0.552308	0.783667
3	0.497500	0.746494	0.897523	0.963031

5.1.3 *Inference from the Response Model: Quality Status Aspect at the Estimated Median Level*

Automobile Companies
- The company having an employee strength between 762–3505 (scale 4) has the high probability of 96.3% to move up, from score of 6.54 (from quality level '0' to '1').
- The company having an employee strength between 475–762 (scale 3) also has the high probability of 89.8% to move up, from quality level '0' to '1'.
- The company having an employee strength between 92–475 (scale 2) still has the probability of 74.6% to move up, from quality level '0' to '1'.
- A small scale company having low employee strength between 0–92 (scale 1) has 49.8% probability of move up, from quality level '0' to '1'.

Electronic Companies
- The company having an employee strength between 762–3505 (scale 4) has the probability of 78.4% to move up, from score of 6.54 (from quality level '0' to '1').
- The company having an employee strength between 475–762 (scale 3), probability to move up, from quality level '0' to '1' reduces to 55.2%.
- The company having an employee strength between 92–475 (scale 2), the probability drops to 29.3% to move up, from quality level '0' to '1'.
- A small scale company having low employee strength between 0–92 (scale 1) has very low probability of 12.2% to move up, from quality level '0' to '1'.

General Companies
- The company having an high employee strength between 762–3505 (scale 4) has low probability of 34.1% to move up, from score of 6.54 (from quality level '0' to '1').
- The company having an employee strength between 475–762 (scale 3), probability to move up, from quality level '0' to '1' reduces to 14.8%.

- The company having an employee strength between 92–475 (scale 2), the probability drops to mere 5.5% to move up, from quality level '0' to '1'.
- A small scale company having low employee strength between 0–92 (scale 1) has near impossible probability of 1.9% to move up, from quality level '0' to '1'.

It can be said that type of company plays a major role in attaining quality status and within same type of company also employee strength plays a major role in attaining higher quality status. As the employee strength can be related to size of the company so as the there is growth in the company it automatically becomes conscious of the quality status. This can also be inferred by a 3D plot in Figure 5.

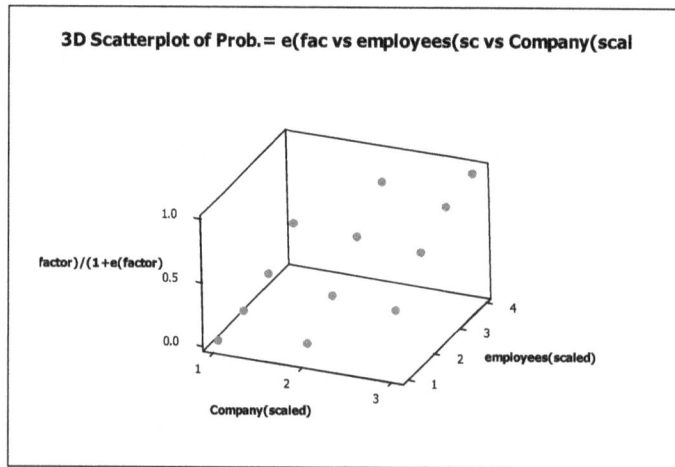

Fig. 5: 3-D Scatter Plot

5.2 Binary Logistic Regression on Quality Status Aspect at the Benchmark Level

The above analysis shows that companies at the median level of 6.54, have significant propensity factors for improvement from level '0' to '1'. To further identify the significant factors at the bench mark level of 7, binary logistic analysis has been carried out and test results are shown below.

5.2.1 *Binary Logistic Regression: Quality Status Aspect at the Benchmark Level*

Response = 7.0 (Significant median score) versus factors X_1, X_2, X_3 and X_4.

Link Function: Logit

Response Information

Variable	Value	Count
Response binary	1	15 (Event)
	0	23
	Total	38

Logistic Regression Table

Predictor	Coef	SE Coef	Z	P
Constant	−7.78692	3.01348	−2.58	0.010
X_1	1.98921	0.860933	2.31	0.021
X_2	−0.114719	0.541450	−0.21	0.832
X_3	0.411130	0.592124	0.69	0.487
X_4	0.529775	0.487806	1.09	0.277

Log-Likelihood = −15.894
Test that all slopes are zero: G = 19.195, DF = 4, P-Value = 0.001

Measures of Association: (Between the Response Variable and Predicted Probabilities)

Pairs	Number	Percent
Concordant	299	86.7
Discordant	40	11.6
Ties	6	1.7
Total	345	100.0

The above binary logistic regression analysis done at the significant benchmark median level of 7.0 shows concordance of 86.7%. The overall p-value = 0.001 for the model predicts high significance of the model. It can be seen that all the regressor coefficients are having positive sign except for X_2 (age). Age has got a negative sign, since it is highly insignificant parameter as indicated by p = 0.832 the variable X_2 was dropped. With the remaining factors rerun of the analysis was done at 10% level of significance. Binary logistic regression performed is shown below.

5.2.2 *Binary Logistic Regression*

Response = 7.0 (Significant median score) versus factors X_1, X_3 and X_4.

Predictor	Coef	SE Coef	Z	P
Constant	−7.92385	2.94611	−2.69	0.007
X_1	2.01941	0.855669	2.36	0.018
X_3	0.513428	0.481582	1.07	0.286
X_4	0.336093	0.471599	0.71	0.476

Log-Likelihood = −15.916
Test that all slopes are zero: G = 19.150, DF = 3, P-Value = 0.000

Measures of Association (Between the Response Variable and Predicted Probabilities)

Pairs	Number	Percent
Concordant	298	86.4
Discordant	39	11.3
Ties	8	2.3
Total	345	100.0

The above analysis table shows that the model has concordance of 86.4% as well as is highly significant (with p-value = 0.00). As X_1 with p = 0.018 is the only significant factors so rerun of the analysis with different combination was made. It was found, that only X_1 comes out to be significant factor and Binary logistic regression performed is shown below.

5.2.3 *Binary Logistic Regression*

Response = 7.0 (Significant median score) versus factor X_1.

Logistic Regression Table

Predictor	Coef	SE Coef	Z	P
Constant	−6.67160	2.78732	−2.39	0.017
X_1	2.42057	0.968900	2.50	0.012

Log-Likelihood = −17.153
Test that all slopes are zero: G = 16.677, DF = 1, P-Value = 0.000

Measures of Association: (Between the Response Variable and Predicted Probabilities)

Pairs	Number	Percent
Concordant	221	64.1
Discordant	8	2.3
Ties	116	33.6
Total	345	100.0

In the above analysis, concordance drops to 64.1%, but there is tie on 33.6% cases, thus concordance may be treated well. The p-value = 0 .0 signifies that model is highly significant. It can be seen that only regress or coefficients X_1 shows propensity to improve in quality status, as well it is a significant parameter as indicated by p = 0.012. Taking the coefficients of X_1 the predicted response model may be given by:

5.2.4 *Response Model: Quality Status Aspect at the Benchmark Level*

$$Y = -6.67 + 2.42 \times X_1 \qquad \qquad \dots (4)$$

The developed model, shows that once the level of bench mark is achieved then only type of company plays role to move up in a quality status scale. Using the developed predictive response model given by equation 4 the calculated predicted probabilities are shown below in Table 14.

Table 14: Predictive Probabilities

Company Type X_1	Probability
1	0.013789
2	0.133542
3	0.629483

5.2.5 *Inference on Response Model: Quality Status Aspect at the Benchmark Level*

- A general nature company has near impossible probability of 1.3% to move up, above the benchmark level of median 7.0.
- Electronic companies have also very low probability of 13.3% to move up, above the benchmark level of median 7.0.
- Lot of automobile companies are already above the benchmark median level of 7.0 but others who have not reached has 62.9% probability to rise up from the benchmark level of 7.0.
- The relative change is also depicted by a scatter graph in Figure 6.

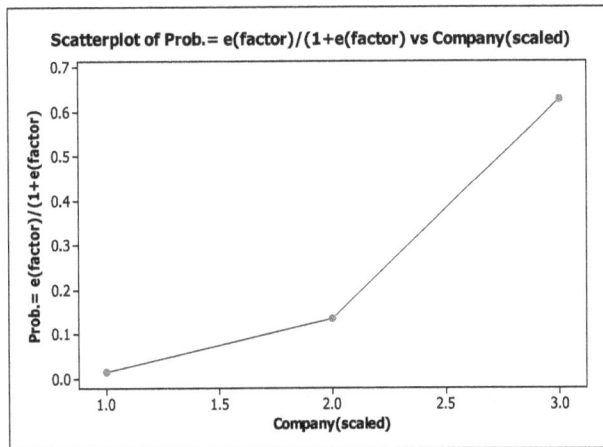

Fig. 6: Scatter Plot

5.2.6 *Comparison between Current and Benchmark Level of Quality Status*

As a company is gearing up to reach benchmark median level of 7 from current median level of 6.54, it is observed that X_4, the employees strength does not play any role. This situation maybe because saturation level of employees has been reached and any increase is just used up in increasing the production levels. At this level only one variable X_1 (type of company) has a significant role to play. So it can be said company profile is more important.

Based on above comparison we move from macro to micro level analysis to study for differentiation among same type of companies. As the research was also specific to automobile industries so we now analyze to see which factor dominates in the automobile sector to attain high quality status.

5.2.7 *Micro Analysis of Quality Status*

In order to analyze this differentiation, data for automobile industries for quality status aspect is studied. Data failed on normality test, so Kruskal Wallis test is applied to see the significance of different facets. As test (Table 15) shows high significance with p = 0.0, appropriate weight age is calculated, accordingly weighted scores are calculated, test results and scores are shown below in Table 16.

Table 15: Kruskal-Wallis Test: Quality Status versus Facets (Automobile sector)

Facets	N	Median	Ave Rank	Z	Weights
1	22	4	28	−3.49	1
2	22	6	34	−2.23	1.21
3	22	8	50	1.17	1.79
4	22	10	66	4.55	2.36
Overall	88		44.5		

H = 29.38 DF = 3 P = 0.000
H = 31.55 DF = 3 P = 0.000 (adjusted for ties)

Table 16: Scores on Quality Status Aspect (Automobile Sector)

Automobile Company	A1	A2	A3	A4	A5	A6	A7	A8	A9	A10	A11
Average Score	7.35	7.37	7.73*	8.68*	7.67*	6.97	8.68*	7.35	8.68*	6.61	7.75*
Automobile Company	A12	A13	A14	A15	A16	A17	A18	A19	A20	A21	A22
Average Score	7.73*	8.49*	5.51	7.73*	7.73*	8.30*	8.49*	7.73*	5.96	6.99	7.35

The Anderson-Darling test performed shows the data for weighted average on quality status does not follow normal (p = 0.08). Accordingly, median test, Wilcoxon-signed rank test is applied to see whether all industries may be ranked equally on quality status or they are significantly different. The results of the test are shown in Table 17.

Table 17: Wilcoxon Signed Rank CI: Weighted Average

			Confidence Interval		
	N	Est. Median	Achieved Confidence	Lower	Upper
Wt. Avg	22	7.643	94.9	7.316	8.014

The test gives the estimated median of 7.64 with 95% C.I. (7.316, 8.014). For this purpose a sensitivity analysis for the 95% C.I. region is done on median of quality status. The result of the test in this region is given below in Table 18.

Table 18: Wilcoxon Signed Rank Test: Wt. Avg

		Test of est. median = 7.643 versus median < 7.643		
	N	Wilcoxon Statistic	P	Est. Median
Wt. Avg.	22	129	0.539	7.643

The test shows that median at level of 7.643 is not significant. However, as it can be seen there are quite a few companies whose comprehensive score is very much more than average median score of 7.64 (marked as * in Table 16). To verify whether these companies are a benchmark companies a sensitivity analysis was carried out and test results are shown below in Table 19.

Table 19: Wilcoxon Signed Rank Test: Weighted Avg.

	N	N for test	Wilcoxon Statistic	P	Est. Median
Test of median = 8.0 versus median < 8.0					
Wt. Avg.	22	22	74	0.046	7.643

The test shows that median at level of 8.0 is significant. A mathematical model developed in sec. 4 equation 1, was used to find out factor propensity to improve w.r.t quality status aspect within automobile companies. Here as are looking within automobile sector so variable X_1 is dropped and we have three factors, the corresponding model reduces to:

$$Y = f (X_2, X_3, X_4) \qquad \qquad ... (5)$$

5.3 Binary Logistic Regression Analysis within Automobile Sector: Quality Status Aspect

The purpose of the analysis is to assess the effects of multiple explanatory variables, on estimated current median score using model in Equation 4. The results of binary logistic regression done w.r.t three factors X_2, X_3 and X_4 is shown below:

5.3.1 *Binary Logistic Regression: Quality Status Aspect within Automobile Sector*

Response = 7.643 (estimated median score) versus factors X_2, X_3 and X_4.

Link Function: Logit

Response Information

Variable	Value	Count
Response binary	1	13 (Event)
	0	09
	Total	22

Logistic Regression Table

Predictor	Coef	SE Coef	Z	P
Constant	−1.54140	2.04232	−0.75	0.450
X_2	−0.451452	0.575977	−0.78	0.433
X_3	0.531474	0.555835	0.96	0.339
X_4	0.529537	0.641696	0.83	0.409

Log-Likelihood = −14.016
Test that all slopes are zero: G= 1.734, DF = 3, P-Value = 0.629

Measures of Association: (Between the Response Variable and Predicted Probabilities)

Pairs	Number	Percent
Concordant	68	58.1
Discordant	41	35.0
Ties	8	6.8
Total	117	100.0

The above analysis shows a low concordance of 58.1%. The overall p-value = 0.629 for the model predicts model is not significant. It signifies that at the current median score none of the factors considered under this project proved to be significant.

To further check for significant factors at the benchmark level of 8.0, binary logistic analysis has been carried out and test results are shown below:

5.3.2 *Binary Logistic Regression*

Response = 8.0 (significant median score) versus factors X_2, X_3 and X_4.

Link Function: Logit

Response Information

Variable	Value	Count
Response binary	1	6 (Event)
	0	16
	Total	22

Logistic Regression Table

Predictor	Coef	SE Coef	Z	P
Constant	−1.17089	2.15496	−0.54	0.587
X_2	−0.154114	0.593236	−0.26	0.795
X_3	−0.215883	0.571161	−0.38	0.705
X_4	0.437605	0.676525	0.65	0.518

Log-Likelihood = −12.612
Test that all slopes are zero: G = 0.558, DF = 3, P-Value = 0.906

Measures of Association: (Between the Response Variable and Predicted Probabilities)

Pairs	Number	Percent
Concordant	53	55.2
Discordant	39	40.6
Ties	4	4.2
Total	96	100.0

The above analysis again shows a low concordance of 55.2%. The overall p-value = 0.906 for the model predicts model is highly insignificant. It signifies that within automobiles to attain higher level in quality status the factors considered under this project does not play significant role.

6. CONCLUSION

In general, in all survey analysis, questions are framed and opinions are sought from various individuals as a method of understanding the realities. In all such responses the tendency is to give weight age to all the sub-questions is prevalent to get the comprehensive score. In this paper, an attempt was made to find the weight age according to criticality of the questions.

In this kind of analysis response are ordinal in nature and hence non-parametric testing methodology and binary logistic regression analysis was followed to develop the models.

REFERENCES

[1] Hosmer, D.W. and Lemeshow, S. (2000). Applied Logistic Regression, John Wiley and Sons, New York.

[2] Kanji, G.K. (2006). 100 Statistical tests, SAGE Publications Ltd., New York.

[3] Montgomery, D.C., Peck, E.A. and Vining, G.G. (2006). Introduction to Linear Regression Analysis John Wiley and Sons, New York.

Fuzzy Multinomial Cumulative Sum and Exponentially Weighted Moving Average Control Charts

Hassen Taleb[1] and Ali Achouri[2]

[1]University of Gafsa
[2]University of Tunis, LARODEC
E-mail: [1]hassentaleb@yahoo.co.uk; [2]ali.achouri@live.fr

ABSTRACT: In this article we propose to apply fuzzy theory to construct cusum and ewma control charts for multinomial processes. Each sample was presented by a single fuzzy set that summarizes the 5 categories of quality that can take the product. Each category is described by a linguistic variable that defines a fuzzy set. Fuzzy addition and multiplication are used for different fuzzy sets to reach one representative value for the entire sample.

Keywords: Statistical Process Control, Fuzzy Set, Linguistic Variable, Multinomial Data, Membership Function, Representative Value.

1. INTRODUCTION

Montgomery[1] described statistical methods useful in quality improvement. Various steps of construction of Shewhart control chart were explained for attribute and variable data and some more advanced Statistical Process Control (SPC) methods, included are the cumulative sum (cusum) and exponentially weighted moving average control chart (ewma). Zadeh[2] proposed fuzzy theory in order to formalize human subjectivity. Expertise of operators was used to model phenomena that are difficult to estimate by conventional logic for lack of quantitative data. Modeling was done with the help of fuzzy sets which reflect the degrees of memberships by assigning 0 for non-membership and 1 for full meembership. Zadeh,[3,4] introduced the concept of linguistic variables. Description of an observed phenomenon, subjectively judged, can be converted to numerical values with the help of fuzzy theory and membership function. Futoshi *et al.*[5] required mathematical methods in the determination of membership functions using human subjectivity and probability. Direct Rating is used and expert is called to answer the question of "How x is u?" Dombi[6] gives an overview of the different kinds of mathematical forms of the membership functions, different approaches are discussed. Previous articles on membership functions are classified in several ways. Some examples of heuristically based membership, membership functions based on reliability concerns with respect to the particular problem and more theoretical demand are presented. Andrs *et al.*[7] exploit the benefits provided by fuzzy technology, desirable characteristics of efficient membership function generating mechanism are proposed. Accuracy, flexibility, Computationally Affordable and Easiness to use are defined as key characteristics that may exist in the membership function. On the other hand, some disadvantages were explained in Taleb and Limam.[8] They suggested two approaches for constructing control charts to monitor multivariate attribute processes when data is presented in linguistic form. Construction of control chart for such multivariate attribute processes is analyzed using fuzzy sets and probability theories (combination of a chi-square statistics). A frozen food example is given to compare the two methods. The disadvantage of fuzzy theory is that it is strongly related to the choice of membership functions and the degree of fuziness. This choice is usually with no theoretical foundation. Taleb and Limam[9] compared several procedures for constructing control charts. Construction of charts was based on fuzzy theory and linguistic variables. Each quality category is made on the basis of fuzzy sets with fuzzy modes. Each sample was represented by the sum of fuzzy modes for all categories. Alipour and Noorossana[10] proposed an application of fuzzy theory on multinomial data to draw ewma control charts, each quality category is summarized by a linguistic variable. Among the methods of converting data, the fuzzy mode was used as the representative value to transform qualitative values to measurable values. The representative value for each sample is obtained by adding fuzzy modes weighted by the proportion corresponding to each category in each sample.

We propose to apply fuzzy theory to linguistic and multinomial data to construct cusum and ewma control charts. Real data is used to compute representative values. In section 1, fuzzy theory and membership functions will be presented, in section 2 cusum and ewma control chart are discussed and in section 3, real data from "Ideale Sanitaire" is taken to apply the new method, then a sensitivity analysis will be provided.

2. FUZZY THEORY AND MEMBERSHIP FUNCTION

Fuzzy logic is based on fuzzy set which is composed by objects with different degrees of belonging in it. The degree of membership is a number included in the interval,[0,1] and fuzzy logic is applied only to objects that we are not sure about their nature accurately and doubtful about the category in which it may be part.

In statistical quality control, 2 methods are proposed which are variable and attribute control charts. For variable control chart, the sample introduces all information with certainty data (weight, diameter, length...) and compares them with the target value in order to judge whether if the product is in or out of control. However, for attribute control chart, characters are not measurable and represented by linguistic variables. These variables are represented by a triplet L, T (L) and U where: "L" is the variable name (size, age...), "T (L)" is the subset of the possible values of the linguistic variable called "term set" and "U " is then the universe of discourse, sets of subsets "T (L)". Data associated to describe these characters may be binomial (consistent vs not consistent or good vs bad) or multinomial (perfect, good, bad, very bad...). Exact values will not be found as the 1st case of control, but fuzzy values reflecting the degree of uncertainty are investigated.

The importance of membership function comes from the fact that it is impossible to precisely identify the class to which belongs objects. The assignment is related to human subjectivity which allows several questions about the logic used for choosing the category. Shapes can be a either triangular, Gaussian or trapezoid. The vagueness, the lack of clearly defined criteria for classifying some objects make it necessary to resort to the fuzzy theory and membership functions to solve the problem of identification. The classic functions take only 2 values 0 or 1 which indicate the certainty of the class that owns the object. If the value assigned is 0 so this object does not belong to this set, else (value = 1), it will be certain that the object is part of the set. However, the membership function can cover a multitude of values between 0 and 1 according to the degree of membership of an object to a class. If the value is different from 1 or 0, so we are doubting about the membership of the object to the class. Membership functions are applications defined from an interval [a, b] to the interval [0.1] where [a, b] is the universe of discussion of the fuzzy object.

$$\mu_A : U = [a,b] \rightarrow \{0,1\}, x \rightarrow \mu_A(x) \qquad \qquad \dots (1)$$

Fuzzy logic is mainly based on expertise, it is from experience and knowledge of the expert that the degree of membership of any element in the universe of discussion is found, this method refers only to the background of the expert who uses his knowledge to assign x to u. To establish this assignment, he is based on a set of questions which guide him to choose the right assignment with the least mistakes. According to the degree of hesitation in the assignment, the possible forms are either: triangular, trapezoid or gaussian (curve).

3. CUSUM AND EWMA CONTROL CHARTS

The basic principle of control charts is to take repeated samples of production at a specified frequency and represent them in charts to judge if the product is in or out of control, the chart is composed by segments that link observations from each other. During the inspection, if there are observations out of limits, then the product is out of control and an investigation should be established to determine the origin of the problem.

3.1 Ewma Control Chart

Ewma control chart is a statistical tool of SPC used to detect small variabilities in production processes. Various observations are taken from the current one to the oldest. It is also applied either for attribute data or variable data, it presents an advantage over the Shewhart control chart that it includes at the same time all previous data added to the current one. 3 main characteristics are presented; first of all, small variabilities at around 1.5σ and less are detected. Then, all information in the sample are considered by examining the current observation to the previous ones, and finally, problem of non normality can be solved with the help of ewma specifications. Normality is a necessary condition for applying statistical control in the case of Shewhart control chart, no control can exist in Xbar R; Xbar S charts ... without checking the normality of data. Ewma is applied on a statistic summarizing all informations from all data collected which solves this issue.

$$Z_i = \lambda \bar{X}_i + (1 - \lambda Z_{i-1}) \qquad \qquad \dots (2)$$

According to the central limit theorem, Xbar of any distribution follows the normal distribution with mean μ and variance σ_2, therefore, X_i si replaced by Xbar and the problem of non normality is solved.

$$Z_t = \lambda \bar{X}_t + (1 - \lambda Z_{t-1}) \qquad \qquad \dots (3)$$

where Xbar is the weighted average of the actual observation with all previous observations, λ shows the weight of the actual observation to determine the statistical value of Z_i, this value is a proportion that summarizes the importance of actual observation compared by the previous ones and $(1 - \lambda)$ is the weight of previous observations in explaining the process. Weighting decreases geometrically with distance away from the current observation because $(1 - \lambda) \subset [0, 1]$. This will be faster if we give a small value for λ because $(1 - \lambda)^2$ will decrease exponentially and tends to 0. Whatever the chosen value of λ, ewma decreases geometrically and gives smaller and smaller weights for observations that are more distant from the current one. Ewma is a weighted average of all the past without neglecting any previous observations and keep then all the information included in the sample, it is based on a statistic Z_i which contains the different weights of all existing observations in the sample, these weights depend crucially on λ values. In short, higher value for λ indicates that the controller takes into account primarily the current observation and consequently a relatively low efficiency of detecting small vari-abilities because λ is high. Weights tends to 0 so the oldest observations are the smaller weighted observations of the sample, then it focus mainly on the observations nearest in time.

3.1.1 *The Choice of λ*

- If $\lambda = 0$, the current observation is removed.
- If $\lambda = 1$, we take into account the current observation and old observations are removed, this case is similar to the Shewhart control chart.
- For small λ, in order of 0.3 or less, Ewma is effective for detecting small variabilities.

$$UCL = \mu_0 + L\sigma\sqrt{\frac{\lambda}{1-\lambda}(1-(1-\lambda)^{2i})}$$

$$CL = \mu_0$$

$$LCL = \mu_0 - L\sigma\sqrt{\frac{\lambda}{1-\lambda}(1-(1-\lambda)^{2i})} \qquad \ldots (4)$$

L is a parameter used in the upper and lower bounds of control to show the order of sigma that we want to affect to our process. L is the multiple of standard deviation σ.

3.1.2 *Cusum Control Chart*

Another technique used to statistical control chart in the production process and referred to detect small variability in the order of 2σ and less is the cusum control chart. The logic of this method is to calculate the C_i as follows:

$$C_i = \sum_{j=1}^{i} (\bar{X}_j - \mu_0) \qquad \ldots (5)$$

where μ_0 is the target value.

C_i is a statistic which aims to compute the sum of deviations between the average of the observations and the target value μ_0, this statistic gives an idea about the quality control but it is not a tool of statistical quality control because the concept of control charts requires the existence of upper and lower limits.

C_i presents the deviation of the averages of observations from the target value μ_0, the process remains under control if C_i fluctuates around the horizontal line of μ_0. If there is a trend upwards or downwards, there is a signal that the process has just changed its average, the value of x would be either higher or lower than μ_0 and does not change orientation. In this case, we must note the probability of existence of an assignable cause that has changed the distribution of observations.

Only control charts cheks if the process is in or out of control.

The Tabular (or Algorithmic) cusum control chart.

Cusum accumulates deviations above or below compared to the target value μ_0. If the deviation is above, so C_{i+} is called one sided upper cusum, else C_{i-} is called one sided lower cusum.

$$C_t^+ = \max\left[0;\ X_t - (\mu_0 + k + C_{i-1}^+)\right] \qquad \dots (6)$$

$$C_i^+ = \max\left[0;\ (\mu_0 + k) - X_i + C_{i-1}^+\right]$$

k is a reference value, where

$$K = \frac{\mu_0 + \mu_1}{2} \qquad \dots (7)$$

C_i^+ and C_i^- are cumulative deviations from the target value $\mu 0$, such that $C_i^{(+/-)} \subset R^+$. If we find a negative difference, so it is essential to bring them to 0. So, the logic is essentially to draw deviations C_i^+ and C_i^- around μ_0, and then see if the process is in control or not by comparing the upper and lower limits. Control limits are defined by H knowing that H is a multiple of σ. This is an important parameter of cusum control chart, it is often chosen as a multiple of 5 times the standard deviation σ.

If C_i^+ or C_i^- exceeds H, so the process is out of control, we must seek the time from which the process becomes out of control. To find the time t source of deviation, N should be computed such that N indicates the number of consecutive periods where C_i is strictly positive. The first period where $N > 0$ is the period from which there is an assignable cause that has triggered the process, the objective is to detect and remove it:

- Assume that the process is out of control at the ith observation.
- Suppose that the last 5 cusum values are positive.

So, the period during it there is the emergence of the assignable cause is the period that coincides with the observation $(J - 5)$. Subsequently, a new statistical control is taking place. For this new control, we associate 0 to the last value of cusum.

4. OVERVIEW OF OUR APPROACH

"Ideal Sanitaire" is a company specializing in sanitary appliance (washbasin, close coupled wc, kitchen sinks, Baths, Showers). Products are classified by items where each item can have 5 possible classes: 1st choice, 2nd choice (standard), 3rd choice, 4th choice or 5th choice. Data are collected about "Saphir" WC's article [Figure 1].

The target item is classified by one expert into 5 possible categories. The assignment refers to the experience and there are no instruments used for this classification. When a product has no defect or a minor defect in a invisible area, it is classified as a first choice, when it shows any minor defect in an invisible area which does not affect the use of the product, it is classified as second choice (standard). If there is a visible defect that does not affect the use of the product is classified as a 3rd choice. Finally, when the use is affected and may be correct, the item is considered to be Repair (Class 4), otherwise waste (class 5). The construction of membership functions is made on the basis of ratings of expert. There are 3 areas in the product [Figure 1], the assignment of product quality varies from one area to another, these areas are mainly:

- Red Zone: that which is opposite: Hyper sensitive.
- Green Zone: the inside of the bowl is less sensitive.
- Black area: one below the bowl the less sensitive.

The x-axis shows the depth of the holes, and the y-axis shows the degree of membership of each depth at each fuzzy subset [Figure 2]. We determine for each product a single membership function for all fuzzy sets.

Table 1: Fuzzy Modes Per Sample

Sample	1	2	3	4	5	6	7	8	9	10
FM	0.22	0.15	0.24	0.17	0.47	0.04	0.29	0.27	0.06	0.10

Sample	11	12	13	14	15	16	17	18	19	20
FM	0.15	0.14	0.24	0.00	0.00	0.35	0.11	0.26	0.31	0.14

Sample	21	22	23	24	25	26	27	28	29	30
FM	0.14	0.25	0.15	0.26	0.29	0.26	0.28	0.21	0.35	0.31

Sample	31	32	33	34	35	36	37	38	39	40
FM	0.23	0.29	0.00	0.04	0.28	0.00	0.00	0.14	0.00	0.00

5. DETERMINATION OF REPRESENTATIVE VALUES

To successfully draw fuzzy multivariate ewma and cusum control charts, we have to construct representative values. Several steps should be followed, collect the ratings of experts to judge the category where belongs the product, transform ratings of experts to membership functions (membership function 1st choice, 2nd choice, 3rd choice, Rework and Gar), then, define a general membership function by adding different fuzzy sets (general membership function 1st choice, 2nd choice ... and Gar), by fuzzy addition of all zones. After that, use general membership functions to construct fuzzy sets of each sample by multiplying each category with it's proportion, make the fuzzy addition of different classes to give only one fuzzy set per sample. Finally, representative value is fuzzy mode [Table 2].

Membership functions are constructed from ratings of experts. Each expert can judge, with a simple direct observation, the category of the product. No instruments of measures are used by the expert, only, its expertise is applied. We proceed to make the measurements corresponding to each rating. The meter is used as a measuring instrument of the depth of the hole in the product. Then, from each measurement, a numerical value is assigned to each qualitative ratings released from the expertise of the expert. By assumption, the membership functions shape is triangular, then, the curves are linear. A simple measurement of boundaries in each category determines the slope of each line. So, fuzzy sets can be plotted. Each fuzzy set is composed of two straight linear, where one is increasing and one is decreasing [Figure 2]. The product is composed by three zones. Each zone has its own membership functions. For each of the three areas, there are 5 possible categories determined from the depth of the hole. The result is 15 fuzzy sets, three in each category. A fuzzy addition is made for each category for all product zones. The result is 5 fuzzy sets per product instead of 15. About 40 observations were collected each day, Fuzzy multiplication will be made for each sample where each category is multiplied by the proportion of matches. Let v is the number of products considered 1st choice, w for the 2nd choice, x for 3rd choice, y for 4th choice and z for 5th choice. Then:

$$RV = FM = v \times \mu_{1C} + w \times \mu_{2C} + x \times \mu_{3C} + y \times \mu_{4C} + z \times \mu_{5C} \qquad \dots (8)$$

Table 2: Cusum Parameters $k = (\mu_1 - \mu_0)$

0	μ_1	σ	$k - (\mu_1 - \mu_0)/2)$	$H = 5 \times \sigma$
0, 0157	0, 182	0, 121	0, 0834	0, 609

Where:
- RV = Representative Value and FM = Fuzzy Mode.
- $1C = 1^{st}$ choice, $2C = 2^{nd}$ choice, $3C = 3^{rd}$ choice, $4C = 4^{th}$ choice and $5C = 5^{th}$ choice

A single representative value is obtained per sample.

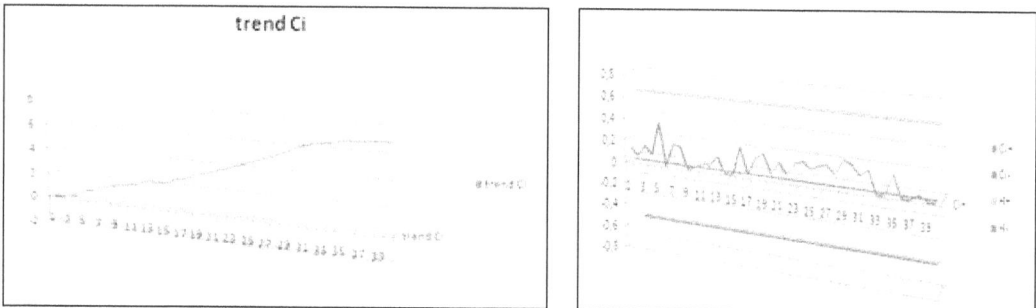

Fig. 1: Fuzzy MCUSUM

6. FUZZY MULTIVARIATE CUSUM CONTROL CHART

Table 1 shows the chosen parameters. Representative values are used to define the specific parameters of cusum control chart. C_i chart has a trend which give an idea that, the process is out of control, however, all observations lie between the limits of control chart. If the trend continues, then the process will soon be out of control.

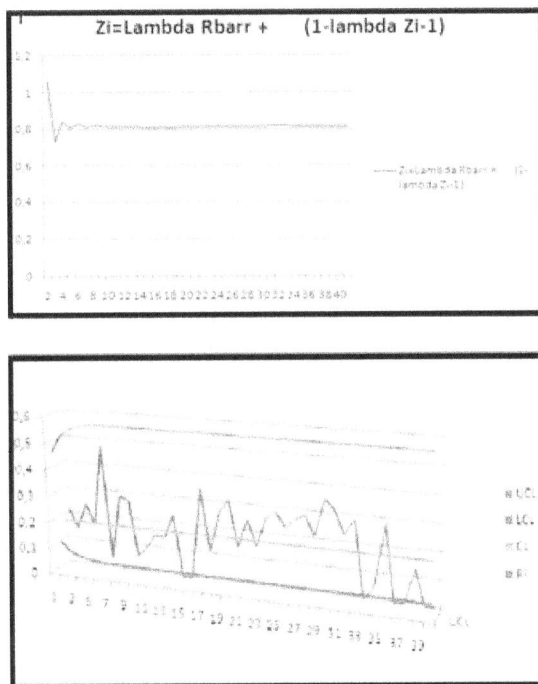

Fig. 2: Fuzzy MEWMA

Table 3: Ewma Parameters

n	λ	μ_0	L	σ
40	0, 3	0, 182	2.7	0.121

Fuzzy multivariate ewma control chart

Table 3 shows the chosen parameters. All observations lie between the upper and lower bounds, then the process is under control.

7. CONCLUSION

Fuzzy multivariate cusum and ewma control charts use current and previous observations to draw quality charts based on linguistic variables and fuzzy sets with triangular shapes and single fuzzy core. Fuzzy sets are multiplied by proportions of quality that characterizes them in the sample and all added after that in order to get a single fuzzy set which represent all products in this sample. Data of "ideale Sanitaire" for the 40 samples are between upper and lower limits which implies that the products are under control. However, the cumulative sum shows a trend which provides that the sample will be soon out of control. Future research should study accurately the exact ways of membership function's identification to reduce subjectivity and improve the use of mathematical tools.

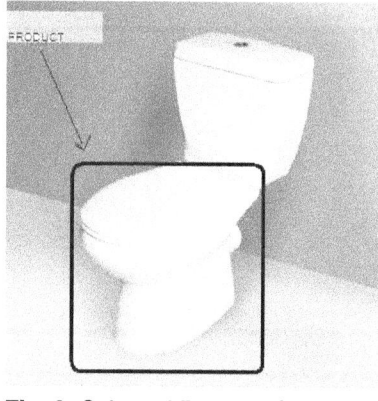

Fig. 3: Selected Product "Saphir wc"

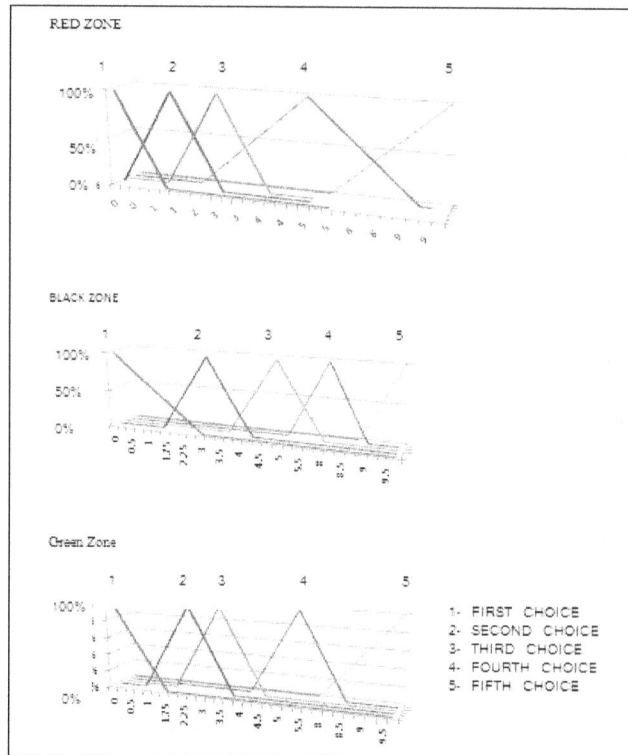

Fig. 4: Fuzzy Sets

REFERENCES

[1] Montgomery, D. (2005). "Introduction to Statistical Quality Control", 5th edition. Zadeh, L.A. (1965). "Fuzzy sets", "Information and control", 8, 338–353.

[2] Zadeh, L.A. (1975). "The Concept of a Linguistic Variable and its Application to Approximate Reasoning-I"; Information Sciences 8, 199–249.

[3] Zadeh, L.A. (1975). "The Concept of a Linguistic Variable and its Application to Approximate Reasoning-II", Information Sciences 8, 301–357.

[4] Futoshi, Tamak; Akihiro, Kanagawa and Hiroshi, Otha (1998). Identification of membership functions based on fuzzy observation data, Fuzzy Sets and Systems, Volume 93, Issue 3.

[5] Dombi, J. (1990). Membership function as an evaluation; Fuzzy Sets and Systems, Vol. 35, Issue 1.

[6] Andrs, L. Medaglia, Shu-Cherng Fang, Henry L.W. Nuttle and James R. Wilson, (2002), "An efficient and flexible mechanism for constructing membership functions, European Journal of Operational Research, Vol. 139, Issue 1.

[7] Taleb, H. and Limam, M.M.T. (2005). "Fuzzy Multinomial Control Charts", Lecture Notes in Computer Science, Vol. 3673/2005, 553–563.

[8] Taleb, H., Limam, MMT (2002). "On fuzzy and probabilistic control charts", Taylor and Francis, 40, 12. 2849–2863.

[9] Alipour, H. and Noorossana, R. (2010). "Fuzzy multivariate exponentially weighted moving average control chart"; 48:10011007 DOI 10.1007/s00170-009-2365-4.

Application of Taguchi's Beta Correction (Partial Correction) Technique for Process Control

Vidya Sagar Akkinapalli[1] and U.H. Acharya[2]

[1]TVS Motor Company, Mysore
[2]Indian Statistical Institute, Bangalore
E-mail: [1]a.vidyasagar@tvsmotor.co.in; [2]uhacharya@gmail.com

ABSTRACT: This case study details the application of Taguchi Beta Correction (Partial Correction) technique on machining process in an automobile industry. During the process of machining called fine boring, of the component Crank shaft, it was being observed that components were getting rejected for oversize defect. Upon observation of the defective components, it was evident that components were rejected due to setting error or size correction error by the machine operator. The process of fine boring was studied for effecting proper controls. The on-line quality control technique viz, Partial correction technique (Taguchi Beta Correction Technique) was applied for adjusting the bore size based on process capability. A ready reckoner was developed and implemented. The process scrap due to setting or adjustment of the bore size was totally eliminated. To sustain the results a reference chart has been prepared and displayed near the machine, and also included in the control plan of the component.

Keywords: Beta Correction Technique, Process Control, Analysis of Variance, Fine Boring, Crank Shaft.

1. INTRODUCTION

During the process of machining in an operation called fine boring of the component, crank shaft (Figure 1); it was being observed that components were getting rejected for oversize defect on internal diameter. The specification for the same was 24.990–25.010 mm.

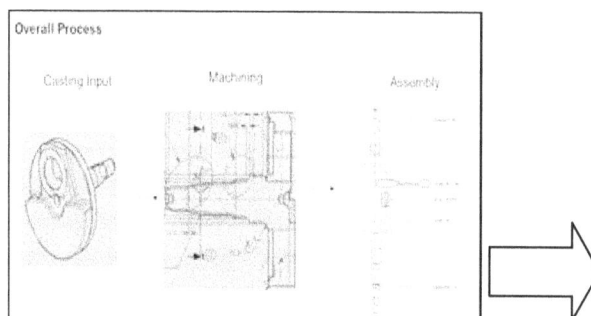

Fig. 1: Process Flow of Crank Shaft Assembly

Average rejection due to bore over size defect was 3665 PPM (Max 6874 and Min 2398 during February 2011 to July 2011 (Figure 2). Upon observation in the defective components, it was evident that components were rejected due to setting error or size correction error by the machine operator. Given below is the trend of bore oversize defect.

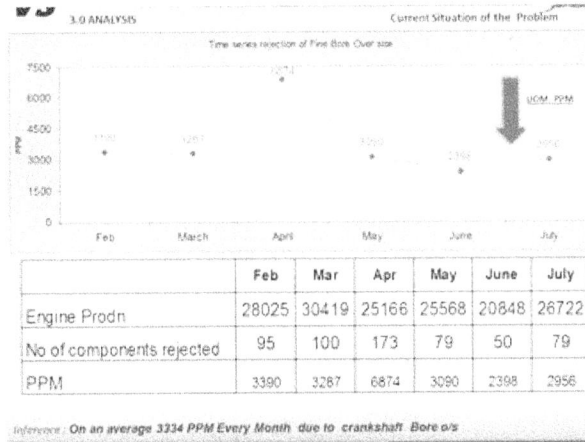

	Feb	Mar	Apr	May	June	July
Engine Prodn	28025	30419	25166	25568	20848	26722
No of components rejected	95	100	173	79	50	79
PPM	3390	3287	6874	3090	2398	2956

Inference : On an average 3334 PPM Every Month due to crankshaft Bore o/s

Fig. 2: PPM Chart

All the process conditions with respect to the 4M i.e. Man, Machine, Method and Material were checked for any assignable cause of variation and found that the given process conditions were being adhered (Table 1).

Table 1

4 'M	Sl.no	Parameter	Standard	Actual	Checking Method	Status
Machine	1	RAM PRESSURE	100 - 150 BAR	120 bar	PRESSURE GAUGE	OK
	2	Hydraulic level	Green band	OK	Visual	OK
	3	Lub oil level	Green band	OK	Visual	OK
Method	4	Locating diameter LH fixture	20/20.05 mm	20.01 mm	Bore Mic.	OK
	5	Locating diameter -RH fixture	18/18.05 mm	18.02 mm	Bore Mic.	OK
	6	Top fixture parallism	80 % blue match	85%	Use blue	OK
	7	Bottom fixture parallism	80 % blue match	85%	Use blue	OK
	8	Tightness of top & bottom fixture	Should be tight	OK	Use allen key	OK
	9	Orientation pin projection	Not allowed	OK	Visual	OK
	10	Orientation pin diameter -RH	24.97/24.95 mm	24.96 mm	Micrometer	OK
Man	11	Operator skill level	L3	L3	Skill matrix	OK
Material	12	Truing value	0.05mm max	0.05mm	Gauge	OK
	13	Checking frequency	100%	100%	Gauge	OK
	14	Con rod float	0.2 /0.4 mm	0.25 mm	Feeler gauge	OK
	15	Assembly width	42 +/ 0.2 mm	42.13 mm	Snap Gauge	OK

2. ANALYSIS STRATEGY

Though the problem could have been studied using conventional control charts, it was not clear as to how often samples are to be evaluated and if adjustment on bore size is found necessary, how much adjustment is to be made. As these questions can easily be answered in the Beta Correction Technique method, it was decided to plan the study through it. The following steps were identified for applying the beta correction technique.

- Step 1: Study the literature of Beta correction technique.
- Step 2: Collect data and study the extent of inherent variation in the process.
- Step 3: Arrive at the correction factor based on Beta correction technique.
- Step 4: Prepare and implement ready reckoner for the operation.

3. LITERATURE STUDY

"β-Correction factor" for on line process control is technique developed by Dr Taguchi of Japan:

1. Taguchi defines quality as a loss $= K\ E(Y - M)^2$ where m is the target and K is a constant. $K = A/\Delta^2$
2. Where, A is the tolerance on one side
3. $L = K\ E(y - m)^2$
   $= K\ E(y + \mu + \mu - m)^2$
   $= K[E(y - \mu)^2 + E(\mu - m)^2]$
   $= K\ [\square^2 + \square^2]$

Best quality products are those whose values are at m. Quality Control should aim for product at target with least possible variation around the target.

Reduction of overall Variation is aimed through the following:

(a) Reduction or elimination of deviation from the aimed at value i.e., $\square = (\mu - m)$;
(b) Controlling σ^2 with σ as inherent standard deviation of the process i.e., maintenance of variation with capability;
(c) Reduction of process capability.

R charts aim for achieving a) and b), whereas β-correction factor aims at a) only β-correction factor is applicable when,

- The characteristic is measurable and adjustable.
- The process is so engineered that the process standard deviation does not require day to day monitoring.
- Process setting requires close monitoring.

The major advantage of β-correction factor is that it not only tells when to control but also tells how much has to be adjusted.

4. DERIVATION OF B-CORRECTION FACTOR

Let μ_0 be the aimed at average, μ be an estimate of population parameter which is over unknown.

Once a process is set by an operator, we want to know whether it has been properly set or not. We take a sample i.e., μ and find out the deviation from is μ_0 i.e., $(\mu - \mu_0)$. Let β be the corrector factor which is assumed to be non-negative. Therefore, the amount of correction needed is $-\beta (\mu_0 - \mu)$.

Thus, μ becomes $\mu - \beta (\mu - \mu_0)$, after correcting which is supposed to be equal to μ_0 (aimed at value). Find out β such that $E (\mu - \beta (\mu - \mu_0) - \mu_0)^2$ is minimum.

Now $E (\mu - \mu)^2 = \sigma^2$

$E (\mu - \mu)^2 = E (\mu - \mu + \mu - \mu_0)^2$

$\quad = E (\mu - \mu)^2 + E (\mu - \mu_0)^2 + 2E (\mu - \mu) (\mu - \mu_0)$

$\quad = \sigma^2 + ((\mu - \mu_0)^2$

$E (\mu - \beta (\mu - \mu_0)^2 - \mu_0)^2 = E (\mu - \beta (\mu - \mu + \mu - \mu_0) - \mu_0)^2$

$\quad = E (\mu - \mu_0) - \beta(\mu - \mu) - \beta(\mu - \mu_0)^2$

$\quad = E (\mu - \mu_0) (1 - \beta) - \beta (\mu - \mu)^2$

$\quad = (1 - \beta)^2 (\mu - \mu_0)^2 + \sigma^2\beta^2$ ... (1)

By differentiating Equation (1) with respect of β and equating to zero

$-2 (1-\beta) (\mu-\mu_0)^2 + 2\beta\sigma^2 = 0$

Or $(1-\beta) (\mu-\mu_0)^2 - \beta\sigma^2 = 0$

Or $(\mu-\mu_0)^2 \beta(\sigma^2 + (\mu-\mu_0)^2) = 0$

i.e., $\beta = (\mu-\mu_0)^2/(\sigma^2 + (\mu-\mu_0)^2$

$\quad = 1 - \sigma^2/(\sigma^2 + (\mu-\mu_0)^2)$

$\quad = 1 - \sigma^2/(\mu - \mu_0)^2$

$\quad = 1 - 1/F$ where $f = ((\mu-\mu_0)^2/\sigma$

Since, β by definition is non negative,

$B = 0$ when $(\mu-\mu_0)^2 \leq \sigma^2$

$\quad = 1 - 1/F$ otherwise

5. CONTROL SCHEME

(a) Estimate σ i.e., prediction error standard deviation as follows:

Let y be the characteristic to be controlled. Let $y_1, y_2, y_a, ..., y_k$ be the observation in order of production with an internal of $\triangle t$. $\triangle t$ could be 5 mts. 10 mts, 15 mts., etc. depending on the rate of production.

$K = 2^n$

Carry out ANOVA as follows (Nested design)

Between 2^{n-1} observations

Between $2^{n-2}/2^{n-1}$ observations

Between $2^{n-4}/2^{n-2}/2^{n-t}$ observations

Find out the smallest significant interval and estimate σ.

(b) For control purpose taken an observation at a time interval $\triangle T$, where $\triangle T$ is the half of the lowest significant interval.

(c) Find out $(\mu - \mu_0)$

If $(\mu - \mu_0) > \sigma$, then

(d) Calculate $F = (\mu - \mu_0/\sigma)^2$

(e) Calculate $\beta = 1 - 1/F$

(f) Amount of correction to be made = i.e., $-\beta (\mu - \mu_0)$

(g) Make ready reckoner for different hypothetical $(\mu - \mu_0)$ i.e., calculate $-\beta (\mu - \mu_0)$ for different value of μ. This will avoid calculations on the shop floor.

6. STUDY OF VARIATION

The extent of variation in the process was studied through histogram and capability analysis as shown in the Figure 3.

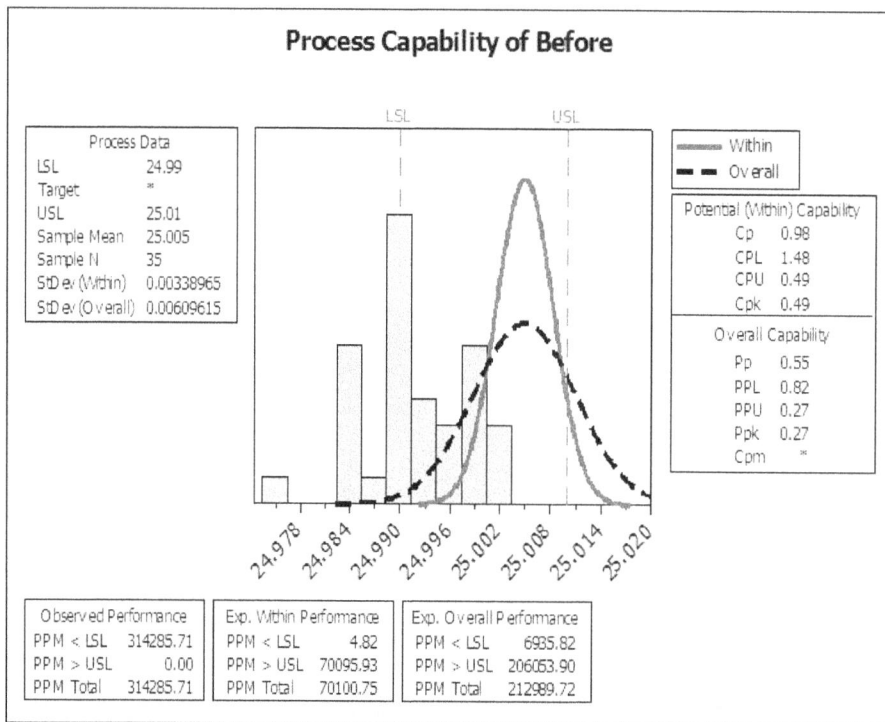

Fig. 3: Process Capability Analysis

7. ARRIVING AT THE CORRECTION FACTOR

The Correction factor was arrived after the above process analysis and the detailed working for arriving at the correction factor has been shown in Table 2.

Table 2: Correction Factor

SL No	Dia Of the Component	X-T	X-T square	F = X-T square/ Sigma Square	1/F	Beta Value = 1-1/F	Beta Correction = (-) Beta(X-T)
1	25.010	0.010	0.0001	11.11111111	0.09	0.91	-0.009
2	25.009	0.009	8.1E-05	9	0.111111111	0.88888889	-0.008
3	25.008	0.008	6.4E-05	7.111111111	0.140625	0.859375	-0.007
4	25.007	0.007	4.9E-05	5.444444444	0.183673469	0.81632653	-0.006
5	25.006	0.006	3.6E-05	4	0.25	0.75	-0.005
6	25.005	0.005	2.5E-05	2.777777778	0.36	0.64	-0.003
7	25.004	0.004	1.6E-05	1.777777778	0.5625	0.4375	-0.002
8	25.003	0.003	9E-06	1	1	7.5939E-14	0.000
9	25.002	0.002	4E-06	0.444444444	2.25	-1.25	-0.003
10	25.001	0.001	1E-06	0.111111111	9	-8	-0.008
11	25	0	0	0	0	0	0
12	24.999	-0.001	1E-06	0.111111111	9	-8	0.008
13	24.998	-0.002	4E-06	0.444444444	2.25	-1.25	0.0025
14	24.997	-0.003	9E-06	1	1	7.5939E-14	-2.27818E-16
15	24.996	-0.004	1.6E-05	1.777777778	0.5625	0.4375	-0.00175
16	24.995	-0.005	2.5E-05	2.777777778	0.36	0.64	-0.0032
17	24.994	-0.006	3.6E-05	4	0.25	0.75	-0.0045
18	24.993	-0.007	4.9E-05	5.444444444	0.183673469	0.81632653	-0.005714286
19	24.992	-0.008	6.4E-05	7.111111111	0.140625	0.859375	-0.006875
20	24.991	-0.009	8.1E-05	9	0.111111111	0.88888889	-0.008
21	24.99	-0.01	0.0001	11.11111111	0.09	0.91	-0.0091

8. PREPARE AND IMPLEMENT READY RECKONER

Operation – Fine Boring

Component – Crank Shaft

Quality Characteristic – Internal Diameter

Dimension – 24.990–25.010 mm (Target value–25.000, Tolerance–0.020 mm)

The ready reckoner was displayed near Fine Boring Machine of Crank Shaft. The machine operators were trained on how to give correction based on ready reckoner.

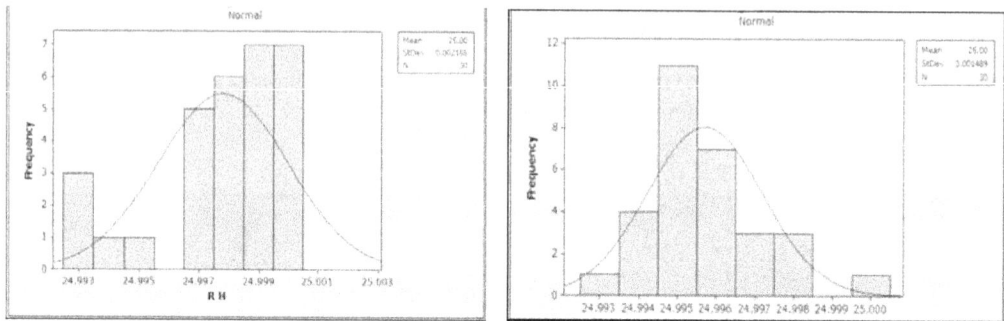

Fig. 4: Histogram

After Implementation of Partial correction technique the process scrap due to setting or adjustment of the bore size was totally eliminated.

Table 3: Implementation

Dia Of the Component	Beta Correction= Beta(X-T)
25.010	-0.009
25.009	-0.008
25.008	-0.007
25.007	-0.006
25.006	-0.005
25.005	-0.003
25.004	-0.002
25.003	
25.002	
25.001	
25.000	No Correction to be carried out
24.999	
24.998	
24.997	
24.996	0.002
24.995	0.003
24.994	0.005
24.993	0.006
24.992	0.007
24.991	0.008
24.990	0.009

7. CONCLUSION

In this paper, the authors discussed a case study details the application of Taguchi Beta Correction technique on fine boring in an automobile industry. The on–line quality control technique viz, Partial correction technique (Taguchi Beta Correction Technique) was applied for adjusting the bore size based on process capability. A ready reckoner was developed and implemented. The process scrap due to setting or adjustment of the bore size was totally eliminated.

Statistical Economic Design of X̄ Control Chart with Pareto In-Control Times

Neelufur

GITAM Institute of Technology, GITAM University, Visakhapatnam–45, India
E-mail: neelufur2000@yahoo.com

ABSTRACT: The design of X̄-chart is heavily influenced by the distributional parameters of the in-control time. This paper deals with the economic statistical design of X̄-chart with Pareto in-control times in order to investigate, in general, the effect on the economic control chart parameters like sample size, time between two successive samples, and the cost per unit time of the distributional assumption. Using the cost model, the sensitivity analysis of the statistical economic design of the X̄-chart with respect to the distributional parameters is studied. This design is extended to the case when the out-of-control times are also random and follows a Weibull distribution. These designs are much useful for analyzing and monitoring the production processes of chemical, fertilizer and oil industries where the in-control times of the process are random and are having a long upper tail distribution.

Keywords: X̄ Control Chart, In-Control Time, Pareto Distribution, Out-of-Control Time, Expected Cycle Cost, Expected Cycle Time.

1. INTRODUCTION

The economic and statistical design of X̄ control chart plays a dominant role in production and manufacturing. Much work has been reported regarding economic design of X̄ control chart. Lorezen and Vance (1986) presented a unified model which establishes a common notation and flexible methodology for economic design of X̄ control chart. Extending this Mc Williams (1989) has considered one of the significant factors effecting the design of the X̄ control chart, namely the in-control times of the process. These in-control times of the process are random due to various reasons like tool wear, raw material quality, operator skill, etc. Al-Oraini and Rahim (2003) have developed economic statistical design of X̄ control chart with the assumption that the in-control time follows a Gamma distribution. However, in production processes such as fertilizers and chemicals the in-control times of a process will have a minimum period of time since the assignable cause arises only after certain period of time of the system operation. One of the probability distribution that can be characterized with this feature is the Pareto distribution. The Pareto distribution characterizes a limiting distribution of the waiting time (time to exceed a specific value of the process character) (Johnson *et al.*). With this motivation, in this article an economic statistical design of X̄ control chart with the assumption that the in-control times of the process are random and follow Pareto distribution has been developed and analyzed.

2. COST MODEL

The production cycle is considered by following the heuristic arguments of Lorenzen and Vance (1986) and shown in Figure 1. The production process is assumed to start in an in-control state. In order to detect a shift in the process mean, a sample of 'n' independent quality characteristic measurements $X_1, X_2, X_3, ..., X_n$ is taken at intervals of 'h' hours. The sample average $\bar{X}$ follows normal distribution with normal cumulative distribution function, 'Φ'.

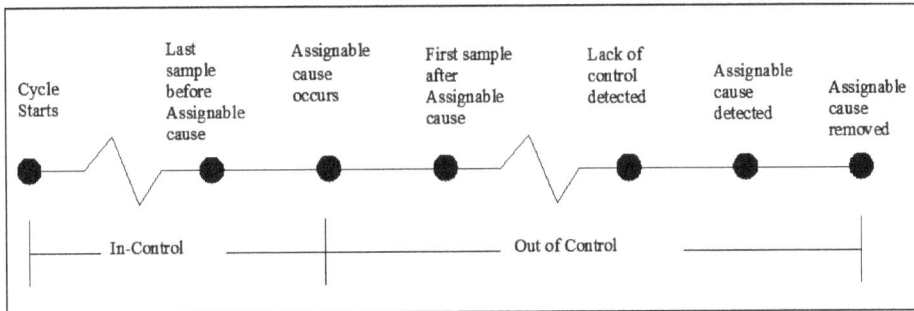

Fig. 1: Production Cycle

The $\bar{X}$ control chart is designed to detect whether the process is out of control or not. The design parameters 'n' and 'h' are chosen to minimize the expected cost per unit time. A quality cycle is defined as the time until the next in-control period. The in-control times in each cycle are identically and independently distributed. Hence, the expected hourly cost E(C/T) equals the ratio of the expected cycle cost to the expected cycle time *i.e.*, E(C)/E(T).

3. EXPECTED CYCLE TIME

From Figure 1.the expected cycle time consists of 4 parts namely, (1) expected time elapsed before assignable cause occurs, (2) expected time between the occurrence of the assignable cause and the next out-of-control signal, (3) expected time 'T$_1$' to identify the assignable cause and (4) expected time 'T$_2$' to repair the process. In this model, it is assumed that the process in-control times follow Pareto distribution with mean, $\left(\frac{c\theta}{c-1}\right)$ and the in-control times in each cycle are independently and identically distributed. Let $\delta_1 = 1$ if production continues during searches and '0' otherwise. Hence, the expected cycle time is,

$$E(T) = (1 - \delta_1).T_0.S.\alpha + \left(S + \frac{1}{(1-\beta)}\right).h + n.T_3 + T_1 + T_2 \qquad ...(1)$$

where, $\alpha = 2\Phi(-k)$ and $\beta = \Phi(k - \Delta\sqrt{n}) - \Phi(-k - \Delta\sqrt{n})$ ...(2)

Here, 'S' is the expected number of samples taken while the process is in control .We have,

$$S = \sum_{i=0}^{\infty} i * \Pr\{\text{assignable cause occurs between the } i^{th} \text{ and } (i+1)^{th} \text{samples}\}$$

$$= \sum_{i=\left[\left(\frac{\theta}{h}\right)+1\right]}^{\infty} i\left[\left(\frac{\theta}{i\,h}\right)^c - \left(\frac{\theta}{(1+i)\,h}\right)^c\right] \qquad ...(3)$$

where, $\left\{ 1 - \left(\frac{\theta}{\imath}\right)^c \right\}$ is the cumulative distribution function of Pareto distribution.

4. EXPECTED CYCLE COST

The cost of the entire cycle includes (1) cost of non-conformities, (2) cost of false alarms, (3) expected cost for sampling and charting the result and (4) cost of repairs, 'W'. Let 'C_0' be the hourly cost due to non-conformities produced while the process is in control and 'C_1' be the hourly cost due to non-conformities produced while the process is out of control, ($C_1 > C_0$). The expected cycle cost is,

$$E(C) = C_0 \cdot \left(\frac{c.\theta}{c-1}\right) + C_1 \left\{ \left(S + \frac{1}{(1-\beta)}\right).h - \left(\frac{c.\theta}{c-1}\right) + n.T_3 + \delta_1 T_1 + \delta_2 T_2 \right\} + C_2.S.a +$$
$$\frac{(a+b.n)}{h} \cdot \left\{ \left(S + \frac{1}{(1-\beta)}\right).h + n.T_3 + \delta_1.T_1 + \delta_2.T_2 \right\} + W \qquad \dots (4)$$

5. OPTIMAL DESIGN PARAMETERS

In this section the optimal design parameters of the $\bar{X}$-chart namely the sample size 'n' and the sampling interval 'h' are obtained by minimizing the expected cost per unit time 'Z', where,

$$Z = \frac{E(C)}{E(T)} \qquad \dots (5)$$

Substituting equations (1) and (4) in equation (5),

$$Z = \frac{\left\{\begin{array}{l} C_0\left(\frac{c.\theta}{c-1}\right) + C_1\left\{\left(S + \frac{1}{(1-\beta)}\right).h - \left(\frac{c.\theta}{c-1}\right) + n.T_3 + \delta_1 T_1 + \delta_2.T_2\right\} + C_2.S.a \\ + \frac{(a+b.n)}{h}.\left\{\left(S + \frac{1}{(1-\beta)}\right).h + n.T_3 + \delta_1.T_1 + \delta_2.T_2\right\} + W \end{array}\right\}}{\left\{(1-\delta_1).T_0.S.a + \left(S + \frac{1}{(1-\beta)}\right).h + n.T_3 + T_1 + T_2\right\}} \qquad \dots (6)$$

The optimum values for 'n' and 'h' are obtained by differentiating 'Z' with respect to 'n' and 'h' and equating them to zero.

Let $R = \frac{\partial S}{\partial h} = \sum_{i=\left[\left(\frac{\theta}{h}\right)+1\right]}^{\infty} \left[i.c.\theta^c.h^{-(c+1)} \cdot \left(\frac{1}{(1+i)^c} - \frac{1}{i^c}\right) \right]$

$\frac{\partial Z}{\partial n} = 0$ implies,

$$\frac{\left\{\begin{array}{l} \left\{(1-\delta_1).T_0.S.a + \left(S + \frac{1}{(1-\beta)}\right).h + n.T_3 + T_1 + T_2\right\}. \\ \left\{C_1.T_3 + \frac{(a+b.n)}{h}.T_3 + \left(\frac{b}{h}\right).\left\{\left(S + \frac{1}{(1-\beta)}\right).h + n.T_3 + \delta_1 T_1 + \delta_2.T_2\right\}\right\} \\ -\left\{\begin{array}{l}C_0\left(\frac{c.\theta}{c-1}\right) + C_1\left\{\left(S + \frac{1}{(1-\beta)}\right).h - \left(\frac{c.\theta}{c-1}\right) + n.T_3 + \delta_1.T_1 + \delta_2.T_2\right\} \\ + C_2.S.a + \frac{(a+b.n)}{h}\left\{\left(S + \frac{1}{(1-\beta)}\right).h + n.T_3 + \delta_2.T_1 + \delta_2.T_2\right\} + W\end{array}\right\}.T_3 \end{array}\right\}}{\left\{(1-\delta_1).T_0.S.a + \left(S + \frac{1}{(1-\beta)}\right).h + n.T_3 + T_1 + T_2\right\}^2} = 0$$
$$\qquad \dots (7)$$

$\frac{\partial z}{\partial h} = 0$ implies,

$$
\frac{
\begin{bmatrix}
\left\{(1-\delta_1).T_0.S.\alpha + \left(S+\frac{1}{1-\beta}\right).h + n.T_3 + T_1 + T_2\right\}. \\
\left(C_1.\left[S+\frac{1}{(1-\beta)}+h.R\right] + C_2.\alpha.R + \frac{(a+b.n)}{h}.\left\{S+\frac{1}{(1-\beta)}+h.R\right\}\right) \\
-\frac{(a+b.n)}{h^2}.\left\{\left(S+\frac{1}{(1-\beta)}\right).h + n.T_3 + \delta_1.T_1 + \delta_2.T_2\right\} \\
-\left\{\left(C_0\left(\frac{c.\theta}{c-1}\right)+C_1\left\{\left(S+\frac{1}{(1-\beta)}\right).h-\left(\frac{c.\theta}{c-1}\right)+n.T_3 + \delta_1.T_1 + \delta_2.T_2\right\} + C_2.S.\alpha\right)\right. \\
\left.+\frac{(a+b.n)}{h}.\left\{\left(S+\frac{1}{(1-\beta)}\right).h + n.T_3 + \delta_1.T_1 + \delta_2.T_2\right\} + W\right\} \\
\left\{S+\frac{1}{(1-\beta)}+h.R+(1-\delta_1).T_0.S.\alpha.R\right\}
\end{bmatrix}
}{
\left\{(1-\delta_1).T_0.S.\alpha + \left(S+\frac{1}{(1-\beta)}\right).h + n.T_3 + T_1 + T_2\right\}^2
} = 0
$$

... (8)

Solving the equations (7) and (8) iteratively using numerical method the optimal sample size (n*) and the optimal sampling interval (h*) can be obtained for given values of the model parameters and cost parameters.

6. NUMERICAL ILLUSTRATION

A fertilizer plant which produces urea, with phosphoric acid as a crucial intermediate product, whose quality has a direct bearing on the quality of the final product is considered. The quality character of this product is measured through the concentration of phosphoric acid in terms of weight percentages. The manufacturer uses $\overline{X}$ chart to monitor the process. Based on the analysis of Quality Control technicians salaries, cost of testing equipment etc., it is estimated that a = ₹ 0.5, b = ₹ 0.1, T_0 = 1 hr, T_1 = 2 hrs, T_2 = 1 hr, T_3 = 0.05 hrs, C_0 = ₹ 10, C_1 = ₹ 20, C_2 = ₹ 50, W = ₹ 200. From the records of the manufacturer, the concentration of phosphoric acid in the fertilizer follows normal distribution with α = 0.01 and β = 0.1. It is also estimated from the records of the manufacturer that the process in-control times follows Pareto distribution with parameters c = 2 and θ = 5. The goodness of fit of Pareto distribution to the in-control times is validated through a chi square test. The manufacturer wishes to design the $\overline{X}$ chart statistically and economically such that 'k' is 3 times standard deviation from the target value of the glass thickness and the cost can be reduced. A computer programme for minimizing the expected total cost per unit time with respect to the design parameters namely sample size (n) and time interval between sampling (h) has been developed using the equations (7) and (8). The optimal values of 'n' and 'h' are obtained as 8 and 5.703 hours respectively with an expected minimum cost of ₹ 26.169 per hour. Using these values of 'n' and 'h', a suitable $\overline{X}$ control chart can be designed.

7. SENSITIVITY ANALYSIS

In this section, sensitivity analysis has been performed to investigate how the parameters of Pareto distribution and the cost parameters effect the optimal values of the design parameters *i.e.*, sample size, 'n', sampling interval, 'h' and the expected cost per hour, 'Z'. The initial parameters of the cost model are set as follows: $c = 2$, $\theta = 5$, $\delta_1 = \delta_2 = 1$, $\alpha = 0.01$, $\beta = 0.1$, $T_3 = 0.05$ hrs, a = ₹ 0.5, b = ₹ 0.1, $T_0 = 1$ hr, $T_1 = 2$ hrs, $T_2 = 1$ hr, C_0 = ₹ 10, C_1 = ₹ 20, C_2 = ₹ 50, W = ₹ 200. Using the equations (7) and (8) and the initial values of the parameters as given above, the optimal sample size (n*) and the optimal interval between the successive samples (h*) are obtained. Substituting these values in the total cost, 'Z', the optimal total cost Z* is computed and all these values are presented in Table 1. The variation in optimal design parameters 'n', 'h' and 'Z' for various values of 'c' and 'θ' are shown in Figures 2 to 7.

Table 1: Optimal Values of n, h and Z for Various Values of c, θ, α, β, C_0, C_1, C_2, W, a and b

c	θ	α	β	C_0 (₹)	C_1 (₹)	C_2 (₹)	W (₹)	a (₹)	b (₹)	n*	h* (hrs)	z* (₹)
3	5	0.01	0.1	10	20	50	200	0.5	0.1	82	5.345	28.486
4										118	5.465	29.245
5										140	5.554	29.65
6										154	5.602	29.92
2	4.7	0.01	0.1	10	20	50	200	0.5	0.1	17	5.163	26.846
	4.8									14	5.34	26.615
	4.9									11	5.52	26.389
	5.1									5	5.891	25.954
2	5	0.01	0.1	10	20	50	200	0.5	0.1	8	5.703	26.169
		0.02								7	5.636	26.207
		0.03								6	5.567	26.245
		0.04								5	5.498	26.285
2	5	0.01	0.01	10	20	50	200	0.5	0.1	20	5.703	26.371
			0.03							17	5.703	26.329
			0.05							15	5.703	26.286
			0.10							8	5.703	26.169
2	5	0.01	0.1	11	20	50	200	0.5	0.1	18	5.506	26.75
				12						26	5.332	27.315
				13						34	5.178	27.866
				14						42	5.039	28.403
2	5	0.01	0.1	10	16	50	00	0.5	0.1	42	5.039	24.403
					17					34	5.178	24.866
					18					26	5.332	25.315
					19					18	5.506	25.75

c	θ	α	β	C_0 (₹)	C_1 (₹)	C_2 (₹)	W (₹)	a (₹)	b (₹)	n^*	h^* (hrs)	z^* (₹)
2	5	0.01	0.1	10	20	10	200	0.5	0.1	9	5.757	26.139
						30				9	5.73	26.154
						70				8	5.676	26.184
						90				7	5.649	26.199
2	5	0.01	0.1	10	20	50	210	0.5	0.1	18	5.506	26.75
							220			26	5.332	27.315
							230			34	5.178	27.866
							240			42	5.039	28.403
2	5	0.01	0.1	10	20	50	200	0.1	0.1	10	5.755	26.099
								0.3		9	5.729	26.134
								0.7		7	5.678	26.204
								0.9		6	5.653	26.239
2	5	0.01	0.1	10	20	50	200	0.5	0.08	29	5.225	26.101
									0.09	18	5.472	26.145
									0.10	8	5.703	26.169

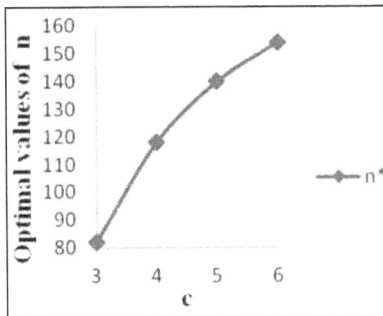

Fig. 2: 'c' Vs Optimal Values of 'n'

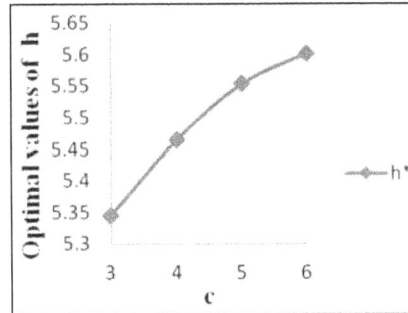

Fig. 3: 'c' Vs Optimal Values of 'h'

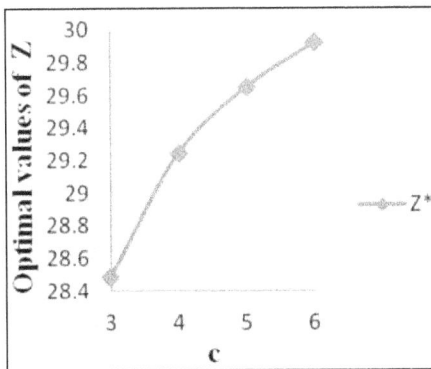

Fig. 4: 'c' Vs Optimal Values of 'Z'

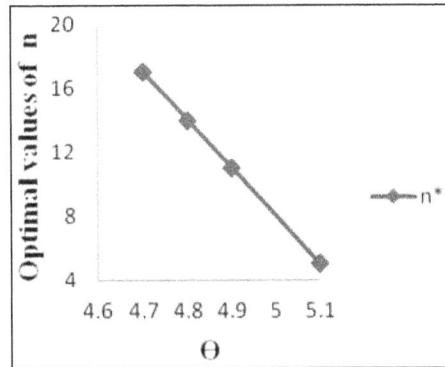

Fig. 5: 'θ' Vs Optimal Values of 'n'

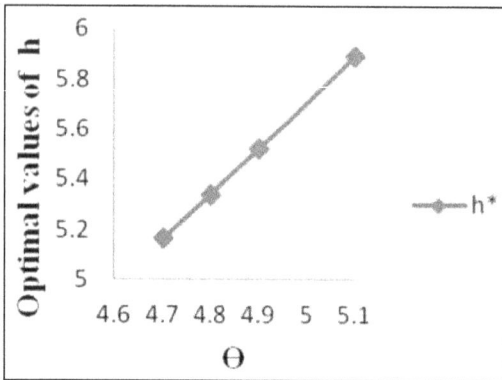

Fig. 6: 'θ' Vs Optimal Values of 'h'

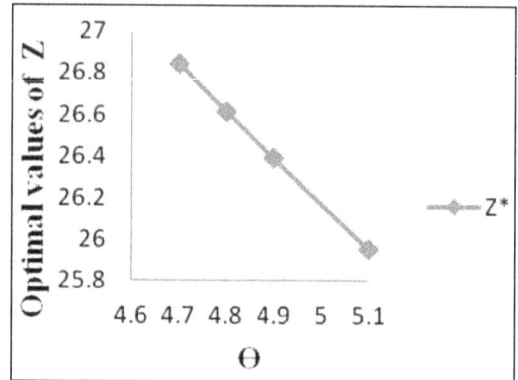

Fig. 7: 'θ' Vs Optimal Values of 'Z'

8. COST MODEL WITH GENERALIZED PARETO IN-CONTROL TIMES AND WEIBULL OUT-OF-CONTROL TIMES

In the earlier sections, it is assumed that 'T_1' and 'T_2' are fixed and known. But in many production processes there are multiple assignable causes such as defective raw materials, faulty setup, untrained operators, the cumulative effect of heat, vibration, shocks, power fluctuations, etc., When the process is governed by multiple assignable causes, 'T_1' and 'T_2' are also random and follow a probability distribution. A suitable distribution for 'T_1' and 'T_2' is a Weibull distribution since it accommodates constant, increasing and decreasing hazard rates. Hence, here it is assumed that 'T_1' and 'T_2' follow Weibull distributions with parameters (λ_1, v_1) and (λ_2, v_2) respectively, 'λ' being the scale parameter and 'v' being the shape parameter.

The expected values of 'T_1' and 'T_2' are:

$$E(T_1) = \frac{1}{\lambda_1} . \Gamma\left(1 + \frac{1}{v_1}\right) \qquad \qquad \dots (9)$$

$$\text{and } E(T_2) = \frac{1}{\lambda_2} . \Gamma\left(1 + \frac{1}{v_2}\right) \qquad \qquad \dots (10)$$

Substituting these values in the equation (6), the expected cost per unit time,

$$Z = \frac{\left\{ C_0\left(\frac{c.\theta}{c-1}\right) + C_1\left\{\left(S + \frac{1}{(1-\beta)}\right).h - \left(\frac{c.\theta}{c-1}\right) + n.T_3 + \frac{\delta_1}{\lambda_1}.\Gamma\left(1 + \frac{1}{v_1}\right) + \frac{\delta_2}{\lambda_2}.\Gamma\left(1 + \frac{1}{v_2}\right)\right\}\right\}}{\left\{(1-\delta_1).T_0.S.\alpha + \left(S + \frac{1}{(1-\beta)}\right).h + n.T_3 + \frac{\Gamma\left(1 + \frac{1}{v_1}\right)}{\lambda_1} + \frac{\Gamma\left(1 + \frac{1}{v_2}\right)}{\lambda_2}\right\}}{\left\{(1-\delta_1).T_0.S.\alpha + \left(S + \frac{1}{(1-\beta)}\right).h + n.T_3 + \frac{\Gamma\left(1 + \frac{1}{v_1}\right)}{\lambda_1} + \frac{\Gamma\left(1 + \frac{1}{v_2}\right)}{\lambda_2}\right\}}$$

$$\dots (11)$$

For obtaining the optimal design parameters of the $\bar{X}$ chart, 'Z' is differentiated with respect to 'n' and 'h' and is equated to zero. $\frac{\partial Z}{\partial n} = 0$, $\frac{\partial Z}{\partial h} = 0$ imply,

$$\frac{\left[\begin{array}{l}\left\{(1-\delta_1).T_0.S.\alpha+\left(S+\frac{1}{(1-\beta)}\right).h+n.T_3+\dfrac{\Gamma\left(1+\frac{1}{v_1}\right)}{\lambda_1}+\dfrac{\Gamma\left(1+\frac{1}{v_2}\right)}{\lambda_2}\right\}.\\ \left\{C_1.T_2+\frac{(a+b.n)}{h}.T_2+\left(\frac{b}{h}\right).\left[\left(S+\frac{1}{(1-\delta)}\right).h+n.T_3+\frac{\delta_1}{\lambda_1}.\Gamma\left(1+\frac{1}{v_1}\right)+\frac{\delta_2}{\lambda_2}.\Gamma\left(1+\frac{1}{v_2}\right)\right]\right\}\\ -\left\{\begin{array}{l}C_0\left(\frac{c.\theta}{c-1}\right)+C_1\left\{\left(S+\frac{1}{(1-\beta)}\right).h-\left(\frac{c.\theta}{c-1}\right)+n.T_3+\frac{\delta_1}{\lambda_1}.\Gamma\left(1+\frac{1}{v_1}\right)+\frac{\delta_2}{\lambda_2}.\Gamma\left(1+\frac{1}{v_2}\right)\right\}\\ +C_2.S.\alpha+\frac{(a+b.n)}{h}.\left\{\left(S+\frac{1}{(1-\beta)}\right).h+n.T_3+\frac{\delta_1}{\lambda_1}.\Gamma\left(1+\frac{1}{v_1}\right)+\frac{\delta_2}{\lambda_2}.\Gamma\left(1+\frac{1}{v_2}\right)\right\}+W\end{array}\right\}.T_2\end{array}\right]}{\left\{(1-\delta_1).T_0.S.\alpha+\left(S+\frac{1}{(1-\beta)}\right).h+n.T_3+\dfrac{\Gamma\left(1+\frac{1}{v_1}\right)}{\lambda_1}+\dfrac{\Gamma\left(1+\frac{1}{v_2}\right)}{\lambda_2}\right\}^2}=0$$

$$\dots (12)$$

$$\frac{\left[\begin{array}{l}\left\{(1-\delta_1).T_0.S.\alpha+\left(S+\frac{1}{(1-\beta)}\right)h+n.T_3+\dfrac{\Gamma\left(1+\frac{1}{v_1}\right)}{\lambda_1}+\dfrac{\Gamma\left(1+\frac{1}{v_2}\right)}{\lambda_2}\right\}.\\ \left\{\begin{array}{l}C_1.\left[S+\frac{1}{(1-\beta)}+h.R\right]+C_2.\alpha.R+\frac{(a+b.n)}{h}.\left\{S+\frac{1}{(1-\beta)}+h.R\right\}\\ -\frac{(a+b.n)}{h^2}.\left\{\left(S+\frac{1}{(1-\beta)}\right).h+n.T_3+\frac{\delta_1}{\lambda_1}.\Gamma\left(1+\frac{1}{v_1}\right)+\frac{\delta_2}{\lambda_2}.\Gamma\left(1+\frac{1}{v_2}\right)\right\}\end{array}\right\}\\ -\left\{\begin{array}{l}C_0\left(\frac{c.\theta}{c-1}\right)+C_1\left\{\left(S+\frac{1}{(1-\beta)}\right).h-\left(\frac{c.\theta}{c-1}\right)+n.T_3+\frac{\delta_1}{\lambda_1}.\Gamma\left(1+\frac{1}{v_1}\right)+\frac{\delta_2}{\lambda_2}.\Gamma\left(1+\frac{1}{v_2}\right)\right\}\\ +C_2.S.\alpha+\frac{(a+b.n)}{h}.\left\{\left(S+\frac{1}{(1-\beta)}\right).h+n.T_3+\frac{\delta_1}{\lambda_1}.\Gamma\left(1+\frac{1}{v_1}\right)+\frac{\delta_2}{\lambda_2}.\Gamma\left(1+\frac{1}{v_2}\right)\right\}+W\end{array}\right\}\\ \left\{S+\frac{1}{(1-\beta)}+h.R+(1-\delta_1).T_0.S.\alpha.R\right\}\end{array}\right]}{\left\{(1-\delta_1).T_0.S.\alpha+\left(S+\frac{1}{(1-\beta)}\right).h+n.T_3+\dfrac{\Gamma\left(1+\frac{1}{v_1}\right)}{\lambda_1}+\dfrac{\Gamma\left(1+\frac{1}{v_2}\right)}{\lambda_2}\right\}^2}=0$$

$$\dots (13)$$

Solving the equations (12) and (13) simultaneously for 'n' and 'h' using numerical technique, the optimal sample size (n*) and the optimal time interval between successive samples (h*) are obtained. To study the effect of the random nature of 'T_1' and 'T_2' on the optimal design parameters the sensitivity analysis for the parameters 'λ_1', 'v_1', 'λ_2' and 'v_2' has been carried out with the initial values of the other parameters as $c = 2$, $\theta = 5$, $\delta_1 = \delta_2 = 1$, $\alpha = 0.01$, $\beta = 0.1$, $T_3 = 0.05$ hrs, a = ₹ 0.5, b = ₹ 0.1, $T_0 = 1$hr, $C_0 = ₹ 10$, $C_1 = ₹ 20$, $C_2 = ₹ 50$, W = ₹ 200, $\lambda_1 = 1.5$, $v_1 = 0.7$, $\lambda_2 = 1$ and $v_2 = 0.5$ and are shown in Table 2.

Table 2: Optimal Values of n, h and Z for Various Values of λ_1, v_1, λ_2 and v_2

λ_1	v_1	λ_2	v_2	n*	h* (hrs)	Z* (₹)
3	0.7	1	0.5	15	5.559	26.374
4.5				16	5.525	26.425
6				17	5.508	26.450
7.5				18	5.498	26.466
1.5	0.5	1	0.5	4	5.789	26.053

λ_1	v_1	λ_2	v_2	n^*	h^* (hrs)	Z^* (₹)
	0.8			11	5.642	26.255
	1.2			12	5.61	26.301
	1.6			13	5.603	26.311
1.5	0.7	1	0.5	10	5.664	26.224
		2		22	5.42	26.585
		3		25	5.342	26.709
		6		29	5.266	26.834
1.5	0.7	1	0.6	16	5.541	26.401
			0.8	20	5.452	26.536
			1.2	22	5.406	26.607
			1.6	23	5.396	26.623

9. CONCLUSION

The economic statistical design of $\overline{X}$ control chart has been developed assuming that the mean of the quality characteristic, $\overline{X}$, follows normal distribution and the in-control time follows Pareto distribution. By fixing the Type-I error and Type-II error probabilities at given levels of 'α' and 'β', the out of control can be identified easily and effectively. Minimizing the expected cost per unit time, the optimal design parameters, namely, sample size and time interval between successive samples, are derived. The numerical values indicate that the effect of Pareto distribution parameters and cost parameters have significant effect on optimal design parameters. Sensitivity analysis carried out indicates that the optimal design parameters and the expected cost per unit time are more sensitive towards the cost parameters than other parameters. These charts are much useful for Quality Control of several manufacturing processes such as fertilizers, chemicals, paints, etc., where the in-control time of the process is characterized by a Pareto distribution.

REFERENCES

[1] Al-Oraini, H.A. and Rahim, M.A. (2003). "Economic statistical design of $\overline{X}$ control charts for systems with gamma (λ, 2) in-control times", *Journal of Applied Statistics*, Vol.30, pp. 397–409.

[2] Lorenzen, T.J. and Vance, L.C. (1986). "The economic design of control charts: A unified approach", *Technometrics*, Vol.28, pp. 3–10.

[3] McWilliams, T.P. (1989). "Economic control chart designs and the in-control time distribution: A Sensitivity study", *Journal of Quality Technology*, Vol. 21, pp. 103–110.

[4] Johnson, N.L., Kotz, S. and Balakrishnan, N. (1994). *Continuous Univariate Distributions*, Vol. 1 and 2, John Wiley & Sons. Inc, New York.

APPENDIX

n	Sample size
h	Time interval between successive samples
k	Number of standard deviations from control limits to centre line
Φ	Normal cumulative distribution function
ARL_0	Average run length when process is in control
ARL_1	Average run length when process is out of control
S	Expected number of samples taken during the in-control state
Δ	Number of standard deviations slip when out of control
δ_1	Indicator variable to indicate whether production continues or not during the assignable cause search, $\delta 1 = 1$ if production continues and '0', otherwise
δ_2	Indicator variable to indicate whether production continues or not during the repair process, $\delta 2 = 1$ if production continues and '0', otherwise
T_0	Expected assignable cause search time for a false alarm
T_1	Expected time to identify the assignable cause
T_2	Expected time to repair the process
T_3	Expected sampling time for one observation
a	Fixed cost per sample
b	Variable cost per sample
C_0	Hourly cost due to nonconformities produced while the process is in control
C_1	Hourly cost due to nonconformities produced while the process is out of control ($C1> C0$)
C_2	Cost per false alarm
W	Cost for locating and repairing the assignable cause
α	Probability that $\overline{X}$ falls outside the control limits when the process is in control
β	Probability that $\overline{X}$ falls within the control limits when the process is out of control
c, ν	Shape parameter
θ	Location parameter
λ	Scale parameter
E(C)	Expected cycle cost
E(T)	Expected cycle time
Z	Expected cost per unit time.

A Hybrid Software Quality Forecasting Model

Arun Duraibalan

Honeywell Technology Solutions, Bangalore, India
E-mail: arun.duraibalan@honeywell.com

ABSTRACT: Software is considered as a product whose production is fundamentally similar to other products and forecasting its quality can be approached using the same basic principles espoused by quality pioneers of hardware and consumer products. Software quality prediction models are based on software engineering measurements hypothesis and often built using historical measurements, metrics and fault data of similar projects or modules of previously developed releases. Practical challenges are the unavailability of such historical metrics and fault data and in precise classification of projects or releases.

To counter this twin fold problem a hybrid of heuristic decision making approach and Support Vector Machine (SVM) based classification is taken up in forecasting the software quality. Primary focus of this paper is on the classification of software product to enable effective SQPM subsequent to classification. The SVM makes a coherent classification of projects based on key parameters and differentiators, employing Linear SVM model and Optimum Separation Hyperplane method. The software specific data are scaled in tune with the training data scaling functions and applying the technical and project engineering expertise in selection on vital few for the software and in establishing standardized scaling and error factors. The domain expertise and scaling enables the specification of the generic functions and leading to SVM classification with less than 4% error. A suitable SQPM can be adapted to suite the current model, that may be selected from an array of Heuristic algorithm and regression techniques defined and analyzed in detail in the suggested references. The historical data or apposite customized fault data, even for a novice project type can be taken from the closest clusters from the SVM to attain precise quality prediction model.

Keywords: Software Quality Forecasting (SQF), Support Vector Machines (SVM), SVM Classification, Optimum Separation Hyperplane (OSH), Linear SVM, Software Quality Prediction Model (SQPM).

1. INTRODUCTION

Software development is being considered just as similar to product development in hardware, mechanical and such industrial approach. This healthy orientation paves way for the Software industry to apply and re-use a wide range of standards and very well structured models prevalent in those industries, developed over decades and fine tuned over the course. The application and adoption of the standards and proven models into Software industry has shown phenomenal business advantage along with cost benefits, especially in the phases of software development, tracking and effective management of the projects. Similarly for quality assurance and defect management the industry practices are being applied, which have shown considerable improvement and are evolving more specifically to suite the software environment.

The software quality assurance is an important part of any software project. The overall complexity and the average size of the software product keep growing whilst the time to market keeps drawing nearer. Assuring whether the desired software quality and reliability is met for a product is equally important and a critical factor as for keeping with the schedule, budget and customer requirements. Software Quality Forecasting or prediction (SQF) techniques are applied to appraise the quality of any software product during conception or during any other phases of SDLC. A Software Quality Prediction Model (SQPM) enables the project team to track and detect potential software defects during development cycle and reduce the rework effort and bring down the Cost of Poor Quality (COPQ) considerably. SQPM is created as regression expression or through trained models based on historical project and process metrics data and helps meeting the objectives of projects reliability and testing initiatives. The basic hypothesis of software quality prediction is that a module currently under development is fault prone if a module with the similar product or process metrics in an earlier project developed in the same environment was fault prone [Khoshgoftaar *et al.*, 1997]. Fault-proneness models are models that are built from information about the code and its faults, and that relate code to faults [Denaro G, Pezze M, 2002]. The ability of SQPM to accurately identify critical components drives for a focused development and verification and allows automated formal analysis methods to be applied. Efficient resource management and risk management practices can be laid on once the critical few are identified and fault-proneness data is available [Briand *et al.*, 1993]. Different types of SQPM models may be constituted based on the type of historical data and project environment, as fault-proneness of a software module is the probability that it contains faults. Distinctively a correlation exists between the fault-proneness of the software and the measurable attributes of the code (i.e. the static metrics) and of the testing (i.e. the dynamic metrics) [Denaro G., 2000]. Early detection of fault-prone software components enables verification experts to concentrate their time and resources on the problem areas of the software system under development. Early lifecycle data includes metrics describing unstructured textual requirement and static code metrics (E.g. Halstead complexity, Cyclomatic complexity, McCabe's complexity, etc.).

Forecasting Software Quality through the techniques and models espoused by quality pioneers of product industries has shown that the results are not applicable widely and are restricted only to a specific type or group of software products or projects. The most important reason for the SQPM based on software engineering measurements hypothesis not being successful is attributed to the fact that these models heavily rely on historical measurements, metrics and fault data (metric repositories constructed from similar projects or modules of previously developed releases). Practical challenges are the unavailability of such historical metrics and fault data and in precise classification of projects or products. In the rest of the sections of this paper, briefly explains the Support Vector Machine (SVM) classification technique followed by its application in software product classification. How the software specific data are scaled and applying the technical and project engineering expertise in selection on vital few for the software and in establishing standardized scaling and error factors. The advantage of such an approach boosts the generic functions and impact on the SVM classification. Concludes with possible SQPM approaches, which can be adapted to suite the current project or product with suggested references.

2. SUPPORT VECTOR MACHINES (SVM)

2.1 SVM Classification

SVM is Kernel-based techniques which are a group of supervised learning methods that can be applied to classification or regression. The SVM algorithm is based on the statistical learning theory and the Vapnik-Chervonenkis (VC) dimension and characterized by the loss function $V = (1 - y(fx))+$. The major hypothesis is that the descriptors capture some important characteristics of the pattern, and then a mathematical function (machine learning algorithm) can generate a mapping (relationship) between the descriptor space and the property. Another hypothesis is that similar objects (objects that are close in the descriptor space) have similar properties. SVM constructs a hyperplane or set of hyperplanes in a high- or infinite-dimensional space, which classifies the data. To achieve better results the functional margin has to be larger (*i.e.* the hyperplane that is farthest to the nearest training data points of any class), this keeps the generalization error minimal. There are many hyperplanes which can separate the data the criterion to choose best hyperplane depends on the data type and classification intended and the same is embedded in an appropriate cost function. It often happens that the sets to discriminate are not linearly separable in that space. For this reason, it was proposed that the original finite-dimensional space be mapped into a much higher-dimensional space, presumably making the separation easier in that space. To keep the computational load reasonable, the mapping used by SVM schemes are designed to ensure that dot products may be computed easily in terms of the variables in the original space, by defining them in terms of a kernel function $K(x, y)$ selected to suit the problem [Press *et al.*, 2007]. Let S be a set of l points xi ϵ IRd Each xi is given a label yi ϵ {−1, 1}. The hyperplane $w \cdot x + b = 0$ is a separating hyperplane. The signed distance di of xi from the separating hyperplane is di $= (w \cdot xi + b)/w$ [Cortes, Vapnik 1995].

Any hyperplane in the space S can be written as,

$$\{x \in S \mid w \cdot x + b = 0\}, w \in S, b \in R$$

The dot product $w \cdot x$ is defined by:

$$w \cdot x = \sum_{i=1}^{n} w_i x_i$$

The hyperplanes in the higher dimensional space are defined as the set of points whose inner product with a vector in that space is constant. The vectors defining the hyperplanes can be chosen to be linear combinations with parameters α_i of images of feature vectors that occur in the data base. Note that if $K(x, y)$ becomes small as y grows further from x, each element in the sum measures the degree of closeness of the test point x to the corresponding data base point xi. In this way, the sum of kernels above can be used to measure the relative nearness of each test point to the data points originating in one or the other of the sets to be discriminated.

2.2 SVM Model Selection

VC dimension measures the capacity of a hypothesis space. Capacity is a measure of complexity and measures the expressive power, richness or flexibility of a set of functions by assessing how wiggly its members can be, based on the definition of VC dimension on a set of real functions [Vapnik, 1999]. The VC dimension of a set of functions is the maximum number of points that can be separated in all possible ways by that set of functions. Hence for a chosen real function based on above SVM kernel and hyperplane equations a right model or hyperplane function has to be selected based on the domain. Once the right set of training data is identified, choose the function that performs well on the training data. Analyze the Empirical and true risk associated and this enables the overall risk reduction with the model and function selected. For a class of classification rules, F, consider the approach of finding a rule fn that minimizes the empirical risk $Rn(f)$ [Binkley, Schach 1998] As the empirical error is a discrete non-convex function of f, actual computation necessary to get the best rule over F could be complex. However, we set aside this computational issue for a moment and focus on the relationship between the empirical error and the true probability of error. An inductive principle for model selection used for learning from finite training data sets, describes a general model of capacity control and provides a tradeoff between hypothesis space complexity and the quality of fitting the training data (empirical error) [Vapnik, 1999]. Vapnik and Chervonenkis showed that an upper bound on the true risk can be given by the empirical risk + an additional term.

Simplification of bound: Test Error = Training Error + Complexity of set of Models.

2.3 Linear SVM

Classifying data is a common task in machine learning. Suppose some given data points each belong to one of two classes, and the goal is to decide which class a new data point will be in. In the case of support vector machines, a data point is viewed as a p-dimensional vector (a list of p numbers), and we want to know whether we can separate such points with a (p − 1)-dimensional hyperplane. This is called a linear classifier. The training set of data is linearly separable when there exists at least one linear classifier defined by the pair (w, b), that correctly classifies all training patterns (Figure 1). This linear classifier is represented by the hyperplane H (w · x + b = 0) and defines a region for class +1 patterns (w · x + b > 0) and another region for class −1 patterns (w · x + b < 0).

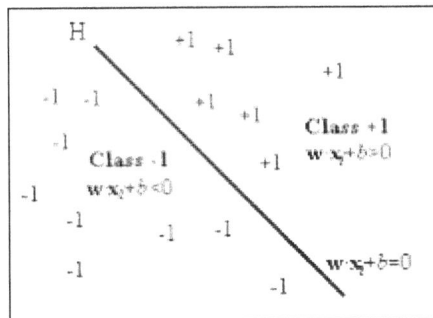

Fig. 1: Linear SVM Classifier

After training, the classifier is ready to predict the class membership for new patterns, different from those used in training. The class of a pattern x_k is determined with the equation:

$$\text{Class } (x_k) = \begin{cases} +1 & \text{if} \quad w \cdot x_k + b > 0 \\ -1 & \text{if} \quad w \cdot x_k + b < 0 \end{cases}$$

Therefore, the classification of new patterns depends only on the sign of the expression w · x + b.

2.4 Maximum Margin Classifier

There are many hyperplanes that might classify the data. One reasonable choice as the best hyperplane is the one that represents the largest separation, or margin, between the two classes. So we choose the hyperplane so that the distance from it to the nearest data point on each side is maximized. If such a hyperplane exists, it is known as the maximum-margin hyperplane and the linear classifier it defines is known as a maximum margin classifier or equivalently, the perceptron of optimal stability. The Optimum Separation Hyperplane (OSH) is the linear classifier with the maximum margin for a given finite set of learning patterns [Freund, Schapire 1998]. The OSH computation with a linear support vector machine is presented in this section.

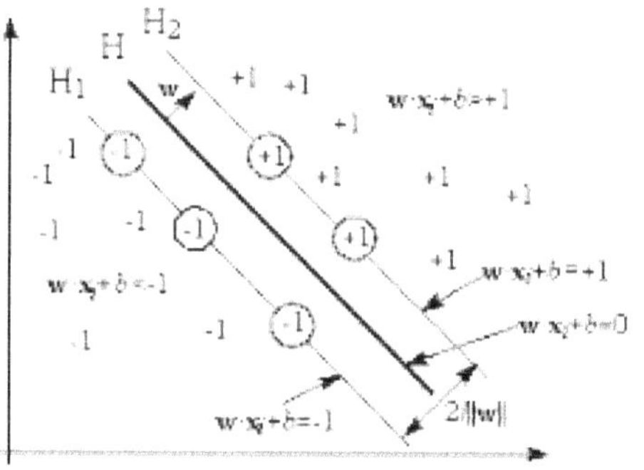

Fig. 2: The Optimum Separation Hyperplane (OSH)

Consider the classification of two classes of patterns that are linearly separable, *i.e.*, a linear classifier can perfectly separate them (Figure 2). The linear classifier is the hyperplane H (w · x + b = 0) with the maximum width (distance between hyperplanes H1 and H2). Consider a linear classifier characterized by the set of pairs (w, b) that satisfies the following inequalities for any pattern xi in the training set:

$$\begin{cases} w \cdot x_i + b > +1 & \text{if} \quad y_i = +1 \\ w \cdot x_i + b < -1 & \text{if} \quad y_i = -1 \end{cases}$$

These equations can be expressed in compact form as,

$$y_i(w \cdot x_i + b) \geq +1$$

Or

$$y_i(w \cdot x_i + b) - 1 \geq 0$$

Because we have considered the case of linearly separable classes, each such hyperplane (w, b) is a classifier that correctly separates all patterns from the training set:

$$\text{Class}(x_i) = \begin{cases} +1 & \text{if} \quad w \cdot x_i + b > 0 \\ -1 & \text{if} \quad w \cdot x_i + b < 0 \end{cases}$$

For all points from the hyperplane H (w · x + b = 0), the distance between origin and the hyperplane H is |b|/||w||. We consider the patterns from the class −1 that satisfy the equality w · x + b = −1, and determine the hyperplane H1; the distance between origin and the hyperplane H1 is equal to |−1−b|/||w||. Similarly, the patterns from the class +1 satisfy the equality w · x + b = +1, and determine the hyperplane H2; the distance between origin and the hyperplane H2 is equal to |+1−b|/||w||. Of course, hyperplanes H, H1, and H2 are parallel and no training patterns are located between hyperplanes H1 and H2. Based on the above considerations, the distance between hyperplanes (margin) H1 and H2 is 2/||w||.

From these considerations it follows that the identification of the optimum separation hyperplane is performed by maximizing 2/||w||, which is equivalent to minimizing ||w||2/2. The problem of finding the optimum separation hyperplane is represented by the identification of (w, b) which satisfies:

$$\begin{cases} w \cdot x_i + b \geq +1 & \text{if} \quad y_i = +1 \\ w \cdot x_i + b \leq -1 & \text{if} \quad y_i = -1 \end{cases}$$

3. HYBRID QUALITY FORECASTING MODEL

3.1 Hybrid Algorithm Overview

The proposed model is a Hybrid algorithm that employs the SVM techniques for Software Product classification [Scholkopf *et al.*, 2000)] and determines the cluster of projects that are in close proximity (level 0 or level 1). The historical data and metrics from the chosen software product (or normalized value from the set of projects in the clusters is taken as applicable). The generalized Heuristic algorithm/regression model (based on current project under consideration), is taken up and the specific adaptation applied for generic correction factors, model parameters. The metrics data (with critical software parameter factor corrections and any other error correction) is taken up and run over the SQPM as depicted in Figure 3.

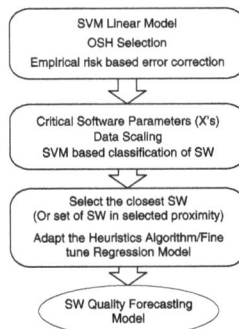

Fig. 3: Hybrid Software Quality Forecasting Model

3.2 Software Product Classification

The critical software parameters are determined and then based on technical and project engineering expert the software product under consideration is suitably graded off against the defined set of parameters or differentiators, with standard predefined set of grading or values. Note the critical software parameters are specific to the domain and type of software and it distinguishes and identifies them, this need not include the quality parameters only [Ebert, 1996]. The parameters can be based on the architecture used, the operating system, the technologies applied, domain type, number of interfaces, complexity and interactions of modules, etc. These data are then subject to Scaling before applying in SVM [David 2011], this is important and significantly improves the performance of the classifier. The main advantage of scaling is to avoid attributes in greater numeric ranges dominating those in smaller numeric ranges. Another advantage is to avoid numerical difficulties during the calculation. Because kernel values usually depend on the inner products of feature vectors and large attribute values might cause numerical problems. [Shawe, Cristianini 2004] Linearly scaling each attribute to the range [−1; +1] or [0; 1]. Note that the same method should be used to scale both training and testing data. For example, suppose that we scaled the first attribute of training data from [−10; +10] to [−1; +1]. If the first attribute of testing data lies in the range [−11; +8], then the testing data needs to be scaled equivalently as [−1:1; +0:8].

The SVM classifier from section 2 is trained with scaled data sets as described above, then the project under consideration is taken up and post equivalent scaling used as test data and subject to classification. The linear SVM with OSH function is deployed for classification, since the classification in this model is initial step and the accuracy to the degree of 0.1 and lower is not desired as the RBF and other polynomial SVM classifiers used in pattern recognition do give such accurate results. The disadvantage is the huge training time and the hyperplane functions to be made very precise and needs multiple analysis and test runs. Linear SVM has trade off and reduces all these factors and classification results are about 7% or more error prone. With proper parameter and factor selection and consistent scaling of data the Linear SVM has performed consistently better.

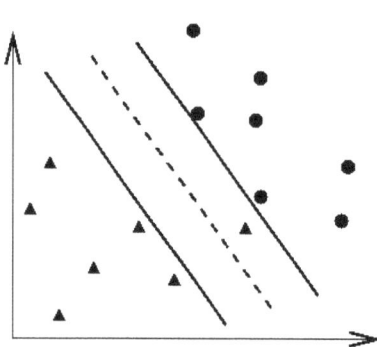

Fig. 4: Training Data SVM Classifier   **Fig. 5:** SVM Classifier with Testing Data

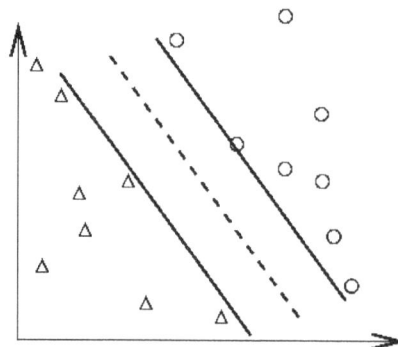

3.3 Application of the Heuristic Algorithm

This SQPM works in subsequent to the SVM classifier as defined in earlier sections. The Linear SVM with lesser accuracy (less than 7% error rate), is acceptable. SVM is employed for the purpose of classification of the concerned software product and identifying the closest cluster [Boetticher, 2004]. The hybrid algorithm selects the projects in close proximity *i.e.* Level 0 or Level 1 for consideration of historical data to be used in the prediction stages. Typically a Level 0 is the narrow band of project clustered around the project in consideration and based on type of project and confidence level of the expert graded values supplied it can be moved to Level 1 (*i.e.* –1 to 1) and this would give a larger cluster. The proximity measurements provide intrinsic measures of similarities between cases. Proximity measurements can also serve as similarity scores which is subsequently used in clustering. The historical data affecting the quality parameters are pulled up from the clusters and either a representative or normalized values are subject to final use in the SQPM. Figures 4 and 5 depicts the classifier response for the training and test data respectively. The final stage of implementing the model is performed through Heuristic algorithm approach or through the generic regression equation developed based on the historical data. The software data and metrics to use for the quality predicting depend on the project spectrum and Critical to Quality (CTQs) of the concerned software product. The detailed analysis on metrics and their implication in fault detection can be referred from [Khoshgoftaar, Seliya 2002], [Hudepohl *et al.*, 1996]. The classification and further assessment from subject matter experts if converges acutely and the error factor is also negligible, then the need for specific tuning of factors or correction factors is non evident. Such cases the use of regression function could satisfactorily provide the necessary prediction data and further analysis as desired. This scenario is applicable for proven and tested categories or like releases from same product category that has passed through quite a few revisions rendering further changes futile. For products that are comparatively novice entrance into the system even if converges on acutely need to cater as divergent ones and may not suffice the application of generic regression function. Hence the Heuristic algorithm need to applied for such instances where the convergence is greater than the optimal minima or when the concerned product is marked as novice. When software engineering metrics are used to predict quality attributes, ranking the importance of measurements is very important because it helps explain the dependencies in the development life-cycle. Having importance measures incorporated in the classification toolset further simplifies the experimental design. Numerous empirical studies confirm that many software metrics aggregated in software quality prediction models are valid predictors for qualities of general interest like maintainability and correctness. Hence, the selection of the quality model has influence on the quality assessment of software based on software metrics. Various SQPM models are analyzed and results compared and detailed out in the following referenced presentations and papers, viz. Rudiger *et al.*, 2010, Mitat, 2009 and Azar D., 2002, which enables in selecting the right approach and model.

4. CONCLUSION

The most important reason for the SQPM based on software engineering measurements hypothesis not being successful is attributed to the fact that these models heavily rely on hhistorical measurements, metrics and fault data (metric repositories constructed from similar projects or modules of previously developed releases). Practical challenges are the unavailability of such historical metrics and fault data and in precise classification of projects or products. The proposed SQFM model is a Hybrid algorithm that employs the OSH based Linear SVM techniques for Software Product classification and determines the cluster of projects that are in close proximity. The historical data and metrics from the identified cluster are taken and the generalized Heuristic algorithm/regression model based on current project under consideration is taken up and the specific adaptation applied for generic correction factors, model parameters. The Linear SVM backed by technical expert data and input is able to achieve excellent classification and in sample runs was able to achieve less than 4% error rate. This enables the Hybrid SQF model to counter the practical challenges and decision making issues. The SVM makes a coherent classification of projects based on key parameters and differentiators. The SQFM deploys the accurate or the best available historical data or apposite customized fault data from the closest clusters. The final predictions are focused and hence the reduction of error bandwidth provides very accurate and dependable results.

REFERENCES

[1] Azar, D., Bouktif, S., K´egl, B., Sahraoui, H. and Precup, D. (2002). Combining And Adapting Software Quality Predictive Models By Genetic Algorithms, *Proc. 17th IEEE International Conference on Automated Software Engineering (ASE2002)*, p. 285.

[2] Binkley, A. and Schach, S., "Validation of the Coupling Dependency Metric as a risk Predictor", *Proceedings in ICSE 98*, 452–455, 1998.

[3] Boetticher, Gary D. (2004). Nearest Neighbor Sampling for Better Defect Prediction, Retrieved June 2005, from http://promise.site.uottawa.ca/proceedings/pdf/7.pdf.

[4] Briand, L.C., Basili, V.R. and Hetmanski, C., "Developing interpretable models for optimized set reduction for identifying high-risk software components," *IEEE Transactions on Software Engineering*, vol. SE-19, no. 11, pp. 1028–1034, 1993.

[5] Cortes and Vapnik, V. (1995). *Support-vector network. Machine Learning*, 20:273–297.

[6] David Meyer (2011). *Support Vector Machines, The Interface to libsvm in package* e1071.

[7] Denaro, G. (2000). Estimating software fault-proneness for tuning testing activities, *Proc. the 2000 International Conference on on Software Engineering (ICSE 2000)*.

[8] Denaro, G. and Pezze, M. (2002). An empirical evaluation of fault-proneness models, *Proc. the 24th Intl Conference on Software Engineering (ICSE'02)*, p. 241.

[9] Ebert, C. (1996). Classification Techniques for Metric-based Software Development, *Software Quality Journal*, 5(4): 255–272.

[10] Freund, Y. and Schapire, R.E. (1998). Large margin classification using the perceptron algorithm, *Proc. 11th Annu. Conf. on Comput. Learning Theory*, 209–217, ACM Press, New York, NY.

[11] Hudepohl, J., Aud, S.J., Khoshgoftaar, T.M., Allen, E.B. and Maryland, J. (1996). Emerald: Software Metrics and Models on the Desktop, *IEEE Software*, pp. 56–60.

[12] Khoshgoftaar, T.M., Allen, E.B., Ross, F.D., Munik oti, R., Goel, N. and Nandi, A. (1997). Predicting Fault-Prone Modules with Case-Based Reasoning, *Proc. the Eighth International Symposium on Software Engineering (ISSRE'97)*, p. 27.

[13] Khoshgoftaar, T.M. and Seliya, N. (2002). Tree-Based Software Quality Estimation Models for Fault Prediction, *Proc. the Eighth IIIE Symposium on Software Metrics (METRICS'02)*, p. 203.

[14] Mitat uysal (2009). Using heuristic search algorithms for predicting the effort of software projects, *applied and computational mathematics*, v. 8, n. 2, p. 251–262.

[15] Press, W.H., Teukolsky, S.A., Vetterling, W.T. and Flannery, B.P. (2007). *Section 16.5. Support Vector Machines. Numerical Recipes: The Art of Scientific Computing* (3rd ed.). New York: Cambridge University Press.

[16] Rudiger Lincke, Tobias Gutzmann and Welf L¨owe (2010). Software Quality Prediction Models Compared, *10th International Conference on Quality Software (QSIC)*.

[17] Scholkopf, B., Smola, A., Williamson, R.C., Bartlett, P. (2000). *New support vector algorithms. Neural Computation*, 12, 1207–1245.

[18] Shawe-Taylor, J. and Cristianini, N. (2004). *Kernel Methods for Pattern Analysis*, Cambridge, UK: Cambridge University Press.

[19] VAPNIK, Vladimir, N. (1999). *The Nature of Statistical Learning Theory*. Second ed. Statistics for Engineering and Information Science. New York: Springer-Verla.

A VIKOR Multiple—Attribute Decision Method for Improving Supply Chain Leanness

P. Parthiban[1], H. Abdul Zubar[2], Richa Tiwari[3] and Surapaneni Bhavana[4]

National Institute of Technology, Tiruchirappalli, India
E-mail: [1]parthee_p@yahoo.com; [2]abdulzubar@rediffmail.com;
[3]bhavanasurapaneni06@gmail.com; [4]rtiwari1234@gmail.com

ABSTRACT: In this highly competitive world, the manufacturers always need to be a step ahead of their competitors. For this they need to shift to cutting-edge technologies. Since, the Toyota system came into existence, the focus of the manufacturing world has shifted to "LEAN "and as the manufacturers started adopting lean, its many advantages came to the fore. In a multiple attribute decision making problem, a decision maker has to choose the enabler satisfying the attributes to the best extent among a set of candidate solutions. The VIKOR method introduces the ranking index based on the "closeness" of the enablers to the "ideal" solution. By linking Lean Attributes and Lean Enablers, this study used VIKOR method for lean enablers to be practically implemented in order to enhance the leanness of the supply chain in the food industry and thus effective selection was made in the industry.

Keywords: Food Supply Chain, Fuzzy Logic, House of Quality (HOQ), Lean Production, Quality Function Deployment (QFD), Triangular Fuzzy Numbers (TFN), Vikor.

1. INTRODUCTION

Material selection is one of the most prominent activities in design process and various researches have been going on it. Damage or failure of an assembly occurs on selection of inappropriate material. Multi Criteria Decision Making (MCDM) came into picture because cheapest is not always the most promising approach in a process. MCDM is about generating alternatives, establishing criteria, evaluation of alternatives and assessment of criteria weights. Every criteria is to obtain an objective in the given decision context. The properties whose higher values are desirable, called positive criteria or beneficial attributes (e.g. strength, and toughness) and those properties whose smaller values are always preferable, named negative criteria, cost criteria or non-beneficial attributes (e.g. density, cost and corrosion rate).

2. REVIEW OF THE LITERATURE

2.1 Food Supply Chain

Better management of supply inputs and timely delivery of products and services at the lowest possible cost are effective practices for achieving sustainable business success (Cox and Chicksand, 2006). Supply chain management integrates suppliers, manufacturers, distributors, and customers to meet final consumer needs and expectations efficiently and effectively (Cox, 1999). Appropriate global supply chain strategies can be developed depending upon market characteristics and by simultaneously seeking to achieve higher levels of customer responsiveness at a lower total cost to the supply chain as a whole (Christopher and Towill, 2002). In this paper, an integrated approach for increasing the leanness of the food supply chain is introduced.

2.2. Leanness in Supply Chain

Supply chain leanness and agility play an important role in all manufacturing processes. Agility is nowadays an important factor in the design of supply chains; this refers to the ability of the supply chain to respond quickly to changes in customer demands (Christopher, 1999). In low-volume highly volatile supply chains, where customer requirements are often unpredictable and supplier capabilities and innovations are difficult to control, a more responsive or agile approach based on innovative products is appropriate operationally (Christopher and Towill, 2000). In contrast, the lean approach operates best when there is high volume and predictable demand with supply certainty, as a result of which functional products can be created (Cox and Chicksand, 2005). Lean manufacturing which has its origins in the Toyota production system, is one of the initiatives that many major businesses all around the world have been trying to adopt in order to remain competitive in the increasingly global market (Schonbergerm, 2007; Womack *et al.*, 1990). The focus of this multi-dimensional approach is on cost reduction by eliminating non-value-added activities and using tools such as Just in Time (JIT), cellular manufacturing, total productive maintenance, and setup reduction to eliminate waste (Abdulmalek and Rajgopal, 2007; Monden, 1998; Nahmias, 2001) not only within the organization but also along the company's supply chain network (Scherrer-Rathje *et al.*, 2009). A key feature of the above mentioned book was that it discussed not only manufacturing operations but also the supply chain (Holweg, 2007). The core thrust of lean production is that the mentioned tools can work synergistically to create a streamlined high-quality system that produces finished products at the pace of customer demand with little or no waste (Shah and Ward, 2003). Lean production promises significant benefits in terms of increased organizational and supply chain communication and integration (Scherrer-Rathje *et al.*, 2009). However, when summarizing the lean evolution, Hines *et al.* (2004) presented comments on approaches that have sought to address some of the gaps in lean thinking. The objective was to provide a framework for understanding not only the evolution of lean production as a concept but also its implementation within an organization (Hines *et al.*, 2004). The lean approach operates best when there is a predictable demand with supply certainty, as a result of which functional products can be created (Cox and Chicksand, 2005). The adoption of lean principles, which place more emphasis on levelling the production schedule (Naylor *et al.*, 1999), leads to a positive outcome with stable and/or increasing profitability (Cox and Chicksand, 2005). Furthermore, managing the supply chain and working closely with suppliers is facilitated by rationalizing the supplier base and focusing on suppliers committed to the ideals of lean production (Kannan and Tan, 2005).The adoption of Table 4 is from (M. Zarei a, M.B. Fakhrzad b, M. Jamali Paghaleh 2010).

3. CASE STUDY

Since the low rate of innovation and high rate of failure for new food products is a cause for concern, Stewart-Knox and Mitchell (2003) carried out an analysis of existing models of product development. They also developed a model for reduced fat food product development and discussed implications for best practices in food product development. Finally, they identified market and consumer knowledge and retailer involvement as key success factors in food product development (Stewart-Knox and Mitchell, 2003).

Table 1: Lean Attributes and Enablers Defined for Lean Supply Chain from Related Literature

Lean Enablers (LEs)	Lean Attributes (LAs)
Supplier management Conformance quality	Design for manufacturing Delivery reliability
Total preventive maintenance Low buffering cost	Pull production Low variability in process time
Eliminate obvious wastes Low variability in deliver times	Variability reduction Low variability in demand rates
Continuous improvement Cost efficiency	JIT manufacturing Delivery speed
Human resource training	Total quality management
Knowledge management	

Table 2: Degree of Relationships, and Corresponding Fuzzy Numbers

Degree of Relationship	Fuzzy Number
Strong (S)	(0.7, 1, 1)
Medium (M)	(0.3, 0.5, 0.7)
Weak (W)	(0, 0,0.3)

3.1 Defuzzification

MCDM requires the comparison of fuzzy numbers in a fuzzy environment. The problem of comparing fuzzy numbers is a tough problem. A good ranking method would be one that takes into account all the factors like shape, spread, height, and relative location on the x-axis. Since a fuzzy number represents many possible real numbers that have different membership values, one will face a difficult problem of comparing two different fuzzy number. In general, two approaches are used: (1) comparison of fuzzy numbers and (2) converting fuzzy number into crisp score (defuzzification).

The kth weighted mean method has been developed to be used as defuzzification. It uses membership function to the power of k as a weighted factor. The crisp value Crisp (N) for the triangular fuzzy number (l, m, r) is determined by the following formula.

$$\text{Crisp (N)} = \int_l^r x\mu^k(x)dx \Big/ \int_l^r \mu^k(x)dx$$

Integrating the integrals the following formula is obtained:

Crisp (N) = (k × m + 1 + r)/(k + 2)

Or

$C = m + (s_r - s_l)/(k+2)$

and

$$\mu(C) = \begin{cases} \frac{k+1}{k+2} + \frac{s_r}{s_r(k+2)} & , \quad C \le m \\ \frac{k+1}{k+2} + \frac{s_l}{s_r(k+2)} & , \quad C \ge m \end{cases}$$

where C = Crisp (N) ; sl m–l and sr = r–m are left and right support (spread), respectively

The parameter (power) k has the impact on defuzzification result as follows:

k = 1: C = (m + 1 + r) /3 or C = m + $(s_r - s_l)$/3 and C ≥ 2/3.

Increasing k (k = 2, 3, . . .) C moves toward m (core) and membership (C) increases; for example, for k = 4: C = (m + (sr – sl)/6); and, $\lim_{n \to \infty} C(k) = m$ and $\lim_{n \to \infty} \mu(C, k) = 1$

3.2 Application of Vikor

Within the VIKOR method, the various J alternatives are denoted as a1, a2,.,aj. For alternative aj the rating of the ith aspect is denotedby fij, *i.e.*, fij is the value of the ith criterion function for the alternative aj; and n is the number of criteria. The compromise ranking algorithm VIKOR has the following four steps (Opricovic and Tzeng [31]).

Table 3: Fuzzy-HOQ of the Case Study

Lean Attributes Lean Enablers	LA1	LA2	LA3	LA4	LA5	LA6	LA7	LA8
Weight (Wi)	(.31)	(.09)	(.06)	(.11)	(.1)	(.11)	(.12)	(.1)
LE1	M	M			W	W	W	M
LE2	S		W				W	
LE3	W		M					W
LE4		M		W				
LE5	W		M				M	
LE6		W		S	M	M		
LE7	M	M					W	M
LE8	M	S	M	M				M
LE9	W						W	
LE10		W						W
LE11				W			W	

Table 4: Fuzzy-HOQ Converted Into Crisp Numbers Taking K=2

Lean Attributes Lean Enablers	LA1	LA2	LA3	LA4	LA5	LA6	LA7	LA8
Weight (Wi)	(.31)	(.09)	(.06)	(.11)	(.1)	(.11)	(.12)	(.1)
LE1	M	M			W	W	W	M
LE2	S		W				W	
LE3	W		M					W
LE4		M		W				
LE5	W		M				M	
LE6		W		S	M	M		
LE7	M	M					W	M
LE8	M	S	M	M				M
LE9	W						W	
LE10		W						W
LE11				W			W	

Step 1: Determine the best f i* and the worst f i⁻ values of all criterion functions, i = 1, 2,., n. If the ith function represents a benefit then fi* = max fij and fi⁻ = min fij, while if the ith function represents a cost fi*= min fij and fi⁻ = max fij (Table 5).

Table 5

	LA1	LA2	LA3	LA4	LA5	LA6	LA7	LA8
f_i^*	1.1	1.1	.45	1.1	.45	.45	.075	.45
f_i^-	0	0	0	0	0	0	0	0

Step 2: Compute the values Sj and Rj,

j = 1, 2, m by the relations

$$S_j = \sum_{i=1}^{n} W_i \, (f_i^* - f_{ij})/(f_i^* - f_i^-)$$

$$R_j = \max_i [W_i \, (f_i^* - f_{ij})/(f_i^\varepsilon - f_i^-)]$$

Step 3: Compute the values Q_j as follows:

$S^- = $ max of S_j

$S^* = $ min of S_j

$R^- = $ max of R_j

$R^* = $ min of R_j

$$Q_j = \frac{v(S_j - S^*)}{(S^- - S^*)} + \frac{(R_j - R^*)(1-v)}{(R^- - R^*)}$$

Where v is introduced as a weight for the strategy of maximum group utility, whereas 1- v is the weight of the individual regret (Table 6).

Table 6: Determining Values as in Step 3 (v = .5)

	LA1	LA2	LA3	LA4	LA5	LA6	LA7	LA8
S_j	8.53	7.72	.47	1.04	.5	.55	.6	.767
R_j	.31	.09	.06	.11	.1	.11	.12	.1
Q_j	1	.078	0	.0675	.082	.105	.1281	.098

Step 4:

Table 7: Rank the Alternatives Sorting by Values S, R and Q in an Ascending Order

	1	2	3	4
By S	A4	A1	A3	A2
By R	A4	A1	A2	A 3
By Q	A4	A1	A3	A2

4. CONCLUSION

In this paper, a VIKOR approach was proposed to enhance the leanness of the food supply chain. The approach showed the applicability of the HOQ of QFD methodology, to identify viable lean enablers for achieving a defined set of LAs.

A numerical example illustrates an application of the VIKOR method to a food industry, aiming to numerical justification. It is an intention to illustrate the conceptual and operational validation of the application of this method in real world problem. The VIKOR method background and comparisons of the results by different methods are presented in order to show the position of this new method in the literature on fuzzy MCDM. Researchers are challenged to provide a guide for choosing the method that is both theoretically well founded and practically operational to solve actual problems.

REFERENCES

[1] Abdulmalek, F.A. and Rajgopal, J. (2007). Analyzing the benefits of lean manufacturing and value stream mapping via simulation: a process sector case study, *International Journal of Production Economics* 107, 223–236.

[2] Christopher, M. (1999). The agile supply chain: competing in volatile markets, *Industrial Marketing Management* 29, 37–44.

[3] Christopher, M. and Towill, D. (2000). Supply chain migration from lean and functional to agile and customized.

[4] Cox, A. (1999). Power, value and supply chain management. Supply *Chain Management: An International Journal* 4 (4), 167–175.

[5] Cox, A. and Chicksand, D. (2005). The limits of lean management thinking. *European Management Journal* 23 (6), 648–662.

[6] Cox, A. and Chicksand, D. (2006). Aligning Brand and Supply Chain Strategies. CPO Agenda, Autumn, pp. 46–50.

[7] Cox, A., Chicksand, D. and Yang, T. (2007). The proactive alignment of sourcing with marketing and branding strategies: a food service case. Supply Chain.

[8] Hines, P., Holweg, M. and Rich, N. (2004). Learning to evolve: a review of contemporary lean thinking. *International Journal of Operations and Production Management* 24 (10), 994–1011.

[9] Holweg, M. (2007). The genealogy of lean production. Journal of Operations.

[10] Kannan, V.R. and Tan, K.C. (2005). Just in time, total quality management and supply chain management: understanding their linkages and impact on business performance. Omega 33, 153–162.

[11] Naylor, J.B., Naim, M.M. and Berry, D. (1999). Liability: integrating the lean and agile manufacturing paradigm in the total supply chain. *International Journal of Production Economics* 62, 107–118.

[12] Perez, C., de Castro, R., Simons, D. and Gimenez, G. (2010). Development of lean supply chains: a case study of the Catalan pork sector. *Supply Chain Management: An International Journal* 15 (1), 55–68.

[13] Prince, J. and Kay, J.M. (2003). Combining lean and agile characteristics: creation of virtual groups by enhanced production flow analysis,. *International Journal of Production Economics* 85, 305–318.

[14] Rusdiansyah, A. and Tsao, D. (2005). An integrated model of the periodic delivery.

[15] Scherrer-Rathje, M., Boyle, T.A. and Deflorin, P. (2009). Lean, take two! Reflections from the second attempt at lean implementation. Business Horizons 52, 79–88.

[16] Schonbergerm, R.J. (2007). Japanese production management: an evolution with mixed success. *Journal of Operations Management* 25, 403–419.

[17] Serafim Opricovic; Fuzzy VIKOR with an application to water resources planning, Expert Systems with Applications.

[18] Shah, R. and Ward, P.T. (2003). Lean manufacturing: context, practice bundles, and performance. *Journal of Operations Management* 21, 129–149.

[19] Stewart-Knox, B. and Mitchell, P. (2003). What separates the winners from the losers in new food product development? Trends in Food Science and Technology 14, 58–64.

Tuning Sprint Goal Achievement in Different Project Scenarios

Vijay Wade

Siemens Information Systems Limited
E-mail: vijay.wade@siemens.com; vijay.wade@gmail.com

ABSTRACT: Sprint Goal Achievement by Agile Scrum means whatever planned is to be achieved in the Sprint to be goal and requires it to be 100% by end of sprint. But with due course of time this Metric is making Scrum behaves the way traditional Lifecycles were being used and hence, losing its shin. My paper highlights on the better use of this metric in different scenarios so that the meaning of scrum is not lost and it is used in the right spirit. This actually relates the metrics like velocity, Capacity available in the sprint, the trend/slope of delivery done sprint by sprint with respect to the expected delivery trend required for planned release. It gives the solutions to for doing required changes in different scenarios of glitches in the way sprint goals are planned for every different scenario in the projects.

Keywords: Sprint, Goal, Achievement, Scrum, Velocity, Capacity.

1. SCOPE

The scope of this white paper is limited for the Agile Scrum Projects. The below content of my paper are useful for the projects using Agile Scrum in their projects specially the projects doing Development and testing however it is not limited to this. When I refer to Metric here I mean the quantitative measurement if it is not being formally called as metric in your scenario.

2. PROBLEM SCENARIOS

1. Sprint Goals defined beyond the Velocity.
2. Unclear User stories accepted in the Sprint (Assuming will get clarity during sprint).
3. Capacity is less for the said sprint but the planned work is not reduced.

Before I go ahead and explain the scenarios in depth and propose about solutions, would like to have some definitions clear.

Sprint Goal

It's the commitment by a scrum team to their Product Owner (PO) for completing all the accepted set of Product Backlog. When we say commitment it can not be 70% or 60% or anything less than 100% of whatever committed.

Sprint Goals are more rigid as far as the sprint duration is concerned as it can not to be changed within Sprint so in any case where goal defined is not realizable or achievable then we have to abort the Sprint. Sprint Goals are expected to be SMART:

- *Specific:* Indicate precisely what is to be done. Avoid vague alternatives;
- *Measurable:* You should be able to quantify your goal;
- *Accepted:* Goals must be accepted as worthwhile, realistic and attainable;
- *Recorded:* Write your goals down. This is the basis of a contract with yourself;
- *Time-constrained:* Set specific time-limits.

Sprint Goal Achievement

It's a metric generally used by Scrum Teams to check effectiveness of scrum practices. Looking at only this metric one can judge the success of scrum team and the seriousness of implementation.

2.1.1 *Problem Scenario 1:* → *Sprint Goal beyond actual Velocity*

While calculating Sprint Goal Achievement the formula used is

$$\text{Sprint Goal Achievement} = \frac{\text{Earned Story Points} \times 100\%}{\text{Planned Story Points}}$$

Or

$$\text{Sprint Goal Achievement} = \frac{\text{Delivered Size} \times 100\%}{\text{Planned Size}}$$

This way of calculating Sprint Goal Achievement works well only up to the point where Planned Story Points for a sprint is equal to Velocity for the sprint, but the problem starts where the Planned Story Points start deviating from the Velocity. It deviates in the following cases:

1. Sprint is planned aggressively and work more than velocity is committed.
2. The buffer to be considered in the planning is not considered and the features relatively takes more time than estimated and the team do not want to re-plan release.

Sprint Goal Achievement metric goes in RED when the commitment is not met which is 100% as per Scrum Methodology.

In the example mentioned below. Teams proven velocity is 9 per sprint, however planned is 12 which is not met eventually and hence the Sprint Goal Achievement is not met as we go with the formula which is mentioned above and hence its in RED mostly.

Table 1

Sprint Number	1	2	3	4	5	6	7	8
Earned Story Points	10	10	12	13	11	11	11	11
Planned Story Points	12	12	12	12	12	12	12	12
Velocity	9	9	9	9	10	10	10	10
Sprint Goal Achievement	83%	83%	100%	108%	92%	92%	92%	92%

The figure below shows the graphical representation of the data mentioned in above Table. This graph shows that the earned story points are well below the planned story points or the actual work done is less than the planned work.

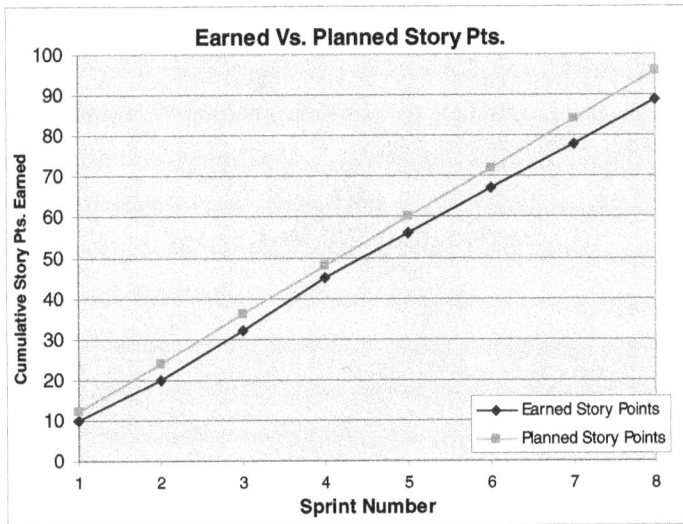

Fig. 1

With the above formula team will in most of the cases not achieve the goal and hence, this metric would always be in RED to demotivate the team. When we generally show that team has met or not met the target why not show them they also overachieve the target every sprint which will keep them motivated.

If we use the following formula as Sprint Goal achievement then if the team does the required work they will see that they have achieved the goal and if the planning is more aggressive then they have over achieved it.

$$\text{Sprint Goal Achievement} = \frac{\text{Earned Story Points} \times 100\%}{\text{Velocity}}$$

Let us use the same data as used in Table 1 and see the difference.

Table 2

Sprint Number	1	2	3	4	5	6	7	8
Earned Story Points	10	10	12	13	11	11	11	11
Planned Story Points	12	12	12	12	12	12	12	12
Velocity	9	9	9	9	10	10	10	10
Sprint Goal Achievement	111%	111%	133%	144%	110%	110%	110%	110%

As you see above the Sprint Goal is achieved and is Green for all the 8 sprints which is 100% achievement. If we compare with the earlier table where it was RED or not achieved in 6 sprints out of 8 which is 75% or is GREEN or achieved Sprint Goals only 25% times. Imagine both datasets shared with team and how it makes a difference on team's mindset. The team velocity as their planned story points will be more positive and will feel more energetic to carry on with more work. Also given in below figure is the graphical representation of the same for all of the 8 sprints. The graph shows team is well above the goal and the gap between earned and planned goal is increasing sprint by sprint which any motivated team would generally like to widen as it progresses with their performance.

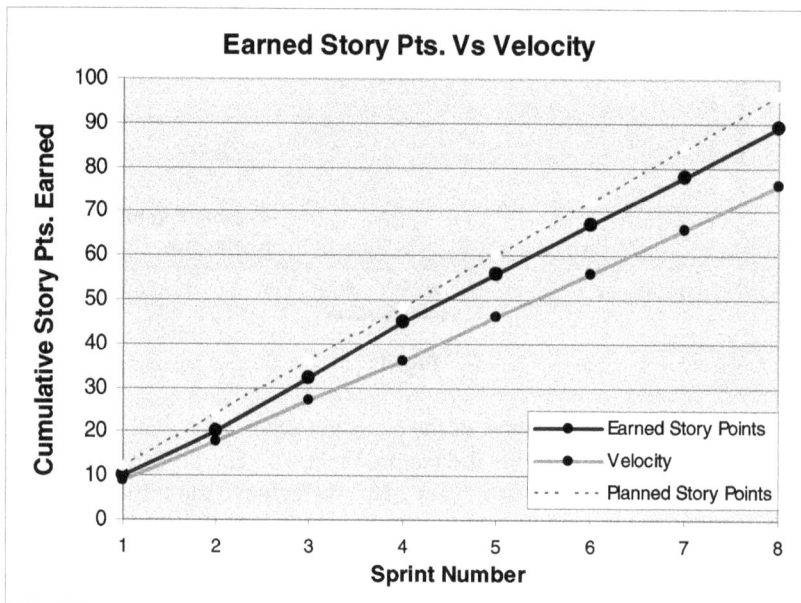

Fig. 2

For aggressive planning too we can have metrics like the one below. This is a bit unusual to use something that shows more than 100% but it's definitely not unethical.

I assume here if my team achieves 10% more than what is expected then it would be great despite my tuned velocity is 100%. This 10% here is just an example and you can use any step up figure as per your expectation. The advisable is continual improvement figure which can be increased on certain definite period or a number of sprints provided you achieve this on numerous occasions to feel confident of upgrading.

$$\text{Sprint Goal Achievement (L1)} = \frac{\text{Earned Story Points} \times 100\%}{\text{Velocity} \times (110/100)}$$

In this case, even if I do not achieve this my team is not going to be demotivated or feel bad however if I meet this then they definitely feel good and motivated to achieve more of it. And let's not make this metric mandatory. If team achieves this then this is bonus and a feel good factor or a moral boost to the team. If team keeps achieving this for say more than 6 Sprints which typically would be anything between 3 to 6 months ideally then we can think of upgrading or revisiting the Velocity.

2.1.2 *Problem Scenario 2:* → *Unclear (Vague) Stories Planned in Sprint*

Many a times there are situations where the unclear stories are to be planned in the Sprint assuming that it will get clarified as the sprint unfolds. Here, clarity also speaks of its dependence on other features or stories which are may be developed by other team or some clarification is pending which is expected in near future. Many a times it happens the dependency is not resolved or the clarification is not sought in time.

Scrum methodology clearly mentions that only clear and implementable stories or features are to be accepted for the sprint which generally happens during sprint planning meetings. If followed religiously does not present any issue but with the real time scenario and off shoring in picture and also the distributed teams these things keep happening and we are in the situation of accepting the unclear (vague) stories in the Sprint. Scrum also speaks of Product Backlog refinement which generally is done in the prior sprint to make sure the clear stories are available for next sprint planning meeting but still we land in the scenarios I am speaking about here.

The Solutions

As mentioned earlier in the Ideal world of Scrum the problems never come but for the scenes beyond it I give following solutions:

1. Tune the Sprint Goal Achievement to the actual velocity and anything achieved comparable to it is considered achieved.
2. The Sprint Goal Achievements LSL to be exactly same as Velocity (Tuned). If this is done then the Solution Description:

In case the vague (unclear) stories are planned in the Sprint because they are high priority then lets have some buffer stories as backup if the planned stories which may be dependent did not get clarified or clear then lets go ahead with back up stories. The Sprint Goal here should be to complete the story points equal to velocity and not all the planned stories as it might be much more than the planned amount. In this case, you will still achieve the goal of the number of stories. The back stories should be clear and may be less in the priority than what was planned as high priority but were vague/unclear.

With this approach you will not have to—Abort the sprint and do re-planning which is suggested by Scum in case of not able to achieve sprint goal.

Below example shows what we discussed above. Here, the planned story points are addition of clear story points and unclear story points. Unclear are the ones which have higher priority and the clear stories which are accepted as backup for the unclear ones may be have the lower priority which have to be completed only in case of unclear stories can not be realized in the running sprint.

Table 3

Sprint Number	1	2	3	4	5	6	7	8
Earned Story Points	9	10	9	10	10	10	10	11
Story Pts. (Clear Stories) (C)	9	9	9	9	10	10	10	10
Story Pts. (Prioritized Vague Stories) (V)	5	4	4	3	3	6	7	7
Planned Story Points (C + V)	14	13	13	12	13	16	17	17
Velocity	9	9	9	9	10	10	10	10
Sprint Goal Achievement	100%	111%	100%	111%	100%	100%	100%	110%

With this approach the Sprint Goal achievement generally goes down due to complexity of handling more than required stories but with this approach you will not:

- Falter on the Sprint Goal Achievement.
- Have to abort the sprint which is advised by Scrum in case of no goal achievement.

Below given is the graphical representation of the Figure 3. If you see the Planned story points line is gone up to increase margin between planned and achieved stories but the achievement is still above or equal to the expectation.

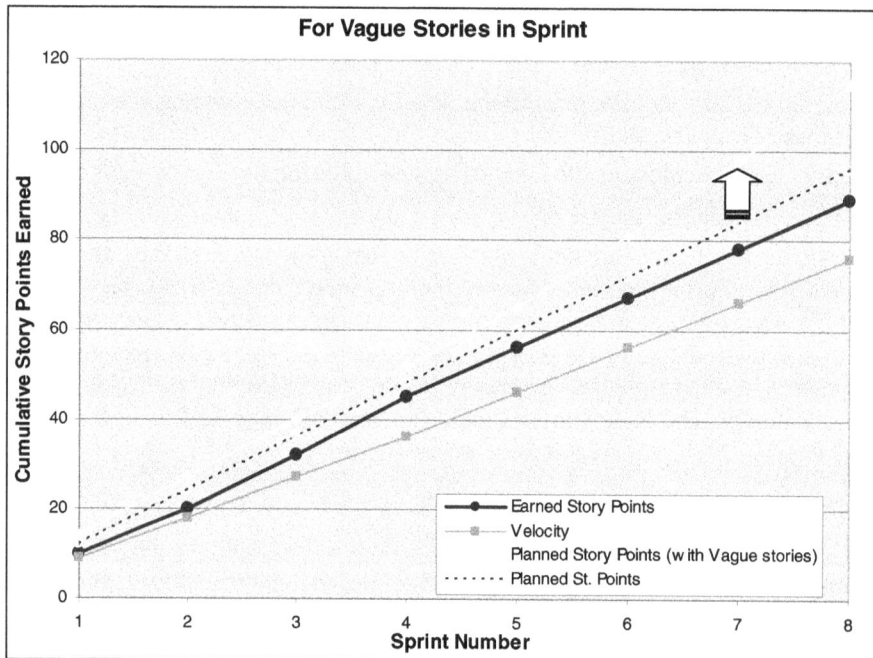

For Vague Stories in Sprint

Legend:
— Earned Story Points
— Velocity
— Planned Story Points (with Vague stories)
······· Planned St. Points

X-axis: Sprint Number
Y-axis: Cumulative Story Points Earned

Fig. 3

2.1.3 *Problem Scenario 3 → Capacity is less but planned work not reduced:*

A. Ideal Scenario: As far as scrum methodology is concerned it says there is no specific role like Developer, Tester or Technical Writer and everybody does the multitasking. This is an ideal situation.

B. The Reality: However, this idea of multitasking is still way beyond the reality in most of the cases. Assuming this scenario practical in ideal world we still have a realistic scenario where we have task specific roles even in Scrum teams. Even today the Sprint teams have specialized members for development and for testing and they do not really perform both the roles simultaneously. We also have different reasons for this.

(a) When a tester joins an organization he/she has already done some testing specific certification and is not very interested in the Programming. Even if I assume that organization forces people to change there ways it would be a forceful conversion which can not last long and also will not be productive. In today's scenario if some one says he wants to have an ideal scrum team then he or she will have to find the most flexible and dexterous people to form this dream team of absolute multitasking Software Professionals, which may be possible but seems difficult.

(b) Larger companies have many role specific people already working with them for years even before the Scrum was introduced to them and hence, its not a days work to convert a tester in to Developer or vice a versa.

(c) Organizations itself does the role specific recruitment and I have never seen a recruitment advertisement where it asks for a Scrum Team Member and not a tester or developer. This may be a reality outside but at least in India it's not the case. I have seen the ads for recruitment where they ask for a Software Tester or Developer who has some specified years of experience in Agile Scrum. This is also not a case of scrum specific role. So when an organization recruits, it recruits Developers or Testers and that's the role recruits would like themselves to operate in.

The Solution

The solution to this problem also is very specific to the available situations at hand which are Ideal scenario and the reality.

A. Ideal Scenario: If any team member is not available in any sprint then its really easy to select less number of stories equal to his productivity in sprint and the team will achieve the Sprint Goal and hence, the our concerned metric will be in green.

B. The Reality: In the real scenario unavailability of either tester or a programmer impacts the collaboration drastically. If I assume a Scrum team of 9 people where 3 are testers and 5 are developer then unavailability of one tester would impact 1.66 developers work which need to be tested and hence the work equal to one team members productivity only can not be reduced as solution but it has to be more than 1.66 times of productivity generally. As other developers as assumed also does the testing in such scenario will definitely take more time than a seasoned tester.

ACKNOWLEDGEMENTS

For my interest in Agile Scrum Methodology is because of some very influential people in my organization and out of it. This white paper content would not be complete without me referring to my colleagues and motivators. The people who contributed to this work someway or the other are:

- Manoj Apte
- R. Ramakrishnan
- L. Abraham
- Hemanth Jagannath
- Pete Deemer for training CSM
- Ken Schwaber whose writing inspired and attracted me for becoming Agile follower.

REFERENCES

[1] http://www.pponline.co.uk/encyc/goal-setting.html

[2] http://www.versionone.com/Agile101/velocity.asp

[3] http://www.scrumalliance.org/articles/39-glossary-of-scrum-terms

[4] Schwaber, Ken written The Enterprise and Scrum, published by Microsoft Press in year 2007.

[5] Cohn, Mike written Succeeding with Agile, published by Pearson in year 2010.

A Bouquet of Unusual Statistical Applications for Consumer Insights

Ranjan Samanta[1], Gireesh Sabari[2] and Jones Joseph[3]

[1]TVS Motor Company
[2,3]IMRB International
E-mail: [1]ranjan.samanta@tvsmotor.co.in; [2]gireesh.sabari@imrbint.com;
[3]jones.joseph@imrbint.com

ABSTRACT: The purpose of this paper is to illustrate some unusual applications of statistical methods to help illuminate the business context and provide direction for better business decisions.

Understanding where to prioritize investments, with limited resources, is a common business challenge. An answer to this lies in understanding consumer trade-offs on various attributes that determine their choices. However, most trade-off analysis is usually undertaken with specific and measurable attributes. For example, aspects like mileage vs power in an automotive context have specific values that can be easily provided to consumers to determine their choices and trade-offs. However, we often grapple with such aspects as whether to over-invest in the aesthetic aspects of a product's design or to emphasize more on its perceived popularity and find it a challenge to determine whether greater popularity over-rules greater aesthetic appeal. Similarly understanding aspects such as whether consumers are willing to pay a premium for having more control over their time vis-à-vis economic or safety factors in the choice of their mode of transport is not easily determined, but important for an investment decision. Examples of choice-based conjoint analysis using perceptual and attitudinal data can help reveal such trade-offs and thereby enlighten the business perspective.

Market share data based on retail sales is a common metric used in assessing brand performance. The trends are usually believed to be helpful in providing early warning indicators. However, this may actually not be the case. However, when combined with consumer brand profiling data can provide great insights by helping to segment markets. Using a statistical tool-CART (Classification and Regression Trees) an approach to market segmentation is arrived. Typically when segmenting markets some judgement is used in determining the relevant variables. Using CART these pre-judgements on segmentation variables are potentially reduced. The segment shares obtained thereof, can be very useful in enlightening the marketing context without any biases or assumptions based on "past experience". An example of the passenger car industry is provided to support this view-point.

Typically Design of Experiments are used in manufacturing for improving process capability. However, here is an interesting example to illustrate whether consumer preference patterns relate to user height, weight and experience in a specific market for consumer durables, namely 2Ws. The results helped determine the direction for product development.

Product development usually faces the challenge of where to create competitive differentiation. On the other hand consumer choices are difficult to comprehend as attributes important to them are varied and interlinked. An example to understand pathways to choice is illustrated using structural equation modelling which help unravel factors and their interdependencies. This in-turn brings focus to product development.

These examples are just a few, but they help throw light on the unusual applications of statistical methods in consumer data analysis, garnering insights for better business decisions.

(For confidentiality reasons we have masked the brands and products and are not be in a position to provide the raw data)

"The purpose of a business is to get and keep a consumer" Theodore Levitt HBS.

1. INTRODUCTION

The purpose of this paper is to illustrate some unusual applications of statistical tools that enlightened the business perspective.

2. USING CONJOINT ANALYSIS

In developing a new product, or in selling it, we often grapple with determining where to over-emphasize the marginal investment with the limited resources at hand. The answer to this really stems from an understanding as to where the marginal value to consumers is and therefore, in doing so, we best influence them on brand choices on aspects that really matter to them.

Determining the marginal value to consumers is really requiring an understanding of consumer trade-offs, on various attributes that determine their choice. Qualitative methods help but they are usually not sufficient to validate results. On the other hand, quantitative methods are fairly limited. A quantitative approach that elicits consumer trade-offs is a conjoint analysis.

However, conjoint analysis is usually undertaken with specific and measurable attributes. For example, aspects like mileage vs power in an automotive context have specific values that can be easily provided to consumers to determine their choices and trade-offs. For example, we may want to know consumer preferences relating to whether they want a 2-Wheeler with greater mileage but lower power or vice-versa. Typically we would design a conjoint experiment where we present combinations of different levels of power with different levels of mileage and assess preferences for such 2 Wheelers. For example, do they want a 2-Wheeler with 100 cc powered engine that gives 65kmpl mileage or one that has a 150cc powered engine but gives only 45 kmpl mileage. Based on this data we would obtain utility values for various combinations of power and mileage and understand their trade-offs and the relative importance of power vs mileage in their preference hierarchy.

However, aspects that really matter to consumers are more perceptual and attitudinal. Our interest may lie in understanding the marginal value in terms of "looks or aesthetics" of a 2-wheeler, for example, vis-a-vis its "perceived popularity or visibility" and find it a challenge to determine whether greater popularity overrules greater aesthetic appeal in consumer preferences. Similarly, is incremental perceived durability seen as providing relatively lesser marginal value vis-à-vis that of increased perceptions on mileage or vis-à-vis that of increased perceptions on popularity or visibility. Such answers help determine the investment focus for developing and selling brands in a competitive context.

We provide here an example of using choice-based conjoint analysis using perceptual and attitudinal data to help reveal such trade-offs.

2.1 Conjoint Analysis Methodology

Attributes that mattered to consumers with respect to the choice of a 2-Wheeler for their use were generated basis qualitative focus groups. These were aspects such as perceived mileage, style or aesthetics, popularity or visibility, power and pick-up, durability, safety or comfort and impressions about technology.

We had to decide how best to determine levels for these attributes. Typically we would have, for example, for mileage levels such as 40 kmpl, 50 kmpl or 60 kmpl. Similarly, we would have power expressed as 100cc engine, 125cc or 150cc. However, for this exercise we wanted to assess attitudinal and perceptual trade-offs. The levels were therefore constructed in a relative perceptual framework. These were at 3 levels-just the same as most brands *i.e.* average, somewhat better than most brands *i.e.* good and much better than most brands *i.e.* excellent.

There were thus 7 attributes or factors that influence choice for a 2-Wheeler each at potentially 3 levels that could create $3 \wedge 7 = 2187$ concepts of various combinations!

Using a traditional conjoint approach with an orthogonal design a total of only 18 concepts were generated that could then be put into a test for obtaining consumer preferences and thereby generating utility values for each of the attribute-level combinations. A representative sample of 200 respondents of our selected target group were administered the conjoint questions. These were as follows:

2.2 Conjoint Card Set

Show the combinations cards (1–18 cards) to the respondents and allow them to go through them and read them carefully:

Q1. *Say:* "here are 18 different motorcycle feature combinations, developed keeping in mind the requirement of users of motorcycles like you. Please go through each of these 18 cards in detail and sort them into 3 separate piles based on your likelihood to buy a motorcycle in case it were to be available in the market today"

Put the cards in the following manner in front of the respondents extreme right hand side: card saying—"most likely to consider buying"; in the middle: card saying—"somewhat likely to consider buying"; extreme left: card saying—"may not be likely to consider buying".

Show the cards and take the respondent through each card and sort them into 3 piles.

Q2. *Say:* "now, i would like you to rank cards in each pile based on how much attractive you feel they are for your needs. Let us begin with the first pile on your right".

Allow the respondent to rank order the cards in the first pile. Record the ranking and repeat the same for the next pile…. Complete all the three piles in this manner.

Code the card id in a continuous order as per their rank in the grid below. E.g.: if respondent selects card 8 as the first ranked among the cards in pile 1, then write '8' against rank 1.

2.3 Conjoint Analysis Findings and Conclusions

What is interesting is that consumers are indeed willing to give up or trade-off greater aesthetics for increased perceptions of popularity or visibility (marginal utility lost 0.77 vs marginal utility got 1.12) (Table 1).

Table 1

Attribute	Utility Value - Average	Utility Value - Good	Utility Value - Excellent	Differential in Utility (Good - Average)	Differential in Utility (Excellent - Good)	Differential in Utility (Excellent - Average)	Relative Importance
Popularity	-0.4761	-0.3267	0.8028	0.1494	1.1295	1.2789	0.15
Mileage	-1.1922	-0.3655	1.5578	0.8267	1.9233	2.75	0.32
Style	-0.5305	-0.1238	0.6543	0.4067	0.7781	1.1848	0.14
Technology	-0.533	0.1411	0.3919	0.6741	0.2508	0.9249	0.11
Safety & Comfort	-0.321	-0.1634	0.4843	0.1576	0.6477	0.8053	0.09
Durability	-0.3369	0.0206	0.316	0.3575	0.2954	0.6529	0.08
Power & Pickup	-0.5289	0.0883	0.4406	0.6172	0.3523	0.9695	0.11
					Total	8.5663	1.00

The old adage of "what gets seen gets sold!" seems to hold true and has more incremental value than that of going from good to great on aesthetics. Also interesting is the diminishing marginal returns on perceived durability. Hence, the incremental investment on durability is of lower value vis-à-vis that of creating greater brand value through higher visibility or popularity. Lastly, what we see, in the relative importance of attributes, is that consumers want a bundle of benefits. In the choice of high-value durable items this is usually the case, but probably in a country like India, it is more so! When you are very hungry you always want to take more on your plate than what you can really consume!!

2.4 Conjoint Analysis Reliability

In passing, we would like to mention that as part of this study we designed this experiment to assess the potential of a new offer in the market. We therefore had both a "Test Panel" and a "Control Panel" both were representative of the target group and randomly matched on certain demographic criteria. The potential of the new offer was assessed by gauging differences in preference shares obtained vis-a-vis the "Test Panel" and "Control Panel". We had administered the conjoint exercise to both Panels independently. Being randomly matched and representative of the same target group we would expect the results of the conjoint exercise to be similar. That indeed was the case! Our belief in the approach was therefore vindicated (Table 2).

Table 2

Control Panel = 200 Attribute	Differential in Utility (Excellent - Average)	Relative Importance	Test Panel=200 Attribute	Differential in Utility (Excellent - Average)	Relative Importance
Popularity	1.2789	0.15	Popularity	1.6751	0.17
Mileage	2.75	0.32	Mileage	2.6105	0.26
Style	1.1848	0.14	Style	1.4598	0.15
Technology	0.9249	0.11	Technology	0.8525	0.09
Safety & Comfort	0.8053	0.09	Safety & Comfort	1.1097	0.11
Durability	0.6529	0.08	Durability	0.8651	0.09
Power & Pickup	0.9695	0.11	Power & Pickup	1.4054	0.14
	8.5663	1.00		9.9781	1.00

2.5 Another Interesting Example of Conjoint Analysis

In the choice of their mode of transport are people willing to pay a premium for saving time, or are they even willing to give up safety for convenience? A conjoint analysis can help answer this!

Consumers were asked to give their preferences for various options of public transport. The factors that influenced their choices and the corresponding levels were generated as follows (Table 3).

Table 3

Attributes	Levels		
Saves Time	Very Fast	Somewhat Fast	Not so Fast
Can Go Anywhere	Fixed routes	Can go anywhere	
Available	Easily Available	Not Easily Available	
Comfort	Very Comfortable	Not very Comfortable	
Safety	Very Safe	Not Very safe	
Affordable	Cheap	Reasonable	Expensive

Interestingly people are willing to give up safety aspects when they are hard-pressed for time or in other words in a hurry! Also more affluent working men are more willing to pay a premium for safety and comfort, while those who are less affluent are willing to give up safety and comfort for cost! Further, men staying far off from their place of work are more willing to give up safety and comfort and to some extent even cost to save time! Women who are not working, in general prefer safety and comfort over time and cost (Table 4).

Table 4

Importance of aspects in choice of public transport	Affluent working men	Middle-class working men	Middle-class working men staying far from place of work	Affluent & non-working women
	%	%	%	%
Attribute				
Can go quickly	7	32	26	21
Can go anywhere	12	3	24	4
Is easily available	24	16	14	21
Saves my time	**44**	**51**	**64**	**46**
Comfort	33	25	21	8
Safe	**16**	**7**	**2**	**29**
Is safe & comfortable	48	31	23	37
Cost	7	18	13	17
Total	100	100	100	100

Who would be more likely to own a car? Who, a 2-wheeler? And, who would be least likely to use a 2-Wheeler? On a lighter vein, when do people give up safety other than being hard-pressed for time or being hard-pressed for money as the above conjoint demonstrated? Possibly when they are enjoying themselves!

3. CART—ING OUT MARKET SEGMENTS

Segmenting and profiling markets has been an age-old business challenge and its importance cannot be over-emphasised. A plethora of approaches exist to solve this issue. However, most approaches assume some degree of judgement on the part of the data analyser or researcher in order to determine the segmentation variables. Further, a typical segmentation exercise is first conducted and thereafter profiled or described in terms of demographics, brand usage, etc. On many occasions however, manufacturers obtain consumer profiles of brands, as and when they are launched in the market, so as to better understand the buyers of these models. They however, often overlook the fact that a brand's consumer profile does not necessarily attract a single segment of consumers and therefore does not reflect the perspective of various segments of the market where the brand could indeed be stronger in some and weaker in others.

Further, the market share of the brand is typically based upon the retail sales data but this still does not reflect the brand's share in various segments of the markets, where differential shares exist for the brand. Retail shares although believed to provide early warning signals are really not so. The fact that different brands are launched at different points in time, and that they are regularly profiled for a better consumer understanding, further blurs the focus and does not really guide targeted actions.

Interestingly, after conducting a brand profiling exercise, using a statistical tool-CART (Classification and Regression Trees) an approach to market segmentation is arrived without any pre-judgements on the part of the researcher in finalizing the segmentation variables. Further the brand's true shares within different segments are determined using retail shares. This can reveal false assumptions that may have been blind-sighted by past experience and helps enlighten the marketing context. An example of a consumer durable context, *i.e.* passenger cars, is provided to support this view-point.

3.1 CART Background and Methodology

This case, confidentiality maintained - so all brand names are masked, relates to the passenger 4-Wheeler market in India. Given the fact that the market is rapidly evolving post de-regulation and liberalization, manufacturers from time to time obtain consumer profiles of users of different brands to understand buyers better. The profile data typically describes the demographics of the user of the brand, reasons for purchase, alternative brands considered, what they like about it including any concerns and their usage patterns of the vehicle.

A randomly selected sample of recent owners and users of lower and mid-end priced cars was recruited. Key brands that constituted the majority shares are given in the Table 5.

Table 5

Recent owners of lower & mid priced cars			
Sample 2261	Sales shares		Sales shares
	%		%
Maruti Swift	36	Hyundai I10	16
Maruti Alto	17	Hyundai I20	7
Maruti Sx4	12	Chevy Beat	4
Maruti Wagon	7	Others	1
Any Maruti	72%	Any Hyundai	23%
	Total	100%	

Using this data a simple CART was run. CART or Tree based method is very suitable when there is large data to explore, the data contains several nominal variables such as gender, occupation, purpose of use, these nominal predictor variables have several levels, the underlying relationship between the predictor variables and the response variable is not linear and the data may contain complex interactions.

The dependent variable was the brand of car recently bought by the main user and the independent variables were Occupation, Age, Monthly Household Income, Driving Experience, Marital Status, Daily distance travelled and Purpose of use of the owner-user.

3.2 CART Findings and Conclusions

The results of the CART are shown below (Figure 1):

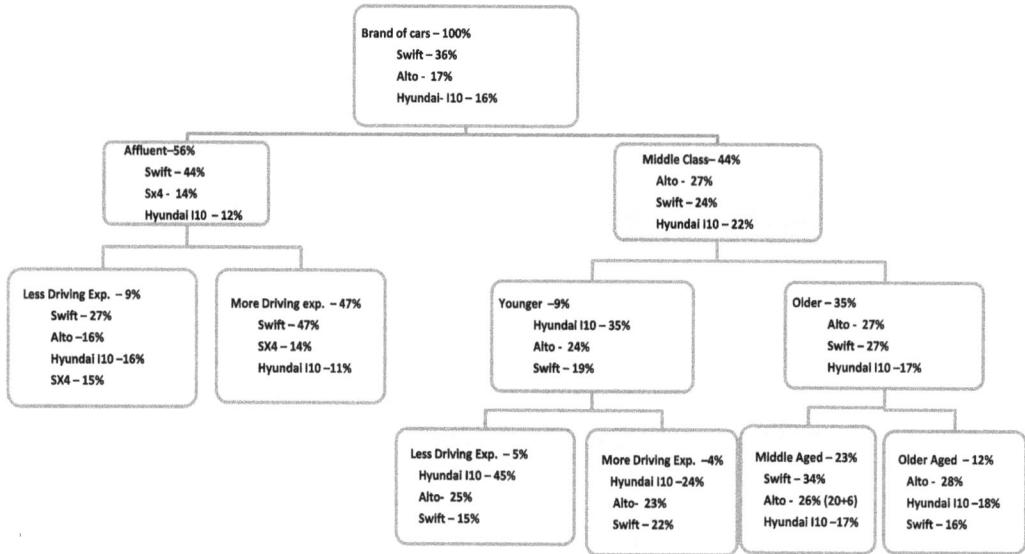

Brand of cars – 100%
Swift – 36%
Alto - 17%
Hyundai- I10 – 16%

Affluent–56%
Swift – 44%
Sx4 - 14%
Hyundai I10 – 12%

Middle Class– 44%
Alto - 27%
Swift – 24%
Hyundai I10 – 22%

Less Driving Exp. – 9%
Swift – 27%
Alto –16%
Hyundai i10 –16%
SX4 – 15%

More Driving exp. – 47%
Swift – 47%
SX4 – 14%
Hyundai I10 –11%

Younger –9%
Hyundai I10 – 35%
Alto - 24%
Swift – 19%

Older – 35%
Alto - 27%
Swift – 27%
Hyundai I10 –17%

Less Driving Exp. – 5%
Hyundai I10 – 45%
Alto- 25%
Swift – 15%

More Driving Exp. –4%
Hyundai I10 –24%
Alto- 23%
Swift – 22%

Middle Aged – 23%
Swift– 34%
Alto - 26% (20+6)
Hyundai I10 –17%

Older Aged – 12%
Alto - 28%
Hyundai I10 –18%
Swift – 16%

Fig. 1

The above construct became revealing. Hyundai-I10, a recently launched competitor, was making serious in-roads amongst the Middle Class Income group who are younger in age. The share of the Hyundai-I10 in this segment was one and a half to two times that of the Maruti Alto, and was already occupying a leadership position here. However, at an overall "All India" level (retail sales share) Hyundai–I10 was actually marginally lower than that of the Alto. Further, the Marketing Director of Maruti seriously believed that the Alto was the brand of choice amongst the Middle Class younger population. There was no other data or approach within the Company that could potentially reveal this picture and disabuse him of this false assumption, which was unfortunately, based on "years of industry experience"!

Lastly, within the profiling exercise, recent owners were asked about their likes and dislikes for owning and using their newly acquired brand. This was elicited by asking them what exactly they liked, helping them to elaborate on this and finally probing them on how it benefitted them. Based on this a laddering of their reasons to believe and benefits derived, was re-constructed for each of the CART segments thereby revealing the relative importance of benefits sought by each segment. The Table 6 below provides a flavour of this.

Table 6

	All recent owners of cars	All recent owners	Affluent- Less Driving exp.	Affluent- More Driving exp.	Middle Class Younger	Middle Class Middle aged	Middle Class Older
Wtd. Base: All Owner	2261	212	1027	233	418	264
Unweighted Base	2261	286	1202	171	279	205
	%	%	%	%	%	%
Ease of use	41	42	44	41	35	43
Comfort	20	20	36	19	17	18
Looks	34	38	27	49	35	32
Affordability	32	34	26	29	39	38

The above analysis showed that the benefits sought for the segment where the Hyundai- I10 was making a disproportional impact related to a need for greater styling and looks. Being younger the need for showing-off is possibly higher!

In the case of, the Hyundai-I10, consumers articulated the exterior body design and shape, the perfectly creased side panels, exterior headlamp styling, the wide front and rear wind-screen glass and the interior dash-board design as strong reasons to believe. This analysis helped give Maruti a good direction for future product planning and in developing a suitable response to this threat.

4. DOE—OUTSIDE MANUFACTURING

Typically Design of Experiments are used in manufacturing for improving process capability. However, here is an interesting example to illustrate whether consumer preference patterns relate to user height, weight and experience in a specific market for consumer durables, namely 2-Wheelers. The results helped determine product development.

4.1 DOE—Methodology

The purpose was to determine whether user height, weight and experience in riding made a difference to choice of a 2-Wheeler and therefore influence product development. A specific target group of recent buyers and users of 2-Wheelers was randomly selected. Each factor was taken at 2 levels. Height, weight and user experience was split into 2 groups each, basis the median levels of these factors. Therefore, there were 8 sub-groups in all ($2 \wedge 3 = 8$ combinations of possibilities). The sample was further matched in terms of age, occupation and affluence, so as to control for such demographics that may influence the outcome.

Users were given 4 vehicles of different formats (shapes and size), each having different performance characteristics. They were asked to ride these and give their preference for the vehicle that most suits their requirements. (The order in which they were provided each of the 4 vehicles was rotated across users within each of the 8 sub-categories as to control for any order biases in the presentation of these vehicles – ideally like that of a Latin square). Post revealing their preferences, they were asked as to why they did so.

4.2 DOE—Findings and Conclusions

The preferences of the 4 vehicles are given (Table 7).

Table 7

	All	Height Tall	Height Short	Difference / 2	Weight Heavy	Weight Light	Difference / 2	Experienced	Inexperienced	Difference / 2
Base: All Respondents	184	94	90		93	91		93	91	
	%	%	%		%	%		%	%	
Vehicle A	20	19	21	-1	17	23	-2	18	22	-2
Vehicle B	50	52	48	2	54	46	4	56	44	6
Vehicle C	25	26	23	1.5	24	25	-0.5	21	29	-4
Vehicle D	5	3	8	-2.5	5	6	-0.5	5	5	0

Interestingly, the user characteristics did not make any significant difference to their preferences. However, we do see that Vehicle B had a majority of the preference. The format and its performance characteristics did make a difference to their ride experience as expressed in their differential vehicle preferences and reasons thereof. Key reasons related to the benefits obtained such as ease of riding, comfort and safety, looks and design, perceived affordability where differences in vehicle performance indeed existed. This insight helped inform product development for replicating format and performance standards of vehicle type B.

5. PATHWAYS TO CHOICE—STRUCTURAL EQUATION MODELLING (SEM)

When a high-value complex product or service item is designed, the performance of this product or service is a function of all the various elements that have gone into it working together. The interactions of various elements of the product can create results that may further interact with each other to create the final consumer experience. In such situations, the design team may want to understand aspects that critically impact the final experience so as to improve the consumer experience further. Typically, it is very difficult to quantify these effects of these elements and trace the direction in which it happens. An attempt was made to understand pathways to choice using structural equation modelling which helped confirm factors and their interdependencies. This provided a rich insight to the product development team for designing the final product experience.

5.1 SEM—Methodology

This case deals with a computer package that can do statistical analysis. Users of data analysis were provided a user experience of this package along with other competitive offers in the market. They were allowed to use each package for a specific period of time. The order of usage was rotated to avoid any order bias. Subsequently, they were asked to give their evaluations on 25 different aspects that related to the package and its overall appeal. A relationship was specified basis the attribute descriptions and their correlations that matched up, basis a confirmatory factor analysis, with broader latent constructs. Basis these specifications a SEM was run to give regressions coefficients and factor loadings. All regression coefficients leading to the overall experience were significant, implying that they were important. The fit of the model was seen to be good with root mean square error at only 14%.

5.2 SEM—Findings and Conclusions

The findings indicated that significant differentiators that led to overall appeal related to the ability of a package to conduct varied sorts of data analysis and to enable the user's comfort. The key aspects relating to data analysis capability were the wide variety of analytical tools available, ability of the package to manage large amounts and varied types of data. On comfort, aspects such as strain to the eyes during usage, speed in doing the analysis, having a comprehensive training or tutorial package and being trouble-free mattered most. On the other hand ease of use and costs were more hygiene aspects to the user experience.

The design team was therefore able to focus on these aspects more to improve the user experience. Path relationships for choice of a data analysis package (SEM) (Figure 2).

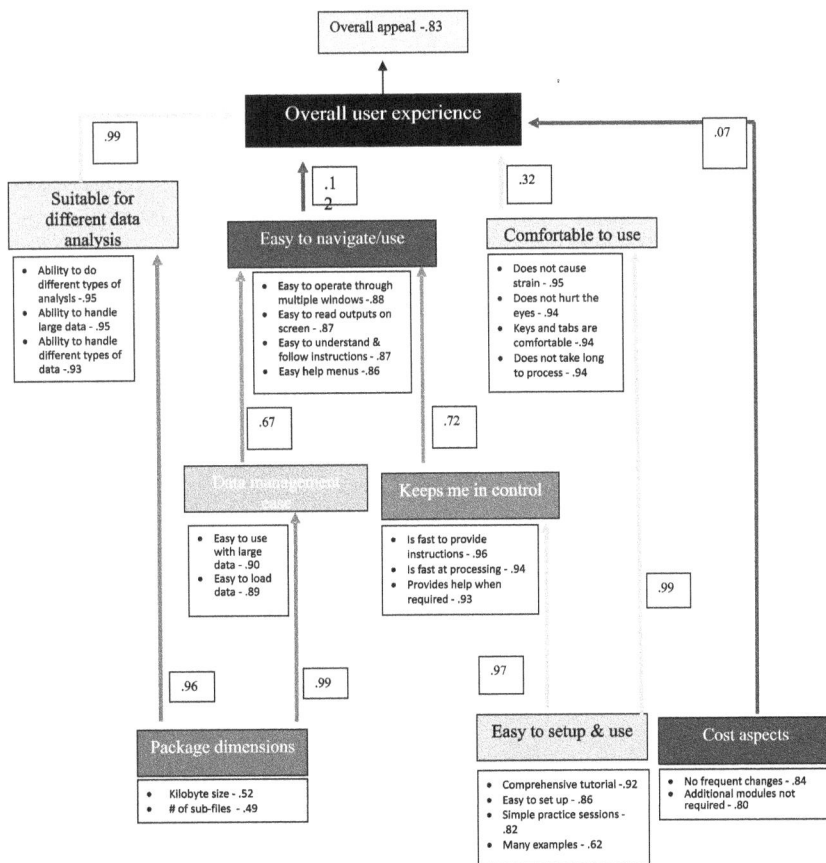

Fig. 2

ACKNOWLEDGMENTS

The authors would like to thank all the well-wishers and supporters at TVS Motor Company, Indian Statistical Institute specially Professor KK Chowdhury, Professor Bobby John and Professor UH Acharya and all at IMRB International including the other anonymous reviewers for their invaluable support.

REFERENCES

[1] Cattin, P. and Wittink, D.R. (1982). Commercial use of conjoint analysis: A survey. *Journal of Marketing,* 46:3, 44–53.

[2] Green, P.E. and Wind, Y. (1973). *Multiattribute decisions in marketing: A measurement approach.* Hinsdale, Ill.: Dryden Press.

[3] Yuan, Y. and Shaw, M.J., Induction of fuzzy decision trees. Fuzzy Sets and Systems 69 (1995), pp. 125–139.

[4] Breiman, Leo; Friedman, J.H., Olshen, R.A. and Stone, C.J. (1984). Classification and regression trees. Monterey, CA: Wadsworth and Brooks/Cole.

[5] Friedman, J.H. (1999). Stochastic gradient boosting. Stanford University.

[6] Horváth, Tamás; Yamamoto, Akihiro, eds (2003). Inductive Logic Programming. Lecture Notes in Computer Science.

[7] Box, G.E., Hunter, W.G., Hunter, J.S. and Hunter, W.G., "Statistics for Experimenters: Design, Innovation, and Discovery", 2nd Edition, Wiley, 2005.

[8] Ghosh, S. and Rao, C.R., ed (1996). Design and Analysis of Experiments. Handbook of Statistics–2.

[9] Goos, Peter and Jones, Bradley (2011). Optimal Design of Experiments: A Case Study Approach. Wiley. ISBN 978-0-470-74461-1.

[10] Bollen, K.A. (1989). *Structural Equations with Latent Variables*, New York: Wiley.

[11] *Structural Equation Models*, ed. K.A. Bollen and Long. S., Newbury Park CA: Sage, 136–162.

[12] Hauser, J. (1991). *Comparison of importance measurement methodologies and their relationship to consumer satisfaction*, M.I.T. Marketing Center Working Paper, Vol. 91–1.

[13] Maddala, G. (1977). *Econometrics*, New York: McGraw-Hill.

A Study on TPM Implementation: Factors and Performance

P.K. Suresh[1] and Mary Joseph[2]

[1]Cochin University of Science and Technology, Kochi
[2]Bharatmatha Institute of Management, Kochi–682021
E-mail: [1]sureshpkputhalath@yahoo.com; [2]miriamjoseph@gmail.com

Keywords: Lean Manufacturing, TPM Pillars, Quality Management Principles, JIPM, Objectives, Training.

1. INTRODUCTION

Globalization in Indian context has started in the first half of 1990's with the opening of Indian economy. Till then, the Indian industry, business and other key vehicles of the economy were well protected by various government rules and regulations. However, with the opening of the economy, the Indian industry faced global challenges of competition and survival. After initial hiccups, the industry soon realized that competitive pricing and quality alone were the means for its survival and growth. Liberalization and opening of the economy to global competition has brought new challenges of the industry in the realism of productivity, reliability and total quality management. The globalization of the economy was putting terrific pressure on industries to have increased adaptability, innovation and process speed.

There is mind-boggling development going on throughout the world. Continuous development and innovations have become order of the day. Technical advancement has made the world today a border less society. Today's competition is ruthless and survival depends upon not only on continuous improvement and invention of new products but also on their availability at low cost, timely delivery, courteous sales and prompts after sales service.

Quality factors, and cost depend upon the quality of the raw material, equipment precision, condition of the methods adopted and on the total involvement and capability of the people who carried out the work. Today's competitive global market helps for availing the quality materials for processing and with the technological advantages helps for sensitive equipments which gives sensitive products and which can operate by a semi skilled operator.

We are now helped by this technological advancement. We have now many precision equipments and devices which help in monitoring and controlling raw material quality and consistency, functional efficiency of equipment and their accuracy, auto operational systems to minimize and or to avoid variations in the methods used and POKA YOKE/mistake proofing system to eliminate human errors. But they are not going to do away with human labour completely. Only output can be automated, maintenance still depends upon human beings.

Upkeep of equipment has become an additional responsibility. The rapid technological advancement, leads to the importance of equipment for market survival. Care for equipment has become a great need. The concept of Total Productive Maintenance (TPM) meets this need.

Quality tools are the buses towards the journey for Total Quality Management (TQM). Quality tools include ISO standards, statistical tools, JIT, Quality circles, TPM, etc.

TPM is a part of Total Quality Management (TQM). Just as Quality Circle activities are carried out in small group, TPM also is carried out as small group activities on a company—wide basis with equipment maintenance as its main aim. This guides the importance of management of equipments, in surviving in the competitive market.

Total Productive Maintenance (TPM) is a very popular manufacturing philosophy and well accepted the world over. TPM aims at creating a system for achieving and maintaining 'Zero breakdown', 'Zero defect', 'Zero accident', and 'Zero pollution'.

Even though basically TPM is meant for manufacturing industries, it can be adapted to service industries also. Many companies in India have adopted TPM and gained lot of benefits.

TPM activity is not an activity restricted to production area alone. It has now a new direction in the function of the entire organization. As mentioned earlier modern tough competition as well as technological advancements has brought in automation and robotization. Now not only quality but also product cost, inventory, safety, health, etc. are dependent on equipment.

2. HISTORICAL BACKGROUND OF TPM CONCEPT

Seiiji Nakajima, a Japanese engineer, came out with this concept in 1971. Till 1950's organizations were carrying out only breakdown maintenance *i.e.*, wherever equipment goes out of order the maintenance crew will attend to that and put it back to operation.

In 1950's a major change came in this method in the form of 'Preventive Maintenance' in USA. This was something revolutionary at that time. Preventive maintenance work was carried out either during holidays or taking a planned shutdown. Preventive maintenance helped to create an awareness and recognition about the importance or reliability and economic efficiency in plant design, etc.

Such awareness led to 'Productive maintenance' concept by 1954. Under this system of day-to-day checkup of the tightness of the nuts and bolts, leakages, etc. were done. This helped to reduce the wear and tear, proper upkeep of the alignment, setup, etc. resulting in minimum breakdown.

By 1957 'Maintainability Improvement' system was developed *i.e.*, improvement of design parameters based on experience, additional control system for timely detection of problems, etc.

All these resulted in considerable reduction in maintenance activities.

3. DEFINITION OF TOTAL PRODUCTIVE MAINTENANCE

TPM is a methodology for maximizing equipment efficiency by establishing a Total system for Productive Maintenance (PM) for the entire life of equipment Participation by all departments, including equipment planning, operating, maintenance departments. Involving all personnel, including top personnel to first line operators. Promoting PM by motivation management, namely by autonomous small group activities.

It can be considered as the medical science of machines. Total Productive Maintenance (TPM) is a maintenance programme which involves a newly defined concept for maintaining plants and equipment. The goal of the TPM programme is to increase production while, at the same time, increasing Employee morale and Job satisfaction.

TPM brings maintenance into focus as a necessary band vitally important part of the business. It is no longer regarded as non-profit activity. Down Time for maintenance is scheduled as a part of the manufacturing day and in some cases and an integral part of the manufacturing process. The Goal is to hold emergency and unscheduled maintenance to a minimum.

4. WHY TPM

TPM was introduced to achieve the following objectives. The important ones are listed below.
- Avoid wastage in a quickly changing economic environment
- Producing goods without reducing product quality
- Reduce cost
- Produce a low batch quantity at the earliest possible time
- Goods send to the customers must be non defective.

5. SIMILARITIES AND DIFFERENCE BETWEEN TQM AND TPM

The TPM programme closely resembles the popular Total Quality Management (TQM) programme. Many of the tools such as employee empowerment, bench marking, documentation etc. used in TQM are used to implement and optimize TPM.

Following are similarities between the two. Total commitment of the programme by upper level of the management is required in both programme.

Employees must be empowered to initiate corrective action and a long range outlook must be accepted as TPM may take a year or more to implement and is an ongoing process. Changes in employee mind set toward their job responsibilities must take place as well.

The difference between TQM and TPM are summarized below.

Category	TQM	TPM
Object	Quality (Output and effects)	Equipment (input and cause)
Means of attaining goal	Systematize the management. It is software oriented	Employees participation and it is hardware oriented
Target	Quality for PPM	Elimination of losses and wastes

6. BENEFITS OF TPM

TPM Targets

P-Performance: Obtain minimum 80% OPE (Overall plant Efficiency)

Obtain minimum 90% OEE (Overall Equipment Effectiveness).

Q-Quality: Operate in a manner so that there are no customer complaints.

C-Cost: Reduce the manufacturing cost by 30%.

D-Delivery: Achieve 100% success in delivering the goods as required by the customer.

S-Safety: Maintain an accident free environment.

M-Mastermind: Increase the suggestions by 3 times. Develop Multi-skilled and flexible workforce.

Direct benefits of TPM:
- Increase productivity and OPE (Overall Plant Efficiency) by 1.5 or 2 times.
- Rectify customer complaints
- Reduce the manufacturing cost by 30%
- Satisfy the customers need by 100% (Delivering the right quantity at the right time, in the required quality)
- Reduce accidents
- Follow pollution control measures.

Indirect benefits of TPM:
- Higher confidence level among employees
- Keep the work place clean, neat and attractive
- Favourable change in the attitude of operators
- Achieve goals by working as a team
- Horizontal deployment of a new concept in all areas of the organization
- Share knowledge and experience
- The workers get a feeling of owning the machine.

7. MOTIVES OF TPM

Adoption of life cycle approach for improving the overall performance of production equipment. Improving productivity by highly motivated workers which is achieved by job enlargement. The use of voluntary small group activities for identifying the cause of failure, possible plant and equipment modifications.

8. UNIQUENESS OF TPM

The major difference between TPM and other concepts is that the operators are also made to involve in the maintenance process. The concept of "I (Production operators) Operate, You (Maintenance Department) fix" is not followed.

9. TPM OBJECTIVES

Achieve zero defects, zero breakdown, and zero accidents in all functional areas of the organization.

Involve people in all levels of organization.

Form different teams to reduce defects and self maintenance.

10. TPM HAVE THE FOLLOWING PILLARS

1. Autonomous Maintenance
2. Focused Improvement
3. Planned Maintenance
4. Quality Maintenance
5. Development Management
6. Education and Training
7. Safety, Health and Environment
8. Office Improvement.

1. *Autonomous Maintenance:* This pillar is geared towards developing operators to be able to take care of small maintenance activities, thus freeing up the skilled maintenance people to spend time on more value added activity and technical repairs.
2. *Focused Improvement:* Target is to achieve and sustain zero losses with respect to minor stops, measurement and adjustments, defects and unavoided down times.
3. *Planned Maintenance:* It is aimed to have trouble free machines and equipments producing defect free products for total customer satisfaction. It breaks maintenance down into 4 families or groups as: a. Preventive maintenance b. Breakdown maintenance c. Corrective maintenance d. Maintenance prevention.
4. *Quality Maintenance:* Quality maintenance activities are to set equipment conditions that preclude quality defects, based on the basic concept of maintaining perfect equipment to main perfect quality products.
5. *Development Management:* Planning project strategies, analyzing the factors influencing project decision. Design new products with customer focus; reduce lead time from design to production to market.
6. *Education and Training:* It is aimed to have multi skilled revitalized employees whose morale is high and who has eager to come to work and perform all required functions effectively and independently.
7. *Safety, Health and Environment:* Focus on targets, actual data and gaps in implementation of systems for ensuring safety, occupational health and clean environment through continuous training.
8. *Office Improvement:* Office TPM must be followed to improve productivity, efficiency in the administrative functions and identify and eliminate losses. This includes analyzing processes and procedures towards increased office automation.

11. RESEARCH PAPERS REFERRED

1. Dr. S. Muthu, Principal, P.S.G. College of Technology, Bharatheeyar University had conducted research work in 2002 under Dr. S.R. Devadasan on the topic 'Design and Exploration of Strategic Maintenance Quality Engineering Model'. This model was designed by integrating the chosen 'Strategic Quality Management model' with TPM principles.

This model recommends:

- Long term and short term Quality Objectives
- Suggestion schemes, small group activities, quality circle and other similar techniques to attain maintenance quality through human knowledge
- System for continuous maintenance quality system
- Programme for continuous approach towards target
- System for continuous transfer of customer feedback
- Programme for continuous training on maintenance quality
- System for continuous maintenance quality failure analysis
- System for continuous control of maintenance quality cost
- System for maintenance quality counseling.

Successful maintenance quality planning, maintenance quality implementation and maintenance quality improvement and maintenance quality control and evaluation is done through:

- Maintenance quality through human knowledge nourishment
- Continuous information system for maintenance quality
- Continuously monitored maintenance quality system
- Approach towards maintenance quality target
- Transfer of customer feedback
- Training and education on maintenance quality
- Maintenance quality through failure analysis
- Maintenance quality costing
- Periodical maintenance quality auditing
- Maintenance quality through counseling.

2. Research paper titled 'The impact of total productive maintenance practices on manufacturing performance' by Mr. Kathleen. E. Mckone, Mr. Roger. G. Schroedar and Mr. Kristy. O. Cua from U.S.A. finds that TPM has a positive and significant relationship with low cost, high levels of quality and strong delivery performance. They are considering the factors influencing TPM as:

- House keeping
- Cross training
- Teams
- Operator involvement
- Disciplined planning
- Information Tracking
- Schedule compliance, etc.

3. Research paper titled as 'Relationship between TQM and TPM implementation factors and business performance of manufacturing industry in Indian context' by Mr. Dinesh Seth and Mr. Deepak Tripathi from NITIE, Mumbai, examines the relationship between factors influencing the implementation of TQM and TPM and business performance.

They are considering the factors as:

- Leadership for improvement
- Strategic planning
- Process management
- Education and training
- Information architecture
- Equipment management
- Performance measurement system.

4. Mr. Shari. M. yusuf and Flaine Aspinwall, University of Birminghan, U.K, conductesd a study on 'TQM implementation issues, Review and Case study' which discusses the various issues confronting small businesses when embarking on TQM. First, reviews the subject of TQM and the quality initiatives undertaken by small businesses (which are treated as Small- to Medium-sized Enterprises (SMEs)) such as ISO 9000 and TQM.

Various issues for effective implementation such as management practices, process of implementation, results and outcomes were investigated.

In terms of structure, processes and people, a small business is in an advantageous position when it comes to adopting a new change initiative, provided that the owner/management has the commitment to, and leadership of, the change process, together with a sound knowledge of it. Lee and Oakes (1995) cited financial and technical resource constraints as being the main problems plaguing small businesses.

Moreno-Luzon (1993) reported on the problems faced by small firms in their attempt to develop a quality culture. Some of the difficulties were:

- Resistance to change;
- Lack of experience in quality management;
- Lack of resources.

5. Mr. Noorliza Karia and Muhamed Hasmi of University of Sains Malaysia had conducted study on 'The effects of Total Quality Management' practices on employee related attitudes.

The results indicate that training and education have a significant positive effect on job involvement, job satisfaction, and organizational commitment. Empowerment and teamwork significantly enhance job involvement, job satisfaction, career satisfaction, and organizational commitment.

Continuous improvement and problem prevention significantly enhance job satisfaction and organizational commitment.

Customer focus does not contribute to job involvement, job satisfaction, career satisfaction, or organizational commitment.

12. AWARDS RELATED WITH QUALITY MANAGEMENT

12.1 IMC Ramakrishna Bajaj National Quality Award Progamme

The Criteria are built upon the following set of interrelated Core Values and Concepts:
- Visionary leadership
- Customer-driven excellence
- Organizational and personal learning
- Valuing employees and partners
- Agility
- Focus on the future
- Managing for innovation
- Management by fact
- Social responsibility
- Focus on results and creating value
- Systems perspective.

12.2 The European Quality Award is the European Equivalent of the Baldrige Award

Special Prizes will be given for:
- Leadership and consistency of purpose
- Customer focus
- Corporate social responsibility
- People development and involvement
- Results orientation.

12.3 Malcolm Baldrige National Quality Award (MBNQA)

Embedded in each of these categories or dimensions—the ideal is defined:
- Leadership
- Strategic Planning
- Customer Focus
- Information and Analysis
- Human Resource Development
- Process Management
- Business Results.

12.4 Deming Prize—The Criteria

Broadly, the following considerations are taken into account for the Deming Application Prize:
- Top Management Leadership, Vision, Strategies
- TQM Frameworks

- Quality Assurance Systems
- Management Systems for Business Elements
- Human Resource Development
- Effective Utilisation of Information
- TQM Concepts and Values
- Scientific Methods
- Organisational Powers (Core Technology, Speed, Vitality)
- Contribution to Realisation of Corporate Objectives.

As per ISO 9000: 2005, eight quality management principles have been identified that can be used by top management in order to lead the organization towards improved performance.

- Customer focus
- Leadership
- Involvement of people
- Process approach
- System approach to management
- Continual improvement
- Factual approach to decision making
- Mutually beneficial supplier relationships.

13. RESEARCH OBJECTIVE

- To assess the degree of relationship between the TPM implementation factors and performance levels.
- To compare the performance of equipments with traditional maintenance practices with equipments have TPM level maintenance practices.
- To assess the difference in the level of implementation of TPM and performance among large scale industries and SME's, among the industries in Kerala and outside Kerala, among public sector and private sector companies, among process plants and production units and the performance with respect to the period of implementation, etc.

14. RELEVANCE OF SECTOR BASED STUDY

In large scale industries the employee strength are comparatively high and in small and medium enterprises, that are generally low. In SME's there is better scope for efficient management control and for providing better training, etc.

In public sector enterprises, there may frequent changes in leadership when comparative to private sector. Similarly, the political influence and other factors influences the decision making process.

Similarly, with respect to equipments, the process equipments are generally continuous running and annual shutdown or similar type maintenance practices are going on. But in production units, it can stop and restart at anytime and routine maintenance can be done very effectively.

When comparative to Kerala and other states, the work culture is generally different. The efficiency and effectiveness of workforce in Kerala is appraised everywhere, but there is some bottlenecks exists in effective utilisation of the force, due to some other issues related with influence of politics and general resistance to change the mindset of the people.

Generally any cultural change will occur with some periods of time. So the performance can be measured with respect to the duration of implementation. So a comparison with duration of implementation is also important.

15. IN THIS RESEARCH THE FOLLOWING IMPORTANT FACTORS ARE CONSIDERED FOR ANALYSIS WHICH INFLUENCES THE SUCCESSFUL IMPLEMENTATION OF TPM

By considering the referred research papers and the factors considered by different award criteria's, the following factors are selected as implementation factors which influences the success in attaining TPM objectives:

- Leadership for improvement
- Strategic Planning
- Human Resources Management, education and Training
- Information architecture
- Equipment and process management
- Monitoring and Analysis.

All these factors results the external and internal customer satisfaction.

The identified results areas are:

- Machine speed
- Machine availability
- Innovative products/processes/equipments
- Zero defects areas/zero customer complaint areas
- Zero breakdown areas
- Plant/Machine flexibility achieved for new product manufacturing/small volume of production of multiple items
- Easiness of operation achieved
- Machine accuracy/process streamlining
- Intellectual property rights/patent applications being obtained
- Reduction in pollution
- Reduction in accidents
- Increase in employee suggestions/small group activities.

The core results of TPM are achieved through the five pillars as addressed below and training, safety, health and environment, which are the plinths of TPM which supports these five pillars.

Office TPM is also considered as one of the plinth supporting the five pillars by providing necessary administrative supports. Here, the related results of five pillars is linked with administrative improvement which results cost reduction by man hours saving and through inventory reduction, etc.

```
                                    ┌──────────────────────────────┐
                                    │ Office (Admn.) Improvement    │
                                    └──────────────────────────────┘
                                                   ↑
┌──────────────────────────────────┬──────────────────────────────┐
│ Leadership                        │ Planned Maintenance          │
│ Strategic Planning                │ Quality Maintenance          │
│ HRM-Education and Training        │ Focused Improvement          │
│ Information Architecture          │ Development management       │
│ Equipment/Plant/Process Management│                              │
│ Monitoring and Analysis           ├──────────────────────────────┤
│                              ⇒     │            ↑                 │
│                                    │ Education and Training       │
│                                    │ Safety, Health and Environment│
└──────────────────────────────────┴──────────────────────────────┘
```

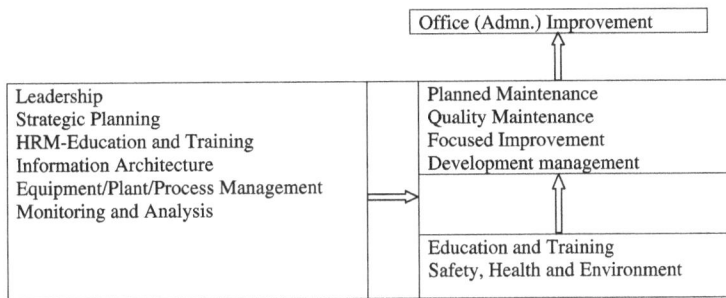

16. ADVANTAGES OF THIS ANALYSIS AND RESULTS

1. This research work will help to scientifically prove the relationship between implementation factors and performance parameters and it will guide to take further necessary steps for effective implementation of the system.

2. The category wise analysis will help to identify the specific nature of implementation and results indifferent categories of industries.

3. This research work will help to identify the importance of leadership and strategic planning which includes the determination of long term and short term objectives and systematic allocation of responsibility and authority, etc.

4. The role of education and training in attaining organisational performance also will be revealed.

5. This research work will prove the importance of effective monitoring and analysis of performance (also through internal audits as recommended by Mr. S. Muthu in Strategic Maintenance Quality model research paper) and the importance of implementation of statistical tools in analysis of performance.

Limitation of the Study:

The study does not consider the financial outcomes through TPM implementation.

17. METHODS ADOPTED FOR DESIGNING QUESTIONNAIRE

Japan Institute of Plant Maintenance (JIPM) have a questionnaire for considering the industries for its different criteria of awards. The questionnaire of JIPM includes the categories from 1. TPM policies and objectives 2. Individual Improvement 3. Autonomous Maintenance 4. Planned Maintenance 5. Quality Maintenance 6. Product and Equipment development and control 7. Training and Development 8. Administrative and Supervisory departments 9. Safety, Sanitation and Environment Control 10. Effects and evaluation of TPM.

Based on the questionnaire of JIPM awards and with referance from advanced books of TPM and from different ISO standards related with Quality, environmental and safety, a questionnaire is designed for collecting necessary data in relation with implementation levels of TPM and the results.

18. SAMPLING METHODS

Industries are to be selected already implemented TPM practices and which are in traditional maintenance practices in a convenient sampling basis. Industries in Kerala and from other southern states and industries belong to different sections as large scale and SME's, Public sector and private sector companies, Process industries and production industries, etc.

From these selected industries different machines or process plants to be selected on a sampling basis to collect data.

From TPM club of India, he membership industries who are in practice of TPM is available. But there are other non member industries who already started TPM initiatives. So, a convenient as well as judgment sampling method will be adopted for collecting necessary data from each sector of industries.

So, the correlation, regression analysis will help to measure the strength of relationship and 't' test will help for the comparison on performance in various categories of industries.

19. IMPROVEMENTS OF STUDY AT THE PRESENT LEVEL

Presently data is collected from industries which in the stage of successful implementation of TPM and data is also collected from industries which are in traditional maintenance practices.

The strength of relation ship in implementation factors and performance is measured in both categories of industries and the difference in implementation and performance is done only with TPM companies and Non TPM companies.

20. RESULTS

- The level of correlation among implementation factors and the achieved results evidenced.
- There is significant difference in performance of TPM implemented. companies with industries which are in traditional maintenance practices. (Tables attached)

21. RECOMMENDATIONS WITH THE PRESENT RESULTS

- Internal audits to be strengthened for monitoring performance (As recommended by Mr. S. Muthu, in Strategic Maintenance Quality model).
- Long term and short term objectives to be fixed and reviewed.
- Implementation results are positive, so the awareness among industries to be improved about the importance of TPM.
- Responsibility and authority to be done properly and performance to be monitored.
- Proper training to be done and individual objectives to be done and monitored for performance measurement for increments and rewards.
- Statistical tools with the support of software will strongly support TPM implementation.

22. FURTHER STUDY

The study will be extended to collect more samples from industries with sector wise, process wise and state wise for more elaborate way.

REFERENCES

[1] Emerald International journal of Quality and Reliability management, Vol. 22, No. 3, 2005.

[2] ISO 9000:2005 International Standard, Third edition 2009-09-15.

[3] Journal of Operations Management 19 (2001), 39–58, Elsevier.

[4] Quality Planning and Analysis, Third edition, by J.M, Juran and Frank M. Gryna of Tata McGraw-Hill Edition, New Delhi-8.

[5] Research work of S. Muthu, Principal, P.S.G. College of Technology, Bharatheeyar University in 2002 under S.R. Devadasan on the topic 'Design and Exploration of Strategic Maintenance Quality Engineering model'.

[6] TPM Reloaded, by Mr. Joel Levitt, Industrial Press INC., New York 10018.

ISO 9001: A Platform to Bring Cultural Changes within the Organization: A Case Study

Sudipta Gouri

IMRB International, Bangalore
E-mail: sudipta.gouri@imrbint.com

ABSTRACT: The ISO 9001 standards provide a basic framework for implementing a quality management culture in which the company's responsibility for its employees and clients and its concern for its impact on society at large are central. By establishing a quality culture, companies create an organization in which activity is not defined so much in terms of the hierarchical structure of the organization as in terms of one's contribution to the team (section, division, company). Improved employee motivation has positive implications for the innovation of a company's processes and quality of activities (*i.e.*, continuous improvement and customer friendliness) and for the company's productivity. Through the implementation of the different requirements of the ISO 9001 standards, a company clearly defines and documents its "way of doing things". Establishing the quality system documentation involves the evaluation of all company operations that may affect quality.

Decision making on quality issues is easier to control and involves more aspects of management. The process of decision making becomes much more formalized in a quality system because of the establishment of the management review committee and quality system indicators. Communication between different functional areas may be greatly improved when the responsibilities of employees, work units and divisions are clearly organized in the quality processes described in the quality manual. Work routines and conditions can become more effective, efficient and safe. If ISO 9001 has been combined with delegation and increased responsibilities, most of the recurring problems are now solved at the implementation level and no longer reach the manager's desk. In the documentation process, the company's inefficiencies are detected and eliminated. Contractual considerations with clients and suppliers may improve substantially through enlarged control and evaluation procedures, especially those required by the ISO 9001 model. Office administration may be greatly improved because responsibilities and work systems are clearly defined. Little time is lost when nonconformities are detected, and repetition of work is avoided.

This paper presents a case study of a service organization in the field of market research field operations. The time bound successful implementation of ISO 9001 QMS in the organization has led to substantial improvement in the decision making process, communication, productivity, client satisfaction and employee satisfaction.

Keywords: ISO 9001, Quality Management System, Service Organization, Continuous Improvement.

1. INTRODUCTION

The international quality management system, ISO 9000 has been developed to provide necessary conceptual and structural inputs for fulfilling customer needs by ensuring desired and consistent product quality. Since the creation of the standards in 1987, thousands of companies across India and abroad have gone for implementing the quality system standards and certification.[1] The reason(s) why companies are looking for certification of ISO 9000 quality management system range from maintaining their positions with other companies in their markets that are already certified to getting a step ahead of their competitors. Rao *et al.*[2] reports, based on an international empirical study, that ISO 9000 standards are effective for infusing quality within organization in the form of higher level of quality leadership, information and analysis, strategic quality planning, human resource development, quality assurance, supplier relationships, customer orientation and quality results. Hurang *et al.*[3] reports that adopting ISO significantly helps organizations in Taiwan in quality improvement, international competitiveness, cost reduction and sales increase.

The results of survey in Indian Industries[4] indicate that ISO 9000 implementation has benefited in better understanding of responsibilities/authorities and linkage across the organization. Lee and Palmer[5] find that the causes of ISO 9000 evidenced improvements are different for small and large firms. In large firms the benefits come about from improvements in flexibility due to the breakdown of formal hierarchies and improved cross-functional communication, whereas improvements observed in small firms are due to increased control and formalization associated with the introduction of operating systems. Many authors[3, 6–7] agree that companies that decide to implement ISO motivated by only external causes without believing in the benefits of the implementation of a quality system will get less impactful results than companies that implement them motivated by internal causes.

Terziovski *et al.*[8] undertook a cross-sectional study in Australia and found that there is a significant and positive relationship between manager's motives for adopting ISO 9000 certification and business performance. Those organizations that pursue certification willingly and positively across a broad spread of objectives are more likely to report improved organizational performance, and the individual element that contributes most to the business performance is the customer focus. Docking and Dowen[9] in a research with North American firms found that market reacts positively to certification of small companies. Investigation of Nicolau and Sellers[10] in Spain show that the value of a firm in the market rises due to acquisition of ISO certification. However, findings of Martinez-Costa and Martinez-Lorente[11] do not conform to it.

Many survey results[7, 12–17] have been reported in literature in the recent past. The authors of these surveys affirm that the main motivation for the organization to implement the standard, in general, is of external type. However, these surveys are mostly carried out on the manufacturing industries.

This paper presents a case study of effective implementation of ISO 9000 quality management system in an Indian service organization in the field of market research data collection to bring cultural changes within the organization. Although there was no external pressure, the top management of the organization decided to pursue the ISO certification keeping customer satisfaction as the key focus area. With team efforts, the entire process of certification *i.e.*

defining and documenting the effective implementation of the ISO quality management system could be completed within a span of 12 months.

2. APPROACH

As highlighted by Liao *et al.*,[18] despite the huge popularity and the urgent demand from customers to implement ISO 9000 QMS standards, some concerns for those organizations that are seeking registration to ISO 9000 include the expensive cost and the lengthy time to implement. Lengthy time for implementation of ISO 9000 standard implies the expected competitive advantages due to the certification remain unachieved for longer period that costs heavily in today's environment of high competition. Moreover, as the time-period increases, associated cost for the system development and implementation becomes larger. Lengthy time for implementation also reduce employee morale. Therefore, once the management of a company takes the strategic decision in going for ISO 9000 QMS certification, the entire process of system designing, documentation, implementation and certification of the QMS should be completed as quickly as possible. Keeping these in mind, at the very beginning of the journey the scope and roadmap were defined after extensive deliberations among top management.

2.1 Defining Scope and the Roadmap

After extensive deliberations, the top management decided to implement ISO 9001 QMS at its 5 Eastern Zone Field Offices first as a pilot study and then to extend the scope to the remaining 10 Field offices located across India. For the completion of entire ISO 9001 QMS certification process, a period of 12 months was targeted and accordingly, the road map (shown in Figure 1) was drawn.

2.2 Appointment of Management Representative (MR) and Formation of Review Committees

Top management's involvement and commitment are essential for successful implementation of the ISO 9000 QMS. Therefore, at the very early stages of the system designing, *i.e.* in the first month, a senior management personal of the company was appointed as the Management Representative (MR) and his responsibilities in terms of system establishment, implementation, maintenance, reporting of system performance and promotion of awareness of customer requirements were defined. To assist the MR in developing and implementing the QMS, Office Head of each location was appointed as Assistant Management Representative (AMR). In addition, two high level committees, namely Management Review Committee (MRC) and Steering Committee (SC) were formed for reviewing and monitoring of the progress and effectiveness routinely. The MRC consisted of the following members: Senior Vice President (Chair Person), Management Representative (Convener) and Office Head. On the other hand, the SC consisted of the following members: Office Head (Chair Person and convener) and Team Leaders.

The MRC was made responsible for reviewing the QMS once in a year in terms of effective implementation of audit finding (internal as well as external), corrective and preventive actions, customer complaints and customer queries. On the other hand, the SC was made responsible for reviewing the QMS in every month. It was decided that the agenda of SC review meeting would be the same as the MRC review meeting.

All the above three activities were completed in the first month itself. Adhering to the roadmap, the top management also formulated the quality policy and objectives keeping parity with the company's business plan and preset practices in the first month itself.

2.3 Drafting Organization Structure and Responsibilities of the Process Owners

The very next important activities were to identify and define various business processes in the company and then drafting the organization structure and defining the responsibilities and authorities of various process owners. Based on extensive brainstorming it was identified that the overall business process consists of several sub-processes, which can be classified into three broad categories, e.g. core processes, enabling processes and ISO specified other mandatory processes. The identified core, enabling and ISO specified mandatory processes are enlisted below:

- *Core processes:* Contract review, Data collection, Freelancers' management, Delivery of collected data, Client servicing and Client satisfaction.
- *Enabling processes:* Recruitment and training, Inspection and testing, Identification and traceability, Management review and preservation of products.
- *ISO specified other mandatory processes:* Control of documents, Control of records, Internal audit, Control of nonconforming products, Corrective actions and preventive actions.

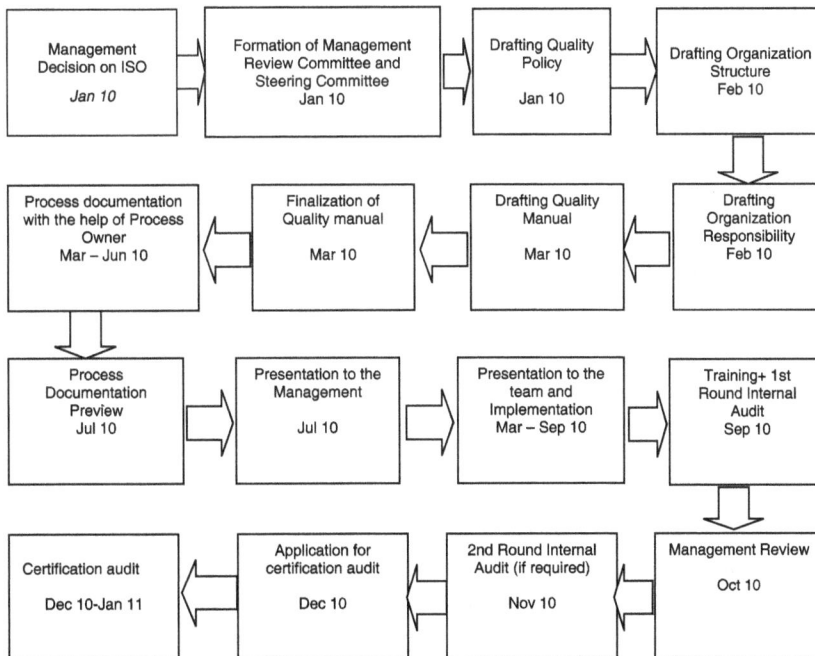

Fig. 1: Roadmap for ISO 9000 QMS Implementation

In order to streamline the responsibility and authority and eliminate the unproductive duplicate activities input-output analysis was carried out for each process. The inputs/outputs of a process were classified into the following categories:

- Tangible input/output
- Instructions and reports
- Information (generally verbal) and
- Routine activities (QC check, training etc.).

The input-output analysis necessarily connected different processes since the outputs of one process were the inputs of another process. The responsibilities of individual process owner were to ensure that the outputs were produced in time. Carrying out input-output analysis, outputs required to come out from each process were identified. Further, the input-output analysis clearly brought out the duplicate activities and the problem of the existing system. It was noted that there was excess paper work in many places. After discussion, excess unproductive activities at each process stage were eliminated. Then, the responsibilities and authorities of the personnel up to the level of process owners were defined clearly and unambiguously. The drafting of the organization chart and responsibilities and authorities of the process owners were completed by the end of the second month.

2.4 Designing and Documentation of the System

The lengthiest part in the ISO 9001 QMS implementation is the designing and documentation of the system. With the aim to accomplish this task successfully and effectively, the following three steps were completed sequentially: (i) documentation of the existing system (ii) comparison of the existing system with the elements of ISO 9001:2008 QMS, and (iii) identification and removal of discrepancies. Then the task of initial documentation of the quality manual and quality procedures were taken up. The quality manual and quality procedures for all the identified core, enabling and ISO specified mandatory processes were documented keeping in mind the present practices and ISO 9001 clause requirements. After preparation of draft documents, those were sent to all the steering committee members to get feedback from them. Based on their feedback the documents were fine tuned and frozen. The entire process for finalization of the documentation took about six months.

2.4.1 *Designing the System for Monitoring Continuous Improvement*

The main focus of the top management was to ensure continuous improvement in the business processes by implementing ISO 9001 QMS. Therefore, special care was taken to design the corrective and preventive action system. The basic requirement for achieving continuous improvement is to identify out the problem areas within the organization. Keeping this in mind, several brainstorming sessions were organized and finally it was identified that the system should focus on the following: 1) minimization of the noncompliance to the ISO 9001 QMS requirements, 2) identification of weak areas of different individuals and enhancement of their skills in those areas and 3) to institute proper methods for analysis of the error data and client query data so that the processes can be improved.

In order to assess the extent of compliance of the company's activities to the ISO 9001 QMS requirements, a system of routine internal audit was established. The internal audit is planned for every 4 months. The key indicator to the extent of compliance to the ISO 9001 QMS requirements and effectiveness of the QMS is the number of Non-Compliance Reports (NCR). It was expected that over the time when the QMS stabilizes in the company the number of NCRs would be reduced substantially.

To improve the skill of each individual, the weak areas of all the company personnel were identified through skill set matrix which was made subject to review every 4 months. Basic IT skills.

With the aim to improve the processes, it was planned to analyze the error data and client query data every month through following steps:

- Carrying out Pareto Analysis of error data and client query data separately and identifying the vital few, *i.e.* most important problem areas in the error data and client query data
- Diagnosing the frequently occurring problems by using Cause and Effect diagram
- Initiating action plans for the identified potential causes.

It was planned to evaluate the effectiveness of the action plans during next monthly meeting.

2.5 Implementation of the System

The documented systems were explained in details to all steering committee members by MR. The valid feedbacks from the steering committee members were incorporated into the documentation as amendments. The steering committee members and the MR explained the documented systems to the personnel at all levels within the organization. Suggestions from the employees were sought to build confidence among the employees. All the valid suggestions were incorporated in the final documentation as amendments. Then all the process owners and team leaders were asked to take direct initiatives to implement the processes as per the documentation with the help and active support from the senior management.

With the aim to encourage the employees in implementing the QMS system, Office Head was asked to justify job allocation among the employees and share the job allocation record with them. In this process, norms were set for each category of work in terms of time taking into account the experience of the employee. An Excel sheet which was the summary of time occupied by an employee was maintained in a common location so that every employee could share the information. The Excel sheet was subjected to updation on continuous basis by the Office Head.

2.5.1 *Monitoring the Implementation Process and Maintaining the System*

A team of internal auditors was formed who were made responsible for monitoring the effective implementation of the documented QMS and ensuring maintenance of the implemented QMS. The MR trained all team members on internal auditing using live examples, and all the successfully trained members were selected as the internal auditors.

Internal audit was planned for once in every 4 months. The AMR was made responsible to prepare the internal audit schedule taking into account the auditor and auditee's availability. Once audit was over, the AMR would prepare the internal audit summary and send it to the MR. It was AMR's responsibility to ensure that the corrective/preventive actions with respect to the internal audit finding were implemented before the agreed completion date between the auditee and auditor.

The internal audit findings, corrective actions and preventive actions and the findings of the client query analysis were subjected to review every month, which was known monthly meeting. The monthly meeting was chaired by the Office Head. It was mandatory that all Team Leaders would take part in this meeting.

Once in a year the unresolved issues of internal audits, corrective actions and preventive actions and client query analysis of the office along with external audit findings were subjected to review by the MRC. The MRC meeting was chaired by the Senior Vice President and convened by the MR. Office heads of all locations also took part in this meeting.

2.6 Certification

The second round of internal audit was completed at the eleventh month, *i.e.* by date and it was found that the ISO 9001 QMS was fully implemented in all Field offices. Therefore, the senior Management decided to go ahead with certification audit by December, 2010 and it decided to invite DEKRA, a Netherlands based certification body for carrying out the certification audit. The DEKRA scheduled the audit for Dec 2010 and Jan 2011. The results of the external audit carried out by DEKRA were satisfactory and so all offices of Abacus Business Operations were recommended for the certification on January 2011.

3. ANALYSIS OF SOME QUALITY MEASURES

A natural interest was to assess the impact of the ISO 9001 QMS on the company's quality frontier. For this purpose, six quality measures were analyzed for the current year (2011). These six quality measures are: 1) Productivity 2) Internal error (cancellation), 3) Client satisfaction (AQMS - Abacus Quality Measurement System), 4) Employee satisfaction, 5) Reduction in inspection (Back Check) and 6) Improvement in process monitoring. The analysis and results with respect to these six quality measures are given below.

3.1 Productivity

Average number of job completed per Monday in a year is considered as the productivity in the year. Figure 2 shows the bar diagram of the associate's increase in productivity in 2011 over 2010.

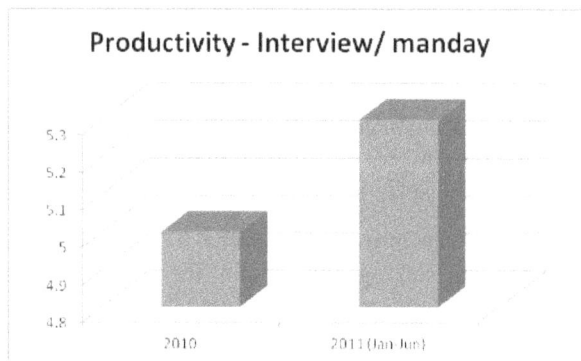

Fig. 2: Bar Diagram of Productivity Per Associate for the Year 2010 and 2011

3.2 Internal Error (Cancellation)

With reference to the company norm a percentage of filled up data collection instrument are subject to inspection. Based on the completeness of data, filled up data collection instruments are subject to acceptance. The rejected filled up data collection instrument is known as cancellation. The percentage of cancellation is calculated as ratio of total number of cancellation and total number of data collection instrument checked. Figure 3 shows the time series plot of % of cancellation for the period September, 2010 – August, 2011. It can be observed from Figure 3 that there is an improvement trend in % cancellation.

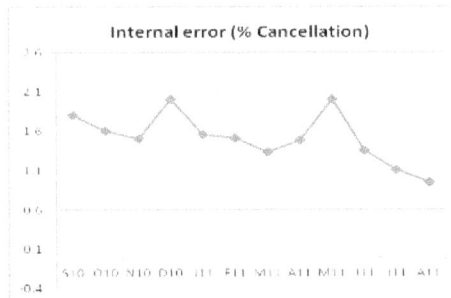

Fig. 3: Monthwise Time Series Plot of Cancellation

3.3 Client Satisfaction

After completion of each project researcher gives their feedback on project on different parameters which is known as Abacus Quality Measurement System (AQMS) in 5 point scale. The parameters are overall responsiveness from field at initial stage, understanding of project requirements, quality of questionnaire Translation, keeping RE informed on the status, quality of scrutiny, open ended probing, quality of translation of OE (Open Ended) responses, maintaining the agreed dispatch schedule, quota Meeting, overall quality of fieldwork/ Recruitments, meeting time target, overall communication - proactiveness, timely and clarity, overall co-operation level of the EIC and overall Satisfaction with the work done. The percentage (%) of compliance is calculated for Top Two Boxes (TTB) and Bottom Two Boxes (BTB). Time series plot of the month-wise % AQMS score (for the period January, 2011 – June, 2011) is shown in Figure 4. Although it is too early, but initial trend indicates (shown in Figures 4 and 5) that there is a improvement in the % AQMS score.

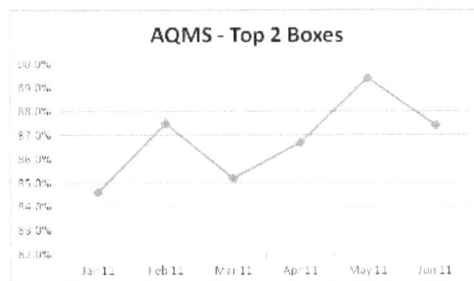

Fig. 4: Monthwise AQMS Score for Top Two Boxes

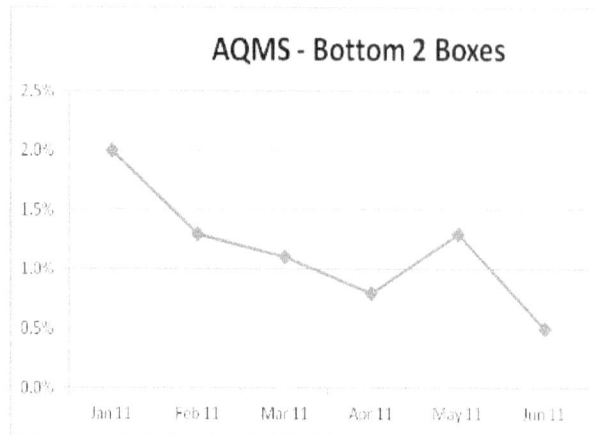

Fig. 5: Monthwise AQMS Score for Bottom Two Boxes

3.4 Employee Satisfaction

Every year KANTAR, parent company, survey employee satisfaction in the month of August on different parameters. In the year 2011, more than 80% IMRB employee participated in this survey. Results show sharp improvement in employee satisfaction in the year 2011 with reference to year 2010 (shown in Figure 6).

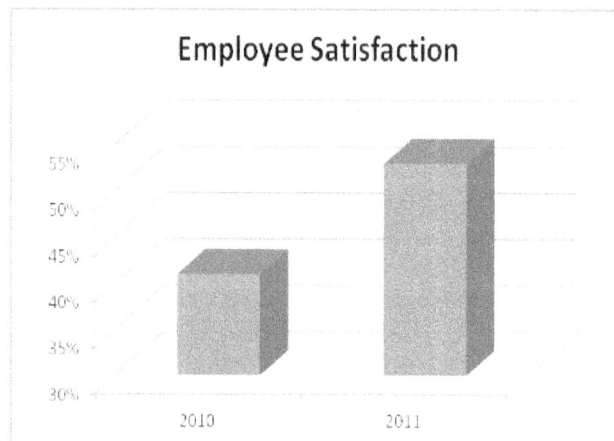

Fig. 6: Bar Diagram on Employee Satisfaction for the Year 2010 and 2011

3.5 Reduction in Inspection

After completion of the data collection at field, filled up questionnaires are checked to confirm the quality of data collection which is known as back check (B/C). At least 10% of the filled up questionnaires need to be back checked with reference to the office norms. Inspection is a non value added activity. Over the time company wants to minimize dependency on back check. Figure 7 clearly indicates that there is a trend in reduction of back-check.

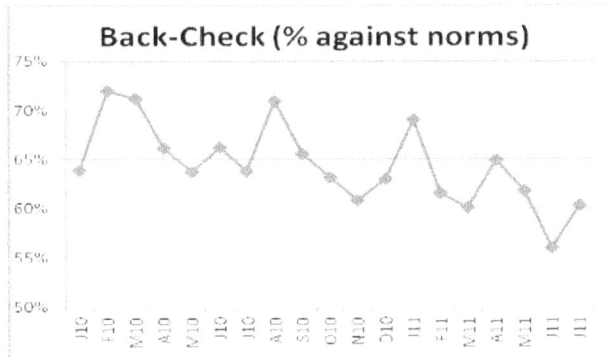

Fig. 7: Monthwise Time Series Plot of Back-Check

3.6 Improvement in Process Monitoring

After project is commissioned, a senior field professional guide the concerned field staff during data collection which is known as accompany check (A/C). At least 2% of the field work need to be accompanied checked with reference to the office norm. Figure 8 clearly indicates that there is trends in improvement in accompany check.

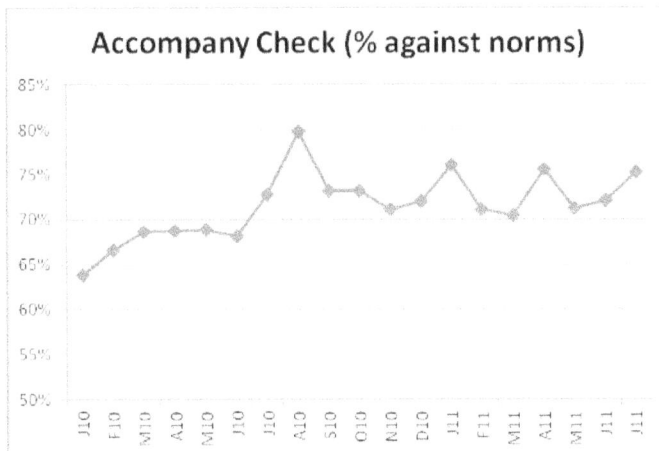

Fig. 8: Monthwise Time Series Plot of Accompany Check

4. CONCLUSION

Many survey studies have found that those organizations that pursue ISO 9001 QMS certification willingly and positively across a broad spread of objectives are more likely to report improved organizational performance. But these surveys are carried out mainly in the manufacturing industries. This paper presents a systematic approach for time bound, effective and successful implementation of ISO 9001: QMS in a service organization engaged in market research data collection services. Since the entire process of system design documentation, implementation and certification could be completed within the planned timeframe, the associated cost of the

ISO 9001 QMS certification process could be maintained at a low level. On the other hand, the organization could achieve substantial improvement in the business processes in the form of increase in productivity, client satisfaction, improved employee satisfaction, decrease in internal error and reduction in dependency on inspection after implementation of the ISO 9001 QMS. However, designing and implementation of the system was not smooth at all. There were several hiccup and barriers. Two important barriers were 1) Difficulty in identification of measurable process parameters and quality characteristics, and 2) Difficulty in separating in-process inspection from process control. However, all the obstacles could be overcome with the active involvement and support of the top management.

REFERENCES

[1] Saraiva, P.M. and Duarte, B. (2003). "ISO 9000: some statistical results for a worldwide phenomenon." *Total Quality Management and Business Excellence* 14, no. 10: 1179–1191.

[2] Rao, S.S., Ragu-nathan, T.S. and Solis, L.E. (1997). "Does ISO 9000 have an effect on quality management practices? An international empirical study." *Total Quality Management* 8, no. 6: 335–346.

[3] Hurang, F., Horng, C. and Chen, C. (1999). "A study of ISO 9000 process, motivation and performance." *Total Quality Management* 10, no. 7: 1009–1025.

[4] Acharya, U.H. and Ray, S. (2000). "ISO 9000 certification in Indian industries: a survey." *Total Quality Management* 11, no. 3: 261–266.

[5] Lee, K.S. and Palmer, E. (1999). "An empirical examination of ISO 9000 registered companies in New Zealand." *Total Quality Management* 10, no. 6: 887–899

[6] Meegan, S.T. and Taylor, W.A. (1997). "Factors influencing a successful transition from ISO 9000 to TQM: the influence of understanding and motivation." *International Journal of Quality and Reliability Management* 14: 100–117.

[7] Huges, T, Williams, T. and Ryall, P. (2000). "It is not what you achieve it is the way you achieve it." *Total Quality Management* 11, no. 3: 329–340.

[8] Terziovski, M., Damien Power, D. and Sohal, A.S. (2003). "The longitudinal effects of the ISO 9000 certification process on business performance." *European Journal of Operational Research* 146: 580–595.

[9] Docking, D.S. and Dowen, R. (1999). "Market interpretation of ISO 9000 registration." *The Journal of Financial Research* 22, no. 2: 147–160.

[10] Nicolau, J.L. and Sellers, R. (2002). "The stock market's reaction to quality certification: empirical evidence from Spain." *European Journal of Operational Research* 142, no. 3: 632–641.

[11] Martinez-Costa, M. and Martinez-Lorente, A.F. (2003). "Effects of ISO 9000 certification on firms' performance: a vision from the market." *Total Quality Management and Business Excellence* 14, no. 10:1179–1191.

[12] Rayner, P. and Porter, L.J. (1991). "BS5750/ISO 9000 – the experience of small and medium-sized firms." *International Journal of Quality and Reliability Management* 18, no. 1: 35–49.

[13] Askey, J.M and Dale, B.G. (1994). "From ISO 9000 series registration to total quality management: an examination." *Quality Management Journal* (July): 67–76.

[14] Brown, A., Van Der Wiele, T. and Loughton, K. (1998). "Small enterprise's experiences with ISO 9000." *International Journal of Quality and Reliability Management* 15, no. 3: 273–285.

[15] Ebrahimpour, M., Withers, B. and Hikmet, N. (1997). "Experiences of US and foreign-owned firms: A new perspective on ISO 9000 implementation." *International Journal of Production Research* 37, No. 2, 567–576.

[16] Anderson, S.W., Daly, J.D. and Johnson, M.F. (1999). "Why firms seek ISO 9000 certification: Regulatory compliance or competitive advantage?." *Production and Operations Management* 8, No. 1, 28–43.

[17] Withers, B. and Ebrahimpour, M. (2000). "Does ISO 9000 affect the dimensions of quality used for competitive advantage?" *European Management Journal* 18, No. 4, 431–443.

[18] Liao, H.T., Enke, D. and Wiebe, H. (2004). An expert advisory system for the ISO 9001 quality system. *Expert System with Applications,* 27(2), 313–322.

Re-Engineering Business Processes—
A Winning Approach to Transform
Businesses Using the BVEM

Senthil Anantharaman

Tata Communications, Chennai
E-mail: senthil681972@gmail.com

ABSTRACT: Business Process Reengineering (BPR) defines processes, analyzes and designs workflows within an organization and is the prime basis for many recent developments in management. It fundamentally rethinks and radically redesigns an organization's existing resources and is more than improving processes. Primarily it reduces costs by aligning to mission of the organization and brings about dramatic improvement in service, quality and speed. A framework and an approach along with a suitable business process reengineering/management tool when properly implemented would make BPR live up to its expectations. The methodology known as Business Value Enhancement Methodology (BVEM) is discussed in detail in this paper. This paper depicts its use and the phenomenal growth and rise in the industry, specifically the telecom sector, where benefits have been reaped to a great extent. The paper also explores in detail, the ten step methodology of BVEM, its advantages and finally its pitfalls without the application of a proper BPM Tool.

Keywords: BPR, BVEM, Re-Engineering, Value Enhancement, Improvement, Process Transformation.

1. INTRODUCTION

Business Process Re-engineering (BPR) fundamentally rethinks and radically re-designs an organization's existing resources. More than just business improvising, it redesigns the way the work is done to support the organization's mission and reduce costs. Re-engineering is the basis for many developments in management. Figure 1 elicits a typical Business Process Re-engineering cycle.

Successful BPR can result in enormous reductions in cost or cycle time. It can also potentially create substantial improvements in quality, customer service, or other business objectives. The promise of BPR is not empty—it can actually produce revolutionary improvements for business operations. Reengineering can help an aggressive company to stay on top, or transform an organization on the verge of bankruptcy into an effective competitor. The successes have spawned international interest, and major reengineering efforts are now being conducted around the world.

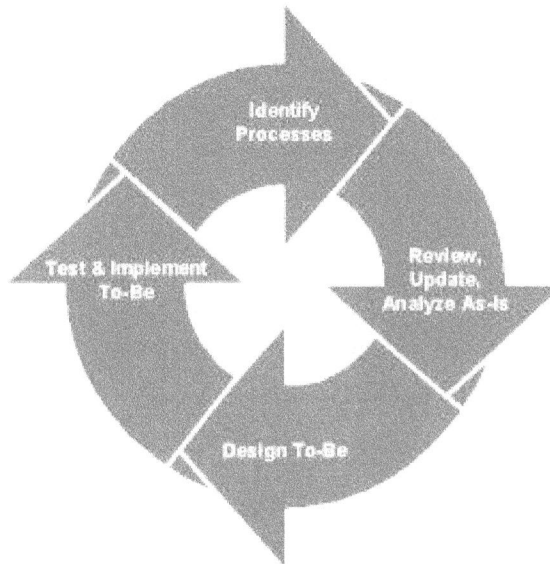

Fig. 1: Business Process Reengineering Cycle

On the other hand, BPR projects can fail to meet the inherently high expectations of reengineering. Recent surveys estimate the percentage of BPR failures to be as high as 70%. Some organizations have put forth extensive BPR efforts only to achieve marginal, or even negligible, benefits. Others have succeeded only in destroying the morale and momentum built up over the lifetime of the organization. These failures indicate that reengineering involves a great deal of risk. Even so, many companies are willing to take that risk because the rewards can be astounding.

Many unsuccessful BPR attempts may have been due to the confusion surrounding BPR, and how it should be performed. Organizations were well aware that changes needed to be made, but did not know which areas to change or how to change them. As a result, process reengineering is a management concept that has been formed by trial and error—or in other words practical experience. As more and more businesses reengineer their processes, knowledge of what caused the successes or failures is becoming apparent.

At Tata Communications Limited, a robust BPR approach and framework known as Business Value Enhancement Methodology (BVEM) along with a BPMS software Case wise has been adopted and implemented which tries to overcome the above mentioned pitfalls and capitalizes on strengths based on enhancing value and create a transformation which provides greater benefits on the whole.

2. BUSINESS VALUE ENHANCEMENT METHODOLOGY

A unique approach developed by KB Unni, Seven Hat Consulting, the Business Value Enhancement Methodology is truly an excellent reengineering framework consisting of the steps shown in Figure 2. Each of the steps is briefly described below.

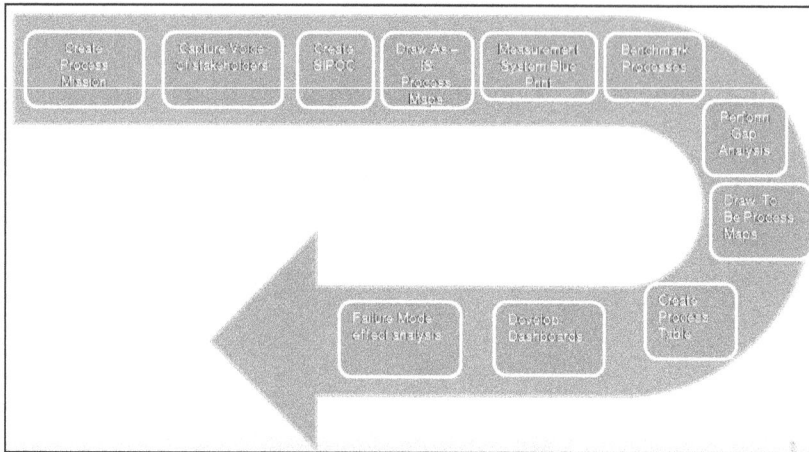

Fig. 2

2.1 Create Process Mission

Creating Process mission is the first step in the process of re-engineering in BVEM Methodology and should answer the following questions:

- Why does **this** process exist?
- For **whom** does it exist?
- What are our own reasons for setting this process up—what are our drivers?

These are not the same as Process Goals. It precedes and is more generic than goals but specific in its scope.

The Characteristics of a good Process Mission:

- Reflects and Represents the values of the Process founders
- Brief, succinct, but clear
- Covers "reason why" the process was set up
- Specific to what you are delivering.

For Example, Process Mission of a traditional restaurant's kitchen: To be the first choice in our catchment area for home-like food cooked using traditional recipes and processes. From this, the Stakeholders will be identified and thus Vision and Goals will be based on how Stakeholders describe their expectations for this process.

2.2 Capturing Voice of Stakeholders

In this step of reengineering, BVEM tries to understand stakeholders by gathering inputs not only on customer's needs but on various stakeholders requirements and translate them into meaningful terms and metrics for the product, process and service by following the steps described below:

- List all Stakeholder groups
- List roles/levels within each

- List individual representatives
- Conduct interview/Record responses
- Convert to Expectations/Needs in the QFD
- Brainstorm metrics
- Look for measures with high correlation to needs.

2.3 Create SIPOC

The next step is to build the SIPOC (or COPIS) so it represents the Customer (stake-holder of the process) view. In this, capture this view formally in defining the Output. The Output defines the sub processes (components) in the Process. The Process then determines the Input. The Output and the Input are tangible, physical….and have attributes that are then measured. The Output coincides with the Start Point. The Input coincides with the End Point.

2.4 Draw As-IS Process Maps

After Creating the SIPOC (or COPIS), draw the As-Is process maps starting from the high level process maps to the detailed L3 level process mapping as shown in Figure 3.0. It starts with the strategic L0 Level to L1 elements to a very detailed L2 which has activities and decision boxes, etc.

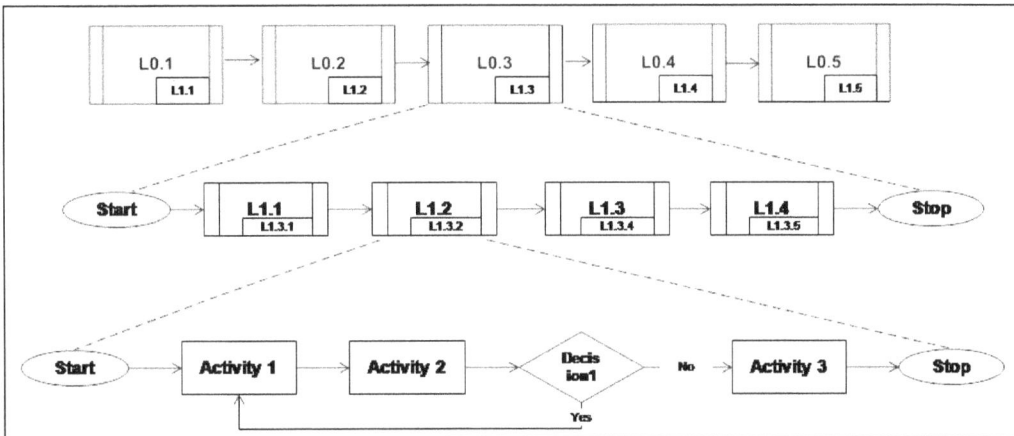

Fig. 3

2.5 Create Measurement System Blueprint

The next important step is measurement because without measuring we cannot improve and hence cannot reengineer the business processes. The following are the list of steps to be performed in creating a measurement system blueprint:

- List all metrics/measures
- Write out operational definition of the metric
- Identify distinct data elements in each metric

- For each data element identify certified data source
- Determine when the data has to be extracted
- Determine the method of collection
- Determine the frequency of collection
- Determine who will collect.

2.6 Conduct Bench Marking

Benchmarking is the process of comparing one's business processes and performance metrics to industry bests and/or best practices from other industries. Dimensions typically measured are quality, time and cost. In the process of benchmarking, management identifies the best firms in their industry, or in another industry where similar processes exist, and compare the results and processes of those studied (the "targets") to one's own results and processes. In this way, they learn how well the targets perform and, more importantly, the business processes that explain why these firms are successful.

Robert Camp (who wrote one of the earliest books on benchmarking in 1989)[4] developed a 12-stage approach to benchmarking.

The 12 stage methodology consists of:
1. Select subject
2. Define the process
3. Identify potential partners
4. Identify data sources
5. Collect data and select partners
6. Determine the gap
7. Establish process differences
8. Target future performance
9. Communicate
10. Adjust goal
11. Implement
12. Review and recalibrate.

2.7 Perform Gap Analysis

In business processes, gap analysis is a tool that helps compare actual performance with potential performance. At its core are two questions: "Where are we?" and "Where do we want to be?" If a company or organization does not make the best use of current resources, or forgoes investment in capital or technology, it may produce or perform below its potential. This concept is similar to the base case of being below the production possibilities frontier. Gap analysis identifies gaps between the optimized allocation and integration of the inputs (resources), and the current allocation level. This reveals areas that can be improved. Gap analysis involves determining, documenting, and approving the variance between business requirements and current capabilities. Gap analysis automatically and naturally flows from benchmarking and other assessments like measurement system blue print. Once the general expectation of performance in the industry is understood, it is possible to compare that expectation with the company's current level of performance. This comparison becomes the gap analysis. Such analysis can be performed at the strategic or operational level of an organization.

2.8 Draw To-Be Process Maps

After performing the gap analysis with various corrective actions, draw the To-Be process maps in a similar way as in step 2.4, but with the corrections.

2.9 Create Process Tables

The next step in BVEM is creation of Process Tables. The process table that we will be creating consists of RACI matrix along with the following elements which are enlisted. The advantage of having a process table is that it provides a single point of information of the accountability, responsibility of the process activities apart from having enlisting the KPI's and other key measures at each step. The elements (columns) of the process table are given below while the rows will be the activities or the process steps.

Process/Sub Process Names—Each Process and sub process has its own documentation. Clear link to the Deployment Flow Charts.

- **Activity id**—In the format 1.1.1.1…The number of places corresponds to the number of levels of drill down.
- **Activity** –Description of specific task…taken as is from Deployment flow chart. Verb-Noun format.
- **TAT**—standard time for task.
- **R**—the role(s) Responsible for this activity/task.
- **A**—the role Accountable –only one and as far down as it will go.
- **C**—the role(s) that need to be consulted (two-way).
- **I**—the role(s) that need to be informed (one way).
- **Input**—define at activity level for those activities that need an Input. Develop format/ instructions, link as hyperlinks.
- **Source**—where the specific input is obtained from.
- **Output**—define at activity level for those activities that need an Output. Develop format/ instructions, link as hyperlinks.
- **Destination**—where the specific input is obtained from.
- **KPI**—For all input and output points examine if a metric is required to ensure success of the process.

2.10 Create Dashboards

The penultimate step of BVEM is creation of dashboards of the relevant matrix and populating them with relevant data, mainly relating to the KPIs described in the Process table. This is done to unearth any patterns in data and will be a resource for future DMAIC/Six sigma/Lean Projects.

2.11 Perform Failure Modes and Effects Analysis (FMEA)

FMEA is a structured approach to:

- Identify the ways in which a process can fail to meet critical customer requirements
- Estimate the risk of specific causes with regard to these failures
- Evaluate the current control plan for preventing these failures from occurring
- Prioritize the actions that should be taken to improve the process.

In short, FMEA identifies ways a product/Process can fail and then plan to prevent those failures. A FMEA cycle can be shown below in Figure 4 with terminologies explained below.

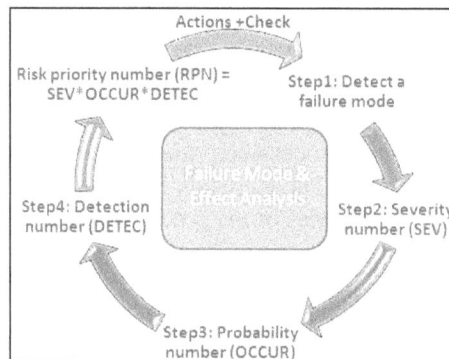

Fig. 4

Failure Mode—For each activity brainstorm all possible ways for the task to fail:

- Failure Effect—For each failure mode define impact/consequence (on stakeholders)
- Severity—Rate the impact/consequence of this effect on a score of 1, 5, 9.
- Cause—For each mode brainstorm the cause.
- Occurrence—Rate the probability of a particular cause occurring on a score of 1, 5, 9.
- Detection—Rate the ability for our systems to detect the occurrence of teach cause, again on 1, 5, 9.
- RPN—Severity score × Occurrence score × Detection score.

3. BRINGING AND PERFORMING THE TRANSFORMATION

No one is ready to transform the organization. One has communicated, strategized, analyzed, reengineered, and blueprinted their ideas for the new process. This is where all of the previous efforts are combined into an actual business system—something one can see and feel and use to enable our organization to meet the market demands of today and tomorrow.

The first step, according to Michael Covert in transforming the organization is to develop a plan for migrating to the new process. We need a path to get from where the organization is today, to where the organization wants to be. Migration strategies include: a full cutover to the new process, a phased approach, a pilot project, or creating an entirely new business unit. Further use of proper BPM software tool is prescribed. Many tools are available in the market. Without the proper utilization of these tools the BPR effect may be felt as incomplete. One of the software tools that can bring about the necessary transformation effectively is Case wise whose modeler effectively visualize, analyze, document and optimize business processes and systems.

Another important point to consider is the integration of the new process with other processes. If only one process is reengineered, then it must interact with the other existing processes. If multiple processes are slated for reengineering, then the new process must not only integrate with existing processes, but also with the newly reengineered processes that will come on line in the near future; therefore, the implementation of the new process must be flexible enough to be easily modified later on.

Successful transformation depends on consciously managing behavioral as well as structural change, with both sensitivity to employee attitudes and perceptions, and a tough minded concern for results. BPR Implementation requires the reorganization, retraining, and retooling of business systems to support the reengineered process.

The new process will probably require a new organization, different in structure, skills, and culture. The new management structure should result in the *control* paradigm being changed to the *facilitation* paradigm. The new process team structure should result in the *managed* paradigm being changed to the *empowered* paradigm. Once the new structures are established, one should map tasks in the process to functional skill levels, and ultimately to workers.

Transforming the workforce will require an array of activities. It begins with an assessment of the current skills or capabilities of the workforce to include soft skills, operational skills, and technical skills. This inventory may require personal evaluations (including areas of interest), peer evaluations, and supervisor evaluations. Feedback should be provided to all personnel to ensure accuracy of current skills and interests for all staff. Armed with the new process skill requirements and a current skills inventory, the gaps can be assessed. Is the new process feasible with the current skill set? Which are the areas to focus on to enhance personnel skills to meet the requirements of the new process? An education curriculum needs to be established to get all employees educated on the business and, most important, on how their jobs relate to the customer.

As with any dramatic change, people will have personal difficulties, to varying degrees, with the paradigm shift that has taken place. Almost all new process implementations are surrounded by confusion, frustration, and sometimes panic. The best transition strategy is one that minimizes, as much as possible, the interference caused to the overall environment. Attempts should be made to keep the new process chaos to a controlled level, to maintain the focus of the reengineering team and the faith of the employees.

Transforming information systems to support the new process may involve retooling the hardware, software, and information needs for the new process. One approach to this transition could be a *controlled introduction*. The method would ensure that each part of the system is operational for a segment of the business before going on to the next module to implement. Although the risk may be low while the bugs in the new system are ironed out, it may be difficult to integrate the hybrid old/new systems in a step-wise manner. The *flash cut* approach is where the entire system is developed in parallel to the existing system, and a complete transition occurs all at once. This may put the organization at a higher risk if the systems do not function properly at first, but it is the more common approach due to the "all-or-nothing" nature of BPR. Most reengineered processes function in an entirely different manner than existing processes; thus, a step-wise introduction would, most likely, not be fully functional until all steps were introduced anyway. An important reason to justify the flash cut approach is that the reengineering benefits can be realized much sooner than with a controlled introduction.

Transitioning the information used to support the old process to become useful in the new process involves reducing some requirements while expanding others. Usually 30 to 40% of the old information can be discarded because it was administrative data needed to tie the old disjointed, linear processes together. On the other hand, the old systems may have poor data integrity, incorrect data, or insufficient data to support the new business needs. In these cases the data must be expanded to fill the gaps in the existing data and supply the new information requirements of the reengineered process. The information blueprints help manage the development of the new information systems.

4. CONCLUSION

While all writers on BPR highlight the importance of the human factors in implementation, we can safely conclude that a proper methodology along with appropriate BPMS software as well as to how to approach the people issues can result in an effective BPR implementation.

REFERENCES

[1] Unni, K.B., Symposium, Seven Hat Consulting, Bangalore, India.

[2] Michael Covert, Successfully Performing BPR, Visible Systems Corporation, MA, USA.

[3] From Wikipedia, the Free Encyclopedia.

Role of Quality in Meeting Global Manufacturing Challenges

K. Balasubramanyam

DRDL Cell
E-mail: balasubkrish@yahoo.com

ABSTRACT: Increase in global competitiveness among production organizations have led to challenges of realizing products at the lowest possible cost while maintaining required delivery schedules and quality standards.

Organizations are now required to focus on activities leading to continuous improvement towards achieving world class. These activities are grouped under core and non core technologies which can lead to minimized cost of manufacturing and decisions to be taken by organizations for make or buy based on the cost of in house manufacturing against cost of outsourcing the products. The competitiveness has led to adopting of world class manufacturing concept.

In the present scenario, the quality is defined by the customer based on his perception of the product as a designer. Quality is required to be built into the product. The concepts such as robust design and six sigma have contributed extensively in achieving the organizational goals by meeting the quality requirements and have universal applications. The author intends to discuss some of the concepts of the application of statistical tools and techniques in manufacturing with the quality benefits derived in R&D manufacturing.

Keywords: Robust Design, Flexible Manufacturing System, Just in Time, Lean Manufacturing, Kaizen, Concurrent Engineering, Six Sigma.

1. INTRODUCTION

To achieve competitiveness a good manufacturing strategy is essential. Superior product quality, innovation and robust after sales service are distinct manufacturing strategies to gain competitive advantages (1). The move towards Flexible Manufacturing System (FMS) has offered many advantages such as to incorporate the changes required by a customer instantly. The cost of manufacturing needs to be controlled. FMS reduces labour cost, lessen manufacturing costs and reduces lead time. It is better to produce items in small batches using FMS and this is particularly suited for R&D manufacturing. For items which have market demand frequently changed concurrent engineering is extremely useful and where the design of the product is simultaneously developed considering the manufacturing implications.

Production processes need to be continuously improved (KAIZEN) to be effective, efficient and adoptable. Improvements can be achieved by waste elimination and non value added activities

(Lean). Statistical measures like Pareto diagram, process flow chart, cause and effect diagram and control charts are used to study and improve processes. The Kaizen principle based on small increments in every function is one of the effective means to improve processes, produce defect-free quality products and services at lowest production costs. The competition to meet the global challenges include apart from cost competition, customer driven quality requirements and realistic delivery schedules in addition to infrastructure development, development of techno-logical strengths, skill training, investments in R&D, manufacturing excellence through product designing and adaptation of strategies for flexibility and benchmarking. If the organization decides on outsourcing, there is a compelling requirement for framing the rules for prevention of leakage of information to competing organizations through vendors. India is recognized as a viable option for outsourcing by the west due to the low cost of manufacturing here.

2. QUALITY CONCEPTS IN R&D PRODUCTION

Quality is a prime focus of business attention. Quality of a product can be ensured if there is an active participation of senior management of a company in the development of the product. If the senior management is not showing interest in the process, then the tendency develops among the actual product developers and vendors to take things for granted and compromise with the quality. Improving quality is always profitable. The eight critical dimensions of quality as identified by Gavin includes performance, features, reliability, serviceability, durability, aesthetic and perceived quality and conformance (6).

Robust design is a technique that is of significance in the development of product that can very well suit R&D work. The design should be able to produce good quality product in spite of variations in the process.

A knowledge based quality index is presented below to bring out the comparison between different types of production systems with reference to the variety in production. It is clear that by adopting the job type production system a variety of jobs can be carried out . This method is particularly useful in the case of R&D type of work which has a variety of jobs but with low production quantities.

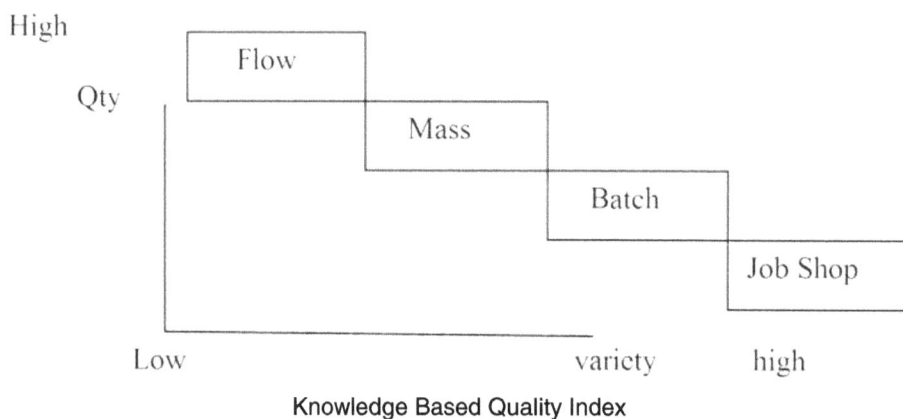

Knowledge Based Quality Index

Technology transfer involving passage of ideas, skills, processes etc especially in R&D work is very beneficial. Certification of ensuring quality all stages increase the confidence level of the customer.

Some of the aspects that are coming to importance in the R&D production include the following:

(a) Concurrent engineering
(b) Design standardization
(c) Quality improvement mechanisms
(d) Dealing with non conformities
(e) Quality Control documentation.

2.1 Concurrent Engineering

While options are available for developing the quality systems for R&D manufacturing such as Zero defect, Six Sigma and Just–in-Time (JIT) , there are clear advantages in applying the Taguchi's Design of Experiments (DOE) in association with the concepts of Concurrent Engineering for R&D manufacturing .The advantages of concurrent engineering are varied including shorter cycle time, reduced cost, optimized design (through robust design techniques), better quality with reduced rejection or scrap and continuous improvement at the shop floor level. The concurrent engineering applications have been applied for this system successfully at the production workcentres.

2.2 Design Standardization

In bulk production there is always a need for design standardization and changes need to be documented. There should also be a mechanism for review of design changes followed by effective documentation. This is very essential from the certification aspect also. As such majority of the organizations carrying out R&D work have been certified by ISO and it is also a requirement of the ISO certification.

2.3 Quality Improvement Mechanisms

In order to meet the quality requirements a continuous and close interaction between design, development agencies and testing agencies, QA teams and integration teams is needed to be established. The quality group also participates in the design review, modification and reporting of non conformities.

The QA controls will consist of raw material control and control of fabrication. This further will require the finalizing the process plans. The quality non conformances are compiled and presented for waiver actions, if found suitable.

Parts may be considered for acceptance as is, or reworking or rejections. Functional tests are required to be carried out as per specified plans and quality control checklists are prepared. The acceptance of the items will be based on the performance tests or functional tests that are specified for each item. All the non conformances that are recorded during the component stage, sub assembly stage and integration stages are recorded for discussion at appropriate forums.

2.4 Dealing with Non-Conformities

Non-conformity is defined as anything that does not conform to the regulations, standards or specifications in addition to any deviation from the prescribed standards for the intermediate processes.

Non conformity is addressed in the waivers which may be either major or minor depending on the performance or functional variations as indicated earlier. A meeting is arranged by the quality department of the work center to discuss the causes of deviations and determine what kind of corrective and preventive measures that are to be taken;

1. Inadequacies leading to non conformities are due to
   – Inadequate verification of the documents
   – inadequate change control
   – not specifying tolerances adequately.
2. Control system for non conformity prevention.

Some of the steps to be taken in this regard are the following:

- implement change control more reliably
- draw up more detailed design review schedules
- establish clear procedures for verification of the documents
- establish clear procedures for carrying out product reviews.

2.5 Quality Control Documentation

The Quality Control documentation indicates the records that are maintained for the following purposes:

- enabling follow up action
- enabling rectification
- repair of non conforming products
- corrective action in preventing the non conformities in design, material specification and production process.

3. CASE STUDY

A typical case study of R&D manufacture of aerospace system is considered for study (ref Figure 1). The technology development for application in aerospace where the drawing/design, process modifications and quality requirement amendments have been effected during the production concurrently.

Any aerospace or missile system is derived from technical specifications that include drawings/ qualification and acceptance test documents, definition dossiers and standards of preparation based on the specific requirements of the order. The production agencies are identified through a series of procedures after which a trial order on development basis is first placed to realize the initial components or sub assemblies. These are evaluated as per the quality plans. After the developmental production items are satisfactorily realized, a series production order may be placed on the production agency in accordance with the users requirements.

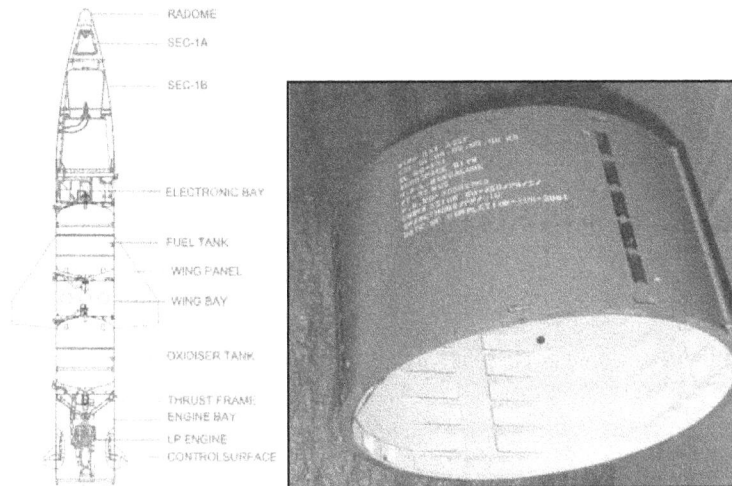

Fig. 1&2: Aerospace Structure and Component for Manufacture

Technical specifications of cast section of aerospace structure (ref Figure 2)

> MATERIAL: AZ 91 C/ML 5/GA 9 magnesium cast alloy
>
> SIZE: 1000 × 550 × 25 mm
>
> WALL THICKNESS – 7.5 TO 10 mm
>
> QUALITY REQUIREMENT – 100% RADIOGRAPHY
>
> To ASTM – B-155
>
> WEIGHT – 100 Kgs as cast 42 Kgs as machined
>
> TESTING – Chemical/Mechanical/HEAT TREATMENT – Homogenization.

A magnesium alloy large size thin walled casting as per specifications above is developed for the airframe structure of a airborne vehicle. This is a complex structure involving the casting of a large size thin walled casting followed by precision machining of cutouts and drilling and tapping for fixing of gaskets as well as hatch covers.

The quality of the casting is assured in the following manner:

- raw material checks inclusive of chemical and mechanical tests
- melt quality checks including chemical and grain size check
- casting quality including integrity through visual inspection followed by 100% radiographic inspection
- inspection of casting stages such as fettling, heat treatment and visual
- mechanical property analysis using separately or integrally cast test bars
- dimensional analysis before and after proof machining
- development and implementation of QC checklists for the component.

During the development of the item, concurrently a number of drawing, process and quality requirements have been identified and effected modifications based on production experience first on a trial basis and later as a permanent requirement.

3.1 Design Tolerance Study and Modification

The design modifications include both the modifications at casting stage and after machining. The casting stage design modification include the design tolerance review and incorporation of appropriate changes in the drawings to foundry specifications that was not earlier made in the development drawings. Apart from this the casting requires to incorporate the weight control strictly and precisely. Hence the machining sequences were modified to include the internal operations like manual scuffing to achieve the minimum wall thickness requirements at various casting zones as well as leveled surface without undulations. The other process requirements that have been brought about concurrently are the fairing strip bonding on the outer surface of the finished casting, fitment of covers and gaskets at casting door and wing pockets with suitable fasteners. The fasteners also have been standardized.

3.2 Design of Experiments on Pouring Parameters

Taguchi techniques have been applied for the optimization of the pouring parameters by experimentation. As a result it was possible to obtain a range of pouring parameters for the maximization of the casting yield has been realized. The alloy AZ 91 C being prone for casting defects and analysis have been carried out and defects have been mapped on a development diagram for analysis and taking corrective actions. The defects have been identified and standardized with the use of ASTM specifications prescribed.

3.3 Casting Defect Identification, Analysis through Mapping and Their Elimination

Magnesium alloy castings are prone for defects as well as corrosion. Long term tests for adequacy of coatings have been introduced as a development concurrently. The author has studied the effects of single as well as multiple (clustered) defects in magnesium casting alloys and identified the Defect Intensity Factor (DIF) for castings in the case of multiple defects with the application of castings inspection standards (ref 9 and 10)

Following are generally used to establish the reasons for the non conformity:

 (a) Cause and effect diagram (Figure 3)
 (b) Pareto chart (Figure 4).

Casting defect types and pattern are shown at Figure 5 and 6 and defect mapping diagram at Figure 7. Casting stage wise rejection analysis at Figure 7 Casting defect types are shown in Figure 8 (all at Annexure 1).

3.4 Process Stage Modifications

The important process modification in the manufacture of magnesium alloy casting is the development of casting protection measures. These include the lanolin protection at the intermediate casting stage, and machining stages and the chromating process and casting impregnation. An additional operation of the fluoride anodizing process was introduced at the casting manufacturing stage to provide protection during casting stage as well as acting at a base for the primer/paint adhesion.

3.5 Quality and Test Requirements

The magnesium casting is subjected to 100% radiographic examination as well as the red dye penetrant testing the latter was eliminated from the process after studies since it was found to be redundant. Above mentioned aspects are some of the key measures that have been carried out concurrently on the magnesium alloy casting during the development and production and the technologies thus developed are transferred to the identified production agency. Summarizing the above activities as follows:

Concurrent Approach	Derived Benefits	QA Check
Design tolerance study	Dimensions retoleranced to casting specifications	Dimensional inspection
	Casting weight control	Weight check
Fluoride Anodizing process introduced	Surface cleanliness on castings	Visual
Defect mapping analysis	Defect location identification on castings	Defect mapping diagram
Corrosion studies	Protection measures developed and tested for adequacy	Visual and corrosion tests
Design of experiments on pouring parameters	Optimization and yield improvement	Pouring at newly developed parameters
Casting defect analysis	Defect identification and measures for elimination	Defect mapping diagram

4. CONCLUSIONS AND RECOMMENDATIONS

The case study discussed indicates some of the aspects of quality that have been followed in realizing the structural components as per the aerospace specifications in the current manufacturing practices.

To meet the future needs of the R&D manufacturing with particular reference to aerospace system following are some of the recommendations that may help in establishing the excellence in manufacturing of which quality plays a prominent role. Manufacturing for aerospace systems denotes one of the typical areas where excellence is of prime consideration for meeting the global manufacturing challenges. Some of the aspects that can contribute substantially for quality improvement include the following:

(a) *Six Sigma Approach:* Six sigma is a methodology by which process improvements are carried out by reviewing and updating the existing processes. Six Sigma was developed with a goal of limiting the manufacturing defects to 3.4 parts per million units produced and a combination of lean and Six Sigma has led to the introduction of Lean Six Sigma. It also helps in decreasing variation and maintains quality .it consists of 5 phases; define, measure, analyse, improve and control (D-M-A-I-C). It concerns with any activity involving quality, cost and time schedules. In R&D manufacturing for improvement of performance, process and measurement systems, quality improvement and customer satisfaction. Many statistical methods such as sample size, confidence interval, variance analysis, correlation and regression are involved.

(b) *ISO 9001:* 2000/AS 9100 and Beyond: For aerospace manufacturing the minimum require-
    ments are specified in Quality Management System (QMS) followed by Business
    Excellence Models (BEM). As 9100 Rev C improves upon ISO 9001 with additional
    requirements including special processes, configuration management, product docu-
    mentation, methods resources recording, inspection and testing, first item inspection and
    so on.

(c) *Performance Management System (PMS):* This is gaining importance with measurement
    systems such as balance score card. The applicability of the same in R&D work needs to
    be considered.

(d) *Just in Time and Total Quality Control:* This aspect is gaining importance with world
    class manufacturing.

(e) *Kaizen for Quality Improvement:* This is an innovative application of Kaizen for quality
    improvement apart from tools and techniques such as six sigma, Kaizen and Activity
    Based Costing (ABC)etc. The organizational survival includes cost competitiveness,
    World class quality systems, lower operational cost, flexible working and so on.

(f) *Lean Manufacturing:* This is already implemented in many large and small organizations. It
    involves the elimination of all non value added activities and waste reduction. Reduces cost,
    production time and floor space requirements and increases productivity as well as product
    quality. Lean manufacturing includes value stream mapping, Kaizen, 5S, set up time
    reduction and Poka yoke.

(g) *ERP Implementation:* To improve productivity and performance in all types of organizations
    engaged in manufacturing. Benefits of ERP include reduction in planning cycle time,
    manufacturing cycle time, inventory, manpower reduction etc.

REFERENCES

[1] Chandra, P. and Shukla, P.R., Manufacturing excellence and global competitiveness challenges
    and opportunities for Indian Industries.

[2] Mahadevan, B., Principles of World Class Manufacturing, The Management Accountant, Sept. 1998.

[3] Economic and Political Weekly, 1994.

[4] Liker, J.L., The Toyota way, Tata Mc Graw Hill, 2004.

[5] Kale, P.T. and others, ERP implementation in Indian SMEs.

[6] Gavin, D.A., Competing on eight dimensions of quality, Unconditional Quality, Harward business
    review,1982.

[7] Gavin, D.A., Quality on line, Unconditional Quality, Harward Business Review, 1982.

[8] Hartley, J.P., Concurrent Engineering, Productivity Press, 1990.

[9] Jain, P.L., Quality Control and Total Quality Management, Tata Mcgraw Hill, 2001.

[10] ASTM-E-155: Radiographic inspection standards for aluminium and magnesium alloy castings.

[11] ASTM-E-505: Inspection of aluminium and magnesium alloy castings.

ANNEXURE 1

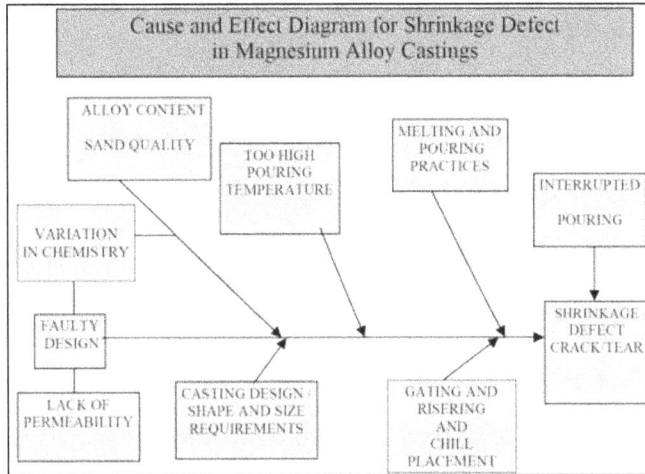

Fig. 3

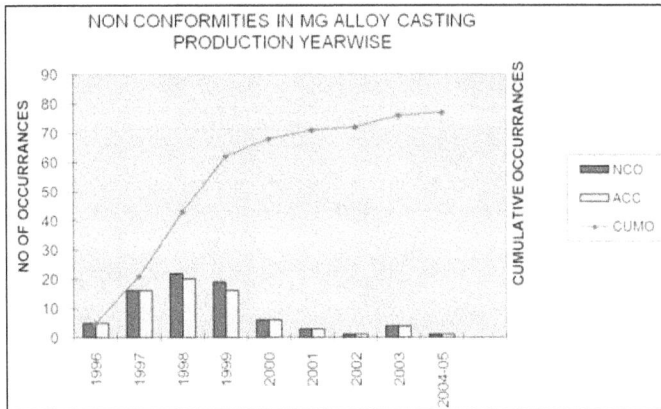

Fig. 4

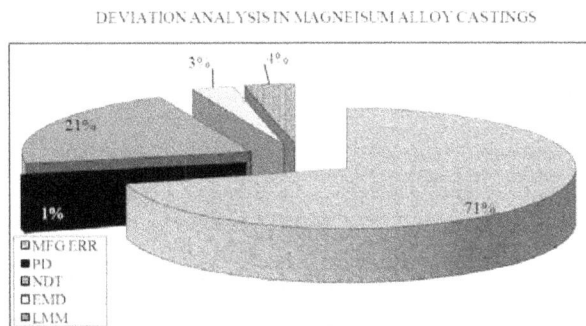

Fig. 5

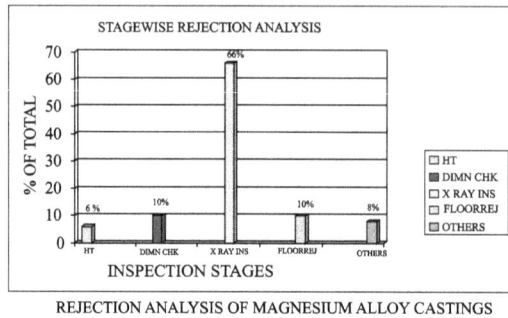

REJECTION ANALYSIS OF MAGNESIUM ALLOY CASTINGS

Fig. 6

Fig. 3–6: Analysis of Non-Conformities (Deviations) in Magnesium Castings

Defect Mapping Diagram
(Casting Zonewise Basis)

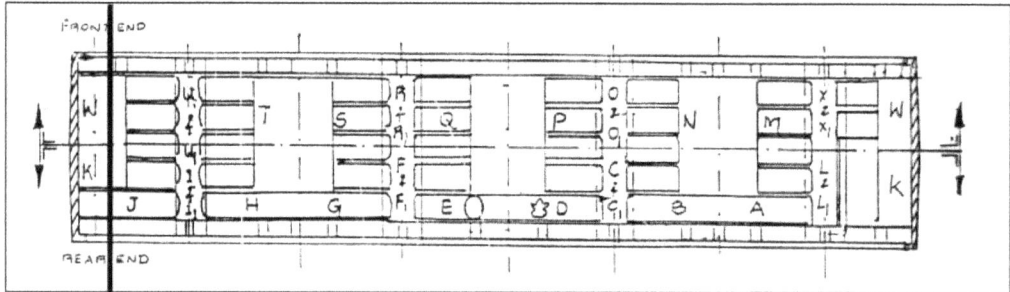

Fig. 7

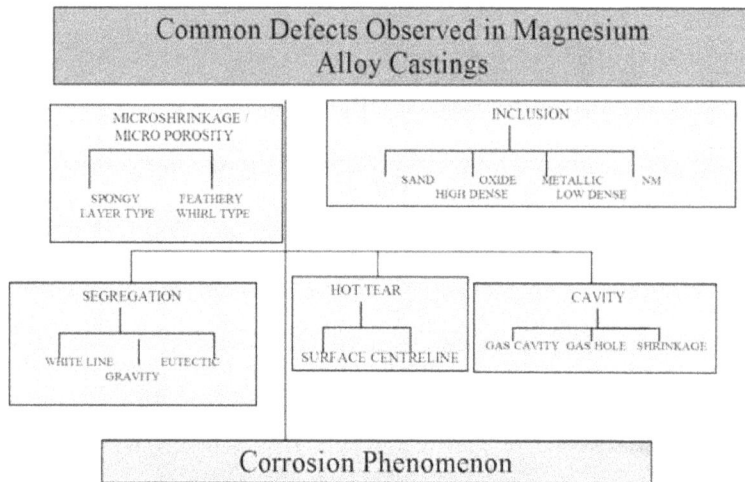

Fig. 8

Fig. 7 and 8: Defect Mapping and Defect Types in Magnesium Alloy Castings

Service Quality of Public Sector Banks in India

Sitaram Vikram Sujir

Rayalaseema University, Kurnool, Andhra Pradesh
E-mail: sitaramsujir@gmail.com

ABSTRACT: The banking industry in India is now running in a dynamic challenge concerning both customer base and performance. Service quality, customer satisfaction, customer retention, customer loyalty and delight are now the major challenges which are gripping the banking sector.

Service quality plays a major role in getting customer satisfaction and creating brand loyalty in banking sector. Human element acts an important role in perceived service quality as well as satisfaction. Public sector banks need to redefine the customer service parameter in order to compete with the private sector banks both in profitability and corporate image.

The Bureau of Indian Standards have also developed a standard for service quality by Public sector organizations IS 15700:2005, though banks do not come within the scope of this standard it has very good practices which could be adopted by the banks.

The committee (By Reserve Bank of India) on customer service in banks headed by Shri. M. Damodaran, former Chairman, SEBI reviewed the existing system of attending to customer service in banks-their approach, attitude and fair treatment to customers. They have come out with their summary of recommendations outlining Customer Service in Banks, Customer Service and Technology, Role of Boards of Banks in Customer Service etc.

This study is just a small step in understanding the multidimensional construct of service quality and its implications in competitive environment. This paper attempts to extract few dimensions of SERVQUAL or service quality as perceived by bank customers and compares with five major dimensions already extracted in past literature.

Keywords: Service Quality, SERVQUAL, Bank, India, RBI, IS 15700:2005.

1. INTRODUCTION TO BANKING IN INDIA

Banks are the most significant players in the Indian financial market. They are the biggest purveyors of credit, and they also attract most of the savings from the population. Dominated by public sector, the banking industry has so far acted as an efficient partner in the growth and the development of the country. Driven by the socialist ideologies and the welfare state concept, public sector banks have long been the supporters of agriculture and other priority sectors. They act as crucial channels of the government in its efforts to ensure equitable economic development.

According to Prof. Hart, "a bank is one who in the ordinary course of business receives money which he repays by honoring cheques of persons from whom or on whose account he receives it" Commercial banks constitute the major portion of the country's credit and banking institutions. In simple term a bank is a dealer in money like a trader in goods. The various functions of a bank are given in Figure 1.

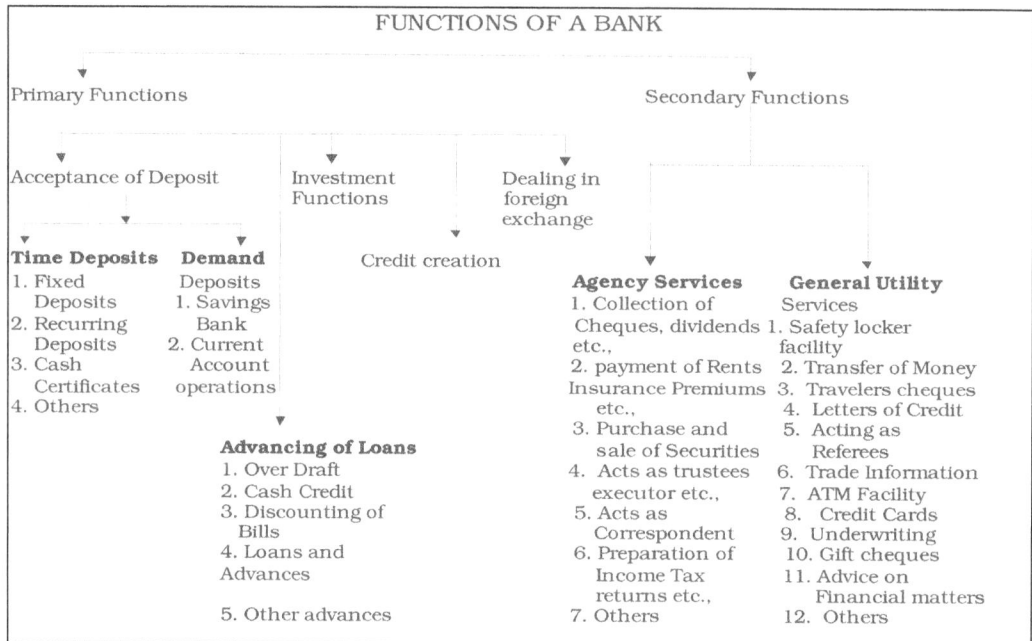

FUNCTIONS OF A BANK

Primary Functions   Secondary Functions

Acceptance of Deposit   Investment   Dealing in
                             Functions   foreign
                                                exchange

Time Deposits Demand   Credit creation   **Agency Services**   **General Utility**
1. Fixed   Deposits   1. Collection of   Services
   Deposits   1. Savings   Cheques, dividends   1. Safety locker
2. Recurring   Bank   etc.,   facility
   Deposits   2. Current   2. payment of Rents   2. Transfer of Money
3. Cash   Account   Insurance Premiums   3. Travelers cheques
   Certificates   operations   etc.,   4. Letters of Credit
4. Others   3. Purchase and   5. Acting as
                                                            sale of Securities   Referees
                Advancing of Loans   4. Acts as trustees   6. Trade Information
                1. Over Draft   executor etc.,   7. ATM Facility
                2. Cash Credit   5. Acts as   8. Credit Cards
                3. Discounting of   Correspondent   9. Underwriting
                   Bills   6. Preparation of   10. Gift cheques
                4. Loans and   Income Tax   11. Advice on
                   Advances   returns etc.,   Financial matters
                                                         7. Others   12. Others
                5. Other advances

Fig. 1

The Indian banking can be broadly categorized into nationalized (government owned), private banks and specialized banking institutions. The Reserve Bank of India acts a centralized body monitoring any discrepancies and shortcoming in the system. Since the nationalization of banks in 1969, the public sector banks or the nationalized banks have acquired a place of prominence and since then seen tremendous progress. The need to become highly customer focused has forced the slow-moving public sector banks to adopt a fast track approach. The Nationalized banks include the fourteen banks nationalized on 19th July 1969 and the 6 banks nationalized on 15th April 1980. They are also scheduled banks. After nationalization these banks render various types of functions by assuming social responsibilities. Through these banks, the government tries to implement its fiscal policies and various welfare schemes. These banks occupy a pivotal place in the Indian banking System. They are also called public sector banks.

The Indian banking has finally worked up to the competitive dynamics of the 'new' Indian market and is addressing the relevant issues to take on the multifarious challenges of globalization. Banks that employ IT solutions are perceived to be 'futuristic' and proactive players capable of meeting the multifarious requirements of the large customer's base. Private Banks have been fast on the uptake and are reorienting their strategies using the internet as a medium The Internet has emerged as the new and challenging frontier of marketing with the conventional physical world tenets being just as applicable like in any other marketing medium.

The Indian banking has come a long way from being a sleepy business institution to a highly proactive and dynamic entity. This transformation has been largely brought about by the large dose of liberalization and economic reforms that allowed banks to explore new business opportunities rather than generating revenues from conventional streams (*i.e.* borrowing and lending). Indian nationalized banks (banks owned by the government) continue to be the major lenders in the economy due to their sheer size and penetrative networks which assures them high deposit mobilization. The Indian banking can be broadly categorized into nationalized, private banks and specialized banking institutions.

The liberalized policy of the Government of India permitted entry of private sector in the banking, the industry has witnessed the entry of new generation private banks. The major differentiating parameter that distinguishes these banks from all the other banks in the Indian banking is the level of service that is offered to the customer. Their focus has always centered around the customer—understanding his needs, preempting him and consequently delighting him with various configurations of benefits and a wide portfolio of products and services. These banks have generally been established by promoters of repute or by 'high value' domestic financial institutions.

The popularity of these banks can be gauged by the fact that in a short span of time, these banks have gained considerable customer confidence and consequently have shown impressive growth rates. Today, the private banks corner almost four per cent share of the total share of deposits. Most of the banks in this category are concentrated in the high-growth urban areas in metros (that account for approximately 70% of the total banking business). With efficiency being the major focus, these banks have leveraged on their strengths and competencies viz. Management, operational efficiency and flexibility, superior product positioning and higher employee productivity skills.

The private banks with their focused business and service portfolio have a reputation of being niche players in the industry. A strategy that has allowed these banks to concentrate on few reliable high net worth companies and individuals rather than cater to the mass market. These well-chalked out and integrated strategy plans have allowed most of these banks to deliver superlative levels of personalized services.

2. RESERVE BANK OF INDIA—THE REGULATOR

Banking sector is the most important segment in the financial system of any country as most of the economic activities are supported and provided funds by the banks. Efficient and profitable functioning of banks are indicators of healthy and vibrant economy of a country. The supervision of banks is generally entrusted to the Central Bank in every country. In India, supervision of commercial banks is carried out by the Department of Banking Supervision (DBS) of RBI. The supervision of co-operative banks and Regional Rural Banks is carried out by NABARD.

The Reserve Bank of India acts as a centralized body monitoring any discrepancies and shortcoming in the system. It is the foremost monitoring body in the Indian financial sector. The nationalized banks (*i.e.* government-owned banks) continue to dominate the Indian banking arena. The banking operations in India are controlled by the Banking Regulation Act 1949 the

Reserve Bank of India Act 1934, the Bankers Book Evidence Act 1981, the Foreign Exchange Regulation Act 1973, etc., The Banking Regulation Act Conferred enormous powers on RBI for controlling the banking companies. The Banking Regulation Act has undergone many changes.

The primary objective of supervision is to ensure safety of funds placed by depositors with banks and to retain the confidence of depositors, investors and clients in the banking systems. Therefore, the objective of supervision is also to ensure the solvency and liquidity of every bank. It also ensures that banks are following the instructions and directives of Central Bank issued on various aspects of banking activities. It is required to identify deficiencies and irregularities in the working of banks and provide early warning system to avert any financial crisis in the banks. In short, the objective of banking supervision is to ensure the overall financial health, liquidity, solvency, and profitable operations of every bank.

3. CUSTOMER SERVICE IN BANKS

Banks cannot exist without customer. The banks operations are intended to serve customer and not vice versa. A customer looks for certain values like product, quality, reliability, superior service state of the art technology, low cost, a premium image, etc while purchasing anything. In service sector like bank, customer service should not only be a critical function but also a way of life. Banks can be said as being customer oriented if its various organizational activities like restructuring, staffing, co-ordination are geared to fulfill customer's needs. Hence total customer satisfaction should be the focal point.

The Reserve Bank of India has constituted a Committee under the chairmanship of Shri M. Damodaran, former Chairman, SEBI to interalia:

1. Review the existing system of attending to customer service in banks—approach, attitude and fair treatment to customers from retail, small and pensioners segment.
2. Evaluate the existing system of grievance redressal mechanism prevalent in banks, its structure and efficacy and recommend measures for expeditious resolution of complaints. The committee may also lay down a suitable time frame for disposal of complaints including last escalation point within that time frame.
3. Examine the functioning of Banking Ombudsman Scheme—its structure, legal framework and recommend steps to make it more effective and responsive.
4. Examine the possible methods of leveraging technology for better customer service with proper safeguards including legal aspects in the light of increasing use of Internet and IT for bank products and services and recommend measures to enhance consumer protection.
5. Review the role of the Board of Directors of banks and the role of Regulators in customer service matter.

The committee submitted its report on July 4th 2011 and has made recommendation on Customer Service in Banks, Grievance Redressal System in Banks, Banking Ombudsman Scheme, Customer Service and Technology, Role of Boards of Banks in Customer Service.

4. DEVELOPING A CUSTOMER SERVICE SYSTEM

Before a bank can measure how well they are providing customer service, they need to have a customer service system in place. Otherwise, a bank will not have a way to measure customer satisfaction. Here is a seven-step approach to developing a successful customer service system followed by some techniques a bank can use to implement customer service and quality improvement.

Step 1: Total Management Commitment

Customer service and quality improvement programs will only succeed when there is total management commitment, and this commitment must begin at the top. The CMD, chairman of the board or owner must develop and communicate a clear vision of what the service quality system is going to be, how it is going to be implemented, what the staff should expect when implementing it, how it will be used to satisfy and retain customers and how it will be supported over time. This process of total management commitment must begin with a vision statement or mission statement related to service quality.

Step 2: Know Your Customers (Intimately)

The bank must do everything possible to get to know its customers intimately and to understand them totally. Some people suggest that the bank needs to know its customers better than they know themselves. This means knowing their likes and dislikes in regards to business; the changes they may want the bank to make; their needs, wants and expectations (now and in the future); what motivates them to buy or change suppliers; what the bank does to satisfy them, retain them and make them loyal.

When the bank has learnt about its customers and it thinks it knows them as well as the bank knows itself, the bank must learn about them all over again. Their needs change daily, even hourly, and you must know to satisfy those needs. Their requirements and expectations change also, and the bank must be able to meet and exceed those expectations. Knowing its customers intimately, and on an ongoing basis, requires that the bank keep in constant contact with them. Call them regularly. Write to them. Invite them for lunch. Find out exactly what they are doing, what they need, and what they want the bank to do for them. This constant contact will help the bank develop the retention and loyalty it needs because they know the bank is interested in them.

Step 3: Develop Standards of Service Quality Performance

Customer service, quality and service quality appear to be intangible items because they are based on perception. However, they do have tangible and visible aspects that the bank can manage and measure. For example, customers dislike waiting for a telephone to be answered or being placed on hold for a long period of time. How long does it take to process a cheque? What is the bank's policy on cheque returned, refunds, exchanges and complaints?

These are all tangible aspects of service quality and they can be measured. If the bank has any doubts about what to measure, just ask your customers. They will tell you (perhaps not directly or exactly) what they are looking for and how they judge service quality. And since service quality and satisfaction only exist in the minds of the customers, the banks should develop its standards and measurement systems to meet their perceptions.

Step 4: Hire, Train and Compensate Good Staff

Superior customer service and quality performance that result in customer satisfaction and retention can only be provided by competent, qualified people. The bank's service quality is only as good as the people who deliver it. If the bank wants its business to be good to people, and that is a requirement for success in today's business environment then it must hire good people.

Once hired, train them extensively to provide superior customer service and do things right the first time. Be sure they understand the bank's standards of service quality performance and the customer's expectations of service quality. Train them in their own jobs and train them in other jobs as well. Let them experience being a customer of a bank, and then have them make suggestions for improving the treatment of customers.

Finally, empower the bank staff to make decisions and do the right thing to satisfy its customers. The staff should not have to look for you or a manager every time a customer asks a question, has a cheque return or a complaint or just needs a problem solved. There are legions of stories about empowered employees who made decisions that were against bank's policy but that satisfied and retained a customer, with the end result of both the business and the customer winning. If the bank is going to place people in customer contact positions, give them the authority that goes with this tremendous responsibility. They must be able to do whatever it takes to satisfy the customer.

Step 5: Reward Service Quality Accomplishments

Always recognize reward and reinforce superior service quality performances, Do this for the bank employees and its customers. Provide psychological, and sometimes financial, incentives for your people. Help them motivate themselves to do even better. Broadcast and make a big deal about all types of service accomplishments that result in more satisfied customers. Recognize and reward even the small wins in a manner similar to what you would use for the major accomplishments.

Also, reward your customers for good customer behavior, everyone wants to be appreciated and made to feel important, especially your customers. Provide them with recognition and appreciation just as you would your employees. This will motivate them to refer more business to the bank and to be more loyal.

Step 6: Stay Close to Your Customers

Even though the bank has got to know its customers intimately in Step 2, it must now do everything possible to stay close to them. Keep in touch with them in any and every way possible. Go visit them. Send them letters, cards, newsletters, published articles that would be of interest to them. Conduct continues research to learn about their changing needs and expectations.

Ask them questions right after they use a service. Ask them why they did not make use of service. Mail them questionnaires and other types of surveys. Call them up on the telephone and ask them how you can do a better job for them. Do whatever it takes to stay close to the bank's customers and continue to build and maintain this valuable relationship.

The bank's relationship with the customer really solidifies after the service is rendered or an account is opened. Let them know the bank cares about them and that it will support their requirements. Make sure they are satisfied, and find out what the bank must do to maintain that satisfaction and loyalty.

Step 7: Work Towards Continuous Improvement

Now that the bank has friendly and accessible customer service systems, have hired and trained the best people for the job and have learned everything it can do about its customers, the bank cannot rest. No system or program is perfect, least of all one that is based on a person's perceptions as is the case for service quality. Therefore, the bank must continually work to improve its customer service and performance quality.

Customers who are initially satisfied with the service will perceive the bank's attempts at continuous service quality improvement as something very positive. They may even want to help. Welcome them with open arms. They are a bank's best source of information about how to get better in their eyes and minds. Plus, when a bank implements their recommendations and suggestions, they perceive that the bank values them even more. The result is that they will do more business with the bank, which leads to more satisfied customers, a happier staff and greater profits.

5. MEASURING THE QUALITY OF SERVICE

The primary benefit of a measurement program is that it provides the bank with immediate, meaningful and objective feedback. The bank can see how they are doing right now, compare it to some standard of excellence or performance, and decide what they must do to improve on that measurement. Have you ever asked yourself why basketball is so popular? It is because the player knows immediately if he or she succeeded. Performance is measured by the ball going through the basket or not, and the player is motivated to try again. This occurs whether or not the shot was successful.

Measurement provides the bank with a sense of accomplishment, a feeling of achievement. Measurements can also form the basis for a reward system that can only be successful if it is based on objective and quantifiable data. How will the bank know which employees or work teams to reward for improving quality and increasing customers' satisfaction if it cannot, or does not, measure their performance?

The benefits of measuring quality of customer service can be summed up in these five items:

1. Measurement provides the bank with a sense of achievement and accomplishment, which will then be translated into superior service to customer.
2. Measurement provides the bank with a baseline standard of performance and a possible standard of excellence which they must try to achieve, which will lead to improved quality and increased customer satisfaction.

3. Measurement offers a performer immediate feedback, especially when the customer is measuring the performer of the bank.

4. Measurement tells the bank what it must do to improve quality and customer satisfaction and how you must do it. This information can also come directly from the customer.

5. Measurement motivates the bank to perform and achieve higher levels of productivity.

6. IS 15700:2005 SERVICE QUALITY IN PUBLIC SERVICE ORGANIZATIONS

Since we are dealing with public sector banks they can adopt the IS 15700:2005 developed by Bureau of Indian Standards as a bench mark. This standard is meant for establishing and maintaining service Quality in all public sector undertakings. Though banks do not come within its scope, it could be used as a reference point while delivering excellence to the customer.

This standard has been specifically designed for the public service organizations, and in its formulation, considerable assistance has been drawn from IS/ISO 9001:2000 'Quality management systems – Requirements', IS/ISO 10002:2004 'Quality management – Customer satisfaction – Guidelines for complaints handling in organizations' and Government of India guidelines on 'Citizens' charter' and 'Public grievance handling'.

This standard can be used by internal and external parties, including certification bodies, to assess the organization's ability to meet customer, regulatory and the organization's own requirements. Below are some of the clauses mentioned in the standard which could be used by the banks.

6.1 The Organization Shall Monitor and Measure

(a) The characteristics of the service and service delivery processes to verify that the service quality objectives and service standards have been met. This shall also be carried out at all stages and locations where the organization has an interface with the customer;

(b) The performance vis-a-vis commitment made in the citizens' charter and complaints handling procedure on a regular basis and report to top management with recommendations for improvement; and

(c) The working of the complaints handling machinery through random checks.

6.2 Customer Satisfaction

The organization shall establish and implement a system for measuring customer satisfaction through suitably designed methodology. This information shall be used for continual improvement.

6.3 Analysis of Data

The organization shall analyze the data collected during monitoring and measurement and customer satisfaction to determine current level of performance and opportunities for continual improvement, particularly where nonconformities are recurring.

6.4 Corrective and Preventive Actions

The organization shall take action to eliminate the cause(s) of non-conformities and potential cause(s) in order to prevent recurrence and occurrence respectively. These shall be appropriate to the effects of the nonconformities encountered and potential problems. Records of action taken and improvements effected shall be maintained

Hence this standard developed by BIS can be used as a benchmark when the PSU banks serve its customers.

7. SERVICE QUALITY OR SERVQUAL TOOL

Though initial efforts in defining and measuring service quality emanated largely from the goods sector, a solid foundation for research work in the area was laid down in the mid-eighties by Parasuraman, Zeithaml and Berry (1985). They were amongst the earliest researchers to emphatically point out that the concept of quality prevalent in the goods sector is not extendable to the services sector. Being inherently and essentially intangible, heterogeneous, perishable, and entailing simultaneity and inseparability of production and consumption, services require a distinct framework for quality explication and measurement. As against the goods sector where tangible cues exist to enable consumers to evaluate product quality, quality in the service context is explicated in terms of parameters that largely come under the domain of 'experience' and 'credence' properties and are as such difficult to measure and evaluate (Parasuraman *et al.,* 1985; Zeithaml and Bitner, 2003).

One major contribution of Parasuraman *et al.* (1988) was to provide a terse definition of service quality. They defined service quality as 'a global judgment, or attitude, relating to the superiority of the service', and explicated it as involving evaluations of the outcome (*i.e.,* what the customer actually receives from service) and process of service act (*i.e.,* the manner in which service is delivered). In line with the propositions put forward by Gronroos (1982) and Smith and Houston (1982), Parasuraman, Zeithaml and Berry (1985, 1988) posited and operationalized service quality as a difference between consumer expectations of 'what they want' and their perceptions of 'what they get.' Based on this conceptualization and operationalization, they proposed a service quality measurement scale called '**SERVQUAL**'. The scale constitutes an important landmark in the service quality literature and has been extensively applied in different service settings.

The foundation for the SERVQUAL scale is the gap model proposed by Parasuraman, Zeithaml and Berry (1985, 1988). With roots in disconfirmation paradigm, the gap model maintains that satisfaction is related to the size and direction of disconfirmation of a person's experience vis-à-vis his/her initial expectations.

The Service quality gap (Figure 2) model provided by Parashuraman *et al.* (1984) outlines the below mentioned gaps in service quality. They are:

1. *Customer Expectation—Management Perception Gap:* This gap exists due to management's inaccurate perception of what customers actually expect. Lack of   proper customer focus by the management generates this gap.

2. *Service Quality Specification Gap:* The gap is created due to the inability on the part of management to translate customer expectations in to service quality specifications. The gap is created due to in adequacy in service design.

3. *Service Delivery Gap:* This gap is created due to in adequate support from the front line staff and variability staff performance.

4. *External Communication Gap:* This gap is created due to communication inadequacy of service providers. In accurate description of service offerings results in difference between expectation and perception.

5. *Expected and Perceived Service Gap (SERVQUAL):* This is how consumers perceive the actual service performance in the context of what they expected.

Parashuraman views that the important element in service to achieve competitive advantage is to have a fine blend of internal marketing, interactive marketing before adoption external marketing straightly.

Internal marketing is marketing to employees within the organization and interactive marketing is employees marketing to customers. Employees are the better ambassadors of an organization. A judicious mix of internal and interactive marketing calls for better relationship and adequate care taken towards employees.

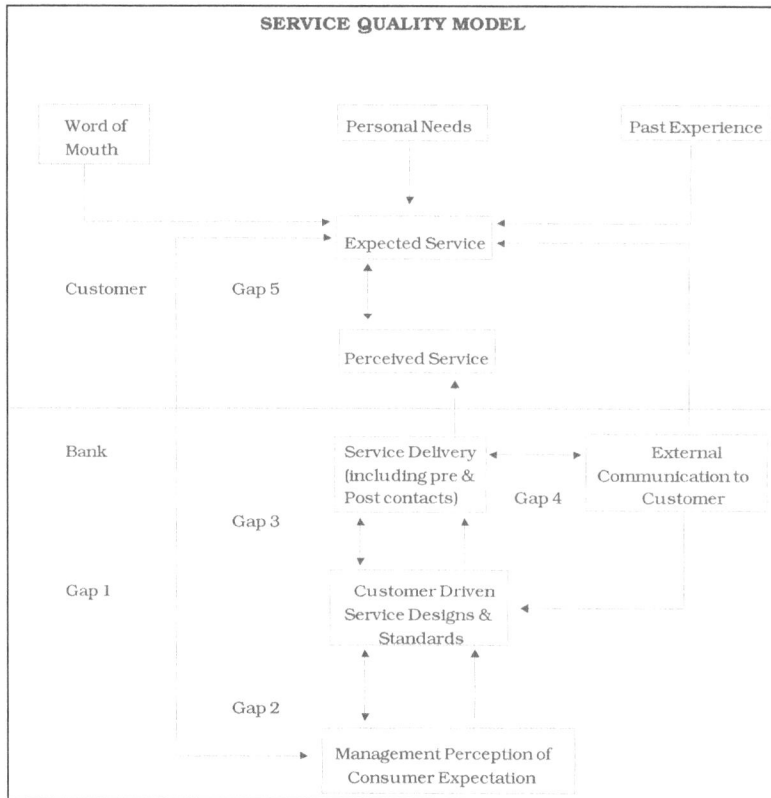

Fig. 2

SERVQUAL (GAP 5) was originally measured on 10 dimensions of service quality: reliability, responsiveness, competence, access, courtesy, communication, credibility, security, understanding or knowing the customer and tangibles. It measures the gap between customer expectations and experience.

By the early nineties the authors had refined the model to the useful acronym RATER:

1. Reliability
2. Assurance
3. Tangibles
4. Empathy
5. Responsiveness.

The above 5 dimensions can be elaborated for banking sector as follows:

7.1 Reliability

- An excellent bank provides its services at the time it promises to do so.
- An excellent bank insists on error-free records.
- When an excellent bank promises to do something by a certain time, it does so.
- Employees in an excellent bank are always willing to help you.

7.2 Assurance

- Behaviors of staff instil confidence in customers.
- Customers feel safe in their transactions.
- Staff having knowledge to answer questions.
- When customers have problems employees in an excellent bank will be sympathetic and reassuring.

7.3 Tangible

- An excellent bank's reception desk employees are neat in appearance.
- An excellent bank's physical facilities are visually appealing.
- Excellent banks have modern looking equipment.
- An excellent bank's credit cards, cheques and similar materials are visually appealing.

7.4 Empathy

- An excellent bank has the customer's best interest at heart.
- The employees of an excellent bank understand customer's specific needs.
- An excellent bank has employees who give customers personal attention.
- An excellent bank has working hours suitable for all customers.

7.5 Responsiveness

- Employees in an excellent bank tell you exactly when the services will be performed.
- Prompt service.
- Staff willingness to help.
- Prompt response from staff.

Banks need to establish quality standards for each of the relevant parameters which are monitored on a regular basis. Designed quality standards will help ensure that quality is planned goal. Some guidelines that could be used to device appropriate standards are:

- Ensure that feedback is regularly sought from customers and acted upon suitably.
- Let the consumers know at one time what documentation is required to process loans.
- Ensure that all letters of credit, guarantees etc are issued on time.
- Ensure that all systems and procedures are scrupulously adhered to and all deviations are promptly brought to the notice of the top management.
- Ensure that systems are in place for benchmarking practices/processes particularly of those in relation to customer service.

8. CURRENT SCENARIO OF PSU OR NATIONALIZED BANKS

Nationalized Banks are in far better position than Private Banks in terms of low Cost of Funds, huge Customer Base and Consumer Trust. Due to these factors Nationalized banks are in better position to offer cheaper services and when someone thinks of bank he thinks of money, be it depositing cash, withdrawing cash, cheques, loans, forex business or Trade activities Hence it's obvious that Nationalized banks are better. Private Banks cannot provide low cost service since they don't have funds to back it, they can only grow if they offer better services and that's the place Nationalized Banks are a little failure. But if you compare Nationalized Banks which have shown willingness to prove themselves, one will come across many such banks. SBI is now offering all services that too online, be it equity trading, insurance or normal banking services.

Public sector banks are favored because of many positive factors. It is the best way to protect the interest of common people and provide the required financial support to the nation. The major points which make if different from Private Banks:

1. *Larger coverage/market penetration:* Nationalized banks have far better reach to remote people than any Private Banks.
2. *Less risk factor or Safe investment:* Money if safe for common low earning people. This is the most important factor for nation like India where people save a lot of their hard earned money; its protection is needed as critical.
3. *Fare customer practices:* Nationalized banks are accountable and government is involved to make sure they don't foul play with customer's money. Private Banks can easily cheat or misguide people.
4. *Cost effective:* Nationalized banks don't charge high for Default charges, Min. Balance, Cheque book, transaction fee etc when compared to private banks.
5. *Competitive technological advancement:* As time has changed and most of the Nationalized banks are Tech Savvy which helped them to serve customers friendlier.

6. ***Changing mindset of working staff:*** The banking staffs understand that if they do not maintain good relations with customer, they will lose business to the competitor because product and services are generic with minimum variations.

9. CONCLUSION

As competition increases, quality will become the only true differentiator. Successful banks will be those that compete on quality. Competing on quality will empower the bank in staying ahead of the competition. Every aspect of the bank's functioning would have to be governed by quality principles. Leveraging technology with various banking operations coupled with appropriate quality initiatives in place can alone keep the banks in good-stead to face the competition and to make them the true leaders in leading quality revolution.

It emerges that in order to ensure customer satisfaction in the banking industry; they have to ensure proper perception management. This perception management would cover both customer's expectation and customer perception of the actual delivery. Systems for constant feedback have to be set up. Multiple channels for feedback, both formal and informal must be organized. The process of service delivery must be analyzed and examined in great detail to ensure that it conforms to the five dimensions of service quality defined earlier.

Delivering customer satisfaction is at the heart of modern marketing, which is a post-service judgement of the consumers. The study on service quality in banks is measured in five dimensions by using the SERVQUAL scale developed by Parsuraman *et al.* (1988). The expectations exceeding performances are clearly visible with Indian PSU banks. But there is significant room for improvement as competition is increasing and products and services remain similar.

REFERENCES

[1] Gerson, Richard F. (2004). *Measuring Customer Satisfaction*, Viva Books Pvt Ltd, New Delhi.

[2] IS 15700: 2005, Indian Standard developed by Bureau of Indian Standards for Service Quality by Public Service Organizations.

[3] Parameswaran, R. and Natarajan, S. (2002). *Indian Banking*, S. Chand and Company Ltd., New Delhi.

[4] Parasuraman, A., Zeithaml, V.A. and Berry, L.L. (1985). *A Conceptual Model of Service Quality and Its Implications for Future Research*, Journal of Marketing, pp. 41–50.

[5] Parasuraman, A., Zeithaml, V.A. and Berry, L.L. (1988). *SERVQUAL: A Multiple Item Scale for Measuring Consumer Perceptions of Service Quality, Journal of Retailing*, 12–40.

[6] RBI report of Shri Damodaran committee on customer service in banks submitted on July 2011.

[7] Varambally, K.V.M. and Shanbhag, M.M. (2003). Service Quality and customer Orientation-Managerial Changes, Varambally, KVM Allied Publishers Pvt. Ltd, New Delhi, pp. 12–14.

[8] Varambally, K.V.M. and Shanbhag, M.M. (2003). Service Quality gaps in Banking Sector A.J, Joshua and Mubarak, V Allied Publishers Pvt. Ltd., New Delhi, pp. 80–83.

[9] www.rbi.org.in, www.bis.org.in, www.iba.org.in

Vendor-Buyer Model Considering Imperfect Items, Trade Credit and Volume Agility under Inflation

V. Gupta[1] and S.R. Singh[2]

[1]Inderprastha Engineering College, Ghaziabad, India
[2]Department of Mathematics, D.N. College, Meerut, India
E-mail: [1]van.gupta@yahoo.co.in; [2]shivrajpundir@yahoo.com

ABSTRACT: In this paper, we extend Liang-Hsuan Chen, Fu-Sen kang (Coordination between vendor and buyer considering trade credit and items of imperfect quality. Int. J. Production Economics 123 (2010) 52–61) by considering an integrated inventory model of a Volume agility manufacturing system for imperfect items due to non-ideal production processes under the effect of trade credit and inflation. Volume Agility is a major component in the present scenario. The manufacturing agility which is capable of adjusting the production rate with the variability in the market demand is known as Volume agility. The unit production cost is taken to be a function of the finite production rate which is treated to be a decision variable. A numerical example is used to illustrate the results for the proposed models.

Keywords: Supply Chain, Volume Agility, Imperfect Items, Stock Dependent Demand, Trade Credit and Inflation.

1. INTRODUCTION

Many studies have investigated integrated vendor-buyer models from the view point of both the buyer and vendor to find the optimal economic order quantity to achieve the minimal total cost and to maximize the profit. It is usually observed that demand rate of any product is always in a dynamic state. Demand of product may vary with time or with price or even with the inventory displayed in a retail shop but as large pile of goods displayed on shelf will lead the customer to buy more and then generate higher demand. These phenomena attract to investigate inventory models related to stock-dependent demand rate. Several studies have investigated vendor-buyer models with stock-level dependent demand rate and variable holding cost; namely, Levin *et al.* (1972), Baker and Urban (1988), Mandal and Maiti (1997, 1999), Balkhi and Bekherouf (2004), and Alfares (2007).

For inventory problems, a number of studies considered the production rate of a machine are to be pre-determined and inflexible. Schweitzer and Seidmann (1991) adopted for the first time, the concept of agility in the machine production rate and discussed optimization of processing rates for a FMS (flexible manufacturing system). Obviously the machine production rate is a decision variable in the case of a FMS and then the unit production cost becomes a function of the production rate. Khouja and Mehrez (1994) and Khouja (1995) extended the Embedded Platform Logistics System (EPLS) model to an imperfect production process with flexible production rates. Sana (2004) developed inventory models with volume agible production for deteriorating items and shortage. Khouja and Mehrez (2005), Husseini *et al.* (2006), Sana *et al.* (2007a and 2007b) also discussed the volume of agiliity policy in production.

In the existing literature, most of the production models unrealistically ignored the presence of the imperfect production process and equipment. Imperfect items are produced due to non-ideal production processes. A number of academics and practitioners have investigated the effect of items with imperfect quality on the EOQ model; namely, salameh and Jaber (2000), Papachristos and Konstantaras (2006), Wee *et al.* (2007), and Maddah and Jaber (2008), Integrated vendor-buyer models that consider defective items have also been presented. Ouyang *et al.* (2006) developed an integrated vendor-buyer model with defective items represented as crisp and fuzzy cases, respectively. Panda *et al.* (2010) determined optimal production inventory policy for defective items with fuzzy time period.

Moreover, the effects of inflation and time value of money are vital in practical environment, especially in the developing countries with large scale inflation. Therefore, the effect of inflation and time value of money cannot be ignored in real situations. As a trade credit policy, a delayed payment for the buyer could be considered in the model from the buyer's viewpoint. In practice, the delay period is usually given to the buyer to encourage the buyer, to increase the ordering quantity. Within the period, the buyer must pay off the account when the payment is due by borrowing from a bank. A number of studies related to trade credit have discussed the issue Goyal (1985), Aggarwal and Jaggi (1995), chung and Huang (2006) also considered items of imperfect quality in their model. Bierman and Thomas (1977) then proposed an EOQ model under inflation that also incorporated the discount rate. Later, Yang *et al.* (2001) established inventory models with time varying demand under inflation. Recently chern *et al.* (2008) proposed partial backlogging inventory lot- size models for deteriorating items with fluctuating demand under inflation. In this model, shortages are allowed at the buyer's part only and the unfulfilled demand is partially backlogged at a constant rate.

This paper combines the above concepts and develop an integrated production inventory model with imperfect production processes, stock dependent demand rate with trade credit under inflation.

The rest of this paper is organized as follows. The notations and assumptions are defined and described in section 2. Vendor-buyer models are formulated in section 3. A numerical example is presented in section 4. Conclusions are given in section 5.

2. ASSUMPTIONS AND NOTATIONS

The following assumptions are used throughout the whole paper:

1. The replenishment rate is infinite and lead time is zero.
2. The demand rate D(t) at time t is,

$$D(t) = \begin{cases} a + bI(t) & I(t) \geq 0 \\ a & I(t) = 0 \end{cases}$$

where a, b are positive constants and I(t) is the inventory level at time t.

3. Shortages are allowed. Unsatisfied demand is partial backlogged. The fraction of shortages backordered is a differentiable and decreasing function of time t, denoted by δ (t), where t is the waiting time up to the next replenishment, and $0 \leq \delta(t) \leq 1$ with $\delta(0) = 1$. Note that if $\delta(t) = 1$ (or 0) for all t, then shortages are completely backlogged (or lost). Partial backlogging is allowed at the buyer's part only.

4. Multiple deliveries per order are considered.

5. The production cost per unit item is a function of the production rate.

6. The production rate is considered to be a decision variable.

7. The vendor proposes a certain credit period M. During that time the account is not settled, the buyer deposits his/her generated sales revenue in an interest-bearing account with rate Ie. At the end of the trade credit period, the account is settled and the buyer starts paying for the interest charges on the items in stock with rate Ip.

8. Constant deterioration rate is considered.

9. Inflation is also considered.

10. A single vendor and single buyer for a single product is considered.

11. The order from the buyer is processed by the vendor once and delivered to the buyer in k no. of lots in equal size.

The following notation are used throughout the whole paper:

T	Time length for each Cycle.
T_1	The length of production time.
T_2	The length of non production time.
C_{1v}	Set up cost of the vendor per production cycle.
C_{2v}	Holding cost of the vendor per unit.
C_v	Item Cost for the vendor per unit.
C_{1b}	Set up cost of the buyer per.
C_{2b}	Holding cost per unit.
C_b	item Cost for the buyer.
C_3	rework cost per unit.
C_{3b}	Shortage cost per unit for the buyer.
Θ	deterioration rate.
C_4	lost sale cost per unit for the buyer.
P	Production rate per unit.
D	percentage of the defective items.
M	credit period offered by the vendor
K	Number of deliveries
I_e	interest earned per unit.
I_p	interest payable per unit.
$\eta(P)$	The production cost per unit item ;
δ	partial backlogging rate.
r	Inflation rate.

In this paper, we assume that the vendor wants to determine the production p, the duration of inventory cycle T, in order to minimize its total cost per unit time. As a result, we have three decision variables

3. MATHEMATICAL FORMULATION AND SOLUTION

In this section, we first establish vendor's model, then buyer's model and finally compare the total cost of the vendor and buyer in two cases of credit period

3.1 Vendor's Inventory Model

In this subsection, the behavior of the inventory in a cycle is shown in Figure 1. We consider a self-manufacturing system in which the items are manufactured in a machine and the market demand is filled by these manufactured items. In this model initially inventory level starts with zero level and increases due to production at a finite rate and decreases at a constant deterioration rate up to time T_1 after this time, production stops and inventory level decreases due to demand and deterioration up to time T_2.

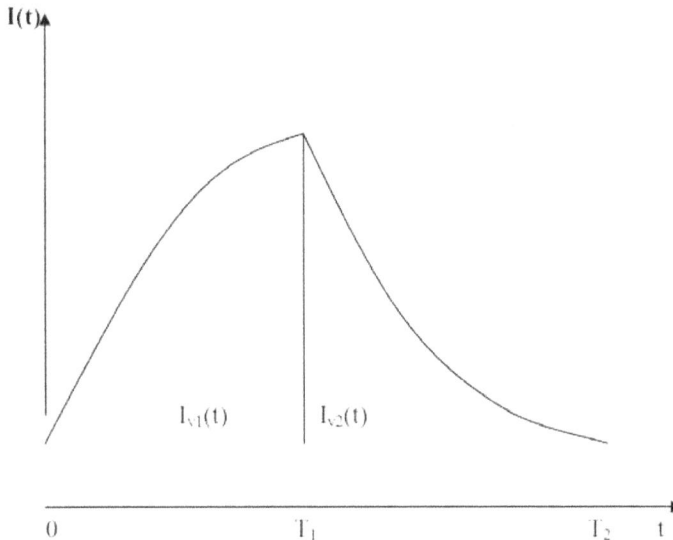

Fig. 1: Vendor's Inventory Model

The differential equations governing the single vendor model for different time durations are as follows:

$$I_{v1}'(t) + \theta I_{v1}(t) = P - [a + bI_{v1}(t)], \ 0 \le t \le T_1 \qquad \text{... (1)}$$

$$I_{v2}'(t) + \theta I_{v2}(t) = -[a + bI_{v2}(t)], \ \ 0 \le t \le T_2 \qquad \text{... (2)}$$

We have boundary conditions

$$I_{v1}(0) = 0, I_{v2}(T_2) = 0$$

The solutions of the above equations by using the above boundary conditions are:

$$I_{v1}(t) = \frac{P-a}{\theta+b}(1-e^{-(\theta+b)t}) \quad 0 \le t \le T_1 \qquad \qquad \ldots (3)$$

$$I_{v2}(t) = \frac{a}{\theta+b}(e^{(\theta+b)T_2} - e^{(\theta+b)t}) \quad 0 \le t \le T_2 \qquad \qquad \ldots (4)$$

Where $T = T_1 + T_2$

Present worth holding cost of the vendor is

$$HC_v = C_{2v}\left[\int_0^{T_1} I_{v1}(t)e^{-rt}dt + e^{-rT_1}\int_0^{T_2} I_{v2}(t)e^{-rt}dt\right]$$

$$HC_v = C_{2v}\left[\begin{array}{l} \dfrac{P-a}{\theta+b}\left\{-\dfrac{e^{-rT_1}}{r} - \dfrac{e^{-(\theta+b+r)T_1}}{\theta+b+r} + \dfrac{1}{r} + \dfrac{1}{\theta+b+r}\right\} \\[3mm] + \dfrac{ae^{-rT_1}}{\theta+b}\left\{-\dfrac{(\theta+b)e^{(\theta+b-r)T_2}}{\theta+b-r} + \dfrac{e^{(\theta+b)T_2}}{r} + \dfrac{1}{\theta+b-r}\right\} \end{array}\right] \qquad \ldots (5)$$

Production cost per unit is

$$\eta(P) = \left(u + \frac{g}{P} + sP\right)$$

The production cost is based on the following factors:
1. The material cost u per unit item is fixed.
2. As the production rate increases, some costs like labor and energy costs are equally distributed over a large number of units. Hence the per-unit production cost $\left(\dfrac{g}{P}\right)$ decreases as the production rate (P) increases.
3. The third term (sP), associated with tool/die costs, is proportional to the production rate.

Now the Production Cost for the inventory is,

$$\eta(P) = \left(\gamma + \frac{g}{P} + sP\right)\int_0^{T_1} Pe^{-rt}dt$$

$$= P\left(\gamma + \frac{g}{P} + sP\right)\frac{(1-e^{-rT_1})}{r} \qquad \qquad \ldots (6)$$

Item cost

$$IT_v = \frac{PC_v}{r}(1-e^{-rT_1}) \qquad \qquad \ldots (7)$$

Present worth set up cost of the vendor is

$$SC_v = C_{1v} \qquad \qquad \dots (8)$$

3.2 Number of Defective Items

There are two cases when the machine turns to out-of-control state. If the machine turns to out-of-control state after the time T_1, then there will be no defective items, but if the machine is in out-of-control state before the time T_1, then there will be defective items as given below:

$$N = \begin{cases} 0 & X \geq T_1 \\ d \int_X^{T_1} P\,dt & X < T_1 \end{cases}$$

$$N = T_1 \begin{cases} 0 & X \geq T_1 \\ dP(T_1 - X) & X < T_1 \end{cases}$$

Therefore, the expected number of defective items in a production cycle is,

$$E(N) = \int_0^{T_1} dP(T_1 - X) f(X)\,dX$$

Rework occurs at $t = T_1$. The rework cost includes the set-up cost, material cost etc. The present worth rework cost can be expressed approximately as,

$$RW = C_3 E(N) e^{-rT_1}$$

$$= C_3 \left\{ \int_0^{T_1} dP(T_1 - X) f(X)\,dX \right\} e^{-rT_1}$$

$$= C_3 dP \left\{ \int_0^{T_1} (T_1 - X) \mu e^{-\mu X}\,dX \right\} \sum_0^{\infty} \frac{(-rT_1)^n}{n!} = C_3 dP\mu \left\{ T_1^2 - (r + \mu)T_1^3 \right\} \quad \dots (9)$$

Present worth average total cost of the producer is the sum of carrying cost, set up cost, item cost, production cost and rework cost.

$$TC_v = \frac{HC_v + IC_v + SC_v + \eta(P) + RW}{T} \qquad \qquad \dots (10)$$

3.3 Buyer's Inventory Model

Buyer has the inventory MI_b. Now buyer's inventory level decreases due to demand and deterioration rate up to time T_3. At time T_3 there is partial backlogging up to time T_4 as shown on Figure 2.

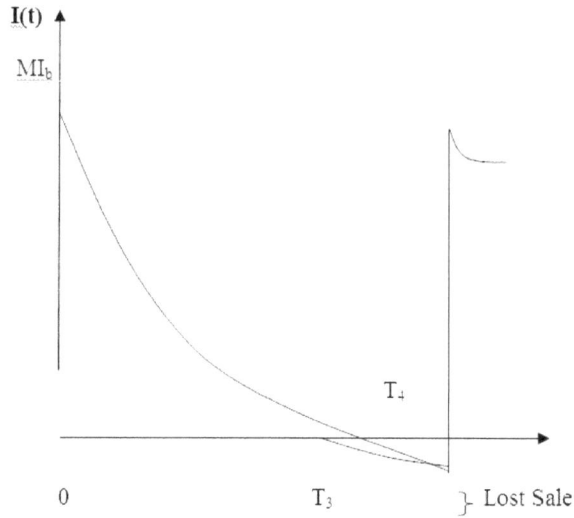

Fig. 2: Inventory System for Buyer

The differential equations governing to the buyer's inventory level are as follows

$$I_{b1}'(t) = -\theta I_{b1}(t) - (a + b I_{b1}(t)) \quad 0 \le t \le T_3 \qquad \qquad \dots (11)$$

After time T_3 partial backlogging occurs and the change in the inventory is directed by the following differential equation,

$$I_{b2}'(t) = -\delta a \quad 0 < t \le T_4 \qquad \qquad \dots (12)$$

We have boundary condition $I_{b1}(T_3) = 0$,

$$I_{b1} = \frac{a}{\theta + b}(e^{(\theta + b)(T_3 - t)} - 1) \quad 0 \le t \le T_3 \qquad \qquad \dots (13)$$

$$I_{b2}(t) = -\delta at \quad 0 < t \le T_4 \qquad \qquad \dots (14)$$

By using the boundary condition $I_{b1}(0) = MI_b$ we have the buyer's maximum inventory level is,

$$MI_b = \frac{a}{\theta + b}(e^{(\theta + b)T_3} - 1) \qquad \qquad \dots (15)$$

The quantity per delivery to the buyer is,

$$Q_b = MI_b + \frac{a}{b}(e^{-\delta b T_4} - 1)$$

$$= \frac{a}{\theta + b}(e^{(\theta + b)T_3} - 1) + \frac{a}{b}(e^{-\delta b T_4} - 1) \qquad \qquad \dots (16)$$

Present worth holding cost of the buyer is,

$$HC_b = C_{2b} \int_0^{T_3} I_{b1}(t)e^{-rt}dt$$

$$= \frac{aC_{2b}}{\theta+b}\left\{e^{-rT_3}\left\{\frac{1}{r}-\frac{1}{\theta+b+r}\right\}+\frac{e^{(\theta+b)T_1}}{\theta+b+r}-\frac{1}{r}\right\} \qquad \dots (17)$$

Present worth set up cost of the buyer is,

$$SC_b = C_{1b} \qquad \dots (18)$$

Shortages occur during the time period (t = 0) to (t = T_4). Therefore the present backlogging cost is,

$$BA = C_4 \int_0^{T_4} -I_{b2}(t)e^{-r(T_3+t)}dt$$

$$= a\delta C_4 e^{-rT_3}\left[\frac{-e^{-rT_4}T_4}{r}+\frac{(1-e^{-rT_4})}{r^2}\right] \qquad \dots (19)$$

Lost sale occurs during the time period 0 to T_4. During this time period, the complete shortage is aT_4 and the partial backlog is $a\delta T_4$. Lost sales are the difference between the complete shortage and the partial backlog.

Thus, the present worth lost sale cost is,

$$LS = C_5 a \int_0^{T_4} (1-\delta)T_4 e^{-r(T_3+t)}dt$$

$$= \frac{C_5 a(1-\delta)e^{-rT_3}T_4(1-e^{-rT_4})}{r} \qquad \dots (20)$$

The item cost includes loss due to deterioration as well as the cost of the item sold.

Therefore, the present worth item cost is,

$$IT_b = aC_b\left[\frac{(e^{(\theta+b)T_3}-1)}{\theta+b}+\frac{(e^{-\delta bT_4}-1)e^{-r(T_3+T_4)}}{b}\right] \qquad \dots (21)$$

Therefore, the present worth total cost per cycle is,

$$TC_b = \frac{(HC_b+IC_b+SC_b+LS+BA)}{T} \qquad \dots (22)$$

There are k deliveries per cycle. The fixed time interval between the deliveries is $T_5 = T_3 + T_4 = T/k$.

$$TC_b = \frac{(HC_b + IC_b + SC_b + LS + BA)}{T} \sum_{i=0}^{k-1} e^{irT_5}$$

$$TC_b = \frac{(HC_b + IC_b + SC_b + LS + BA)}{T} \left(\frac{1 - e^{rT}}{1 - e^{rT_5}} \right) \qquad \qquad \dots (23)$$

The average total cost of the models TC which is the sum of vendor's cost (TC_v) and the buyer's cost (TC_b) can be derived for two possible cases as follows:

Case I: When $T_3 > M$

Interest earned per cycle per unit time is,

$$= c_b I_e \int_0^{T_3} [a + bI_{b1}(t)]e^{-rt} dt$$

$$= \frac{c_b I_p}{\theta + b} \left[-\frac{a(e^{-rT_3} - 1)}{r} + \frac{ab}{\theta + b} \left\{ -\frac{e^{-rT_3}}{\theta + b + r} + \frac{e^{-rT_3}}{r} - \frac{1}{r} + \frac{e^{-(\theta+b)T_3}}{\theta + b + r} \right\} \right] \qquad \dots (24)$$

Interest payable per cycle per unit time is,

$$= c_b I_p \int_M^{T_3} I_{b1}(t)e^{-rt} dt$$

$$= c_b I_p \left\{ \frac{e^{-rM} - e^{-rT_3}}{\theta + b + r} + \frac{e^{-rT_3} - e^{-rM}}{r} + \frac{e^{-(\theta+b)T_3} - e^{-(\theta+b)M}}{\theta + b + r} \right\} \qquad \dots (25)$$

$$TC_1 = TC_v + TC_b + \text{interest paid- interest earned} \qquad \dots (26)$$

Case II: When $T_3 \le M$

Interest earned up to time T_3 is $c_b I_e \int_0^{T_3} (a + bI_{b1})e^{-rt} dt$

$$= \frac{c_b I_p}{\theta + b} \left[-\frac{a(e^{-rT_3} - 1)}{r} + \frac{ab}{\theta + b} \left\{ -\frac{e^{-rT_3}}{\theta + b + r} + \frac{e^{-rT_3}}{r} - \frac{1}{r} + \frac{e^{-(\theta+b)T_3}}{\theta + b + r} \right\} \right]$$

Interest earned during $(M - T_3)$ is,

$$= \frac{c_b I_p}{\theta + b} \left[-\frac{a(e^{-rT_3} - 1)}{r} + \frac{ab}{\theta + b} \left\{ -\frac{e^{-rT_3}}{\theta + b + r} + \frac{e^{-rT_3}}{r} - \frac{1}{r} + \frac{e^{-(\theta+b)T_3}}{\theta + b + r} \right\} \right]_{T_3}^{M} \int e^{-rt} dt$$

$$= \frac{c_b I_p}{\theta+b}\left[-\frac{a(e^{-rT_3}-1)}{r}+\frac{ab}{\theta+b}\left\{-\frac{e^{-rT_3}}{\theta+b+r}+\frac{e^{-rT_3}}{r}-\frac{1}{r}+\frac{e^{-(\theta+b)T_3}}{\theta+b+r}\right\}\right]\left(\frac{e^{-rT_3}}{r}-\frac{e^{-rM}}{r}\right)$$

Total Interest earned per unit time is,

$$= \frac{c_b I_p}{\theta+b}\left[-\frac{a(e^{-rT_3}-1)}{r}+\frac{ab}{\theta+b}\left\{-\frac{e^{-rT_3}}{\theta+b+r}+\frac{e^{-rT_3}}{r}-\frac{1}{r}+\frac{e^{-(\theta+b)T_3}}{\theta+b+r}\right\}\right]\left(1+\frac{e^{-rT_3}}{r}-\frac{e^{-rM}}{r}\right) \quad \dots (27)$$

$$TC_2 = TC_v + TC_b \text{ -total interest earned} \qquad \dots (28)$$

In order to find optimal values of p, T_1 and T_3, we have to solve nonlinear equations:

$$\partial TC(P,T_1,T_3)/\partial P = 0, \partial TC(P,T_1,T_3)/\partial T_1 = 0 \text{ and}$$
$$\partial TC(P,T_1,T_3)/\partial T_3 = 0$$

4. NUMERICAL EXAMPLE

In this section, we use software MATHEMATICA version 5.2 to obtain the optimal solutions for both TC_1 and TC_2.

If a = 50, b = 0.077, r = 0.04, C_{1v} = $2000, C_{1b} = $1000, C_{2v} = $ 0.50, C_{2b} = $1.0, C_v = $0.15, C_b = $14, C_3 = $50, C_4 = $0.9, C = $150, g = 500, s = 0.027, γ = 26.5, I_e = 0.1, I_p = 0.02,

θ = 0.001, δ = 0.1, percentage of defective items d = 0.05, An elapsed time until shift is exponentially distributed with μ = 0.001 and T = 365 days.

When $T_3 > M$ (M = 48 days)

K	T_1	T_2	T_3	T_4	P	TC_1
1	218.207	146.793	304.2951	60.7049	48.3186	10520.7
2	221.234	143.766	128.8433	53.6567	79.1352	5152.65

Now we have shown graphically, the total cost, when production is **79.1352**

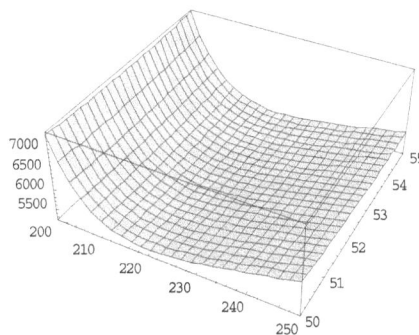

When $T_3 > M$ (M = 70 days).

K	T_1	T_2	T_3	T_4	P	TC_2
1	222.938	142.062	311.2389	53.7611	188.512	5111.13*
2	221.241	143.759	128.828	53.6725	79.2043	5155.84

Now we have shown graphically, the total cost, when production is **188.512**.

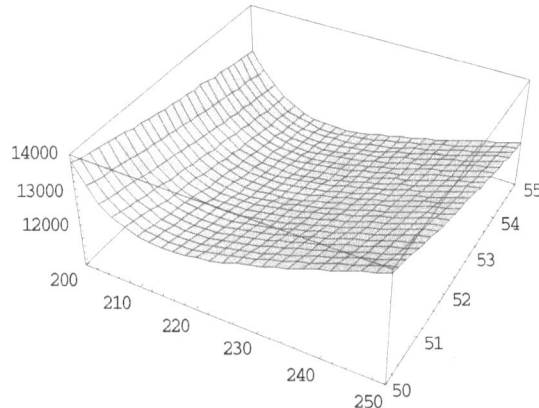

5. CONCLUSION

This paper considers stock dependent demand rate, trade credit, imperfect quality and volume agility in the inflationary environment and develops model to determine the optimal solutions of vendor and buyer's optimal replenishment period and the total cost. Shortages are allowed at the buyer's part only and the unfulfilled demand is partial backlogged. In this research, we assume that in the vendor's production process, the imperfect quality items are reworked after the production is stopped and the vendor must bear the repair cost. This new proposed model shows a different thought on inventory modeling. The expected joint total annual cost function has been derived. Then, by analyzing this derived function, we can obtain the unique procedure to determine the total cycle time, order quantity and the number of shipments per production runs from the vendor to the buyer. The results obtained from the example reveal that successful results in an integrated inventory model leads to impressive cost reduction.

A numerical example is given to illustrate the theoretical results and a conclusion is made that the model is quite stable and suitable to realistic situations. All these facts together make this study very unique and matter of fact. In future, we can extend our approach and thoughts to the supply chain models with more innovative ideas, such as models with price discount.

REFERENCES

[1] Alfares, H.K. (2007). "Inventory model with stock-level dependent demand rate and variable holding cost", International Journal of Production Economics, 108 (1–2), 259–265.

[2] Aggarwal, S.P. and Jaggi, C.K. (1995). "Ordering policies of deteriorating item under Permissible delay in Payment", Journal of the operational Research society, 46, 658–662.

[3] Baker, R.C. and Urban, T.L. (1988). "A deterministic inventory system with an inventory level-dependent demand rate", *Journal of the Operational Research Society*, 39 (9), 823–831.

[4] Balkhi, Z.T. and Benkherouf, L. (2004). "On an inventory model for deteriorating items with stock dependent and time-varying demand rates", Computers and Operations Research, 31 (2), 223–240.

[5] Bierman, H. and Thomas, J. (1977). "Inventory decisions under inflationary conditions", Decision Sciences 8, 151–155.

[6] Chang, H.C. (2004). "An application of fuzzy sets theory to the EOQ model with imperfect quality items", Computers and Operations Research, 31 (12), 2079–2092.

[7] Chung, K.J. and Huang, Y.F. (2006). "Retailer's optimal cycle times in the EOQ model with imperfect quantity and a permissible credit period", Quality and Quantity, 40(1), 59–77.

[8] Chern, M.S., *et al.* (2008). "Partial backlogging inventory lot -size models for deteriorating items with fluctuating demand under inflation", *European Journal of Operational Research,* 191, 127–141.

[9] Goyal, S.K. and Ca´rdenas-Barro´ n, L.E. (2002). "Note on: Economic production quantity model for items with imperfect quality—a practical approach". *International Journal of Production Economics*, 77(1), 85–87.

[10] Husseini, S.M. *et al.* (2006). "A Method to Enhance Volume Flexibility in JIT Production Control", *International Journal of Production Economics*, 104(2), 653–665.

[11] Khouja, M. (1995). "The Economic Production Lot Size Model Under Volume Flexibility", Computer and Operation Research, 22(5), 515–523.

[12] Khouja, M. and Mehrez, A. (1994). "An Economic Production Lot Size Model with Imperfect Quality and Variable Production Rate", *Journal of Operational Research Society*, 45, 12, 1405–1417.

[13] Levin, R.I. *et al.* (1972). "Production/Operations Management: Contemporary Policy for Managing Operating Systems", McGraw-Hill, New York.

[14] Liang-Hsuan Chen and Fu-Sen Kang (2010). "Coordination between vendor and buyer considering trade credit and items of imperfect quality", *Int. J. Production Economics,* 123, 52–61.

[15] Mandal, M. and Maiti, M. (1997). "Inventory model for damageable items with stock-dependent demand and shortages", *Opsearch*, 34(3), 155–166.

[16] Maddah, B. and Jaber, M.Y. (2008). "Economic order quantity for items with imperfect quality: revisited", *International Journal of Production Economics*, 112(2), 808–815.

[17] Ouyang, L.Y. *et al.* (2006). "Analysis of optimal vendor-buyer integrated inventory policy involving defective items", *International Journal of Advanced Manufacturing Technology*, 29 (11–12), 1232–1245.

[18] Papachristos, S. and Konstantaras, I. (2006). "Economic ordering quantity models for items with imperfect quality", *International Journal of Production Economics,* 100(1), 148–154.

[19] Panda, D. (2008). "A single period inventory model with imperfect production and stochastic demand under chance and imprecise constraints", *European Journal of Operational Research*, 188, 121–139.

[20] Salameh, M.K. and Jaber, M.Y. (2000). "Economic production quantity model for items with imperfect quantity", *International Journal of Production Economics*, 64 (1–3), 59–64.

[21] Sana, S. (2004). "On a Volume Flexible Inventory Model", *Advanced Modeling and Optimization*, 6(2), 1–15.

[22] Sana, S. *et al.* (2007a). "On a Volume Flexible Inventory Model for Items with an Imperfect Production System", *International Journal of Operational Research*, 2(1), 64–80.

[23] Sana, S. *et al.* (2007b). "An Imperfect Production Process in a Volume Flexible Inventory Model", *International Journal of Production Economics*, 105, 548–559.

[24] Schweitzer, P.J. and Seidmann, A. (1991). "Optimizing Processing Rate for Flexible Manufacturing Systems", *Management Science*, 37, 454–466.

[25] Wee, H.M., Yu, J. and Chen, M.C. (2007). "Optimal inventory model for items with imperfect quality and shortage backordering Omega", 35(1), 7–11.

[26] Yang, P.C. *et al.* (2001). "Deterministic inventory lot-size models under inflation with shortages and deterioration for fluctuating demand", *Naval Research Logistics*, 48, 144–158.

Process Monitoring through Application of Principal Component Analysis in a Process Industry

S.M. Subhani

Indian Statistical Institute, Hyderabad–500 007
E-mail: smsubhani@rediffmail.com

ABSTRACT: A demonstration study was conducted at a Cigarette manufacturing company to render hands on introduction to Principal Component Analysis, so as to enable the process engineers monitor process variables on a routine basis. The company has several ABC types of machines which are among the fastest rated Cigarette Making machines in the world. The important Quality Characteristics of Cigarette are Weight, Circumference and Pressure Drop. Monitoring different process variables has been a challenging task for the management, as these need to be adjusted according to the moisture content and other quality characteristics of incoming material (cut tobacco). The study was carried out on ABC-1 Machine. Of the five process variables considered on ABC-1 machine the first three Eigenvectors were found to contribute to 96% of the variation. Since the plant personnel are novices to statistical methodology, layman friendly tables and simple graphical tools were used for interpretation. The studies have helped the process engineers gain proficiency on the application of PCA using Minitab statistical software and in depth understanding of the process behavior. Further, the possible root causes for instability in the dominant drivers were identified.

Keywords: Time Series Plot, Matrix Plot, Scree Plot, Score Plot, Eigenvector Weights (loadings) Plot.

1. INTRODUCTION

The plant has several machines of a different technology namely ABC. After consultations with the management personnel PCA study was initiated on ABC machine.

2. BACKGROUND

The ABC machines are controlled by monitoring several process variables mostly related to pressure levels at different locations. A brainstorming session was conducted by the plant personnel and five process variables $X_t = (x_1, x_2, x_3, x_4, x_5)'$ are identified to be the most important and are felt to be correlated with each other contemporaneously. Here an attempt is made focusing primarily on contemporaneous correlation among the above multivariates.

The process variables x_1, x_2, x_3, x_4, x_5 are monitored simultaneously to demonstrate Principal Components Analysis (PCA) method to examine them graphically. Coded data on the 47 consecutive observations of the five variables are presented in Annexure-I.

3. PRINCIPAL COMPONENTS METHOD

Here, a set of orthogonal eigenvectors of the correlation or covariance matrix of the variables are found first. The first principal component accounts for the largest percent of the total data variation. The second principal component accounts the second largest percent of the total data variation, and so on. The goal is to explain the maximum amount of variance with the fewest number of components.

4. EIGENVECTORS

Eigenvectors, which are comprised of coefficients corresponding to each variable, are the weights for each variable used to calculate the principal components scores.

4.1 Eigenvalue

The eigenvalues are the variances of the principal components $Z = A\, Y$.

4.2 Proportion

The proportion of variance explained by the kth principal component $= \dfrac{\gamma k}{\gamma_1 + \gamma_2 + ... + \gamma_p}$

where γ_k is the kth eigenvalue.

4.3 Cumulative Proportion

The cumulative proportion of variance explained by the first k principal components $=$
$\dfrac{\gamma_1 + \gamma_2 + ... + \gamma k}{\gamma_1 + \gamma_2 + ... + \gamma_p}$

Where, γ_k is the kth eigenvalue

The Time Series plots of x_1, x_2, x_3, x_4, x_5 are shown in Annexure-II. It is apparent that none of the series are stable. The plots of x_2 and x_3 show the same trend and hence may be correlated. The matrix plot of the five variables is presented in Annexure-III.

It may be noted from the Matrix plot and the above table that there is a very strong positive correlation between X_2 and X_3, and between X_1 and X_2, and also between X_1 and X_3.

The Box Plot of the five variables is shown in Annexure-IV. It is apparent that x_1, x_2 and x_3 show more variability than x_4 and x_5. This is because these three variables have a downward trend.

Now, Principal Component Analysis is carried out and the summary is shown in Table 2. The Coefficients of x_1, x_2, x_3, x_4, x_5 for the Linear equation of Principal Components are presented in Table 3.

Table 1: Correlations between x_1, x_2, x_3, x_4, x_5 along with Associated
p-Values of ABC-1 Machine Data

	x_1	x_2	x_3	x_4
x_2	0.765			
	0.00			
x_3	0.762	0.992		
	0.00	0.00		
x_4	0.286	−0.019	−0.021	
	0.051	0.901	0.889	
x_5	−0.305	−0.404	−0.392	0.359
	0.037	0.005	0.006	0.013

Table 2: Summary of Principal Component Analysis on x_1, x_2, x_3, x_4, x_5 of ABC-1 Machine

Eigen value	2.898	1.342	0.543	0.209	0.008
Proportion	0.58	0.268	0.109	0.042	0.002
Cumulative	0.58	0.848	0.957	0.998	1

The Scree Plot, Score Plot, Loading Plot and Biplot of x_1, x_2, x_3, x_4, x_5 are shown in Figures 1–4 respectively. The 3D Scatter Plot of PC_1, PC_2 and PC_3, and Surface Plot of PC_1, PC_2 and PC_3 are shown in Figures 5 and 6 respectively. The chart of Eigenvector weights is shown in Figure 7.

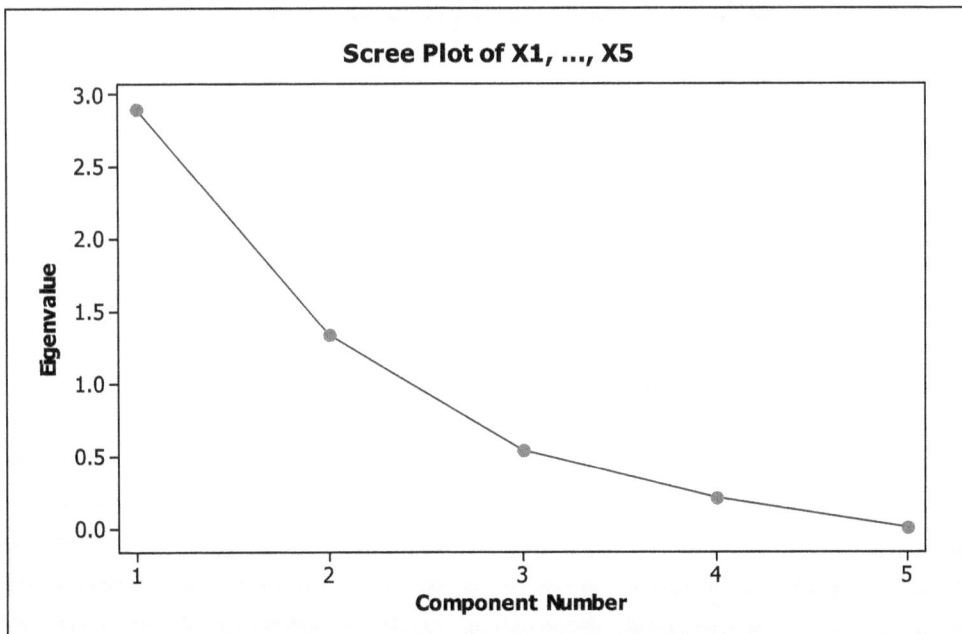

Fig. 1: Scree Plot of x_1, x_2, x_3, x_4, x_5 of ABC-1 Machine

Table 3: Coefficients of x_1, x_2, x_3, x_4, x_5 for the Linear Equation of Principal Components of ABC-1 Machine

Variable	PC1	PC2	PC3	PC4	PC5
X1	0.507	−0.282	0.165	−0.798	0
X2	0.567	−0.02	−0.28	0.31	−0.709
X3	0.566	−0.022	−0.3	0.305	0.705
X4	0.005	−0.796	0.471	0.381	0.006
X5	−0.318	−0.535	−0.763	−0.171	−0.013

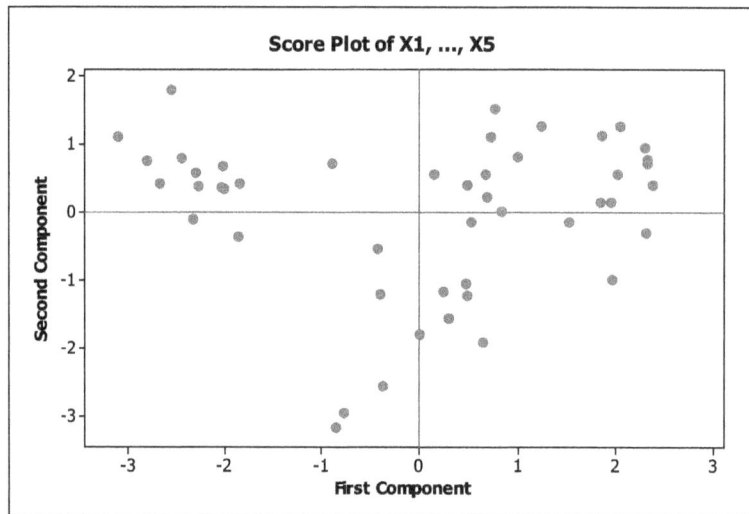

Fig. 2: Score Plot of x_1, x_2, x_3, x_4, x_5 Projected Down the First Two Components of ABC-1 Machine

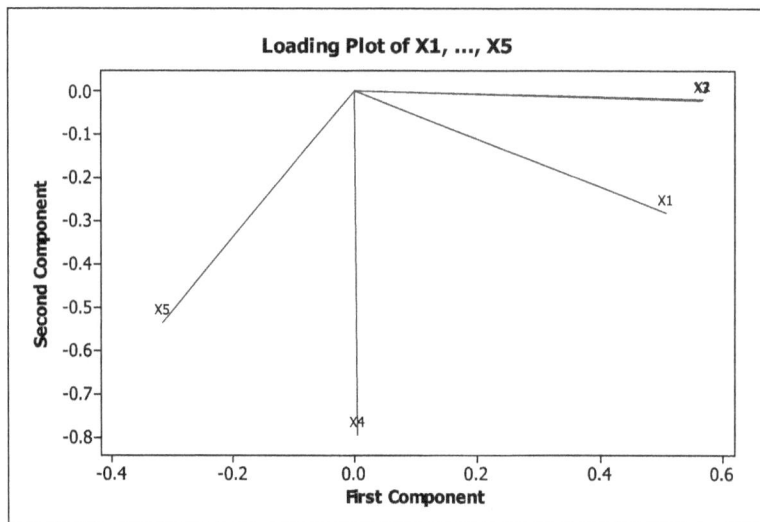

Fig. 3: Loading Plot of x_1, x_2, x_3, x_4, x_5 of ABC-1 Machine

The first, second and third eigenvalues are 2.898, 1.342 and 0.543 and their respective contributions to the variability are 58%, 26.8% and 10.9%. The first three eigenvalues together contribute to 95.7% of the total variability. In other words the data are essentially three dimensional. The Linear equations of the first three principal components are given below.

First Principal component $PC_1 = 0.507 \, X_1 - 0.567 \, X_2 - 0.566 \, X_3 + 0.005 \, X_4 - 0.318 \, X_5$

Second Principal component $PC2 = -0.282 \, X_1 - 0.020 \, X_2 - 0.022 \, X_3 - 0.796 \, X_4 - 0.535 \, X_5$

Third Principal component $PC3 = 0.165 \, X_1 - 0.280 \, X_2 - 0.300 \, X_3 + 0.471 \, X_4 - 0.763 \, X_5$

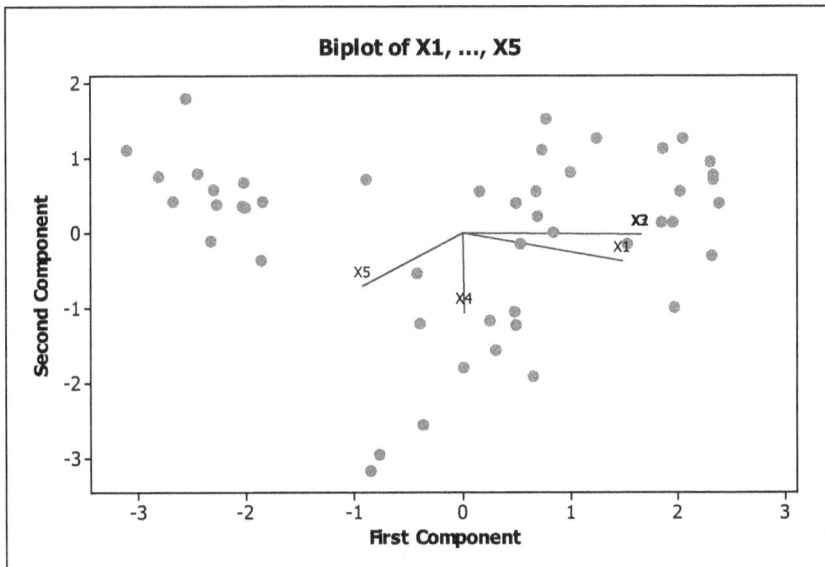

Fig. 4: Biplot of x_1, x_2, x_3, x_4, x_5 of ABC-1 Machine

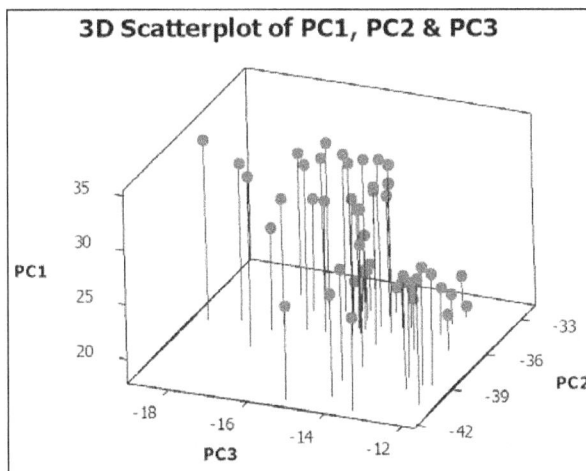

Fig. 5: Three Dimensional Scatter Plot of PC_1, PC_2 and PC_3 of ABC-1 Machine

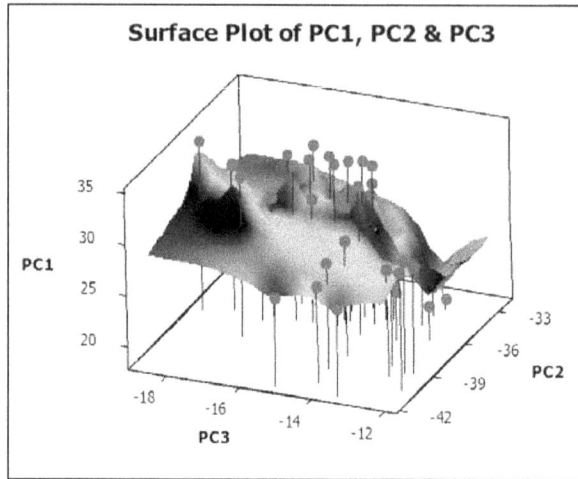

Fig. 6: Three Dimensional Surface plot of PC1, PC2 and PC3 of ABC-1 Machine

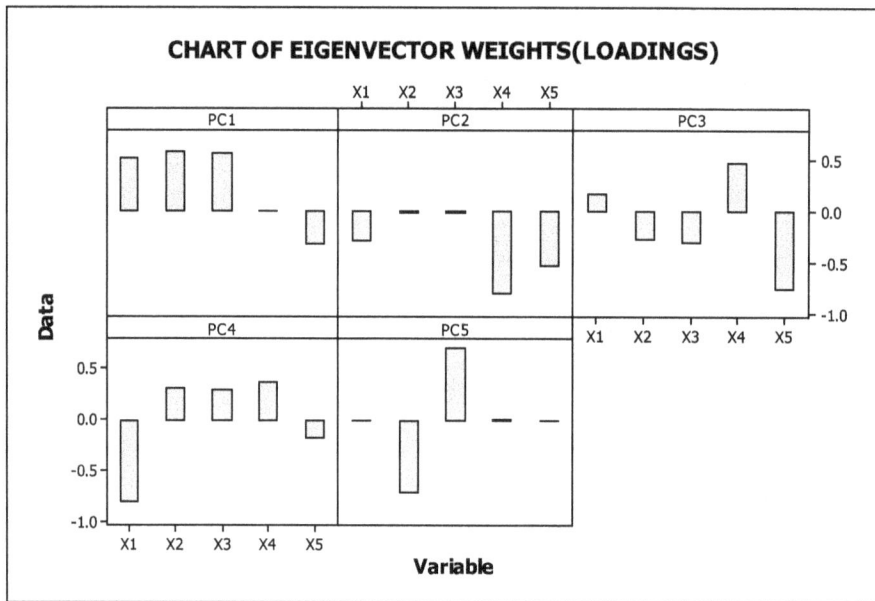

Fig. 7: Plot of the Coordinates of the Five Eigenvectors of ABC-1 Machine

It may be noted from the above chart that the First Principal Component (PC1) is primarily driven by the contrast: Contrast-1 is $(X_1 + X_2 + X_3)/3$.

Similarly, the Second Principal component (PC2) is primarily driven by the following contrast: Contrast-2 is $-(X_4 + X_5)/2$.

Similarly, the Third Principal component (PC3) is primarily driven by the following contrast: Contrast-3 is $-(X_2 + X_3 + X_5 - X_4)/2$.

The values of the five principal vectors and the first three contrasts are given in Annexure-V.

The Scatter plots of Contrast-1 vs PC1, Contrast-2 vs PC2, and Contrast-3 vs PC3 are shown in Annexure-VI. The Scatter plots show a strong correlation between the Contrasts and respective Principal Components.

The correlations observed among the first three Principal components and the respective Contrasts are significant.

Now, we will examine correlations between the three Principal Components and the respective Contrasts.

Corr.(PC_1, Contrast-1) = .996
Corr.(PC_2, Contrast-2) = −.896
Corr.(PC_3, Contrast-3) = −.832

To gain more insight, the Time Series plots of the first three Principal Components are prepared and are presented in Annexure-VII. Let us compare the Time Series plots of x_1, x_2, x_3, x_4, x_5 shown in Annexure-II and the plots in Annexure-VII.

Now let us recall that x_2, x_3, x_1 are the main drivers of the first Principal Component PC_1 are quite similar in appearance with PC_1 showing a significant downward trend. Though x_1 does not have downward trend initially, PC_1 it is dominated by the more dominant drivers x_2 and x_3. Thus the first Principal component representing 58% of the variability, promotes downward trend that is most visible and very similar in x_2, x_3 and x_1. Now the process engineer has looked into the causes for such a downward trend in x_2 and x_3 and why these two process variables are highly unstable compared to other process variables. Lack of control was found to be the root cause.

Similarly, the second principal component PC_2 representing 26.8% of the variability is dominated by the drivers x_4 and x_5 shows their average trend. Now the process engineer has to look into the cause(s) for sudden upward shift of x_4 between S No. 20 and 30 and exceedingly unstable x_5.

Similarly, the third principal component PC_3 representing 10.9% of the variability is driven by x_5 and is similar in appearance. Here too, the process engineer needs to look into the cause(s) for unstable x_5. The process engineers have developed Trouble shooting matrix to initiate appropriate corrective actions.

5. CONCLUSION

The study has helped the process engineers to gain hands on experience on the application of PCA using Minitab statistical software. The layman friendly simple tools and graphical methods used for interpretation have facilitated them in understanding the PCA concepts faster. The probable root causes for instability of the dominant drivers were identified. The process engineers have acquired indepth understanding of the contemporaneous correlations among the dominant drivers, system behavior and further developed Trouble shooting matrix to solve process problems.

6. BENEFITS OF STUDY

The process engineers have started monitoring the process by focusing mainly on the first three Principal Components. The process control has become relatively easy and more effective. The study has resulted in substantial tangible and intangible savings.

REFERENCES

[1] Johnson, R.A., Wichern, D.W. (2002). *Applied Multivariate Statistical Analysis*, 5th ed. Upper Saddle River, NJ: Prentice-Hell.

[2] Soren Bisgaard *et al.* (2006). Quality Quandaries: *The Application of Principal Components Analysis for Process Monitoring,* Quality Engineering, Vol. 18, No. 1, pp. 95–103.

ANNEXURE – I

Coded Data of Pressures on ABC-1 Machine

S. No.	X1	X2	X3	X4	X5
1.	28.0	21.7	20.1	16.2	22.6
2.	30.3	24.4	22.6	17.9	24.2
3.	30.1	22.9	20.5	17.5	24.3
4.	31.2	23.2	21.4	19.5	24.3
5.	30.6	22.5	20.7	18.3	22.8
6.	31.4	22.1	20.2	17.4	22.4
7.	31.4	21.6	19.4	16.4	22.0
8.	31.2	22.1	20.1	17.9	21.0
9.	31.9	21.6	20.1	18.2	21.0
10.	32.5	20.7	19.3	18.5	22.1
11.	31.7	21.8	20.5	17.9	21.4
12,	31.2	22.5	20.8	18.6	21.6
13.	29.7	23.0	20.2	16.7	21.7
14.	29.4	20.3	19.0	17.7	24.5
15.	29.4	20.9	18.5	18.4	23.4
16.	29.0	19.9	18.4	17.7	22.8
17.	27.7	20.2	18.0	16.9	21.4
18.	27.4	20.1	18.4	17.9	21.6
19.	28.7	19.9	18.6	18.7	22.7
20.	31.5	19.9	17.4	16.8	22.5
21.	32.3	18.8	16.9	19.6	24.1
22.	31.1	17.0	15.3	20.5	24.0
23.	32.9	18.6	16.6	22.4	23.4
24.	30.8	18.8	16.7	22.1	23.5
25.	30.4	18.1	16.7	21.8	22.8
26.	30.8	19.2	16.6	21.9	22.9
27.	29.6	19.4	17.2	21.7	24.7
28.	28.9	19.0	17.2	23.1	25.4
29.	29.3	17.8	16.4	23.8	25.7
30.	29.1	18.5	16.9	23.0	26.9
31.	28.4	19.2	17.4	17.8	25.5
32.	27.1	20.0	18.4	17.4	23.6
33.	28.9	19.7	17.8	18.0	22.9
34.	26.9	17.7	15.8	17.1	23.7
35.	26.4	15.4	14.1	17.0	24.8
36.	25.4	15.8	14.2	16.8	24.4
37.	24.8	15.5	14.0	16.8	24.7
38.	24.9	15.3	13.0	16.6	25.3
39.	24.8	14.1	12.7	16.6	24.6
40.	23.9	16.2	14.4	18.8	23.7
41.	23.7	16.7	14.7	20.8	23.6
42.	24.5	15.2	13.0	20.0	23.5
43.	25.3	15.7	13.7	18.8	23.1
44.	25.8	14.8	13.3	17.8	24.2
45.	25.5	14.1	12.4	17.5	23.4
46.	23.4	13.1	11.9	17.4	23.3
47.	24.5	14.0	12.3	16.0	22.8

ANNEXURE-II

Time Series Plots on ABC-1 Machine

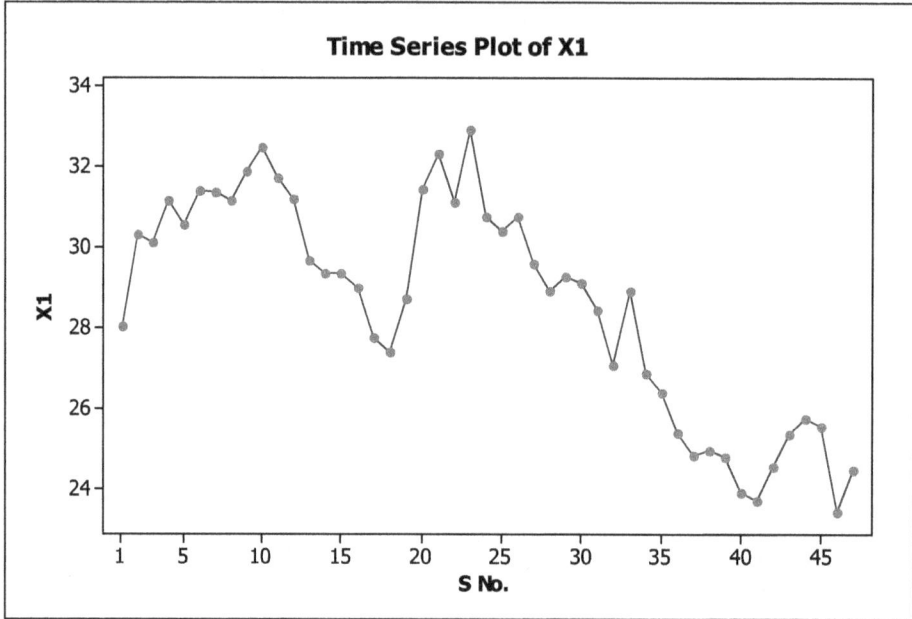

Time Series Plot of X1

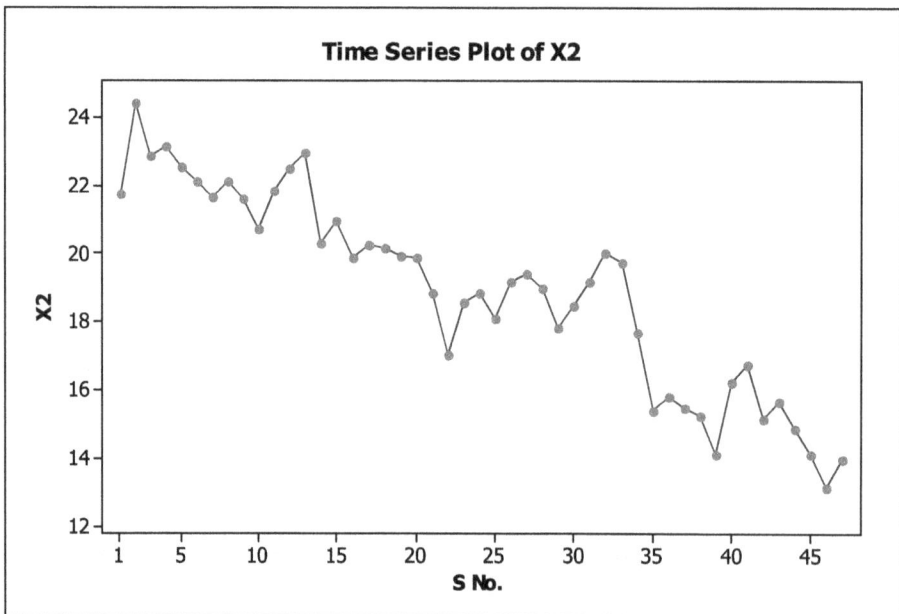

Time Series Plot of X2

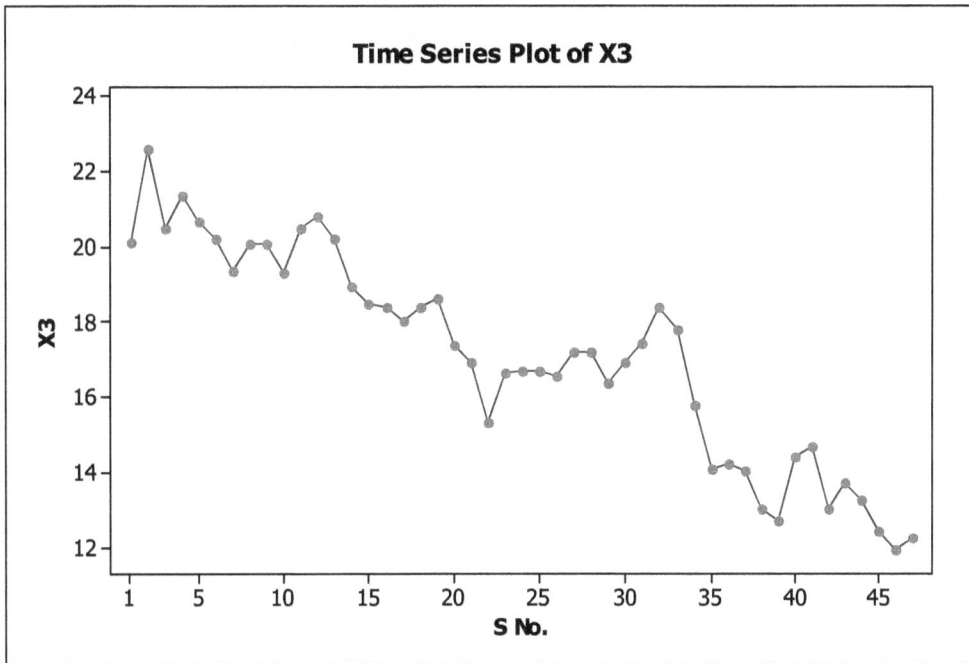

Time Series Plot of X3

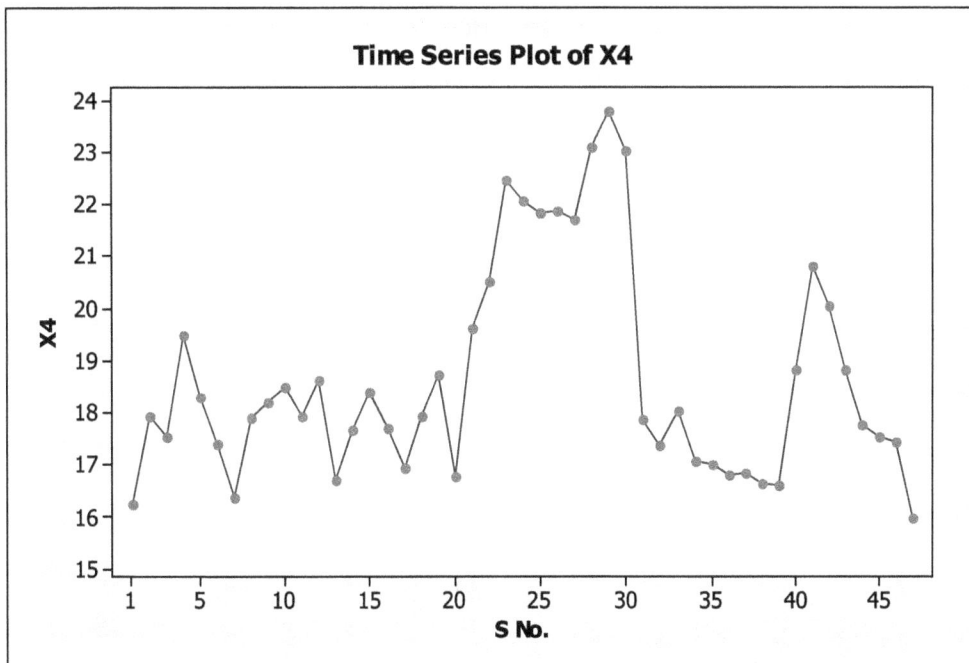

Time Series Plot of X4

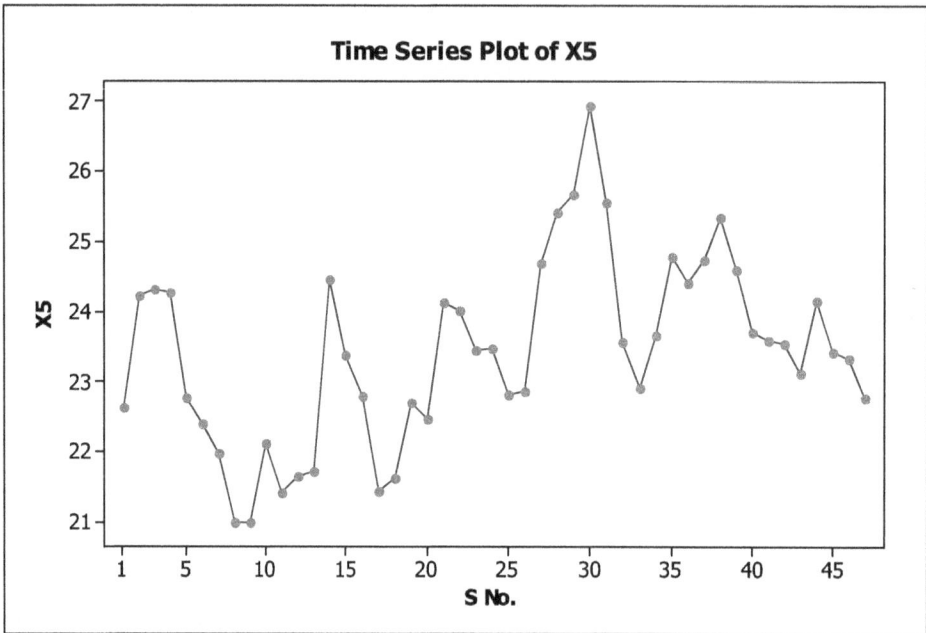

Time Series Plot of X5

ANNEXURE-III

Matrix Plot of X_1, X_2, X_3, X_4, X_5 on ABC-1 Machine

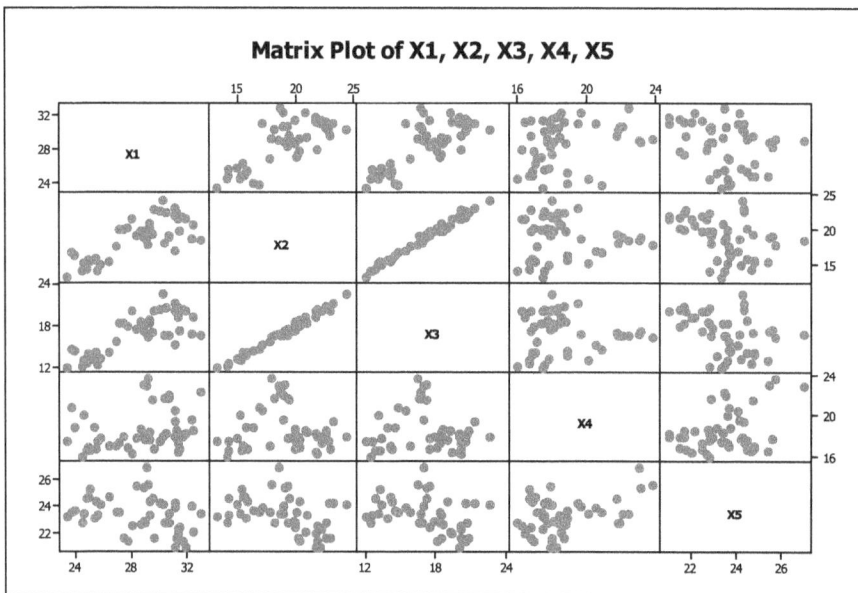

Matrix Plot of X1, X2, X3, X4, X5

ANNEXURE-IV

Box Plot of the Five Variables x_1, x_2, x_3, x_4, x_5 on ABC-1 Machine

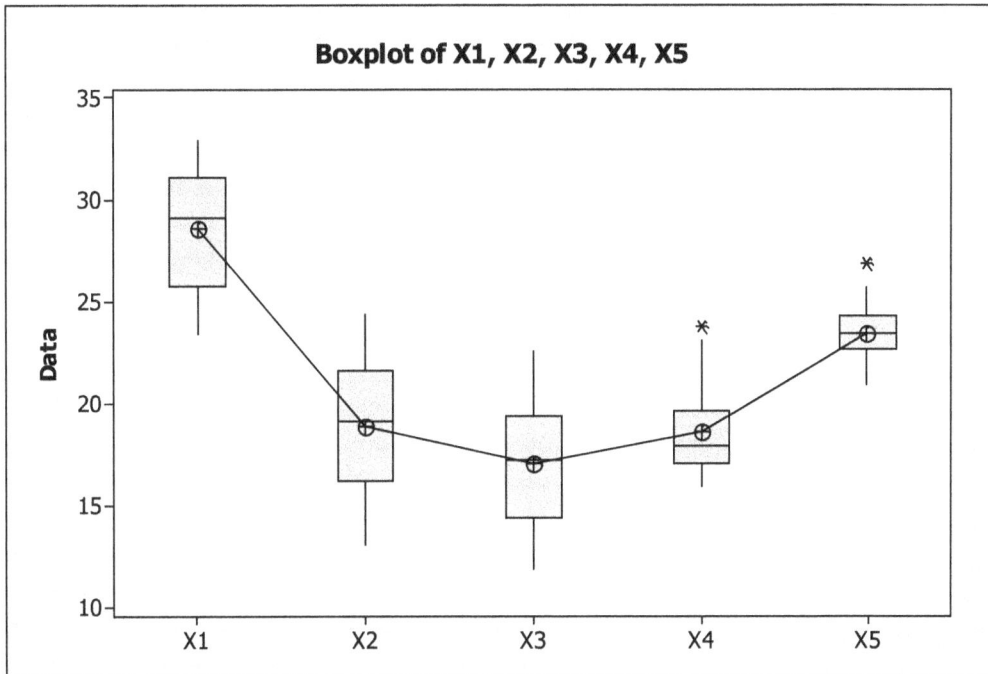

Boxplot of X1, X2, X3, X4, X5

ANNEXURE-V

The Five Principal Vectors and the First Three Contrasts

PC1	PC2	PC3	PC4	PC5	Contrast-1	Contrast-2	Contrast-3
30.8080	−33.8112	−17.1197	−7.1809	−1.40528	23.2956	19.4320	24.1275
34.3822	−36.7458	−18.6468	−7.0498	−1.55245	25.7694	21.0628	26.6477
32.1965	−36.3579	−17.8892	−8.1758	−1.94859	24.4974	20.9191	25.0865
33.4086	−38.2036	−17.0997	−7.9081	−1.55391	25.2323	21.8687	24.6641
32.8308	−36.2598	−16.2377	−8.0386	−1.58848	24.5956	20.5254	23.8558
32.8581	−35.5827	−15.9682	−9.2433	−1.62104	24.5738	19.9042	23.6586
32.2304	−34.4952	−15.7458	−9.9409	−1.86508	24.1257	19.1734	23.3019
33.1272	−35.1565	−14.6569	−8.6887	−1.66090	24.4596	19.4443	22.6422
33.2084	−35.5793	−14.2640	−9.2819	−1.31864	24.5306	19.5920	22.2498
32.2033	−36.5562	−14.3726	−10.3632	−1.24349	24.1674	20.3019	21.8001
33.3447	−35.5564	−14.9341	−9.1257	−1.17318	24.6856	19.6697	22.9186
33.5983	−36.1252	−15.1266	−8.1963	−1.43139	24.8560	20.1352	23.1664

PC1	PC2	PC3	PC4	PC5	Contrast-1	Contrast-2	Contrast-3
32.7033	−34.1887	−16.2853	−7.7669	−2.22647	24.2940	19.2041	24.0890
29.4165	−36.2343	−16.8622	−8.8141	−1.22366	22.8598	21.0531	23.0131
29.8816	−36.2529	−15.7397	−8.3163	−2.00319	22.9316	20.8823	22.2063
29.2123	−35.2570	−15.3533	−8.5262	−1.30449	22.4164	20.2438	21.6777
29.0187	−33.5810	−14.8826	−7.5723	−1.80826	22.0075	19.1912	21.3907
28.9618	−34.3621	−14.6965	−6.8849	−1.47672	21.9922	19.7642	21.1313
29.2357	−35.9377	−14.9286	−7.8214	−1.17013	22.4004	20.7046	21.2483
29.9911	−35.0096	−14.8274	−11.1022	−2.04826	22.9011	19.6108	21.4713
29.0688	−38.3961	−14.1791	−11.4470	−1.60777	22.6923	21.8770	20.1249
26.5612	−38.6375	−12.8880	−11.1754	−1.45313	21.1519	22.2698	17.9180
29.3124	−40.4410	−12.0754	−10.9145	−1.60083	22.7219	22.9472	18.1092
28.3681	−39.5491	−12.7136	−9.2344	−1.77777	22.0959	22.7726	18.4575
27.9801	−38.8776	−12.1883	−9.1619	−1.19175	21.7331	22.3154	17.8919
28.6668	−39.0489	−12.4009	−9.1377	−2.07289	22.1566	22.3579	18.3559
27.9791	−39.5867	−14.3246	−8.3200	−1.81256	22.0548	23.1913	19.7865
27.1908	−40.8938	−14.2079	−7.4769	−1.51706	21.6966	24.2558	19.2419
26.1536	−41.6405	−13.4565	−8.1867	−1.25412	21.1488	24.7298	18.0333
26.3702	−41.7008	−15.1423	−8.1919	−1.36013	21.5078	24.9831	19.6451
27.1330	−36.6631	−16.9947	−9.0051	−1.51735	21.6884	21.6970	22.1591
28.0699	−34.8786	−16.4553	−7.2042	−1.44805	21.8234	20.4723	22.2990
28.7438	−35.5425	−15.1023	−8.5742	−1.62984	22.1598	20.4663	21.2316
25.1191	−34.5067	−15.2611	−8.6929	−1.62919	20.0962	20.3564	20.0161
22.2874	−34.8449	−15.0967	−9.7534	−1.20623	18.6260	20.8889	18.6445
22.2020	−34.2203	−15.1992	−8.8108	−1.40987	18.4691	20.6013	18.8029
21.5382	−34.2713	−15.3948	−8.5570	−1.29478	18.1231	20.7945	18.7125
20.7085	−34.4414	−15.5700	−9.1904	−1.84885	17.7519	20.9985	18.5036
20.0341	−33.9306	−14.6258	−9.4185	−1.28505	17.2103	20.6002	17.4153
22.0333	−35.0359	−14.1524	−6.5400	−1.54720	18.1815	21.2563	17.7691
22.4163	−36.5228	−13.3915	−5.3557	−1.67525	18.3736	22.1992	17.1108
21.0313	−36.0595	−12.6293	−7.2984	−1.73585	17.5797	21.7921	15.8485
22.2395	−35.1221	−13.1075	−7.9478	−1.62070	18.2447	20.9801	16.8520
21.3884	−34.9157	−13.9533	−9.2752	−1.38396	17.9550	20.9602	17.2524
20.6183	−34.2499	−13.0890	−9.5367	−1.43675	17.3586	20.4836	16.2211
18.7290	−33.4815	−12.9729	−8.3425	−1.08437	16.1535	20.3762	15.4671
20.1098	−32.3424	−13.4060	−9.2632	−1.45442	16.9024	19.3691	16.5190

ANNEXURE–VI

Scatter Plots of Contrasts vs Principal Components on ABC-1 Machine

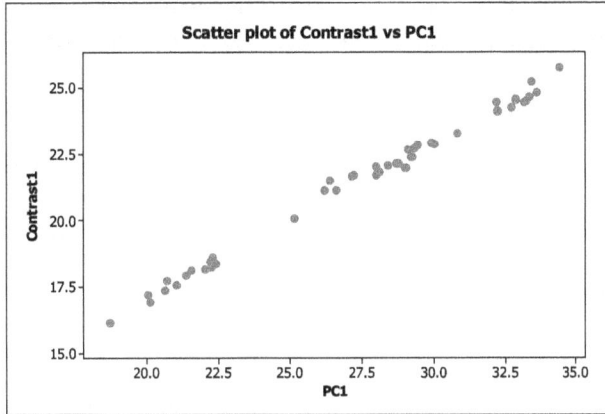

Scatter plot of Contrast1 vs PC1

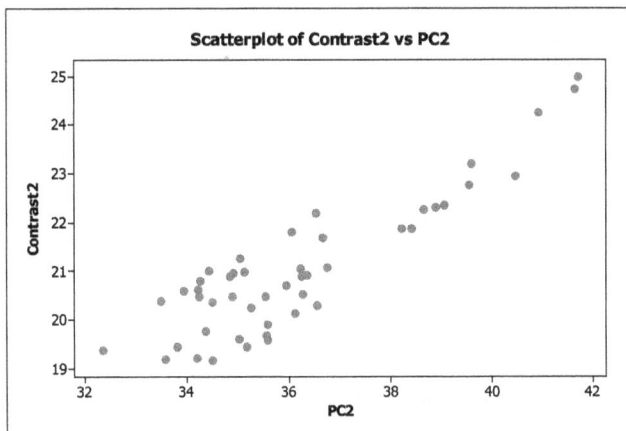

Scatterplot of Contrast2 vs PC2

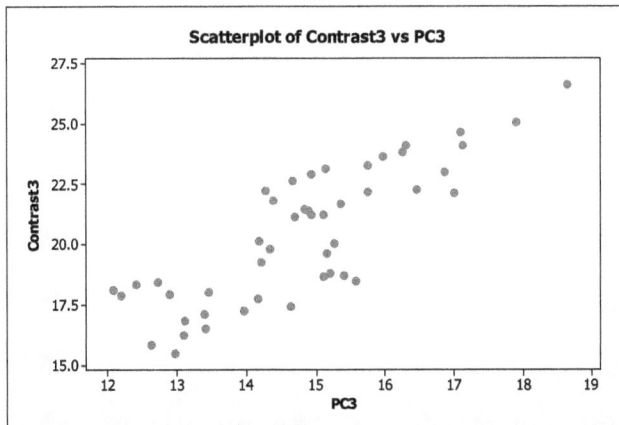

Scatterplot of Contrast3 vs PC3

ANNEXURE-VII

TIME SERIES PLOTS OF FIRST THREE PRINCIPAL COMPONENTS ON ABC-1 MACHINE

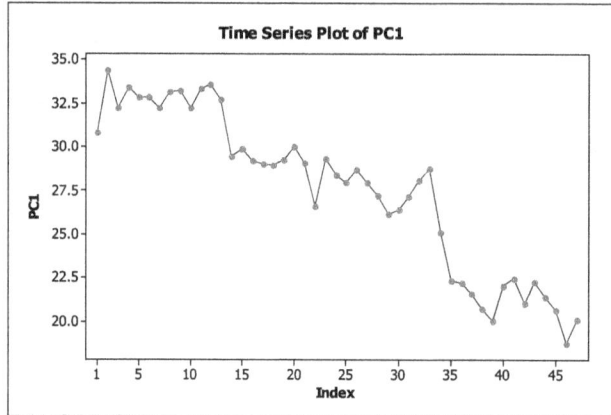

Time Series Plot of PC1

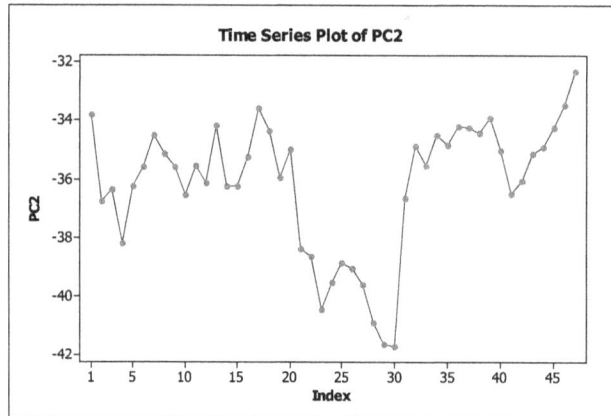

Time Series Plot of PC2

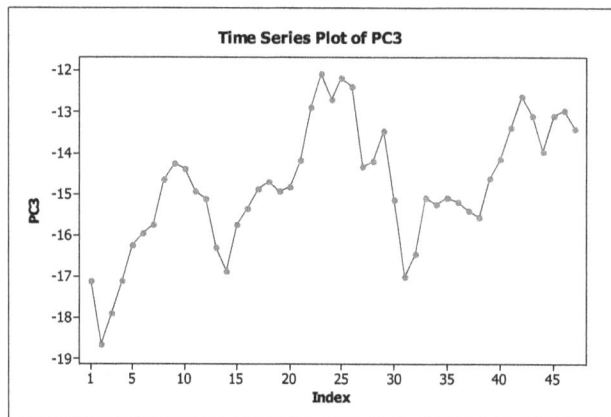

Time Series Plot of PC3

Some Properties of CG (u, v)

Moutushi Chatterjee[1] and Ashis Kumar Chakraborty[2]

Indian Statistical Institute, Kolkata

E-mail: [1]tushi.stats@gmail.com; [2]akchakraborty123@rediffmail.com

ABSTRACT: Multivariate process capability indices have been proposed by many authors for symmetric specification region. Chakraborty and Das (2007) have proposed a superstructure of multivariate process capability indices and denoted it by CG (u, v). We studied here some important properties of CG (u, v) which are useful for understanding and interpretation of the value obtained for CG (u, v). This makes the superstructure more suitable for practical use.

Keywords: Super-Structures of Process Capability Indices, Multivariate Quality Characteristics, Relationship, Comparison.

1. INTRODUCTION

Process Capability Index (PCI) is a single valued measure used to assess the ability of a process to produce products within a pre-assigned specification region. Kane (1986) in his pioneering research paper explained the importance of using PCI's specially in manufacturing industries. Kotz and Johnson (2002), in their review paper, have mentioned the existence of about 200 research papers mostly dealing with univariate process capability indices.

Rodriguez (1992) has rightly pointed out that a stable predictable process distribution is a prerequisite for capability analysis. Also, most of the PCI's available in the literature assume normality of the underlying process distribution. Some possible logic behind this normality assumption may be that normal distribution possesses some useful properties which are easily adoptable in the study of properties of PCI's and also most of the processes found in manufacturing industries are prone to follow normal distribution. Moreover, a process needs to be stable before its capability being assessed since, an unstable process is highly unpredictable and hence it is difficult to judge whether the process is capable or not if it is not at all stable.

Under the assumption of normality of the underlying univariate process distribution, the four classical PCI's are:

1. $C_p = \dfrac{USL - LSL}{6\sigma}$

2. $C_{pk} = \dfrac{d - |\mu - M|}{3\sigma}$

3. $C_{pm} = \dfrac{d}{3\sqrt{\sigma^2 + (\mu - T)^2}}$

4. $C_{pmk} = \dfrac{d - |\mu - M|}{3\sqrt{\sigma^2 + (\mu - T)^2}}$

where, USL and LSL are respectively the upper and lower specification limits of a process, $d = \dfrac{\text{USL} - \text{LSL}}{2}$, $M = \dfrac{\text{USL} + \text{LSL}}{2}$ and 'T' is the target of the process.

Among these four indices, Cp measures the potential process capability. However, since it only incorporates process standard deviation in its definition, Cpk is defined to take into account the process average also. Further, Cpm was defined to establish the relationship between squared error loss and the process capability indices. Finally, Cpmk, 'the third generation PCI' (1992), was constructed from Cpk and Cpm to increase the sensitivity (of a PCI) to departure of the process mean 'μ' from the target value 'T'. Vannman (1995) introduced the following super-structure of univariate process capability indices with bi-lateral specification interval:

$$C_p(u, v) = \dfrac{d - u|\mu - M|}{3\sqrt{\sigma^2 + v(\mu - T)^2}}, \qquad u, v \geq 0 \qquad \qquad \dots (1)$$

It can be easily observed that Cp(u, v) incorporates all the four basic indices as special cases, viz., Cp(0, 0) = Cp; Cp(1, 0) = Cpk; Cp(0, 1) = Cpm and Cp(1, 1) = Cpmk. However, as has been pointed out by Taam *et al.* (1993) among others, in most of the practical situations, a manufactured product has more than one quality characteristics to describe its features, geometric shape and design intent. The prevailing practice of most of the industries is to calculate suitable univariate PCI for each of the quality characteristic and then multiply them to have a single valued capability measure of the process. However, this ignores the interdependence *i.e.* the correlation among the various quality characteristics and as such may, often, be misleading. This necessitates the use of 'Multivariate Process Capability Index (MPCI)'.

Despite the high demand in practical field, there are very few MPCI's [see Taam *et al.* (1993), Chen (1994), Wang *et al.* (2000), Polansky (2001), Kirmani and Polansky (2009), Shariari and Abdollahzadeh (2009), Shinde and Khadse (2009) and so on] which are not adequate to address all the types of problems encountered in practice. In the present paper, some properties of a new super-structure of MPCI's have been discussed and also its threshold value has been suggested so as to make this index more suitable for practical purpose. Finally, the plug-in estimator of the super-structure as well as its expectation is derived which may be used for further investigation of the properties of the super-structure, specially, when the consequence of sampling fluctuation is taken into account. Note that we have used bold faced letters to denote vectors for the remaining part of the article.

2. A NEW SUPER-STRUCTURE OF MULTIVARIATE PROCESS CAPABILITY INDICES

It has been mentioned earlier that, the four classical PCI's, for processes with bi-lateral specification limits and univariate normal process distribution, covers most of the practical situations under the univariate scenario. Hence, the multivariate counterparts of these indices should deliver the good for processes having more than one quality characteristics. Also, their properties can be easily compared to those of the univariate ones, on which substantial research works are available in literature [see Pearn *et al.* (1992), Vannman and Kotz (1995)]. For these reasons, Chakraborty and Das (2007) proposed the following multivariate analogue of Cp(u, v) :

$$C_G(u, v) = \frac{1}{3} \sqrt{\frac{(d - uD)' \Sigma^{-1} (d - uD)}{1 + v (\mu - T)' \Sigma^{-1} (\mu - T)}} \qquad \dots (2)$$

where, $\mathbf{D} = (|\mu_1 - M_1|, |\mu_2 - M_2|, \dots, |\mu_p - M_p|,)'$; $\mathbf{d} = (d_1, d_2, \dots, d_p)'$ with $d_i = (USL_i - LSL_i)/2$; $\mathbf{T} = (T_1, T_2, \dots, T_p)'$; $\mathbf{M} = (M_1, M_2, \dots, M_p)'$ with $M_i = (USL_i + LSL_i)/2$; $\mu = (\mu_1, \mu_2, \dots, \mu_p)'$.

Here, Ti is the target value, Mi is the nominal value and μ_i is the mean of the ith characteristic of the item, for i = 1(1)p ; 'p' denotes the number of characteristics under consideration; μ is the mean vector and Σ is the variance-covariance matrix of the random vector '**X**' representing the quality characteristics under consideration; u and v are the scalar constants that can take any non-negative integer value.

Following assumptions are taken into consideration:

1. Underlying process distribution is multivariate normal with mean vector μ and dispersion matrix Σ.
2. The process has hyper-rectangular specification region.
3. For each process variable specification limits are symmetric about its mean.

It is interesting to see that for u = 0, 1 and v = 0, 1, this MPCI gives us the multivariate PCI's which are analogous to the four basic univariate PCI's viz, Cp; Cpk; Cpm and Cpmk. Similar to Vannman (1995), here also, for analytical purpose, it is assumed that M = T.

3. RELATIONSHIP BETWEEN DIFFERENT INDICES OBTAINED BY VARYING 'U' AND 'V' OF C$_G$(U, V)

Relationship Between C$_G$(0, 0) and C$_G$(u, 0)

We know that, $C_G(0, 0) = \frac{1}{3} \sqrt{d' \Sigma^{-1} d}$

which implies, $C_G^2(0, 0) = \frac{1}{9} d' \Sigma^{-1} d$

Also, $C_G(u, 0) = \frac{1}{3} \sqrt{(d - uD)' \Sigma^{-1} (d - uD)}$

i.e. $C_G^2(u, 0) = \frac{1}{9} tr [\Sigma^{-1} (d - uD)(d - uD)']$ $\qquad \dots (3)$

Now, $(d - uD)(d - uD)' = dd' - uDd' - udD' + u^2 DD'$.

Since, **D** involves 'modulus function' and hence is positive, so,

$d \geq (d - uD)$

i.e., $dd' \geq (d - uD)(d - uD)'$ $\qquad \dots (4)$

i.e., $\frac{1}{9} tr[\Sigma^{-1}dd'] \geq \frac{1}{9} tr[\Sigma^{-1}(d-uD)(d-uD)']$

and hence, $C_G(0, 0) \geq C_G(u, 0)$, for $u \geq 0$ ... (5)

However, it can be easily noticed from (4) that if one can make at least any one of `u' and `D' large enough to make $(d-uD) < 0$, such that $dd' < (d-uD)(d-uD)'$, then the inequality of (5) will be reversed. But this situation is not desirable for a process due to the following reasons:

1. Under the normality assumption the process is symmetric about its mean and is 'just capable' with respect to its rectangular specification region of $\mu \pm 3\sigma$. Hence **D** can not be very large as by definition, $Di = |\mu_i - M_i|$, $i = 1(1)p$.

2. For univariate case, Vannman (1995) has shown, by simulation study, that the index gives the most desirable result for small values of 'u' and 'v' (especially for u = 0 and v = 4). Hence assuming 'u' to be very large is not desirable.

Relationship Between $C_G(0, 0)$ and $C_G(0, v)$

We know that,

$$C_G^2(0,v) = \frac{1}{9} \cdot \frac{d \Sigma^{-1} d}{1 + v(\mu-T)'\Sigma^{-1}(\mu-T)}$$

$$= \frac{C_G^2(0,0)}{1 + v(\mu-T)'\Sigma^{-1}(\mu-T)}$$

... (6)

Now, $(\mu-T)'\Sigma^{-1}(\mu-T)$ is a quadratic form and hence is always non-negative. Therefore, denominator of (6) is greater than or equal to 1. Hence, $C_G(0, v) \leq C_G(0, 0)$, for $v \geq 0$.

Relationship Between $C_G(u, 0)$ and $C_G(u, v)$

We have,

$$C_G^2(u,v) = \frac{1}{9} \cdot \frac{(d-uD)'\Sigma^{-1}(d-uD)}{1 + v(\mu-T)'\Sigma^{-1}(\mu-T)} = \frac{C_G^2(u,0)}{1 + v(\mu-T)'\Sigma^{-1}(\mu-T)}$$

Therefore, $C_G(u, v) \leq C_G(u, 0) \leq C_G(0, 0)$, for all $u \geq 0$ and $v \geq 0$.

Relationship Between $C_G(0, v)$ and $C_G(u, v)$

$$C_G^2(0,v) = \frac{1}{9} \cdot \frac{d \Sigma^{-1} d}{1 + v(\mu-T)'\Sigma^{-1}(\mu-T)}$$

Now, we have already shown that $dd' < (d-uD)(d-uD)'$

Hence $C_G(u, v) \leq C_G(0, v) \leq C_G(0, 0)$, for all $u \geq 0$ and $v \geq 0$.

However, similar to the univariate situation (Kotz and Johnson, 2002), here also relationship between CG(u; 0) and CG(0; v) is not clear cut.

4. COMPARISON OF THE PERFORMANCE OF $C_G(U, V)$ TO THAT OF $C_P(U, V)$

As has been pointed out by Grau (2009), for a PCI to be easily acceptable to industry, it should ideally posses natural properties and interpretations identical to the most commonly used PCI's (univariate) viz., Cp, Cpk, Cpm and Cpmk and in general Cp(u, v) so that even a shop floor person can easily understand its purpose. On the basis of our previous discussion on various properties of CG(u, v), we make the following comparison between the natural properties of CG(u, v) to those of Cp(u, v) to establish the purpose of defining CG(u, v) for processes with multiple quality characteristics.

Property 1: When $\mu = M$ [or, $\mu = M$, for multivariate case], both C_p and $C_G(0, 0)$ directly relate to proportion of non-conformance.

Proof: For univariate processes, if $\mu = M$, then the probability of producing a non-conforming item is $p = 2\Phi(-3C_p)$. We have to check whether similar property holds for $C_G(u, v)$ as well. Now, the general formulation for the proportion of non-conformance is,

P[Producing non-conforming items] = $1 - P[L \leq \mathbf{X} \leq U]$

$= 1 - P[\mathbf{X} \leq U] + P[\mathbf{X} \leq L]$

$= 1 - P_1 + P_2$, say

$= 2(1 - P_1)$, since the specification region is assumed to be symmetric about μ.

Hence, $\quad P_1 = P[\mathbf{X} \leq U]$

$$= P\left[(X - \mu)\,\Sigma^{-1}(X - \mu) \leq d\,\Sigma^{-1}d\right]$$

$$= P\left[(X - \mu)\,\Sigma^{-1}(X - \mu) \leq 9\,C_G^2\,(0,0)\right]$$

$$= P\left[Y \leq 9\,C_G^2\,(0,0)\,\middle|\,Y \sim \chi_p^2\right]$$

Since the values of the Cumulative Density Function (CDF) of `Chi-squared distribution' are tabulated in any standard statistical table, this probability can be easily obtained. As such, since by definition CG(u, v) is always non-negative, similar to Cp, CG(u, v) can also be expressed as a one-to-one function of the proportion of non-conformance.

Property 2: For any univariate process,

$C_{pmk} \leq C_{pk} \leq C_p$

$C_{pmk} \leq C_{pm} \leq C_p$.

Similarly, for any multivariate process,

$C_G(1, 1) \leq C_G(1, 0) \leq C_G(0, 0)$

$C_G(1, 1) \leq C_G(0, 1) \leq C_G(0, 0)$.

Proof: This has already been proved in section 3.

Property 3: Both Cp(u, v) and CG(u, v) take into account the deviation of the process mean from the mid-point of the specification region (viz., 'M' for univariate case and '**M**' for the multivariate case) in a more or less similar manner—the difference, if any, being in the geometrical approach of considering such deviation *i.e.* linear or quadratic way.

Proof: Grau (2009) have already shown that:

1. Cpk takes into account the deviation of 'μ' from 'M' linearly such that it assumes '0' value when the process mean coincides with the specification limits.
2. Cpm, when T = M, takes into account such deviation in quadratic way such that it never assumes the value '0' even if 'μ' approaches either of the specification limits. However, it assumes value close to '0' for large shift of the mean from 'M'.
3. Cpmk being derived from both Cpk and Cpm, allows to take the deviation in a quadratic way [similar to Cpm] and assumes `0' value when the process mean reaches either of the specifications [similar to Cpk].

Now we check whether such characteristics are also valid for C_G (u; v). For this we start with $C_G(1, 0)$. Let us assume that $\mu = U$. Then, $D_i = | \mu_i - M_i | = d_i$, i = 1(1)p. *i.e.*, $D = d$ and hence $C_G(1, 0) = 0$. Similar results can also be proved for the case when $\mu = L$. However, unlike Cpk, $C_G (1, 0)$ takes into account the deviation of μ from M in quadratic way.

From the definition of $C_G(0, 1)$, it is obvious that, similar to Cpm, $C_G(0, 1)$ also takes into account the deviation in quadratic way. Now, let us assume that $\mu = U$. Then, $d \Sigma^{-1} d \neq 0$ and hence, $C_G(0, 1) \neq 0$. However, the denominator of $C_G(0, 1)$ *i.e.* $1 - (\mu - T) \Sigma^{-1} (\mu - T)$ is greater than '1' and increases with increasing distance between μ and T. Therefore, $C_G(0, 1)$ assumes value close to '0' for large shift of 'μ' from 'T'.

Finally, $C_G(1, 1)$, which is analogous to Cpmk, allows to take the deviation in a quadratic way and assumes '0' value when 'μ' equals either of the specification limits. This can be proved directly from our previous discussion on $C_G(1, 0)$ and $C_G(0, 1)$.

5. CONCLUSION

$C_G(u, v)$ was proposed, primarily, to provide a multivariate analogue of Cp(u, v), the famous super-structure of PCI's for univariate processes. In this paper, we have studied the inter-relationship between the MPCI's obtained from CG(u; v) and found that they hold similar relationships compared to Cp(u, v). We have also made a comparative study to establish that CG(u, v) also possesses some interesting properties similar to Cp(u, v) and that makes this super-structure for MPCI's easy to compute and easier to interpret. However, since in practice, the computed value of CG(u, v) is likely to be affected by sampling fluctuation, the properties of the plug-in estimator of the MPCI also needs to be studied. In this context, it is worthy to mention that the presence of the 'modulus- function' in the numerator of CG(u, v) necessitates the use of multivariate analogue of the so-called folded normal distribution. The authors are studying the properties of multivariate folded normal distribution.

REFERENCES

[1] Chen, H. (1994). "A multivariate process capability index over a rectangular solid tolerance zone", Statistica Sinica, Vol. 4, pp. 749–758.

[2] Chakraborty, A.K. and Das, A. (2007). "Statistical analysis of multivariate process capability indices", M.Tech Thesis, Indian statistical Institute, Kolkata, India.

[3] Grau, D. (2009). "New process capability indices for one-sided tolerances", Quality Technology and Quantitative Management, Vol. 6, No. 2, pp. 107–124.

[4] Kane, V.E. (1986). "Process capability index", *Journal of Quality technology,* Vol. 18 No. 1, pp. 41–52.

[5] Kirmani, S. and Polansky, A.M. (2009). "Multivariate process capability via lowner ordering", Linear Algebra and Its Applications, Vol. 430, pp. 2681–2689.

[6] Kotz, S., Johnson. N. L. (2002), "Process capability indices a review, 1992–2000", Journal of Quality technology, Vol. 34 No.1, pp. 02–19.

[7] Pearn, W.L., Kotz, S. and Johnson, N.L. (1992), "Distributional and inferential properties of process capability indices", *Journal of Quality technology,* Vol. 24, No. 4, pp. 216–231.

[8] Polansky, A.M. (2001). "A smooth nonparametric approach to multivariate process capability", Technometrics, Vol. 43, No. 2, pp. 199–211.

[9] Rodriguez, R.N. (1992). "Recent developments in process capability analysis", *Journal of Quality technology*, Vol. 24, No. 4, pp. 176–187.

[10] Shahriari, H. and Abdollahzadeh, M. (2009). "A new multivariate process capability vector", Quality Engineering, Vol. 21, No. 3, pp. 290–299.

[11] Shinde, R.L. and Khadse, K.G. (2009). "Multivariate process capability using principal component analysis", Quality and Reliability Engineering International, Vol. 25, pp. 69–77.

[12] Taam, W., Subbaiah, P. and Liddy, W. (1993). "A note on multivariate capability indices", *Journal of Applied Statistics*, Vol. 20, pp. 339–351.

[13] Vannman, K. (1995). "A unified approach to capability indices", *Statistica Sinica*, Vol. 5, pp. 805–820.

[14] Vannman, K. and Kotz, S. (1995). "A superstructure of capability indices-distributional properties and implications", *Scandinavian Journal of Statistics*, Vol. 22, No. 4, pp. 477–491.

[15] Wang, F.K., Huele, N.F., Lawrence, F.P., Miskulin, J.D. and Shahriari, H. (2000). "Comparison of three multivariate process capability indices", *Journal of Quality Technolog y*, Vol. 32, No. 3, pp. 263–275.

PART–II

Reliability Engineering

Goodness-of-Fit Comparisons of Change-Point Models for Software Reliability Assessment

Shinji Inoue[1] and Shigeru Yamada[2]

Tottori University, Tottori, Japan
E-mail: [1]ino@sse.tottori-u.ac.jp; [2]yamada@sse.tottori-u.ac.jp

ABSTRACT: Software testing manager usually observes a change of the software failure-occurrence phenomenon in an actual testing phase due to some factors being related to the software reliability growth process. Testing-time when behavior of the software failure-occurrence time interval notably changes is called change-point. Such phenomenon influences accuracy of reliability assessment based on a software reliability growth model. This paper discusses software reliability growth modeling with the influence of the change-point by using the environmental factors. Then, we check goodness-of-fit of our change-point models to actual data by comparing with the existing non change-point and single change-point models.

Keywords: Software Reliability Model, Change-Point, Nonhomogeneous Poisson Process, Environmental Function, Goodness-of-Fit.

1. INTRODUCTION

Software reliability growth models (abbreviated as SRGMs) (Yamada and Osaki, 1985; Musa *et al.*, 1987; Pham, 2000) are known as mathematical models for quantitative software reliability assessment. Ordinarily, SRGMs are developed by treating the software failure-occurrence time or the fault-detection time intervals as random variables. And, it is assumed that the stochastic characteristics for these quantities are same throughout the testing phase in the usual software reliability growth modeling. However, such assumptions do not enable us to reflect an actual software failure-occurrence phenomenon to software reliability growth modeling because we often observe a change of the stochastic behavior for the software failure-occurrence time interval due to changing some factors being related to the software failure-occurrence phenomenon, e.g., changing of fault target, changing of testing-effort expenditure, and so forth. Testing-time when such phenomenon is observed is called change-point (Zhao, 1993). It is known that occurrence of the change-point influences accuracy of SRGM-based software reliability assessment. Under the background, software reliability growth modeling with the influence of the change-point has been discussed so far (Zhao, 1993; Huang, 2005; Zou, 2003; Zhao *et al.*, 2006).

However, it is very difficult to find research results, which discuss software reliability growth modeling with a relationship between the software failure-occurrence time intervals before the change-point and those after the change-point. In an actual testing phase, it might be natural to consider that there exists the relationship between the time-intervals before change-point and those after change-point because the same software product is tested even if the change-point is occurred during the testing-phase for the software product. And it is very important to know how the stochastic characteristic of the software failure-occurrence time-interval changes at the change-point from the point of view of software development management. This paper discusses a framework for software reliability growth modeling with the effect of the change-point. Concretely speaking, we incorporate the relationship between the software failure-occurrence time intervals before change-point and those after change-point into the usual modeling framework by using a testing-environmental function. Further, we check goodness-of-fit of our change-point models, which is developed by using our change-point modeling framework, for actual data by comparing with the existing non change-point models.

2. BASIC MODELING FRAMEWORK

Basically, it is known that almost all of the SRGMs in which the total number of detectable faults is finite are developed under the following basic assumptions (Langberg and Singpurwalla, 1985; Miller, 1986; Raftery, 1987; Joe, 1989):

1. Whenever a software failure is observed, the fault which caused it will be detected immediately and no new faults are introduced in the fault-removing activities.

2. Each software failure occurs at independently and identically distributed random times with the probability distribution, $F(t) \equiv \Pr\{T \leq t\}$, where $\Pr\{A\}$ represents the probability of event A. And the probability density function is denoted by $f(t)$.

3. The initial number of faults in the software, $N_0 (> 0)$, is a random variable, and is finite.

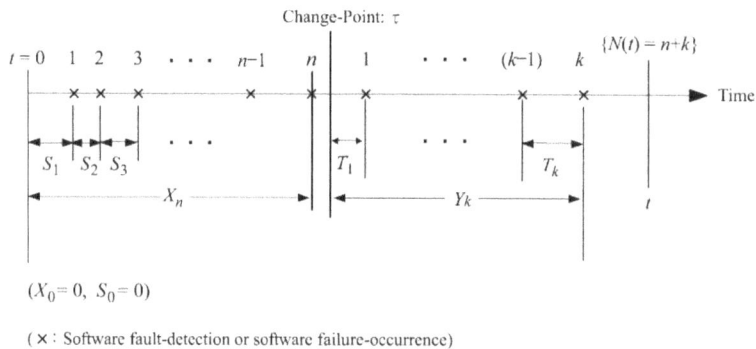

Fig. 1: Stochastic Quantities for the Software Failure-Occurrence Phenomenon with Change-Point

Now, let $\{N(t), t \geq 0\}$ denote a counting process representing the total number of faults detected up to testing-time t. From the basic assumptions above, the probability that m faults are detected up to testing-time t is derived as,

$$\Pr\{N(t) = m\} = \sum_n \binom{n}{m} \{F(t)\}^m \{1 - F(t)\}^{n-m}$$
$$\times \Pr\{N_0 = n\} \quad (m = 0, 1, 2, \cdots). \qquad \ldots (1)$$

As a well-known result, if we assume that the initial fault content, N_0, follows a Poisson distribution with mean ω, the counting process $\{N(t), t \geq 0\}$ in Eq. (1) can be rewritten as,

$$
\begin{aligned}
\Pr\{N(t) = m\} &= \sum_n \binom{n}{m} \{F(t)\}^m \{1 - F(t)\}^{n-m} \frac{\omega^n}{n!} \exp[-\omega] \\
&= \exp[-\omega] \frac{\{\omega F(t)\}^m}{m!} \sum_n \frac{\{\omega(1 - F(t))\}^{n-m}}{(n-m)!} \qquad \cdots (2) \\
&= \frac{\{\omega F(t)\}^m}{m!} \exp[-\omega F(t)] \quad (m = 0, 1, 2, \cdots).
\end{aligned}
$$

Eq. (2) is equivalent to a non-homogeneous Poisson process (NHPP) with mean value function $\omega F(t)$. We need to give a suitable software failure-occurrence times distribution to develop a specific NHPP model. For an example, we obtain an exponential SRGM (Goel and Okumoto, 1979), $E[N(t)] = \omega(1 - e^{-\lambda t})$, which is one of the representative NHPP models, if we assume that the software failure-occurrence times distribution follows an exponential distribution with parameter λ.

3. EXISTING CHANGE-POINT MODELING APPROACH

We discuss an existing framework for software reliability growth modeling with change-point based on the basic modeling framework (Inoue and Yamada, 2007). Extending the assumptions on basic modeling framework (2), we describe a difference between the software hazard rates before and after the change-point as,

$$
z(t) = \begin{cases} z_1(t) & (0 \leq t \leq \tau) \\ z_2(t) & (t > \tau), \end{cases} \qquad \cdots (3)
$$

where $\tau \, (0 < \tau < T)$ indicates change-point. From Eq. (3), the software failure-occurrence times distributions before and after change-point is derived as,

$$
F(t) = \begin{cases} 1 - \exp[-\int_0^t z_1(x)dx] & (0 \leq t \leq \tau) \\ 1 - \exp[-\int_0^\tau z_1(x)dx - \int_\tau^t z_2(x)dx] & (t > \tau), \end{cases} \qquad \cdots (4)
$$

respectively. In Eq. (4), the software failure-occurrence times distribution over $(0 \leq t \leq \tau)$ is denoted as $F_B(t)$ and over $(t > \tau)$ as $F_A(t)$, respectively. Then, we can develop a mean value function with change-point, $H(t)$, as

$$
\begin{aligned}
H(t) &= \omega F(t) \\
&= \omega\{F_B(t)U_B(\tau - t) + F_A(t)U_A(t - \tau)\} \qquad \cdots (5) \\
&= \begin{cases} H_B(t) = \omega F_B(t) & (0 \leq t \leq \tau) \\ H_A(t) = \omega F_A(t) & (t > \tau) \end{cases}
\end{aligned}
$$

where $U_1(x)$ and $U_2(x)$ represent the step functions:

$$U_1(x) = \begin{cases} 0 & (x < 0) \\ 1 & (x \geq 0), \end{cases} \qquad U_2(x) = \begin{cases} 0 & (x \leq 0) \\ 1 & (x > 0), \end{cases} \qquad \dots (6)$$

respectively. From Eqs. (5) and (6), we can develop an SRGM with change-point by assuming the software failure-occurrence rate functions for the software failure-occurrence times distributions before and after the change-point, respectively.

4. OUR CHANGE-POINT MODELING APPROACH

Existing modeling approach mentioned above is developed by focusing on the change of the software hazard rate at the change-point. However, this modeling approach has not been discussed the relationship between the software failure-occurrence time-intervals before and those after change-point. In an actual testing phase, it might be natural to consider that there exists such relationship since we test an identical software system in an actual testing phase even if the change-point occurs.

Now we define the following stochastic quantities being related to our modeling approach in this paper:

X_i = the i-th software failure-occurrence time before change-point ($X_0 = 0$, $i = 0, 1, 2, \cdots$),

S_i = the i-th software failure-occurrence time-interval before change-point

$$(S_i = X_i - X_{i-1}, S_0 = 0, i = 0, 1, 2, \cdots),$$

Y_i = the i-th software failure-occurrence time after change-point ($Y_0 = 0$, $i = 0, 1, 2, \cdots$),

T_i = the i-th software failure-occurrence time-interval after change-point ($T_i = Y_i - Y_{i-1}, T_0 = 0$, $i = 0, 1, 2, \cdots$).

Figure 1 depicts the stochastic quantities for the software failure-occurrence or fault-detection phenomenon with change-point. We assume that the stochastic quantities before and those after the change-point have the following relationships:

$$\begin{cases} Y_i = \alpha(X_i), \\ T_i = \alpha(S_i), \\ K_i(t) = J_i(\alpha^{-1}(t)), \end{cases} \qquad \dots (7)$$

respectively, where $\alpha(\cdot)$ is a test-environmental function representing the relationship between the software failure-occurrence times or time-intervals before change-point and those after change-point, $J_i(t)$ and $K_i(t)$ the probability distribution functions with respect to the random variables S_i and T_i, respectively.

We assume that the testing-environmental function is given as $\alpha(t) = \alpha t$ ($\alpha > 0$) (Okamura *et al.*, 2000), where α is the proportional constant representing the relative magnitude of the effect of change-point on the software reliability growth process. Suppose that n faults have been detected up to change-point and their fault-detection times from the test-beginning ($t = 0$) have been observed as $0 < x_1 < x_2 < \ldots < x_n \leq \tau$. Then, the probability distribution function of T_1, a random variable representing the time-interval from change point to the ($n+1$)-st software failure-occurrence, can be derived as,

$$
\begin{aligned}
\overline{K}_1(t) &\equiv \Pr\{T_1 > t\} \\
&= \frac{\Pr\{S_{n+1} > \tau - x_n + t/\alpha\}}{\Pr\{S_{n+1} > \tau - x_n\}} \qquad\qquad\qquad \ldots (8)\\
&= \frac{\exp[-\{M_B(\tau + t/\alpha) - M_B(x_n)\}]}{\exp[-M_B(\tau) - M_B(x_n)]},
\end{aligned}
$$

where $\overline{K}_1(t)$ indicates the cofunction of the probability distribution function $K_1(t) \equiv \Pr\{T_1 \leq t\}$, *i.e.*, $\overline{K}_1(t) \equiv 1 - K_1(t)$, and $M_B(t) (\equiv \omega J(t))$ represents the expected number of faults detected up to change-point, *i.e.*, a mean value function for the NHPP before change-point. From Eq. (8), the expected number of faults detected up to $t \in (\tau, \infty]$ after change-point, $M_A(t)$, can be formulated as,

$$
\begin{aligned}
M_A(t) &= -\log \Pr\{T_1 > t - \tau\} \\
&= -\log \overline{K}_1(t - \tau) \qquad\qquad\qquad \ldots (9)\\
&= M_B(\tau + \frac{t - \tau}{\alpha}) - M_B(\tau).
\end{aligned}
$$

Then, an NHPP model with change-point, $\Lambda(t)$, can be derived as,

$$
\Lambda(t) = \begin{cases}
\Lambda_B(t) = M_B(t) & (0 \leq t \leq \tau) \\
\Lambda_A(t) = M_B(\tau) + M_A(t) & \\
\quad = M_B(\tau + \dfrac{t - \tau}{\alpha}) & (t > \tau)
\end{cases}
\qquad \ldots (10)
$$

From Eq. (10), we can see that several types of NHPP-based SRGM with change-point can be developed by assuming a suitable probability distribution function for the software failure-occurrence times before change-point.

5. GOODNESS-OF-FIT COMPARISONS

5.1 Models

We develop change-point models for software reliability assessment by using the modeling frameworks in Eqs. (5) and (10). First of all, following the modeling framework in Eq. (5), we derive the following change-point models:

$$
H_1(t) = \begin{cases}
H_B(t) = \omega(1 - \exp[b_1 t]) & (0 \leq t \leq \tau) \\
H_A(t) = \omega(1 - \exp[b_1 \tau - b_2(t - \tau)]) & (t > \tau)
\end{cases}
\qquad \ldots (11)
$$

and

$$H_2(t) = \begin{cases} H_B(t) = \omega(1 - \exp[-\dfrac{b_1}{2}t^2]) & (0 \le t \le \tau) \\[2ex] H_A(t) = \omega(1 - \exp[-\dfrac{b_1}{2}\tau^2 - \dfrac{b_2}{2}(t^2 - \tau^2)]) \\[2ex] \qquad (t > \tau) \end{cases} \qquad \ldots (12)$$

by assuming that the software hazard rates are $z_1(t) = b_1$ and $z_2(t) = b_2$ in Eq. (11) and $z_1(t) = b_1 t$ and $z_2(t) = b_2 t$, respectively. Further, we derive the following our change-point models

$$\Lambda_1(t) = \begin{cases} \Lambda_B(t) = \omega(1 - \exp[-b_1 t]) & (0 \le t \le \tau) \\[2ex] \Lambda_A(t) = \omega(1 - \exp[-b_1(\tau + \dfrac{t-\tau}{\alpha})]) & (t > \tau) \end{cases} \qquad \ldots (13)$$

and

$$\Lambda_2(t) = \begin{cases} \Lambda_B(t) = \omega(1 - (1 + b_1 t)\exp[-b_1 t]) & (0 \le t \le \tau) \\[2ex] \Lambda_A(t) \\[1ex] = \omega(1 - (1 + b_1(\tau + \dfrac{t-\tau}{\alpha}))\exp[-b_1(\tau + \dfrac{t-\tau}{\alpha})]) \\[2ex] \qquad (t > \tau) \end{cases} \qquad \ldots (14)$$

by using our modeling framework in Eq. (10). We assume that the software failure-occurrence phenomenons follow an exponential distribution with parameter λ in Eq. (13) and a gamma distribution with parameters $(2, b_1)$, respectively. We note that Eqs. (11) and (13) are assumed constant software failure rates and Eqs. (12) and (14) are assumed increasing software failure rate, respectively.

5.2 Comparison Criteria

We use the mean square error (abbreviated as MSE) (Pham, 2000) and the predicted relative errors (Musa et al., 1987) as the model comparison criteria in this paper, the MSE is calculated by dividing the sum of squared vertical distance between the observed and estimated cumulative numbers of faults, y_i and $\hat{y}(t_i)$, detected during the time-interval $(0, t_i)$, respectively, by the number of observed data pairs. That is, supposing that N data pairs (t_i, y_i) $(i = 1, 2, \cdots, N; 0 < t_1 < t_2 < \cdots < t_N)$ are observed, we can formulate the MSE as,

$$MSE = \frac{1}{N}\sum_{i=1}^{N}(y_i - \hat{y}(t_i))^2. \qquad \ldots (15)$$

The model having the smallest MSE fits best to the observed data set. And, the predicted relative error, $PRE[t_e]$, is calculated as,

$$PRE[t_e] = \frac{\hat{y}(t_e;t_q) - n_q}{n_q}, \qquad \qquad \qquad \dots (16)$$

where $\hat{y}(t_e;t_q)$ is the estimated value of the number of detected faults at the testing-termination time, t_q, by using the observed data collected up to the arbitrary testing-time, t_q $(0 < t_e \le t_q)$, and n_q the observed number of faults detected up to the termination time of the testing. Then, we can say that the SRGM whose value of the predicted relative errors at each testing-period or testing-time is closed to zero has better performance on the predictive accuracy.

5.3 Results

We arrange the five data sets, DS1-DS5, which were collected from actual testing phase for the Windows version software. DS3 and DS5 show exponential software reliability growth curves and the remainder (DS1, DS2, DS4) show S-shaped software reliability growth curves, respectively. Table 1 shows the results of model comparisons of our change-point models ("Model 1" in Eq. (13) and "Model 2" in Eq. (14)) with existing SRGMs (an exponential (Goel and Okumoto, 1979), a delayed S-shaped SRGMs (Yamada and Osaki, 1985), and existing change-point models in Eqs. (11) and (12) named as "Existing CP Model 1" and "Existing CP Model 2", respectively) based on the MSE. The parameter estimations are obtained by the method of maximum likelihood. From Table 1, we can say that Existing CP Model 1 and Model 1 has the best fitting performance among other SRGMs for the actual data showing exponential software reliability growth curves, which are DS3 and DS5. And we see that MSEs of Existing CP Model 1 and Model 1 are same because these two models have the same model structure essentially. However, we should note that we can get to know the relative magnitude of the effect of the change for the software failure-occurrence phenomenon by using our modeling approach. On the other hand, Model 2 does not have the best performance for the data showing S-shaped software reliability growth curves. However, we can say that Model 2 has better fitting performance for DS1 and DS4 compared with the corresponding existing non change-point model, *i.e.*, the delayed S-shaped SRGM, and the existing change-point model, Existing CP Model 2. For DS4, Existing CP Model 1 and Model 1 fit well to the actual data regardless of the shape of the software reliability growth curve because DS4 is not saturated with respect to the expected initial fault content and might be collected the middle of the software reliability growth process.

Table 1: Results of Goodness of Fit Comparisons Based on MSE

	Exponential SRGM	Delayed S-shaped	Existing CP Model 1	Existing CP Model 2	Our Model 1	Our Model 2
DS1 (S-shpaed)	16.0848	7.21958	18.4096	69.1692	21.6153	**6.12918**
DS2 (S-shaped)	32.0834	**19.3189**	58.3482	191.827	58.3482	22.1304
DS3 (Exponential)	2.40089	6.87789	**2.19263**	50.2619	**2.19263**	6.05874
DS4 (S-shaped)	0.348901	1.70385	**0.331126**	16.0309	**0.331126**	1.60806
DS5 (Exponential)	2.47239	7.63319	**2.2943**	96.12	**2.2943**	7.49945

Table 2: Results of Goodness of Fit Comparisons Based on Predicted Relative Errors for DS2

Testing Progress Ratio (%)	Exponential SRGM	Delayed S-shaped	Existing CP Model 1	Existing CP Model 2	Our Model 1	Our Model 2
85	1.279E-01	6.845E-02	1.947E-15	5.840E-16	5.431E-02	**4.826E-02**
90	8.158E-02	4.352E-02	3.957E-04	**9.467E-16**	2.047E-02	1.951E-02
95	2.048E-02	1.083E-02	1.159E-03	**-9.467E-16**	3.422E-03	3.380E-03
100	-1.947E-16	1.947E-16	0	-3.893E-16	-7.787E-16	-3.893E-16

Table 3: Results of Goodness of Fit Comparisons Based on Predicted Relative Errors for DS4

Testing Progress Ratio (%)	Exponential SRGM	Delayed S-shaped	Existing CP Model 1	Existing CP Model 2	Our Model 1	Our Model 2
85	**1.988E-02**	-2.939E-02	3.492E-02	-4.278E-02	3.492E-02	-3.666E-02
90	2.542E-02	**-8.470E-03**	3.643E-02	-1.558E-02	3.643E-02	-1.388E-02
95	**-6.345E-03**	-1.864E-02	-7.578E-03	-2.103E-02	-7.578E-03	-2.099E-02
100	-2.757E-05	-4.263E-16	1.421E-16	2.842E-16	2.842E-16	0

Further, we conduct calculating the predicted relative errors of these SRGMs for DS2 and DS4 for checking the predictive performance of Model 2 because Model 2 does not have nice performance on MSE for DS2 and DS4, respectively. Table 2 shows the results of goodness-of-fit comparisons based on the predicted relative errors. In Table 2, we started calculating the predicted relative errors from 85% of the testing progress ratio because software reliability assessment is ordinarily conducted after 60-80% of the testing progress ratio. From Table 2, we cannot see the usefulness for the predictive performance of Model 2. However, Model 2 often has better predictive performance than Existing CP Model 1 and 2. From these results of model comparisons, we can say that our modeling approach contributes to improve accuracy of the software reliability assessment based on existing SRGMs. However, we see that our models have some weak point in terms of the predictive performance.

6. CONCLUSION

We considered the relationship of the software failure-occurrence phenomenon before and after the change-point by using a testing-environmental function in our change-point modeling. By our modeling approach, we can get to know the quantitative aspect of the change for the software failure-occurrence phenomenon at the change-point. Further, this paper conducted goodness-of-fit comparisons of our models with existing SRGMs in terms of the MSE and the predicted relative errors by using actual data. From the goodness-of-fit comparisons, we checked that our modeling approach contributes to improve software reliability assessment accuracy based on existing SRGMs. However, we see that our models have some weak point especially in terms of the predictive performance. In further studies, we need to check more software reliability assessment performance of our models with existing SRGMs by using actual data for check the usefulness of our change-point modeling approach, and overcome the weak point on the predictive performance of our models by improving our modeling framework.

ACKNOWLEDGMENT

This work was supported in part by the Grant-in-Aid for Scientific Research (C), Grant No. 22510150, from the Ministry of Education, Sports, Science, and Technology of Japan.

REFERENCES

[1] Goel, A.L. and Okumoto, K. (1979), "Time-dependent error-detection rate model for software reliability and other performance measures," *IEEE Transactions on Reliability*, vol. R-28, no. 3, pp. 206–211.

[2] Huang, C.Y. (2005), "Performance analysis of software reliability growth models with testing-effort and change-point," *The Journal of Systems and Software*, vol. 76, no. 2, pp. 181–194.

[3] Inoue, S. and Yamada, S. (2007), "Software reliability measurement with change-point," in *Proceedings of the Fifth International Conference on Quality and Reliability*, Chiang Mai, Thailand, pp. 170–175.

[4] Joe, H. (1989) "Statistical inference for general-order-statistics and nonhomogeneous-Poisson-process software reliability models," *IEEE Transactions on Software Engineering*, Vol. 15, No. 11, pp. 1485–1490.

[5] Langberg, N. and Singpurwalla, N.D. (1985), "A unification of some software reliability models," *SIAM Journal on Scientific Computing*, Vol. 6, No. 3, pp. 781–790.

[6] Miller, D.S. (1986), "Exponential order statistic models of software reliability growth," *IEEE Transactions on Software Engineering*, vol. SE-12, No. 1, pp. 12–24.

[7] Musa, J.D., Iannio, D. and Okumoto, K. (1987), *Software Reliability: Measurement, Prediction, Application*. McGraw-Hill, New York.

[8] Okamura, H., Dohi, T. and Osaki, S. (2000), "A reliability assessment method for software product in operational phase? Proposal of an accelerated life testing model" (in Japanese), *IEICE Transactions on Fundamentals of Electronics, Communications and Computer Sciences*, Vol. J83-A, No. 3, pp. 294–301.

[9] Pham, H. (2000), *Software Reliability*. Springer-Verlag, Singapore.

[10] Raftery, A.E. (1987), "Inference and prediction for a general order statistic model with unknown population size," *Journal of the American Statistical Association*, vol. 82, no. 400, pp. 1163–1168.

[11] Yamada, S. and Osaki, S. (1985), "Software reliability growth modeling: Models and applications," *IEEE Transactions on Software Engineering*, Vol. SE-11, No. 12, pp. 1431–1437.

[12] Zhao, J., Liu, H.W., Cui, G. and Yang, X.Z. (2006), "Software reliability growth model with change-point and environmental function," *The Journal of Systems and Software*, Vol. 79, No. 11, pp. 1578–1587.

[13] Zhao, M. (1993), "Change-point problems in software and hardware reliability," *Communications in Statistics. Theory Methods*, Vol. 22, No. 3, pp. 757–768.

[14] Zou, F.Z. (2003), "A change-point perspective on the software failure process," *Software Testing, Verification and Reliability*, Vol. 13, No. 2, pp. 85–93.

Impact of Complexity of Object-Oriented Design on Software Reliability

Shrihari A. Hudli[1] and Anand V. Hudli[2]

[1]M.S. Ramaiah Institute of Technology, Bangalore, India
[2]ObjectOrb Technologies, Bangalore–560 010, India
E-mail: [1]shrihari@hudli.com; [2]anand.hudli@objectorb.com

ABSTRACT: Software Reliability theory is an important approach for predicting software failures in the field. A number of reliability models have been developed, but they do not address the complexity introduced by Object-Oriented Design. The object-oriented approach has many significant advantages, such as reusability, extendibility, interoperability, and reliability. However, testing object-oriented software has to deal with new problems resulting from the very features that make these advantages possible. Powerful features, such as encapsulation, inheritance, polymorphism, and dynamic binding, make new types of defects possible. We propose a model for estimating the reliability of object-oriented software based on a complexity measure. The complexity is derived from various metrics, such as the number of encapsulated variables, the number of methods in a class, the number of polymorphic methods, the depth of inheritance, etc. The complexity measure of an object-oriented design is shown to have an impact on the reliability of the resulting software.

Keywords: *Object-Oriented Design, Software Complexity Metrics, Reliability Theory, Reliability Models, Test Effectiveness.*

1. INTRODUCTION

1.1 Object-Oriented Software

The quality of a software product depends on many factors. It depends on the process being used to develop the product, the design of the product, the tools used, the skill levels of the staff who design the product, and the characteristics of the implementations. To improve the quality of software products, we should be able to identify locations of bad designs or programming practices in a software system. One way to identify problem areas in programs is to define metrics that can characterize programs. Using the metrics, the programs can be designed to get good characteristics that will result in a quality product. The metrics provide some confidence in the estimation of a whether a software product has been constructed according to some desired standards of good designs. Moreover, the metrics relate directly to reliability, maintainability, extensibility, and adaptability, all of which are essential requirements for quality. Software Reliability Theory (Lyu, 1996) is an important approach developed to predict the likelihood of failures in the field. However, the assumptions made by software reliability measurement models do not address the complexities of most kinds of software. There is a need to consider software complexity, test effectiveness, and other factors (Whittaker and Voas, 2000, Xu and Xu, 2010) in software reliability research.

Object-Oriented Programming (OOP) is a programming paradigm that is based on abstractions of object types in the application (Jacobson *et al.,* 1992, Rumbaugh *et al.*, 1991). The key difference between object-oriented programming and structured programming is that the former identifies the object types in the application, while the latter models the application as a set of functions. In OOP, object types are identified, and methods/functions that act on the objects are then defined. The object-oriented approach has a number of significant advantages, such as reusability, extendibility, interoperability, and reliability. However, testing object-oriented software has to deal with new problems resulting from the very features that make these advantages possible. Powerful features, such as encapsulation, inheritance, polymorphism, and dynamic binding contribute to the complexity of the software and hence present new problems in testing (Perry and Kaiser, 1990). This indicates that a stronger emphasis on testing is needed for object-oriented software. What is important to note is that this also indicates that these features have an impact on software reliability. In this paper, we propose to incorporate the effect of object-oriented design complexity into software reliability models with a view of making them more accurate for object-oriented software.

1.2 Software Reliability Measurement

The following formula is typically used to measure or predict the reliability $R(t)$ of software under test:

$$R(t) = e^{\lambda t} \qquad \qquad \dots (1)$$

Where λ is the failure rate of the system, typically obtained from testing data. $R(t)$ is the probability that a failure occurs after time t, given that there is no failure before time t. Here an assumption of constant failure rate is made.

A simple method of estimating the reliability of a software system just based on the error removal rate over a given period of time is illustrated by means of an example. Suppose previous life tests have shown that software fails at a constant rate. If a program was tested for 1000 hours and 4 defects were found, under the assumption of a constant error removal rate, we may obtain the failure rate of the software as $\lambda = 4 \times 10^{-3}$. Hence we may derive the reliability of this software using equation (1).

Although this simple approach using equation (1) works for hardware systems and most kinds of software systems as well, it does not work well for object-oriented software systems. As a consequence, the equation for reliability is not accurate for such systems. In this paper, we propose to improve the accuracy of the expression for reliability by taking into account the complexity of the design.

In order to measure the software design complexity, we need to introduce design metrics. These metrics are based on standard features of object-oriented software and are easily calculated for a given program. The paper is organized as follows. Section 2 contains a discussion of object-oriented design metrics and test effectiveness. Section 3 presents experimental results. Section 4 contains conclusions and future directions.

2. COMPLEXITY OF OBJECT-ORIENTED SOFTWARE

2.1 Metrics for Object-Oriented Design

Our presentation of object-oriented design metrics is based on earlier work by researchers (Hudli *et al.*, 1994).The metrics are not only aimed at evaluating an object-oriented design but also help in choosing an optimal design among alternatives.

1. *Data Encapsulation:* One of the goals of object-oriented design is to treat classes as "black boxes", with operations defined on the internals of the black box. The internal details are hidden from the clients of the class. The client sees only the public methods of the class. The instance variables of a class should be manipulated only by the methods of the class. This is possible only if all the instance variables are private members of the class. The number of instance variables that are private members of the class is a measure of the complexity of the class and the encapsulation of its data members. Two metrics are defined for encapsulation:

   (a) *Number of Encapsulated Variables (EV):* This is defined as the count of instance variables that are private members of the class. The higher the number the better the encapsulation is for the class.

   (b) *Number of Non-encapsulated Variables (NEV):* This is defined as the number of instance variables that are public members of the class. The higher the number, the worse the encapsulation is for the class.

2. *Number of Methods (NM):* The complexity of a class is indicated by the number of operations (methods) that the class can support. This gives us an idea of the programming effort involved in the development of the class, and also the functionality of the class. A paper that estimates the effort and time required to complete a programming project using the object-oriented uses the number of methods as a parameter. Number of Methods (NM) of a class is a metric that we define to denote the complexity of the interface of the class.

3. *Depth of Inheritance Hierarchy (DI):* The base classes of a class themselves may be derived from other classes, which may be subclasses of other classes and so on. The inheritance structure of classes can be represented as a directed acyclic graph. The nodes of the graph represent the classes. There is a directed edge from each class to its super class node. The Depth of Inheritance (DI) of a class is the length of the longest path in the graph that originates at the node representing the graph. The DI is an indication of the dependence of the class on the class hierarchy.

4. *Lines of Code per Class (SLOC):* The number of lines of code for the methods is an indication of the size of the class interface. This is a measure of the programming effort involved in implementing the class. This measure in conjunction with the number of methods can be used to compare the complexity of implementation of two or more classes. We use SLOC to indicate the number of source lines of code for the methods of a class.

5. *Number of Subclasses (NSC):* The amount of reuse of a class is determined by the number of derived classes the class has. The class can have subclasses if the class is designed to be generic. A class that is very specialized cannot have subclasses. However, in most

applications, many classes are so specialized that they cannot have subclasses. The number of sub-classes usually decreases as the depth of inheritance increases for classes.

6. *Coupling Between Classes (CBC):* The classes in an object-oriented design will have relations between other classes. Such relationships between pairs of classes are captured in the design. Two classes are coupled when methods of one class use the methods or instance variables of the other. CBC for a class is the number of other classes it is coupled with.

7. *Lack of Cohesion of Methods (LCOM):* This metric indicates the extent to which methods of a class form a cohesive set in the sense they use common instance variables. Consider two methods of a class M_i and M_j. If the two methods share one or more instance variables then we say the methods are cohesive, else they lack cohesion. The count LCOM for a class is the difference between the number, P of pairs of methods that are cohesive and the number, Q of pairs of methods that are not cohesive. If Q is greater than P, then the count LCOM is set to zero. The underlying idea here is that if there is a lack of cohesion of methods, the class can be split into two or more classes that have the same functionality as the original class.

Let the number of classes in the program be N.

Definition 1: The complexity, C_0 of an object-oriented program is defined by:

$$C_O = \ln \sum_{i=1}^{N} (NM_i - DI_i - SLOC_i - CBC_i - LCOM_i)$$

Definition 2: The complexity factor λ_c is defined as:

$$\lambda_c = k \left(1 - e^{-\frac{C_O}{\ln(N_{LOC})}} \right)$$

where N_{LOC} is the number of lines of code in the program. The constant k can be adjusted in computation of the complexity of different programs. In practice, we choose $k = 1$.

In the definition of the complexity factor λ_c, all factors that influence the complexity of a program have been considered. Some metrics such as the number of subclasses and the number of encapsulated and non-encapsulated variables could also have been considered but were left out for reasons of simplicity. As we will see, lower values of λ_c are preferred to higher values.

In calculating the complexity, C_0, we can first compute the terms, $\sum_{i=1}^{N} NM_i$, $\sum_{i=1}^{N} DI_i$, $\sum_{i=1}^{N} SLOC_i$, $\sum_{i=1}^{N} CBC_i$, and $\sum_{i=1}^{N} LCOM_i$ which represent the summation of metrics over all classes for a given program. We can next compute C_0, and λ_c easily using definitions 1 and 2.

2.2 Measuring Reliability of Object-Oriented Software

Once the complexity factor, λ_c of the object-oriented software has been found, we need to modify equation (1) for software reliability so as to accommodate λ_c. We propose the following modification:

$$R^*(t) = e^{-(\lambda t + \lambda_c)} \qquad \qquad \dots (2)$$

Thus the effect of the complexity factor λ_c is to reduce the reliability by multiplying it with a factor $e^{-\lambda_c}$.

2.3 Testing Object-Oriented Software

There are several considerations during the testing of object-oriented software that need to be kept in perspective. For example, encapsulation with inheritance which intuitively ought to reduce testing problems, compounds them instead. Multiple inheritance is recognized as both a blessing and a curse. If testing is not done taking into consideration such concerns, then there will likely be an adverse impact on the reliability of the software. This naturally leads us to the concept of test effectiveness in the context of object-oriented software. Test effectiveness can be represented by the symbol λ_e, which can take values from 0 to 1. $\lambda_e = 0$ represents the case when testing is not done and $\lambda_e = 1$ represents the case when testing is fully effective. The reliability equation (2) can now be written, taking into account the testing effectiveness:

$$R^{**}(t) = e^{-(\lambda t + (1 - \lambda_e)\lambda_c)} \qquad \qquad \dots (3)$$

when $\lambda_e = 0$, the impact of software complexity on reliability is fully felt, and when $\lambda_e = 1$, there is no impact of software complexity on reliability and the original equation (1) holds. There are two approaches to calculate the value of λ_e. One approach is to use the software fault injection technique and calculate the test effectiveness λ_e as the ratio of injected faults found to the total number of injected faults for the software under test. Another approach is to calculate λ_e using historical data from similar projects.

3. EXPERIMENTAL RESULTS

To apply the results from the previous section, we considered four software projects in the healthcare domain. These projects range from hospital billing applications to complex claims processing. Table 1 below summarizes the impact of software complexity.

Next, we list the failure rate λ and the test effectiveness λ_e and calculate the reliability $R(t)$ at t = 200, and the reliability $R^{**}(t)$ at t= 200, considering the software complexity factor and test effectiveness.

The results are shown in Table 2.

Table 1: Calculation of C_0 and λ_C

	Project A	Project B	Project C	Project D
Classes (N)	75	356	747	1031
$N_{LOC} = \sum_{i=1}^{N} SLOC_i$	20140	105730	198680	151830
$\sum_{i=1}^{N} NM_i$	675	2492	11952	12372
$\sum_{i=1}^{N} DI_i$	150	712	2241	3093
$\sum_{i=1}^{N} CBC_i$	225	1424	11205	22682
$\sum_{i=1}^{N} LCOM_i$	38	107	299	516
Complexity, C_0	9.96305279	11.61245	12.32108	12.15737
λ_C	0.63406753	0.633511	0.63577	0.639049

Table 2: Calculation of $R(t)$ and $R^{\sim\sim}(t)$

	Project A	Project B	Project C	Project D
λ	0.001531	0.000513	0.001358	0.000545
λ_e	0.644269	0.82659	0.741335	0.72973
λ_C	0.63406753	0.633511	0.63577	0.639049
$R(t)$	0.736246	0.90252	0.762133	0.896753
$R^{\sim\sim}(t)$	0.587576	0.808624	0.646562	0.754507

With the knowledge of the original or unadjusted reliability $R(t)$, the failure rate λ, the complexity factor λ_C, and the test effectiveness λ_e, we can calculate the new reliability $R^{\sim\sim}(t)$ using equation (3), which takes into account the complexity of object-oriented software and test effectiveness. Consider, for example, the software represented by Project B which has a failure rate of 0.000513 as per the pre-release testing data. According to the original definition, the reliability at t = 200 is 0.90252. However, with adjustments made to incorporate the effects of software complexity $\lambda_C = 0.633511$, and testing effectiveness $\lambda_e = 0.82659$, the new reliability is 0.808624. This shows that we need to consider such adjustments for more accuracy.

The new reliability figures give a better and more realistic picture of the software. They could also be used to improve test effectiveness.

4. CONCLUSIONS AND FUTURE WORK

Reliability models for software should consider the complexity introduced by the object-oriented programming paradigm which is commonly used in the industry. Also, there is a need to take into consideration the test effectiveness, since it is a well-known fact in the industry that software testing cannot ensure a total absence of defects. These and other such considerations that are peculiar to software will result in more accurate reliability models. They could also help the software designer to identify the right design and test engineers to improve the test effectiveness that could result in a more reliable software system.

REFERENCES

[1] Hudli, R.V., Hoskins, C.L. and Hudli, A.V. (1994), "Software Metrics For Object-oriented Designs" *in Proceedings of IEEE International Conference on Computer Design: VLSI in Computers and Processors, ICCD'94,* Cambridge, MA, pp. 492–495.

[2] Jacobson, I, Christerson, M., Jonsson, P. and Overgaard, G. (1992), *Object-oriented Software Engineering A Use Case Driven Approach,* Addison Wesley, Reading, MA.

[3] Lyu, M. (1996), *Handbook of Software Reliability Engineering*, IEEE CS Press, Los Alamitos, Calif. and McGraw-Hill New York.

[4] Perry, D.E. and Kaiser, G.E. (1990), "Adequate testing and Object-Oriented Programming", *Journal of Object-Oriented Programming*, Jan–Feb. 1990, pp. 13–19.

[5] Rumbaugh, J., Blaha, M., Premerlani, W., Eddy, F. and Lorenson, W.(1991)*, Object-Oriented Modeling And Design*, Prentice Hall, New Jersey.

[6] Whittaker, J.A. and Voas, J. (2000), "Toward a More Reliable Theory of Software Reliability", *IEEE Computer*, Dec. 2000, pp. 36–43.

[7] Xu, P. and Xu S. (2010), "A Reliability Model for Obect-Oriented Software", in *Proceedings of the 19th IEEE Asian Test Symposium,* Shanghai, China, pp. 65–70.

Bayesian Accelerated Life Testing Under Competing Exponential Causes of Failure

Soumya Roy* and Chiranjit Mukhopadhyay

Indian Institute of Science, Bangalore
*E-mail: soumya@mgmt.iisc.ernet.in

ABSTRACT: Consider an Accelerated Life Testing (ALT) experiment for a *J*-component series system with independent exponentials as component lives. The existing literature is first extended by considering multiple stresses instead of a single stress variable. A full Bayesian methodology is then developed by letting the parameters of the exponential component lives depend on the stress variables through a class of stress translation functions, which contains the commonly used stress translation functions in the literature as special cases. The priors on all the model parameters, called the stress coefficients, are assumed to be Normal and independent of each other. The univariate conditional posterior of each parameter given the rest, has been shown to be log-concave. This fact is then used to generate samples from the joint posterior of the model parameters using Gibbs sampling. The samples generated from the joint posterior are then used to obtain the Bayesian point and interval estimates of the system reliability at usage condition.

Keywords: Series System, Stress Translation Function, Gibbs Sampling, Log-Concave.

1. INTRODUCTION

Consider a J-component series system which is put on ALT experiment. A system is called a J-component series system if it fails as soon as one of its *J* components fails. Nelson (1990, Chapter 7) has listed a number of real life examples of such systems. The competing risk model introduced by Moeschberger and David (1971) can be used to model the lifetime of series systems in an accelerated life test. The focus of this study is to make statistical inference about certain characteristics of lifetime distribution of series systems at usage condition based on the observations collected in an accelerated life test.

Klein and Basu (1981, 1982a, 1982b) assume that the component lifetimes have either independent exponential or Weibull distributions with equal or unequal shape parameters. They apply the method of maximum likelihood to estimate model parameters for Type-I, Type-II and progressively censored ALT observations. Standard maximum likelihood asymptotics are used to estimate lifetime characteristics at usage stress. Nelson (1990, Chapter 7) obtains the likelihood function for a *J*-component series system subjected to ALT involving *K* stress variables. The techniques to obtain the standard errors and confidence intervals for the estimators are also discussed under this set up. He also discusses methods to estimate lifetime parameters when certain failure modes are eliminated through redesign.

The literature cited above use maximum likelihood asymptotics to infer about the lifetime characteristics at usage stress. Since ALT is performed on expensive systems, the available sample size is expected to be small. In such situations, Bayesian methods are expected to outperform the method of maximum likelihood, since it does not require any asymptotic result. Furthermore, the main objective of ALT being prediction of the lifetime characteristics at usage stresses, the predictive inference can be made more coherently in the Bayesian set-up.

There are quite a few studies on Bayesian analysis of ALT observations when the system has single failure mode. Pathak *et al.* (1991) provide a Bayesian inference of ALT data having single failure mode under the assumption that the lifetime distributions are exponential. Van Dorp and Mazzuchi (2004, 2005) present Bayesian analysis for accelerated life testing assuming failure times at each stress level are either exponential or Weibull random variables. The Bayesian analysis presented by the authors incorporate variation in the testing procedures, namely regular life testing, fixed-stress testing, step-stress testing, profile stress testing for interval censored as well as Type-I censored data. Tojeiro *et al.* (2004) have modified the power rule stress translation function for the exponential distribution that introduces the concept of a threshold stress. The Bayesian analysis for the proposed model proceeds with proper diffuse priors for the model parameters.

Thus though there is considerable amount of literature available on Bayesian inference of ALT data, the literature on Bayesian analysis of ALT observations for a system having multiple failure modes is limited. Bunea and Mazzuchi (2006) assume that the component life distributions are exponential and present a Bayesian analysis for series systems, with gamma priors on the hazard rates with the hyper-parameters depending on the stress level. They then use the method of least squares to estimate the lifetime characteristics at the usage stress using the posterior expectation of the hazards. Tan *et al.* (2009) consider Bayesian analysis of incomplete ALT observations on systems with multiple failure modes. Xu and Tang (2011) develop a Bayesian analysis of ALT observations for series systems with independent Weibull as component lives, and a general stress translation function with a single stress variable. They develop Jeffreys' and Berger-Bernardo reference prior for the corresponding Bayesian analysis. They use the method of least squares to estimate the life characteristics at usage stress.

In this paper, the Bayesian analysis for ALT observations is carried out with the assumption that the component lifetimes are exponential random variables. Though the basic framework of the study is similar to that of Bunea and Mazzuchi (2006), the approach adopted here is quite different as briefly outlined below. The hazards of the exponential distribution are assumed to depend on the stress variable through a class of stress translation functions of which the standard stress translation functions such as Arrhenius, Power Rule, log-linear are special cases. The parameters of the stress translation function are assumed to have independent Normal priors. Resulting conditional posterior of each of the individual model parameters given the rest, is shown to be log-concave. This enables the Bayesian analysis to be conducted using Gibbs sampling from the joint posterior. The lifetime characteristics at the usage stress are assessed using Bayesian predictive inference techniques. One clear advantage of this method of prediction using the full Bayesian approach over the approach of Bunea and Mazzuchi (2006) or for that matter the frequentist approach in general for this problem, is that it accommodates the uncertainty in the knowledge of the model parameters through their posterior distribution, resulting in a completely coherent set of predictions at usage stresses. Apart from this, the present study contributes to the literature in another direction. In the existing literature, mostly, the ALT has been assumed to be performed with only one stress variable. However, in reality, there may be more than one stress variable (sometimes one stress variable and a number of engineering variables) present in ALT (Nelson (2004), p. 98)) and here the methodology has been developed for general K stresses.

The rest of the paper is organized as follows. The basic model for the J-component series system has been presented in Section 2. In Section 3, the likelihood function has been obtained, followed by the posterior analysis in Section 4. In Section 5, the developed methodology has been illustrated with the help of numerical examples. Finally, Section 6 concludes the paper.

2. MODEL

Suppose a J-component series system is put on ALT experiment involving K stress variables. Let the non-negative random variable X_j denote the lifetime of the j-th component for $j = 1, ..., J$. It is assumed that X_j's are mutually independent, and X_j has a probability density,

$$f_j(x \mid \lambda_j) = \lambda_j \, e^{-\lambda_j x}$$

with the survival function,

$$\bar{F}_j(x \mid \lambda_j) = e^{-\lambda_j x}$$

and the hazard rate,

$$h_j(x \mid \lambda_j) = \lambda_j .$$

Now, by the definition of a series system, if system lifetime is denoted by the non-negative random variable T, then $T = \min\{X_1, ..., X_J\}$. Let $f_T(t \mid \lambda)$, $\bar{F}_T(t \mid \lambda)$ and $h_T(t \mid \lambda)$ denote the probability density function, survival function and hazard function of T respectively, where $\lambda = (\lambda_1, \cdots, \lambda_J)'$ denotes the vector of lifetime parameters. Let the discrete random variable I taking values in $\{1, ..., J\}$ denote the component causing a system failure. Then, clearly $I = \text{argmin}_j X_j$. One typically observes (T, I) for a series system that has failed after being put in life test. This necessitates derivation of the joint distribution of (T, I), which is given by,

$$\lim_{\Delta t \to 0} \frac{1}{\Delta t} P(t \leq T \leq t + \Delta t, I = j \mid \lambda) = p_j(t \mid \lambda), \text{say}$$

where,

$$p_j(t \mid \lambda) = f_j(t \mid \lambda_j) \prod_{j' \neq j} \bar{F}_{j'}(t \mid \lambda_{j'}) = h_j(t \mid \lambda_j) \prod_{j'} \bar{F}_{j'}(t \mid \lambda_{j'}) = \lambda_j \exp\left(-\sum_{j'=1}^{J} \lambda_{j'} t \right) \qquad (1)$$

On the other hand, a system that is still functioning at time t after being put on life test at time 0, its lifetime T is censored, and the corresponding probability is given by,

$$P(T > t | \lambda) = \prod_{j=1}^{J} \bar{F}_j(t|\lambda) = \exp\left(-\sum_{j=1}^{J} \lambda_j t\right) = \bar{F}_T(t|\lambda).$$

Let $Z = (Z_1, \dots, Z_K)'$ represent the vector of K stress variables acting on a system, where Z_k is k-th stress variable. For $k = 1, \dots, K$, consider an arbitrary transformation $g_k(\cdot)$ of the k-th stress variable Z_k, and denote the transformed stress variable by $S_k = g_k(Z_k)$. By denoting $S = (S_1, \dots, S_K)'$, consider the following general log-linear stress translation function given by,

$$\ln \lambda_j(\boldsymbol{\theta}_j, S) = \sum_{k=1}^{K} \theta_{kj} g_k(Z_k) = \sum_{k=1}^{K} \theta_{kj} S_k = \boldsymbol{\theta}_j' S \qquad (2)$$

where $\boldsymbol{\theta}_j = (\theta_{1j}, \dots, \theta_{Kj})'$, $j = 1, \dots, J$, is the vector of parameters of the stress translation function and λ_j, the hazard rate of the j-th component is now a function of the applied stresses S and the stress translation function parameters $\boldsymbol{\theta}_j$. $\boldsymbol{\theta}_j$'s will henceforth be referred to as stress coefficients. Note that the stress coefficients θ_{kj}, for $k = 1, \dots, K$, $j = 1, \dots, J$, do not depend on applied stresses. The stress translation function is written in the above general form for a couple of reasons.

First of all, note that the stress translation function given in (2) includes the standard stress translation functions as special cases. For example, for $K = 2$ and $Z = (1, Z)'$, by choosing $g_1(\cdot) = 1$, $g_2(z) = \frac{1}{z}$, one gets the Arrhenius model. Similarly, the power rule model and the log-linear model can be obtained by choosing $g_2(z) = \ln z$ and $g_2(z) = z$ respectively, together with $g_1(\cdot) = 1$. Thus for most standard stress translation functions $S_1 \equiv 1$. Also, if the original number of stress variables Z_k's is $K' \leq K$, and one wants to model $\ln \lambda_j$ as some non-linear functions of the Z_k's, like for example if one has just one stress variable Z and wants to model $\ln \lambda_j$ as $\theta_{1j} + \theta_{2j} z + \theta_{3j} z^2$, K and $g_k(\cdot)$'s may be appropriately chosen to accommodate that. However, for the sake of brevity, from now on S will be referred to as the vector of stress variables without any further reference to Z, the original stress variables.

Let $s = (s_1, \dots, s_K)'$ denote the observed value of S. Then given $S = s$, the density function, survival function and hazard rate of X_j is given by,

$$f_j\left(x \mid \lambda_j(\theta_j, s)\right) = \exp\left(\theta_j' s\right) \exp\left[-\left(e^{\theta_j' s}\right) x\right],\tag{3}$$

$$\bar{F}_j\left(x \mid \lambda_j(\theta_j, s)\right) = \exp\left[-\left(e^{\theta_j' s}\right) x\right] \text{ and}\tag{4}$$

$$\lambda_j(\theta_j, s) = e^{\theta_j' s},$$

respectively. Similarly, the density function, survival function and the hazard rate of the system lifetime T at stress $S = s$ is given by,

$$f_T\left(t \mid \lambda(\theta, s)\right) = \left(\sum_{j=1}^{J} \exp\left(\theta_j' s\right)\right) \exp\left(-\sum_{j=1}^{J}\left(e^{\theta_j' s}\right) t\right),\tag{5}$$

$$\bar{F}_T\left(t \mid \lambda(\theta, s)\right) = \exp\left(-\sum_{j=1}^{J}\left(e^{\theta_j' s}\right) t\right) \text{ and}\tag{6}$$

$$\lambda(\theta, s) = \sum_{j=1}^{J} e^{\theta_j' s},$$

respectively, where $\theta = (\theta_1, \ldots, \theta_J)'$ is the complete set of model parameters.

3. OBSERVED DATA AND LIKELIHOOD FUNCTION

Suppose n' systems are put on ALT experiment and it is assumed that they behave independently of each other. Let the number of failures and survivors be denoted by n and m respectively. Then, $n + m = n'$. Further, let the number of systems that have failed due to component j be denoted by n_j. Obviously, $\sum_{j=1}^{J} n_j = n$.

For $= 1, \ldots, n_j, j = 1, \ldots, J$, let t_j^i denote the observed system lifetime and $s_j^i = \left(s_{1j}^i, \ldots, s_{Kj}^i\right)'$ denote the corresponding applied stress to the i-th system which failed due to failure of the j-th component. Let the m censored observations be denoted by $t_+^1, \ldots, t_+^m$ and the corresponding applied stresses by $s_+^1, \ldots, s_+^m$, where $s_+^l = \left(s_{1+}^l, \ldots, s_{K+}^l\right)'$, for $l = 1, \ldots, m$. Thus let $\mathcal{D} = \left\{\left(\left(t_1^1, s_1^1\right), \ldots, \left(t_1^{n_1}, s_1^{n_1}\right)\right), \ldots, \left(\left(t_J^1, s_J^1\right), \ldots, \left(t_J^{n_J}, s_J^{n_J}\right)\right), \left(\left(t_+^1, s_+^1\right), \ldots, \left(t_+^m, s_+^m\right)\right)\right\}$ denote the observed data set. The structure of these observations is listed in Table 1.

Table 1: Life Times and Corresponding Stresses

Component	Observations	O
1	$\left(t_1^1, s_1^1\right), \left(t_1^2, s_1^2\right), \dots, \left(t_1^{n_1}, s_1^{n_1}\right)$	
2	$\left(t_2^1, s_2^1\right), \left(t_2^2, s_2^2\right), \dots, \left(t_2^{n_2}, s_2^{n_2}\right)$	
.	.	
.	.	
.	.	
J	$\left(t_J^1, s_J^1\right), \left(t_J^2, s_J^2\right), \dots, \left(t_J^{n_J}, s_J^{n_J}\right)$	
Right Censored	$\left(t_+^1, s_+^1\right), \left(t_+^2, s_+^2\right), \dots, \left(t_+^m, s_+^m\right)$	

By using (1), (2) and (6), the likelihood function of θ is given by

$$L(\theta|\mathcal{D}) = \left[\prod_{j=1}^{J} \prod_{i=1}^{n_j} \exp\left(\theta_j s_j^i\right) \exp\left(-\sum_{j=1}^{J}\left(e^{\theta_j s_j^i}\right) t_j^i\right) \right]$$
$$\times \left[\prod_{i=1}^{m} \exp\left(-\sum_{j=1}^{J}\left(e^{\theta_j s_+^i}\right) t_+^i\right) \right]. \quad (7)$$

Now, by using standard competing risk algebra it can be shown that,

$$L(\theta|\mathcal{D}) = \prod_{j=1}^{J} L_j\left(\theta_j|\mathcal{D}\right).$$

where,

$$L_j\left(\theta_j|\mathcal{D}\right) = \exp\left(\sum_{i=1}^{n_j} \theta_j s_j^i - \sum_{j=1}^{J} \sum_{i=1}^{n_j} \left(e^{\theta_j s_j^i}\right) t_j^i - \sum_{i=1}^{m} \left(e^{\theta_j s_+^i}\right) t_+^i \right) \quad (8)$$

Note that $L_j\left(\theta_j|\mathcal{D}\right)$ is a function of θ_j alone and does not depend on any other $\theta_{j'}$, for $j' \neq j$.

4. BAYESIAN ANALYSIS

Given data $\mathcal{D}$, the Bayesian analysis proceeds with the posterior distribution of θ denoted by $\pi(\theta|\mathcal{D})$, which is proportional to the product of the likelihood function of θ and prior of θ. Since the likelihood function has already been derived in the last section, this now requires specification of a prior distribution on θ. The next problem is that of marginalization in terms of obtaining the marginal posteriors of each of the scalar θ_{k_j}'s, for $k = 1, \dots, K$ and $j = 1, \dots, J$, for drawing inference on them. Furthermore, the main interest after performing an ALT is to assess the lifetime distribution of both the components and the systems at usage stress $S = s_u = (s_{u1}, ..., s_{uK})'$, say. As is evident from (3), (4), (5) and (6), these distributions depend on the θ_{kj}'s and in the Bayesian paradigm this particular problem is resolved by computing the predictive distribution of $X_j | S = s_u$ and $T | S = s_u$. Since these schemes of things are somewhat elaborate, this section is divided into three subsections. The prior distributions are introduced in Subsection 4.1. The problem of marginalization is solved via Gibbs sampling in Subsection 4.2 and finally the problem of assessing the life distributions at usage stress is addressed via Bayesian predictive inference in Subsection 4.3.

4.1 Prior Distribution

In the absence of any conjugacy, the priors assumed for the stress coefficients θ_{kj}, for $k = 1, ..., K$ and $j = 1, ..., J$, are independent and Normal. The mean and standard deviation of the Normal prior of θ_{kj} will be denoted by $\mu_{\theta_{kj}}$ and $\sigma_{\theta_{kj}}$ respectively, *i.e.*, it is assumed that apriori $\theta_{kj} \sim N\left(\mu_{\theta_{kj}}, \sigma_{\theta_{kj}}^2\right)$, $k = 1, ..., K$ and $j = 1, ..., J$, and they are mutually independent.

A few remarks about the choice of this prior are in order. First, note that the stress-coefficients θ_{kj}'s, which are similar to regression coefficients, are unrestricted real numbers and this calls for their priors having the entire real line as their support and for this, the most natural choice is the Normal distribution. Next, the easiest way to elicit one's subjective uncertainty regarding a parameter's value is through a guess value and one's confidence on this guess value. Supplying the guess value as mean and confidence through the standard deviation of a Normal distribution is the most obvious choice of expressing one's thus elicited subjective belief as a prior distribution. One's vague prior knowledge of a parameter can also be modeled using a Normal distribution with a very large standard deviation.

Apart from the above mentioned intuitive justifications, there is another solid statistical reason behind the selection of Normal priors. As mentioned by Berger (1985, pp. 90–93), in any Bayesian analysis, the priors used should be as noninformative as possible even in the presence of partial information. One way of finding a noninformative prior is to maximize the entropy of the prior distribution subject to a set of restrictions derived from the partial information. As it is well known that given the first two moments, it is the Normal distribution that maximizes the entropy for distributions with real line as their support, and here since the θ_{kj}'s are unrestricted to take any value in the real line, and one way of eliciting prior information is through a guess value and one's confidence in it in terms of the mean and standard deviation, Normal distributions appear to be the most optimal choice as priors for the θ_{kj}'s from the point of view of maximizing entropy.

4.2 Posterior Draws

Based on the likelihood function given in (7), and independent normal priors on the KJ model parameters θ_{kj}, for $k = 1, \ldots, K$ and $j = 1, \ldots, J$, one has to find the posterior distribution of θ_{kj}. Since the joint posterior $\pi(\boldsymbol{\theta}|\mathcal{D})$ is proportional to the product of the likelihood function $L(\boldsymbol{\theta}|\mathcal{D})$, which equals $\prod_{j=1}^{J} L_j(\boldsymbol{\theta}_j|\mathcal{D})$, and the product of KJ Normal densities with parameters $\left(\mu_{\theta_{kj}}, \sigma_{\theta_{kj}}^2\right)$, $k = 1, \ldots, K$ and $j = 1, \ldots, J$, *i.e.*, it is given by,

$$\pi(\boldsymbol{\theta}|\mathcal{D}) \propto L(\boldsymbol{\theta}|\mathcal{D}) \prod_{j=1}^{J} \prod_{k=1}^{K} \phi\left(\frac{\theta_{kj} - \mu_{\theta_{kj}}}{\sigma_{\theta_{kj}}}\right)$$

$$= \prod_{j=1}^{J} \left[L_j(\boldsymbol{\theta}_j|\mathcal{D}) \prod_{k=1}^{K} \phi\left(\frac{\theta_{kj} - \mu_{\theta_{kj}}}{\sigma_{\theta_{kj}}}\right) \right] \propto \prod_{j=1}^{J} \pi_j(\boldsymbol{\theta}_j|\mathcal{D}). \quad (9)$$

where $\phi(\cdot)$ denotes the density of standard Normal distribution and $\pi_j(\boldsymbol{\theta}_j|\mathcal{D})$ denotes the joint posterior of $\boldsymbol{\theta}_j$ given data $\mathcal{D}$ and is given by,

$$\pi_j(\boldsymbol{\theta}_j|\mathcal{D}) \propto L_j(\boldsymbol{\theta}_j|\mathcal{D}) \prod_{k=1}^{K} \phi\left(\frac{\theta_{kj} - \mu_{\theta_{kj}}}{\sigma_{\theta_{kj}}}\right). \quad (10)$$

Thus the joint posterior of $\boldsymbol{\theta}$ equals the product of the posteriors of $\boldsymbol{\theta}_j$ implying aposteriori for $j \neq j'$, $\boldsymbol{\theta}_j$ is independent of $\boldsymbol{\theta}_{j'}$, hence the marginal distribution of θ_{kj}, $k = 1, \ldots, K$, does not depend on any $\boldsymbol{\theta}_{j'}$, $j' \neq j$. That is by virtue of (9), the posterior analysis for $\boldsymbol{\theta}$ may be carried out parallely and independently for $\boldsymbol{\theta}_1, \ldots, \boldsymbol{\theta}_J$, which can be generically carried out for an arbitrary $\boldsymbol{\theta}_j$, for $j \in \{1, \ldots, J\}$, in terms of analyzing $\pi_j(\boldsymbol{\theta}_j|\mathcal{D})$ given in (10) above.

From (10) and the expression of $L_j(\boldsymbol{\theta}_j|\mathcal{D})$, given in (8), it is evident that trying to analytically obtain the marginal posteriors of θ_{kj} from $\pi_j(\boldsymbol{\theta}_j|\mathcal{D})$ is difficult, if not impossible. However, this problem can be computationally solved if one can obtain a sample from $\pi_j(\boldsymbol{\theta}_j|\mathcal{D})$. Towards this end one can employ the ubiquitous Gibbs sampling technique provided one can easily draw samples from the univariate conditional posteriors of each component of $\boldsymbol{\theta}_j$ given the rest. These univariate densities have the same functional form as that of $\pi_j(\boldsymbol{\theta}_j|\mathcal{D})$, given in (10), viewed as a function (of one variable) of only the component of interest and treating the remaining components as fixed. Again by looking at equation (8), these univariate conditional posteriors do not resemble any standard probability distribution yielding an immediate way of sampling from them.

Let the conditional posterior of each θ_{kj}, given the remaining $(K-1)$ parameters $N_{kj} = \Omega_j \setminus \{\theta_{kj}\}$, be denoted by $\pi(\theta_{kj}|N_{kj}, \mathcal{D})$, where $\Omega_j = \{\theta_{1j}, ..., \theta_{Kj}\}$, denotes the set of all parameters in θ_j and $N_{kj} = \{\theta_{1j}, ..., \theta_{k-1j}, \theta_{k+1j}, ..., \theta_{Kj}\}$ denote the set of all parameters of θ_j, excluding θ_{kj}. Gibbs sampling involves sample generation from each of the these K univariate $\pi(\theta_{kj}|N_{kj}, \mathcal{D})$'s. It has been shown in Appendix that all these $\pi(\theta_{kj}|N_{kj}, \mathcal{D})$'s are log-concave. This fact now facilitates drawing samples from the $\pi(\theta_{kj}|N_{kj}, \mathcal{D})$ using rejection method. Samples from the $\pi_j(\theta_j|\mathcal{D})$ are drawn using Gibbs sampling as follows. Start with an initial value of $\left(\theta_{1j}^{(0)}, ..., \theta_{Kj}^{(0)}\right)$ and denote the value of the parameter vector $\left(\theta_{1j}, ..., \theta_{Kj}\right)$ at the v-th step of iteration by $\left(\theta_{1j}^{(v)}, ..., \theta_{Kj}^{(v)}\right)$. The sampling procedure describes below how $\left(\theta_{1j}^{(v+1)}, ..., \theta_{Kj}^{(v+1)}\right)$ is obtained given the values $\left(\theta_{1j}^{(v)}, ..., \theta_{Kj}^{(v)}\right)$.

- Draw an observation $\theta_{1j}^{(v+1)}$ from the conditional posterior $\pi(\theta_{1j}|N_{1j}, \mathcal{D})$ with $\theta_{2j} = \theta_{2j}^{(v)}, ..., \theta_{Kj} = \theta_{Kj}^{(v)}$, using the rejection method.

- Draw an observation $\theta_{2j}^{(v+1)}$ from the conditional posterior $\pi(\theta_{2j}|N_{2j}, \mathcal{D})$ with $\theta_{1j} = \theta_{1j}^{(v+1)}, \theta_{3j} = \theta_{3j}^{(v)}, ..., \theta_{Kj} = \theta_{Kj}^{(v)}$, using the rejection method.

  .
  .
  .

- Draw an observation $\theta_{Kj}^{(v+1)}$ from the conditional posterior $\pi(\theta_{Kj}|N_{Kj}, \mathcal{D})$ with $\theta_{1j} = \theta_{1j}^{(v+1)}, ..., \theta_{K-1j} = \theta_{K-1j}^{(v+1)}$, using the rejection method.

The above steps complete one step in the transition of the Markov chain wherein one transits to $\left(\theta_{1j}^{(v+1)}, ..., \theta_{Kj}^{(v+1)}\right)$ from $\left(\theta_{1j}^{(v)}, ..., \theta_{Kj}^{(v)}\right)$. Now if the chain is run for a sufficiently long time, in the limit, it will produce an observation from the target joint posterior, $\pi_j(\theta_j|\mathcal{D})$, which is the stationary distribution of the above MC. Now further observations are drawn from $\pi(\theta_{1j}, ..., \theta_{Kj})$ till the required sample size N is achieved as described below.

There are essentially two ways by which using Gibbs sampling one can obtain a sample of size N from the joint posterior $\pi_j(\theta_j|\mathcal{D})$. The first option is to run N parallel chains and eventually collect the end products of each chain as a sample of size N from $\pi_j(\theta_j|\mathcal{D})$. The second option is to run one long chain and suitably sample from the chain in order to obtain N observations on $\pi_j(\theta_j|\mathcal{D})$. Here the second approach, which is computationally cheaper has been adopted. In this single chain approach, a few initial observations (say, b) are discarded as burn-in, a proper thinning interval (say, h) is decided and then a long chain of length L is run where $L = b + (N-1)h + 1$. Denote the generated chain by
$$\left\{\theta_j^{(1)},...,\theta_j^{(b)},\theta_j^{(b+1)},...,\theta_j^{(b+h+1)},...,\theta_j^{(b+2h+1)},...,\theta_j^{(b+(N-1)h+1)}\right\}.$$ Then the final samples of size N from the joint posterior of θ_j is given by
$$\left\{\theta_j^{(b+1)},\theta_j^{(b+h+1)},...,\theta_j^{(b+(N-1)h+1)}\right\}.$$ Thus determination of b and h becomes critical issue in the single chain approach.

As the Gibbs iterative procedure eventually produces a sample from the target density in the limit, convergence of the Gibbs sampler is an important issue and the values of b and h are determined based on the convergence diagnostics mentioned below. The review of the convergence diagnostic tools by Cowles and Carlin (1996) recommends the tests suggested by Raftery and Lewis (1992), Geweke (1992) and Heidelberger and Welch (1983) for ensuring the convergence of the Markov chain. To examine the run length of the chain, Raftery and Lewis diagnostics and Heidelberger and Welch convergence diagnostic are used. Raftery and Lewis diagnostics applied on a short pilot run of a Markov chain provides an assessment of the number of burn-in iterations b to be discarded at the beginning of the chain and the thinning interval h. In addition, the diagnostics for autocorrelation is also taken into consideration while conforming the above thinning interval. Finally Heidelberger and Welch convergence diagnostics and Geweke's convergence diagnostic tests are used to confirm the convergence of the Markov chain. N observations from $\pi_j(\theta_j|\mathcal{D})$, may now be drawn from a Markov chain that satisfies the above mentioned convergence criteria. Now this process of generating N observations from $\pi_j(\theta_j|\mathcal{D})$ are repeated for all $j = 1,...,J$, yielding a sample of size N from $\pi(\theta|D)$. Thus based on this sample of size N from the joint posterior $\pi(\theta|D)$ of the model parameters θ, one can now plot the joint or marginal density estimates of the parameters, obtain point estimates using the mean values (Bayes' estimate under Squared Error Loss), or obtain interval estimates in the form of Highest Posterior Density Credible Sets (HPDCS).

4.3 Prediction

The main purpose of performing an ALT is to make inferences on the lifetime distribution of the components as well as that of the system at the usage stress from the observations collected at the accelerated conditions. To be precise, one is interested in the density of X_j at $S = s_u$, given by $f_j\left(t|\lambda_j(\theta_j, s_u)\right)$, defined in (3), the survival function of X_j at $S = s_u$, given by $\bar{F}_j\left(t|\lambda_j(\theta_j, s_u)\right)$, defined in (4), the density of T at $S = s_u$, given by $f_T(t|\lambda(\theta, s_u))$ defined in (5), and the survival function of T at $S = s_u$ given by $\bar{F}_T\left(t|\lambda(\theta, s_u)\right)$, defined in (6).

Note that the above quantities of interest at $S = s_u$, are functions of either t and θ_j, or, t and θ. However, for the sake of brevity, let us denote these functions as $v(t, \theta)$, where $v(\cdot)$ is either $f_j(\cdot)$, $f_T(\cdot)$, $\bar{F}_j(\cdot)$ or $\bar{F}_T(\cdot)$. Note that $t \in [0, \infty)$ and θ is random. Hence $v(\cdot)$ is a continuous time stochastic process. The probability characterization of the stochastic process $v(\cdot)$ over a prefixed time interval of interest $[0, \tau)$, where τ is sufficiently large, can be done if one can find out joint posterior distribution of $(v(\tau_0), \ldots, v(\tau_P))$, where $\{0 = \tau_0, \ldots, \tau_P = \tau\}$ constitutes a grid of equispaced points in the time interval of interest $[0, \tau)$. Hence one needs to further discuss how to obtain the joint posterior of $(v(\tau_0), \ldots, v(\tau_P))$, given $\{\theta^{(l)}\}$, $l = 1, \ldots, N$, a sample of size N, from the joint posterior $\pi(\theta|\mathcal{D})$. Each of these quantities is computed for $t \in \{0 = \tau_0, \ldots, \tau_P = \tau\}$ for each of the generated $\{\theta^{(l)}\}$, $l = 1, \ldots, N$. These yields a sample from the joint posterior density of $(v(\tau_0), \ldots, v(\tau_P))$ denoted by $\left\{\left(v^{(1)}(\tau_0), \ldots, v^{(1)}(\tau_P)\right), \ldots, \left(v^{(N)}(\tau_0), \ldots, v^{(N)}(\tau_P)\right)\right\}$, where $v(\cdot)$ is either $f_j(\cdot)$, $f_T(\cdot)$, $\bar{F}_j(\cdot)$ or $\bar{F}_T(\cdot)$. These N observations drawn from the joint posterior of $(v(\tau_0), \ldots, v(\tau_P))$ may be used to obtain the Bayes' estimate and 95% HPDCS of the respective quantities. For $p = 1, \ldots, P$, the Bayes' Estimate at $S = s_u$ of $v(\tau_p)$ is given by $\frac{1}{N}\sum_{l=1}^{N} v^{(l)}(\tau_P)$. Using $\left(v^{(1)}(\tau_p), \ldots, v^{(N)}(\tau_p)\right)$, one can obtain a density estimate of the posterior distribution of $v(\tau_p)$ at $S = s_u$, based on which one can compute the HPDCS.

5. NUMERICAL EXAMPLE

In this section, the proposed Bayesian methodology is illustrated with the help of numerical examples. The basic framework for the simulation study has been adopted from Klein and Basu (1982a). A two component series system is subjected to an ALT having one stress variable with three stress levels. The stress variable has three stress levels 10, 15 and 20, with the usage stress 5. A power rule stress translation function has been used to relate the hazard of the j-th component ($j = 1, 2$) to stress variable, with the stress coefficients given by $\theta_{11} = -1$, $\theta_{21} = 1$, $\theta_{12} = -3$ and $\theta_{22} = 2$. The censoring mechanism is assumed to be random, with failure probability being 0.6, 0.8 and 0.9 at three stress levels respectively. It is also assumed that the censoring random variable has an exponential distribution. Then, for generating an observation (either a failure or censored), one generates three exponential random variates and the minimum of these three is considered as an observation. A total of 30 observations is generated, with 10 observations at each stress level. Under the above general set up, Bayesian analysis is presented for the following two schemes:

- **Scheme 1:** Uniform Priors
- **Scheme 2:** Normal Priors with Mean equal to true parameter values, Variance = 25.

Table 2: Posterior Summary Measures for Scheme 1

Summary Statistics	θ_{11}	θ_{21}	θ_{12}	θ_{22}
Mean	−4.016	2.236	−2.357	1.550
Median	−3.906	2.213	−2.227	1.517
Mode	−3.858	2.259	−1.827	1.379
SD	3.126	1.137	3.273	1.206
95% HPDCS	(−9.794,2.489)	(−0.046,4.429)	(−8.923,3.945)	(−0.779,3.978)

A sample of size 5000 (*i.e.*, $N = 5000$) is generated from the joint posterior density $\pi(\theta|\mathcal{D})$ of θ. This sample is then utilized to draw posterior inference on each θ_{kj} ($k = 1, 2$, $j = 1, 2$). The posterior summary measures are presented in Table 2 and Table 3 for Scheme 1 and Scheme 2 respectively. It is observed that even in the absence of any subjective prior knowledge about the priors, the Bayesian point estimates are fairly close to the true value of the parameters. While the 95% HPDCS are fairly wide for uniform priors, they are narrower for informative priors as expected. The marginal densities of the parameters are shown in Figure 1. The marginal posterior densities for each parameter are unimodal and appear to be symmetric in both the cases. Notice that the spread of marginal posterior densities for all the parameters is fairly narrow for informative priors.

Table 3: Posterior Summary Measures for Scheme 2

Summary Statistics	θ_{11}	θ_{21}	θ_{12}	θ_{22}
Mean	−3.021	1.878	−2.526	1.616
Median	−2.948	1.865	−2.481	1.624
Mode	−3.021	1.684	−2.132	1.616
SD	2.526	0.925	2.169	0.967
95% HPDCS	(−8.090,1.777)	(−0.008,3.598)	(−7.678,2.565)	(−0.200,3.577)

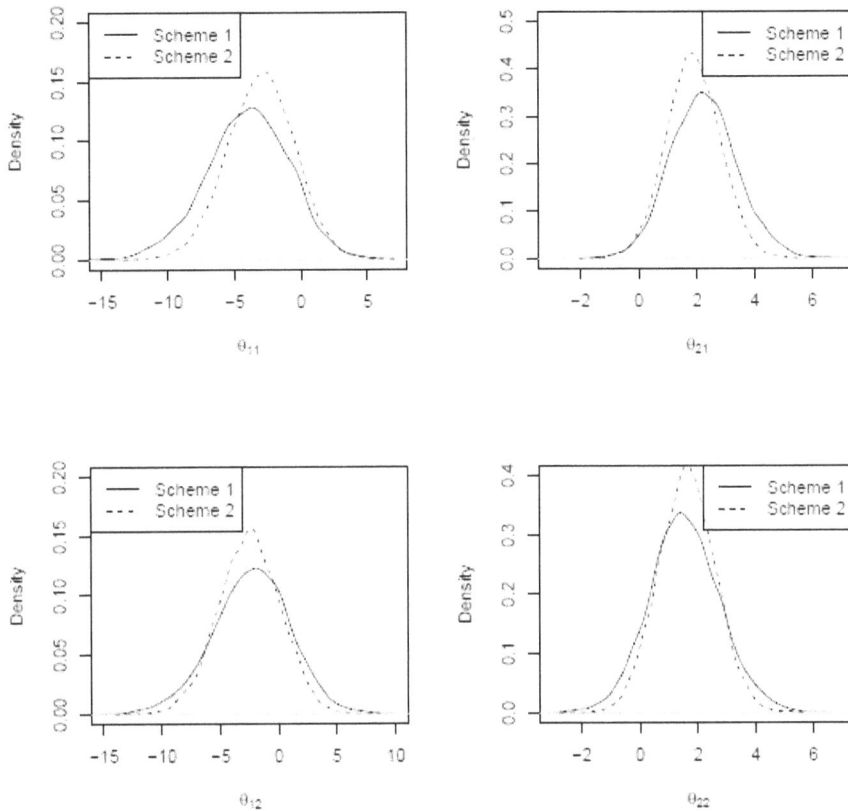

Fig. 1: Density Plots of the Parameters

To obtain the joint posterior of for $\left(v(\tau_0), \ldots, v(\tau_p) \right)$, for $t \in \{0 = \tau_0, \ldots, \tau_p = \tau\}$, at usage stress, a time interval $[0, 2]$ has been considered and divided into 200 equal parts, where $v(\cdot)$ is either $f_j(\cdot)$, $f_T(\cdot)$, $\bar{F}_j(\cdot)$ or $\bar{F}_T(\cdot)$. Based on the $N = 5000$ observations from the joint posterior density of θ, a sample of size $N = 5000$ has been obtained from the joint posterior $\left(v(\tau_0), \ldots, v(\tau_p) \right)$. Based on the N = 5000 observations from $v(\tau_p)$, for a fixed p, the posterior mean of $v(\tau_p)$ has been computed. This is the Bayes' Estimate under squared error loss. Next for each fixed p, the posterior density of $v(\tau_p)$ is estimated based on the sample of size 5000. This density estimate is then used to compute the 95% HPDCS of $v(\tau_p)$, just as it was done for each θ_{kj}. Since all these densities are unimodal, the HPDCS simply boils down to a lower limit (LL) and a upper limit (UL) of $v(\tau_p)$. These two limits (LL and UL) along with the Bayes' estimate of $v(\tau_p)$, are then plotted against τ_p, for $p = 0, \ldots, 201$.

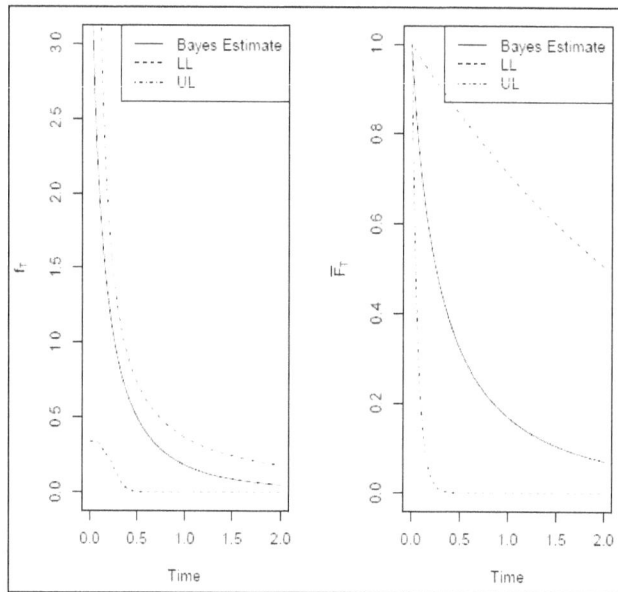

Fig. 2: Bayes' Estimate and HPDCS of T for Scheme 1

As evident from Figures 2 and 3, the HPDCS are fairly wide for the system predictive survival functions, while they are narrower for the system predictive densities. However, the HPDCS are narrower for Scheme 2, justifying that the use of informative priors can improve the prediction when the sample size is small. The predictive survival functions and the predictive density functions for both the components have been obtained similarly, and the figures, which are omitted to save space, are similar to Figures 2 and 3.

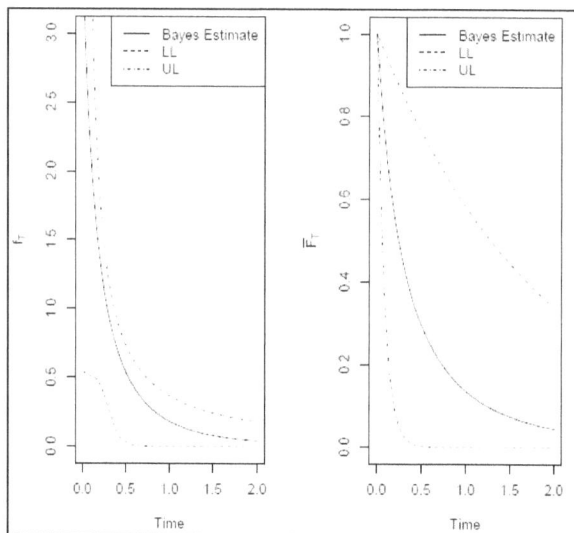

Fig. 3: Bayes' Estimate and HPDCS of T for Scheme 2

6. CONCLUSION

In this paper, a Bayesian analysis has been presented for series systems in ALT involving multiple stresses, assuming the component lifetimes are exponential. A general stress translation function has been introduced, which includes the standard stress translation functions as special cases. Bayesian analysis is presented using both non-informative and informative priors for small sample sizes. Gibbs sampling is used for generating samples from the joint posterior. The samples generated from the joint posterior density of the model parameters are used to obtain the predictive densities at the usage stress. The samples generated can also be used to estimate as well as make certain probability statements for the component and system reliability at the usage stress. The proposed Bayesian analysis can be performed by assuming that the component lifetimes follow lognormal or Weibull distribution.

REFERENCES

[1] Bunea, C. and Mazzuchi, T.A. (2006), "Competing failure modes in accelerated life testing", *Journal of Statistical Planning and Inference*, Vol. 136, pp. 1608–1620.

[2] Cowles, M.K. and Carlin, B.P. (1996). "Markov Chain Monte Carlo Convergence Diagnostics: A Comparative Review", Journal of the American Statistical Association, Vol. 91, pp. 883–903.

[3] Geweke, J. (1992), "Evaluating the Accuracy of Sampling-Based Approaches to the Calculation of Posterior Moments", Bayesian Statistics, J.M. Bernardo, J.O. Berger, A.P. Dawid and A.F.M. Smith (Eds.), Oxford University Press, pp. 169–193.

[4] Heidelberger, P. and Welch, P.D. (1983), "Simulation Run Length Control in the Presence of an Initial Transient", Operations Research, Vol. 31, pp. 1109–1144.

[5] Klein, J.P. and Basu, A.P. (1981), "Weibull accelerated life tests when there are competing causes of failure", *Communications in Statistics—Theory and Methods,* Vol. 10, pp. 2073–2100.

[6] Klein, J.P. and Basu, A.P. (1982a), "Accelerated Life Testing Under Competing Exponential Failure Distributions", IAPQR Transactions, Vol. 7, pp. 1–20.

[7] Klein, J.P. and Basu A.P. (1982b), "Accelerated life tests under competing Weibull causes of failure", Communications in Statistics—Theory and Methods, Vol. 11, pp. 2271–2286.

[8] Moeschberger, M.L. and David, H.A. (1971), "Life Tests under Competing Causes of Failure and the Theory of Competing Risks", Biometrics, Vol. 27, pp. 909–933.

[9] Nelson, W. (1990), *Accelerated testing: Statistical Models, Test Plans and Data Analysis*, Wiley, New York.

[10] Pathak, P.K., Singh, A.K. and Zimmer, W.J. (1991), "Bayes estimation of hazard and acceleration in accelerated testing", IEEE Transactions on Reliability, Vol. 40, pp. 615–621.

[11] Raftery, A.E. and Lewis, S. (1992), "How many iterations in the Gibbs Sampler?", Bayesian Statistics, J.M. Bernardo, J.O. Berger, A.P. Dawid and A.F.M. Smith (Eds.), Oxford University Press, pp. 763–773.

[12] Tan, Y., Zhang, C. and Chen, X. (2009), "Bayesian Analysis of Incomplete Data from Accelerated Life Testing with Competing Failure Modes" in 8th International Conference on Reliability, Maintainability and Safety, pp. 1268–1272.

[13] Tojeiro, C.A.V., Louzada-Neto, F. and Bolfarine, H. (2004), "A Bayesian Analysis for Accelerated Lifetime Tests Under an Exponential Power Law Model with Threshold Stress", Journal of Applied Statistics, Vol. 31, pp. 685–691.

[14] Van Dorp, J.R. and Mazzuchi, T.A. (2004), "A General Bayes Exponential Inference Model for Accelerated Life Testing", Journal of Statistical Planning and Inference, Vol. 119, pp. 55–74.

[15] Van Dorp, J.R. and Mazzuchi, T.A. (2005), "A General Bayes Weibull Inference Model for Accelerated Life Testing", Reliability Engineering and System Safety, Vol. 90, pp. 140–147.

[16] Xu, A. and Tang Y. (2011), "Objective Bayesian analysis of accelerated competing failure models under Type-I censoring", *Computational Statistics and Data Analysis*, Vol. 55, pp. 2830–2839.

APPENDIX

In this Appendix, we shall show that the univariate conditional posterior density of any model parameter, given the rest of the model parameters, is log-concave. Without loss of generality, we shall show that the conditional posterior of θ_{11}, given $\theta_{k1} \neq \theta_{11}$, is log-concave.

The conditional posterior of θ_{11}, given $\theta_{k1} \neq \theta_{11}$, $k = 2,3,\ldots,K$, is given by

$$\pi(\theta_{11}|N_{11},D) \propto L(\theta_{11}|D) \times \pi(\theta_{11})$$

where $\pi(\theta_{11})$ is prior density of θ_{11}, which has been assumed to be Normal in this case and $L(\theta_{11}|D)$ is same as $L_1(\theta_1|D)$ in the equation (8), when $L_1(\theta_1|D)$ is viewed as a function of θ_{11} only.

Since the Normal density is known to be log-concave, it is sufficient to show that the likelihood function $L(\theta_{11}|D)$ is log-concave function of θ_{11} respectively. Equivalently, we can show that $\log L(\theta_{11}|D)$ is concave function of θ_{11}.

By using (8), we have the log-likelihood function,

$$l(\boldsymbol{\theta}) = \log L(\boldsymbol{\theta}) = \sum_{j=1}^{J} l_j(\boldsymbol{\theta}_j),$$

where,

$$l_j(\boldsymbol{\theta}_j) = \log L_j(\boldsymbol{\theta}_j)$$

$$= \sum_{i=1}^{n_j} \boldsymbol{\theta}_j' s_j^i - \sum_{j=1}^{J} \sum_{i=1}^{n_j'} \left(e^{\theta_j' s_j^i}\right) t_j^i - \sum_{i=1}^{m} \left(e^{\theta_j' s_+^i}\right) t_+^i.$$

Note that $l_j(\theta_j)$ is a function of θ_j only and is independent of $\theta_{j'}$, $j' \neq j$. So, in order to prove that $\log L(\theta)$ is a concave function of θ_{11}, it is enough to concentrate on only $l_1(\theta_1)$. Now, we have,

$$l_1(\theta_1) = \sum_{i=1}^{n_1} \theta_1' s_1^i - \sum_{j=1}^{J} \sum_{i=1}^{n_j'} \left(e^{\theta_1' s_j^i} \right) t_j^i - \sum_{i=1}^{m} \left(e^{\theta_1' s_+^i} \right) t_+^i.$$

Hence,

$$\frac{\delta l_1(\theta_1)}{\delta \theta_{11}} = \sum_{i=1}^{n_1} s_{11}^i - \sum_{j=1}^{J} \sum_{i=1}^{n_j'} \left(e^{\theta_1' s_j^i} \right) s_{j,1}^i t_j^i - \sum_{i=1}^{m} \left(e^{\theta_1' s_+^i} \right) s_{1+}^i t_+^i.$$

which implies,

$$\frac{\delta^2 l_1(\theta_1)}{\delta \theta_{11}^2} = -\sum_{j=1}^{J} \sum_{i=1}^{n_j'} \left(e^{\theta_1' s_j^i} \right) \left(s_{j,1}^i \right)^2 t_j^i - \sum_{i=1}^{m} \left(e^{\theta_1' s_+^i} \right) \left(s_{1+}^i \right)^2 t_+^i < 0.$$

Thus since that the second derivative of $l_1(\theta_1)$ with respect to θ_{11} is negative, $\pi(\theta_{11}|N_{11}, D)$ is log-concave.

Reliability in Medical Devices: An Experience

Saraswathi Deora

Philips Electronics India Limited, Bangalore
E-mail: Saraswathi.Deora@philips.com

ABSTRACT: Reliability is one of the key parameters that need to be addressed for a high customer satisfaction. Reliability is defined as probability of a device performing its purpose adequately for the period of time intended under stated operating conditions. In a regulated industry like medical devices, it is more important to ensure high degree of Quality and Reliability as they touch human lives. Standards like IEC 60601–1, 3rd edition also mentions that "Reliability of functioning is regarded as a safety aspect (for life-supporting) ME equipment and where interruption of an examination or treatment is considered as a hazard for the patient". In case of medical devices reliability becomes synonymous with safety and effectiveness of the device. Due to increased complexity of medical devices with advancements in technology it becomes all the more important to design for reliability right from the conception phase. Using time proven analytical techniques like Early Design FMEA, identifying and monitoring essential requirements, objective evaluation of design alternatives, design consideration checks etc., would help improving system reliability by finding potentially hidden failures in the design during early stages of development.

Keywords: Design for Reliability, FMEA, Critical to Quality Parameters, Design Consideration Checks.

1. INTRODUCTION

Philips Innovation Campus (PIC) Bangalore is part of Philips Electronics India Limited which is owned by Royal Philips Electronics N.V., The Netherlands. HealthCare sector in PIC is involved in developing multiple medical devices under different Business Units (BU). Nuclear Medicine (NM) BU is involved in developing NM Application suites, designed to streamline NM workflows to meet productivity demands of NM imaging departments. The development involved teams from United States and Bangalore. One of the key business goals was to ensure product quality and reliability.

2. APPROACH

Typically "Design for Reliability" approach will begin with a "Reliability Goal" in terms of MTBF or MBTSE and so on. In the case of medical devices safety and effectiveness of the device is of at most importance as they deal with human lives. Hence, Design for Reliability should focus on "Design for Safety and Effectiveness" in Medical devices.

2.1 Identifying Essential Requirements

Detailed requirement analysis was done to understand the essential requirements for the product. Essential requirements are identified as safety related or satisfies risk mitigations where the native risk was classified as unacceptable. This is the primary design input to ensure safety and effectiveness of the product. Identifying essential performance is a requirement in IEC60601–1, 3rd edition.

2.2 History Review

One of the important steps that we followed in our projects for ensuring product reliability is doing the "History Review" as shown in figure 1. This is a step where we analyze the defects of legacy product, study any device recalls from similar medical devices as well as previous corrective and preventive actions (CAPA). This provides design inputs to the device. This is also one of the expectations from Quality System Regulations (QSR part 820).

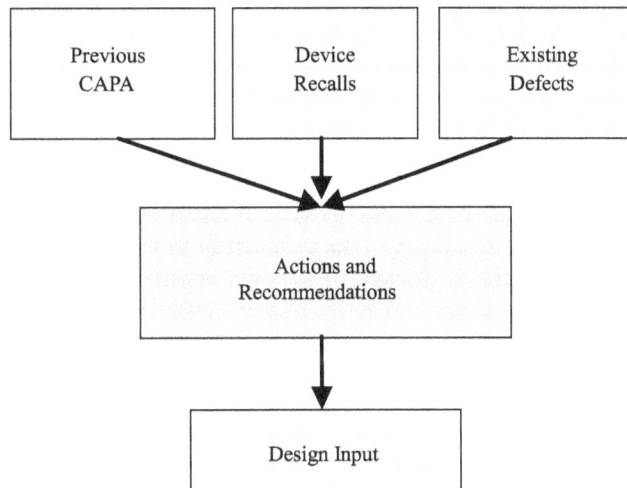

Fig. 1: History Review

The device recalls information can be downloaded from FDA website.

2.3 Safety Risk Assessments (SRA)

Risk management planning, analysis, evaluation and control followed by effectiveness checks and residual risk evaluations are the essential steps mandated by standards like ISO14971. Major steps involved in risk analysis are:

- Identifying the Hazards
- Identifying the resulting hazardous situations
- Estimating the risks resulting from the hazardous situations.

It is extremely important to have a good understanding of the terms used in safety risk assessment as shown in Table 1.

Table 1: Terms Used in SRA

Hazard	Potential source of harm
Harm	Physical injury or damage to the health of people or damage to property or the environment
Hazardous situation	Circumstance in which user or environment is exposed to hazards
Risk	Product of probability of occurrence and the severity of the harm

The following are a few best practices that we followed during safety risk assessment:

- Use of a Cross-Functional team.
- Identifying both normal and fault conditions.
- Use of a broad range of sources of input, including data from FDA medical device report database, output of history review, etc.
- Begin at the earliest stage of the project and refine them throughout life cycle of the product.
- Use of analytical tools like FMEA (Failure Mode and Effects Analysis) and FTA (Failure Tree Analysis) while doing hazard analysis.
- Evaluation of third party elements and their impact on safety and effectiveness of the device.

2.4 Early Design FMEA

An Early Design FMEA (EDFMEA) was conducted to ensure that the design is robust and safe. The output of safety risk assessment was considered as an input to early design FMEA as shown in figure 2. EDFMEA is an analysis tool used to anticipate and plan actions to resolve potential failure modes as soon as a design concept has begun to be developed. They help in influencing the design before it has been implemented.

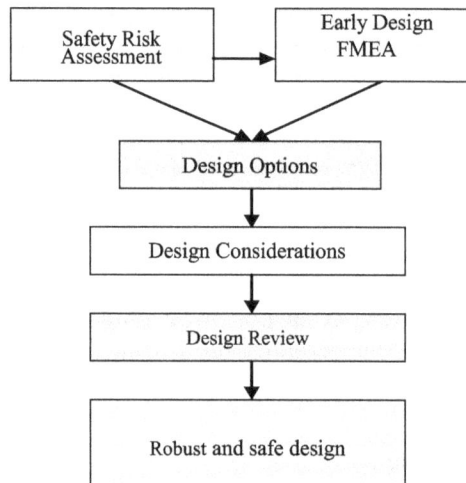

Fig. 2: Design Approach

2.5 Design Consideration Checks

Various parameters need to be considered for evaluation of design. Some of the essential parameters that we considered during the design phase are as follows:

- Intended use or foreseeable misuse ➢ Mitigation of key hazards
- Extremes of humidity, voltage, temperature, etc.
- Designed to meet expected life.
- Design correctly addresses resource management (memory, files, etc.).
- Design ensures compliance with applicable regulatory standards.
- Design makes correct use of architecture facilities/infrastructures.
- The Design is consistent with overall architecture.

2.6 Design Alternatives

In order to ensure that the best design option is considered, we used objective evaluation of design alternatives with the help of Pugh matrix as shown in Figure 3.

	Pugh Matrix - Design approach			
No.	Evaluation Criteria	Significance Rating (1 to 5)	Relative Rating (1 to 5)	
			Design Option 1	Design Option 2
1	Reliability	3	3	3
2	Maintainability	2	3	3
3	Safety Asepct	4	4	3
4	Complexity	3	2	1
5	Interoperability	2	4	3
6	Testability	2	3	2
Net weighted score			**51**	**40**
Stakeholder(s)		Engineering		
Recommendation		Design Option-1		
Rationale for the Decision		Option 1 does not carry significant technical risk and also has advantages in multiple parameters		

Fig. 3: Pugh Matrix-Design Approach

2.7 Design Reviews

Design reviews are systematic assessment of design considerations and selection and results. Focus is on controlling and monitoring design decisions and ensuring optimum quality and reliability of the device. In order to get maximum benefits it is highly recommended to involve cross functional teams during formal design reviews.

2.8 Defect Analysis

Defects found during reviews (review comments), and early integration testing, module integration testing, etc., were analyzed for effectiveness of defect prevention actions. Additional actions were triggered to improve effectiveness.

2.9 Effectiveness Checks

Verification testing included review of all design elements to ensure implementation and effectiveness. Effectiveness testing is the focused testing by the clinical workflow team on the elements of SRA. This is done to verify the effectiveness of the Risk Mitigation.

3. RESULTS AND BENEFITS

By following the "Design for Reliability" approach as explained in the paper, we were able to make significant improvement in the Quality and Reliability of the product. We were able to significantly improve product quality and reduce post release defect density as well as rework effort as shown in Figure 4. Intangible benefits also included high customer satisfaction which was evident after the release of the device. Benefits also included a cost savings of more than 153k Euros to the Business Unit.

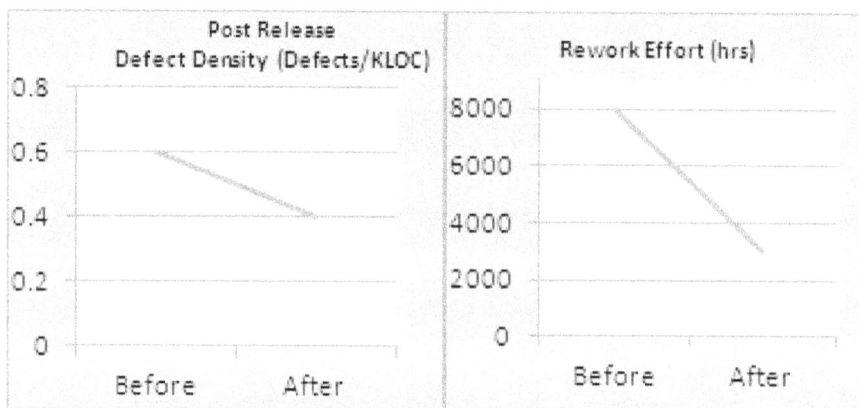

Fig. 4: DfR Approach Results

These practices help in meeting expectations from various regulations (example, ISO14971, QSR CFR part 820 from FDA, IEC 60601-1, etc.).

ACKNOWLEDGEMENTS

The author would like to acknowledge the support provided by PIC Quality Management and Nuclear Medicine Business Unit and NM Team in PIC for the active participation and deployment of DfR approach.

REFERENCES

[1] C.A. Ericson- Hazard Analysis Techniques for System Safety; 2005. Wiley-Interscience.

[2] FDA 21 CFR QSR Part 820 - Food and Drug Administration, Code of Federal Regulations Title 21, Part 820 Quality System Regulation-www.fda.gov.

[3] IEC 60601-1 3rd edition- International Standard applies to the Basic Safety and Essential Performance of Medical Electrical Equipment and Medical Electrical Systems.

[4] ISO 14971:2007- Application of risk management to medical devices.

[5] Poncelin, G.; Derain, J.-P.; Cauvin, A.; Dufrene, D- Development of a design-for-reliability method for complex systems.

Stochastic Analysis of a Complex System with Inspection in Different Weather Conditions

Beena Nailwal[1] and S.B. Singh[2]

G.B. Pant University of Agriculture and Technology, Pantnagar, India
E-mail: [1]bn4jan@gmail.com; [2]drsurajbsingh@yahoo.com

ABSTRACT: This paper investigates reliability analysis of two unit cold standby system under two weather conditions (normal and abnormal). The considered system consists of two units: main and standby. The units are non-identical and each has two modes—normal and total failure. The main unit is k-to-l-out-of-n: F while standby unit is k-out-of-n: G configurations. A single repairman is available with the system who plays the dual role of inspection and repair. Repairman inspects the unit at its failure to see the feasibility of repair. If repair of the unit is not feasible, it is replaced by new one. Inspection activities are stopped in abnormal weather but system remains operative. Failures rates of both units and the rate of change of weather conditions are constant, whereas the repair rates are general. With the help of Supplementary variable technique, Laplace transformations and copula methodology different reliability characteristics and numerical results for special cases are evaluated.

Keywords: Reliability, M.T.T.F., k-to-l-out-of-n: F system, k-out of-n: G system, Gumbel-Hougaard Copula, Supplementary Variable Technique.

1. INTRODUCTION

Recently k-out-of-N: G system has become more important due to its demand in various systems. It is widely used in both industrial and military systems. For example, the multiengine systems in an airplane, the power generation system, the multipump system in a hydraulic control system, data processing system, communication system etc. Many authors (Pham and Upadhyaya, 1988; Shao and Lamberson, 1991) have studied k-out-of-N: G system in different configurations. (Heidtmann, 1981) introduced k-l-out-of-N system which is called non-coherent system in which no fewer than k and no more than 1 out of n units are required to function for the successful operation of a system. The system fails if less than k or more than l out of n units function simultaneously. This type of system is found in many areas, viz. multiprocessor, communication, transportation systems etc. Due to wide applicability of this system, several authors (Jain and Ghimire, 1997; Upadhyaya and Pham, 1993) motivated to study k-Z-out-of-N: G system. In the present study we have focused on k-to-l-out-of-n: F system.

Further the power system such as transformer, generator etc. are operated under physical environment. The weather environment has a great impact on the system performance. The stress on the system due to environment changes is much higher in the bad weather conditions. The weather conditions that are responsible for highly increasing failure rate of components are generally infrequent and have short duration. During these periods the failure rates of system components increase and the probability of overlapping failures is greater than the probability of overlapping failures in normal weather. The phenomenon of component failure due to abnormal weather is known as bunching effect (Cha *et al.*, 2005). The weather condition can be subdivided into two categories: normal and abnormal weather. In abnormal weather the failure rate of a system component increases and reduces the operational ability of the system. Many accidents occur in the abnormal weather conditions due to highly increasing failure rate of system components. In most of the real life situations inspection is carried out on the spot and chances of occurrence of accidents exist which is higher when it is carried out in abnormal weather condition because the system becomes more sensitive due to abnormal weather. So the inspection activities should be stopped in the abnormal weather conditions to reduce the accidents. The study of weather effect on the system performance has drawn the interest of various researchers in the field of reliability theory. (Goel, 1985) used the idea of two weather conditions (normal and abnormal) in a single unit system. (Gupta and Goel, 1991) analyzed two-unit cold standby system with different weather conditions.

Keeping above facts and realistic situations in view, here a reliability model for a system having two units: main and standby with inspection under different weather conditions is developed. Each unit of the system has two modes—good and failed. The main unit is k-to-l–out-of- *n*: F (in which fewer than k and more than l out of n units are required to function for the successful operation of a system) and standby unit is k-out-of-*n*: G. There is a single repairman available for repair who attends the system immediately when required and plays the dual role of inspection and repair. Repairman inspects the unit at its failure to see the feasibility of repair. If repair of unit is not feasible it is replaced by new one so that unnecessary expenses on repair may be avoided and if the repair is feasible then it sends for repair. Inspection activities are stopped in abnormal weather but the system remains operative. Further, in real life problem we can have situations in which a system can be repaired in two different ways. This feature is also incorporated in this study. At states S_{10}, S_{17}, S_{25}, S_{36} and S_{44} both the units are completely failed and being repaired with two different repair rates. We have used Gumbel-Hougaard family of copula (Nelson, 2006; Sen, 2003) to find joint distribution of repair rate from state S_{10} to S_0, S_{17} to S_0, S_{25} to S_0, S_{36} to S_0 and S_{44} to S_0. Failure rates of units and rates of change of weather conditions are assumed to be constants in general whereas the repairs follow general distribution in all the cases. By using Supplementary variable technique, Laplace transformation and copula following characteristics of the system have been analyzed:

1. Transition state probabilities of the system.
2. Asymptotic behaviour of the system.
3. Various reliability measures such as reliability, availability, M.T.T.F., cost effectiveness and sensitivity analysis of the system.

Some numerical examples have been used to illustrate the model mathematically. The transition state diagram of the system is shown in Figure 1.

2. SYSTEM DESCRIPTION AND ASSUMPTIONS

- The considered system consists of two units namely main and standby unit. The main unit is k-to-l-out-of-n: F and standby unit is k-out-of-n: G configurations.
- Both units are non-identical. Initially main unit is operating and the standby unit is kept as cold standby.
- The system fails if neither less than k nor more than l units function simultaneously *i.e.* for the successful operation fewer than k and more than l out of n units are required to function properly.
- There is a single repairman who visits the system immediately whenever required and plays the dual role of inspection and repair.
- Repairman inspects the unit at its failure to see the feasibility of repair. If repair is not feasible, it is replaced by new one.
- Inspection activities are stopped in abnormal weather but the system remains operative.

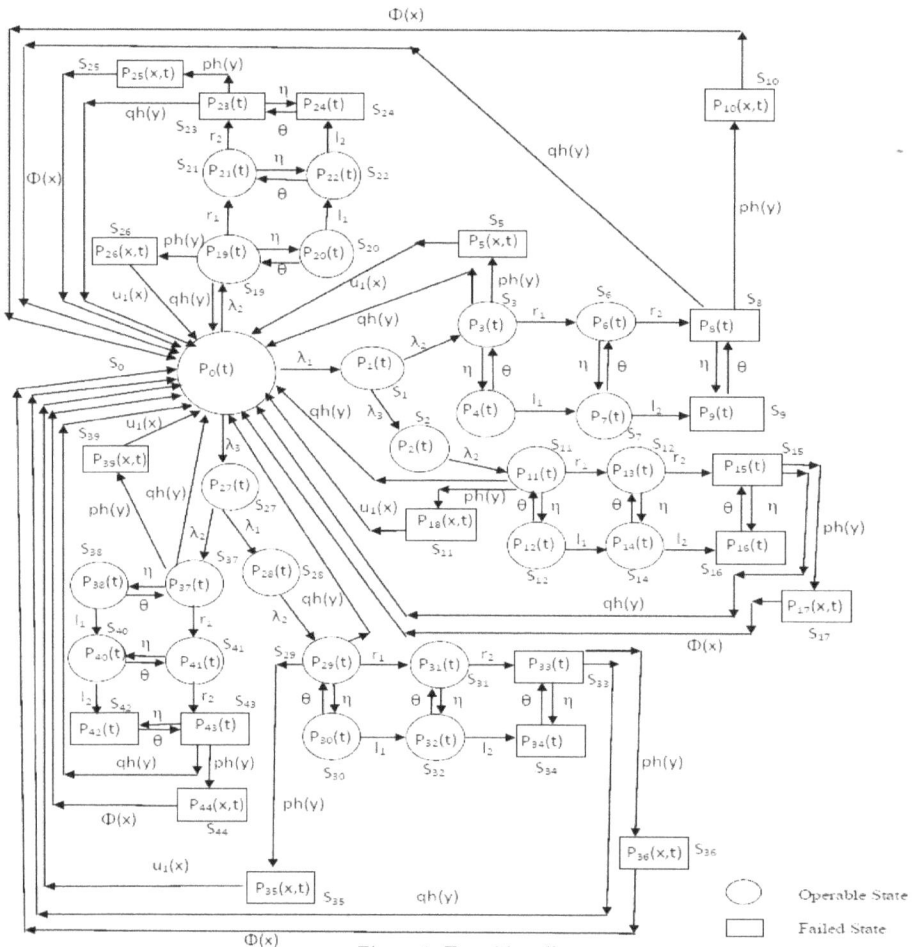

Figure 1. Transition diagram

- The rates of change of weather are constant, η from normal to abnormal and θ from abnormal to normal.
- Each unit has two modes—normal and total failure.
- The repair of the failed unit is perfect. After repair each unit is as good as new.
- Transition from the completely failed state S_{10} to S_0, S_{17} to S_0, S_{25} to S_0, S_{36} to S_0 and S_{44} to S_0 can occur in two ways and their joint probability distribution of repair rate is computed by Gumbel-Hougaard family of copula.

3. STATE SPECIFICATION

Table 1: State Specification

States	S_1	S_2	S_3	S_4	S_5	S_6	S_7	S_8	S_9	S_{10}	S_{11}	S_{12}	S_{13}	S_{14}	S_{15}
Main unit	O	O	f_{ui}	f_{wiA}	f_r	f_{wi}	f_{wiA}	f_{ui}	f_{wiA}	f_r	f_{ui}	f_{wiA}	f_{wi}	f_{wiA}	f_{ui}
Standby unit	S	S	O	O_A	S	O	O_A	f_{ui}	f_{wiA}	f_r	O	O_A	O	O_A	f_{ui}
System state	G	G	G	G_A	F_r	G	G_A	F_{ui}	F_{wiA}	F_r	G	G_A	G	G_A	F_{ui}

States	S_{17}	S_{18}	S_{19}	S_{20}	S_{21}	S_{22}	S_{23}	S_{24}	S_{25}	S_{26}	S_{27}	S_{28}	S_{29}	S_{30}	S_{31}
Main unit	f_r	f_r	f_{ui}	f_{wiA}	f_{wi}	f_{wiA}	f_{ui}	f_{wiA}	f_r	f_r	O	O	f_{ui}	f_{wiA}	f_{wi}
Standby unit	f_r	S	O	O_A	O	O_A	f_{ui}	f_{wiA}	f_r	S	S	S	O	O_A	O
System state	F_r	F_r	G	G_A	G	G_A	F_{ui}	F_{wiA}	F_r	F_r	G	G	G	G_A	G

States	S_{32}	S_{33}	S_{34}	S_{35}	S_{36}	S_{37}	S_{38}	S_{39}	S_{40}	S_{41}	S_{42}	S_{43}	S_{44}
Main unit	f_{wiA}	f_{ui}	f_{wiA}	f_r	f_r	f_{ui}	f_{wiA}	f_r	f_{wiA}	f_{wi}	f_{wiA}	f_{ui}	f_r
Standby unit	O_A	f_{ui}	f_{wiA}	S	f_r	O	O_A	S	O_A	O	f_{wiA}	f_{ui}	f_r
System state	G_A	F_{ui}	F_{wiA}	F_r	F_r	G	G_A	F_r	G_A	G	F_{wiA}	F_{ui}	F_r

G: Good state, F_r: Failed system under repair, f_r: Failed unit under repair, O: Operating, O_A: Operating under abnormal weather, f_{ui}: Failed unit under inspection, f_{wi}: Failed unit waiting for inspection, f_{wiA}: Failed unit waiting for inspection due to abnormal weather, F_{ui}: Failed system under inspection, F_{wiA}: Failed system waiting for inspection due to abnormal weather: G_A: Good under abnormal weather

4. NOTATIONS

$\lambda_1 / \lambda_2 / \lambda_3$:	Failure rates of main unit, where
	$$\lambda_1 = \sum_{i=1}^{k-1} \lambda_i, \ \lambda_2 = \sum_{j=k}^{l} \lambda_j \text{ and } \lambda_3 = \sum_{l+1}^{n} \lambda_k$$
r_1 / r_2 :	Failure rates of standby unit in normal weather where $r_1 = \beta_i$ ($i = 1, 2, .. $ k–1) and $r_2 = \beta_j$ ($j = k+1..n$)
l_1 / l_2 :	Failures rates of standby unit in abnormal weather
η :	constant rate of change from normal to stormy weather
θ :	constant rate of change from stormy to normal weather
$u_1(x)$:	Repair rate of main unit
$u_2(x)$:	Repair rate of standby unit
$\phi(x)$:	Repair rate of standby unit
x:	Elapsed repair time
$ph(y)/qh(y)$:	Rate of change of failed unit under inspection to see the feasibility of repair / replacement where y is elapsed inspection time
$P_i(t)$:	Probability that the system is in S_i state at instant t for $i = 1$ to $i = 44$
$\overline{P}_i(s)$:	Laplace transform of P_i (t)
$P_i(x, t)$:	Probability density function that at time t the system is in failed state
	S_i and the system is under repair, elapsed repair time is x
$E_p(t)$:	Expected profit during the interval (0, t]
K_1, K_2:	Revenue per unit time and service cost per unit time respectively
$\overline{S}_n(x)$:	Laplace transform of $S_n(x) = \int_0^\infty \eta(x) \exp(-sx - \int_0^x \eta(x)dx)dx$
If $v_1 = u_1(x)$, $v_2 = u_2(x)$ then the expression for the joint probability according to Gumbel-	
Hougaard family of copula is given by	
$\phi(x) = \exp[-\{(-\log u_1(x))^{\theta_1} + (-\log u_2(x))^{\theta_1}\}^{1/\theta_1}]$	

5. FORMULATION OF MATHEMATICAL MODEL

By elementary probability and continuity arguments, one can obtain the following set of integro-differential equations for the considered system

$$\left[\frac{d}{dt} + \lambda_1 + \lambda_2 + \lambda_3\right] P_0(t) = qh(y)P_{19}(t)$$

$$+ qh(y)P_{23}(t) + qh(y)P_3(t) + qh(y)P_8(t) + qh(y)P_{11}(t)$$

$$+ qh(y)P_{15}(t) + qh(y)P_{29}(t) + qh(y)P_{33}(t) + qh(y)P_{37}(t)$$

$$+ qh(y)P_{43}(t) + \int_0^\infty u_1(x)P_{26}(x,t)dx + \int_0^\infty \phi(x)P_{25}(x,t)dx$$

$$+ \int_0^\infty u_1(x)P_5(x,t)dx + \int_0^\infty \phi(x)P_{10}(x,t)dx + \int_0^\infty u_1(x)P_{18}(x,t)dx$$

$$+ \int_0^\infty \phi(x)P_{17}(x,t)dx + \int_0^\infty u_1(x)P_{35}(x,t)dx + \int_0^\infty \phi(x)P_{36}(x,t)dx +$$

$$+ \int_0^\infty u_1(x)P_{39}(x,t)dx + \int_0^\infty \phi(x)P_{44}(x,t)dx \qquad \ldots(1)$$

$$\left[\frac{d}{dt} + \lambda_2 + \lambda_3\right] P_1(t) = \lambda_1 P_0(t) \qquad \ldots(2)$$

$$\left[\frac{d}{dt} + \lambda_2\right] P_2(t) = \lambda_3 P_1(t) \qquad \ldots(3)$$

$$\left[\frac{d}{dt} + qh(y) + ph(y) + r_1 + \eta\right] P_3(t) = \lambda_2 P_1(t) + \theta P_4(t) \qquad \ldots(4)$$

$$\left[\frac{d}{dt} + \theta + l_1\right] P_4(t) = \eta P_3(t) \qquad \ldots(5)$$

$$\left[\frac{\partial}{\partial t} + \frac{\partial}{\partial x} + u_1(x)\right] P_5(x,t) = 0 \qquad \ldots(6)$$

$$\left[\frac{d}{dt} + r_2 + \eta\right] P_6(t) = r_1 P_3(t) + \theta P_7(t) \qquad \ldots(7)$$

$$\left[\frac{d}{dt} + l_2 + \theta\right] P_7(t) = l_1 P_4(t) + \eta P_6(t) \qquad \ldots(8)$$

$$\left[\frac{d}{dt} + qh(y) + ph(y) + \eta\right] P_8(t) = r_2 P_6(t) + \theta P_9(t) \qquad \ldots(9)$$

$$\left[\frac{d}{dt} + \theta\right] P_9(t) = \eta P_8(t) + l_2 P_7(t) \qquad \ldots(10)$$

$$\left[\frac{\partial}{\partial t}+\frac{\partial}{\partial t}+\phi(x)\right]P_{10}(x,t)=0 \qquad\qquad\qquad\text{...(11)}$$

$$\left[\frac{d}{dt}+qh(y)+ph(y)+r_1+\eta\right]P_{11}(t)=\theta\,P_{12}(t)+\lambda_2 P_2(t) \qquad\qquad\text{...(12)}$$

$$\left[\frac{d}{dt}+\theta+l_1\right]P_{12}(t)=\eta P_{11}(t) \qquad\qquad\qquad\text{...(13)}$$

$$\left[\frac{d}{dt}+r_2+\eta\right]P_{13}(t)=r_1 P_{11}(t)+\theta\,P_{14}(t) \qquad\qquad\text{...(14)}$$

$$\left[\frac{d}{dt}+\theta+l_2\right]P_{14}(t)=l_1 P_{12}(t)+\eta P_{13}(t) \qquad\qquad\text{...(15)}$$

$$\left[\frac{d}{dt}+qh(y)+ph(y)+\eta\right]P_{15}(t)=r_2 P_{13}(t)+\theta\,P_{16}(t) \qquad\qquad\text{...(16)}$$

$$\left[\frac{d}{dt}+\theta\right]P_{16}(t)=\eta P_{15}(t)+l_2 P_{14}(t) \qquad\qquad\qquad\text{...(17)}$$

$$\left[\frac{\partial}{\partial t}+\frac{\partial}{\partial x}+\phi(x)\right]P_{17}(x,t)=0 \qquad\qquad\qquad\text{...(18)}$$

$$\left[\frac{\partial}{\partial t}+\frac{\partial}{\partial x}+u_1(x)\right]P_{18}(x,t)=0 \qquad\qquad\qquad\text{...(19)}$$

$$\left[\frac{d}{dt}+qh(y)+ph(y)+r_1+\eta\right]P_{19}(t)=\theta\,P_{20}(t)+\lambda_2 P_0(t) \qquad\qquad\text{...(20)}$$

$$\left[\frac{d}{dt}+l_1+\theta\right]P_{20}(t)=\eta P_{19}(t) \qquad\qquad\qquad\text{...(21)}$$

$$\left[\frac{d}{dt}+\eta+r_2\right]P_{21}(t)=r_1 P_{19}(t)+\theta\,P_{22}(t) \qquad\qquad\text{...(22)}$$

$$\left[\frac{d}{dt}+l_2+\theta\right]P_{22}(t)=l_1 P_{20}(t)+\eta P_{21}(t) \qquad\qquad\text{...(23)}$$

$$\left[\frac{d}{dt}+qh(y)+ph(y)+\eta\right]P_{23}(t)=r_2 P_{21}(t)+\theta\,P_{24}(t) \qquad\qquad\text{...(24)}$$

$$\left[\frac{d}{dt}+\theta\right]P_{24}(t)=\eta P_{23}(t)+l_2 P_{22}(t) \qquad\qquad\qquad\text{...(25)}$$

$$\left[\frac{\partial}{\partial t} + \frac{\partial}{\partial x} + \phi(x) \right] P_{25}(x,t) = 0 \qquad \qquad \ldots(26)$$

$$\left[\frac{\partial}{\partial t} + \frac{\partial}{\partial x} + u_1(x) \right] P_{26}(x,t) = 0 \qquad \qquad \ldots(27)$$

$$\left[\frac{d}{dt} + \lambda_1 + \lambda_2 \right] P_{27}(t) = \lambda_3 P_0(t) \qquad \qquad \ldots(28)$$

$$\left[\frac{d}{dt} + \lambda_2 \right] P_{28}(t) = \lambda_1 P_{27}(t) \qquad \qquad \ldots(29)$$

$$\left[\frac{d}{dt} + qh(y) + ph(y) + r_1 + \eta \right] P_{29}(t) = \lambda_2 P_{28}(t) + \theta P_{30}(t) \qquad \qquad \ldots(30)$$

$$\left[\frac{d}{dt} + \theta + l_1 \right] P_{30}(t) = \eta P_{29}(t) \qquad \qquad \ldots(31)$$

$$\left[\frac{d}{dt} + r_2 + \eta \right] P_{31}(t) = r_1 P_{29}(t) + \theta P_{32}(t) \qquad \qquad \ldots(32)$$

$$\left[\frac{d}{dt} + \theta + l_2 \right] P_{32}(t) = l_1 P_{30}(t) + \eta P_{31}(t) \qquad \qquad \ldots(33)$$

$$\left[\frac{d}{dt} + qh(y) + ph(y) + \eta \right] P_{33}(t) = r_2 P_{31}(t) + \theta P_{34}(t) \qquad \qquad \ldots(34)$$

$$\left[\frac{d}{dt} + \theta \right] P_{34}(t) = \eta P_{33}(t) + l_2 P_{32}(t) \qquad \qquad \ldots(35)$$

$$\left[\frac{\partial}{\partial t} + \frac{\partial}{\partial x} + u_1(x) \right] P_{35}(x,t) = 0 \qquad \qquad \ldots(36)$$

$$\left[\frac{\partial}{\partial t} + \frac{\partial}{\partial x} + \phi(x) \right] P_{36}(x,t) = 0 \qquad \qquad \ldots(37)$$

$$\left[\frac{d}{dt} + qh(y) + ph(y) + \eta + r_1 \right] P_{37}(t) = \lambda_2 P_{27}(t) + \theta P_{38}(t) \qquad \qquad \ldots(38)$$

$$\left[\frac{d}{dt} + \theta + l_1 \right] P_{38}(t) = \eta P_{37}(t) \qquad \qquad \ldots(39)$$

$$\left[\frac{\partial}{\partial t} + \frac{\partial}{\partial x} + u_1(x) \right] P_{39}(x,t) = 0 \qquad \qquad \ldots(40)$$

$$\left[\frac{d}{dt}+\theta+l_2\right]P_{40}(t)=\eta P_{41}(t)+l_1 P_{38}(t) \qquad\qquad\qquad \text{...(41)}$$

$$\left[\frac{d}{dt}+r_2+\eta\right]P_{41}(t)=r_1 P_{37}(t)+\theta P_{40}(t) \qquad\qquad\qquad \text{...(42)}$$

$$\left[\frac{d}{dt}+\theta\right]P_{42}(t)=\eta P_{43}(t)+l_2 P_{40}(t) \qquad\qquad\qquad \text{...(43)}$$

$$\left[\frac{d}{dt}+qh(y)+ph(y)+\eta\right]P_{43}(t)=r_2 P_{41}(t)+\theta P_{42}(t) \qquad\qquad \text{...(44)}$$

$$\left[\frac{\partial}{\partial t}+\frac{\partial}{\partial x}+\phi(x)\right]P_{44}(x,t)=0 \qquad\qquad\qquad \text{...(45)}$$

Boundary conditions:

$$P_5(0,t)=ph(y)P_3(t) \qquad\qquad\qquad\qquad\qquad\qquad\qquad \text{...(46)}$$

$$P_{10}(0,t)=ph(y)P_8(t) \qquad\qquad\qquad\qquad\qquad\qquad\qquad \text{...(47)}$$

$$P_{17}(0,t)=ph(y)P_{15}(t) \qquad\qquad\qquad\qquad\qquad\qquad\qquad \text{...(48)}$$

$$P_{18}(0,t)=ph(y)P_{11}(t) \qquad\qquad\qquad\qquad\qquad\qquad\qquad \text{...(49)}$$

$$P_{25}(0,t)=ph(y)P_{23}(t) \qquad\qquad\qquad\qquad\qquad\qquad\qquad \text{...(50)}$$

$$P_{26}(0,t)=ph(y)P_{19}(t) \qquad\qquad\qquad\qquad\qquad\qquad\qquad \text{...(51)}$$

$$P_{35}(0,t)=ph(y)P_{29}(t) \qquad\qquad\qquad\qquad\qquad\qquad\qquad \text{...(52)}$$

$$P_{36}(0,t)=ph(y)P_{33}(t) \qquad\qquad\qquad\qquad\qquad\qquad\qquad \text{...(53)}$$

$$P_{39}(0,t)=ph(y)P_{37}(t) \qquad\qquad\qquad\qquad\qquad\qquad\qquad \text{...(54)}$$

$$P_{44}(0,t)=ph(y)P_{43}(t) \qquad\qquad\qquad\qquad\qquad\qquad\qquad \text{...(55)}$$

Initial condition:

$P_0(t)=1$ at t = 0 and all other probabilities are zero initially.   ...(56)

6. SOLUTION OF THE MODEL

Taking Laplace transformation of (1) to (56) and on further simplification, one can obtain transition state probabilities of the system as:

$$\overline{P}_0(s)=\frac{1}{A(s)} \qquad\qquad\qquad\qquad\qquad\qquad\qquad\qquad \text{...(57)}$$

$$\overline{P}_1(s) = \frac{\lambda_1}{s + \lambda_2 + \lambda_3} \overline{P}_0(s) \qquad \qquad ...(58)$$

$$\overline{P}_2(s) = \frac{\lambda_3}{s + \lambda_2}.\overline{P}_1(s) \qquad \qquad ...(59)$$

$$\overline{P}_3(s) = \frac{\lambda_2(s + \theta + l_1)}{(s + \theta + l_1)(s + qh(y) + ph(y) + r_1) + \eta(s + l_1)} \overline{P}_1(s) \qquad ...(60)$$

$$\overline{P}_4(s) = \frac{\eta}{(s + \theta + l_1)} \overline{P}_3(s) \qquad \qquad ...(61)$$

$$\overline{P}_5(s) = ph(t)\overline{P}_3(s)\left[\frac{1 - \overline{S}_{u_1}(s)}{s}\right] \qquad \qquad ...(62)$$

$$\overline{P}_6(s) = \frac{r_1(s + \theta + l_2)}{(s + \theta + l_2)(s + r_2) + \eta(s + l_2)} \overline{P}_3(s) + \frac{l_1\theta}{(s + \theta + l_2)(s + r_2) + \eta(s + l_2)} \overline{P}_4(s) \qquad ...(63)$$

$$\overline{P}_7(s) = \frac{\eta}{(s + \theta + l_2)} \overline{P}_6(s) + \frac{l_1}{(s + \theta + l_2)} \overline{P}_4(s) \qquad ...(64)$$

$$\overline{P}_8(s) = \frac{r_2(s + \theta)}{(s + qh(y) + ph(y))(s + \theta) + \eta s} \overline{P}_6(s) + \frac{l_2\theta}{(s + qh(y) + ph(y))(s + \theta) + \eta s} \overline{P}_7(s) \qquad ...(65)$$

$$\overline{P}_9(s) = \frac{\eta}{(s + \theta)} \overline{P}_8(s) + \frac{l_2}{(s + \theta)} \overline{P}_7(s) \qquad ...(66)$$

$$\overline{P}_{10}(s) = ph(t)\overline{P}_8(t)\left[\frac{1 - \overline{S}_\phi(s)}{s}\right] \qquad \qquad ...(67)$$

$$\overline{P}_{11}(s) = \frac{\lambda_2(s + \theta + l_1)}{(s + qh(y) + ph(y) + r_1)(s + \theta + l_1) + \eta(s + l_1)} \overline{P}_2(s) \qquad ...(68)$$

$$\overline{P}_{12}(s) = \frac{\eta}{(s + \theta + l_1)}.\overline{P}_{11}(s) \qquad \qquad ...(69)$$

$$\overline{P}_{13}(s) = \frac{r_1(s + \theta + l_2)}{(s + r_2)(s + \theta + l_2) + \eta(s + l_2)} \overline{P}_{11}(s) + \frac{\theta l_1}{(s + r_2)(s + \theta + l_2) + \eta(s + l_2)} \overline{P}_{12}(s) \qquad ...(70)$$

$$\overline{P}_{14}(s) = \frac{l_1}{(s + \theta + l_2)} \overline{P}_{12}(s) + \frac{\eta}{(s + \theta + l_2)} \overline{P}_{13}(s) \qquad ...(71)$$

$$\overline{P}_{15}(s) = \frac{r_2(s + \theta)}{(s + qh(y) + ph(y))(s + \theta) + \eta s} \overline{P}_{13}(s) + \frac{l_2\theta}{(s + qh(y) + ph(y))(s + \theta) +} \qquad ...(72)$$

$$\overline{P}_{16}(s) = \frac{\eta}{(s+\theta)}\overline{P}_{15}(s) + \frac{l_2}{(s+\theta)}\overline{P}_{14}(s) \qquad \qquad \text{...(73)}$$

$$\overline{P}_{17}(s) = ph(t)\overline{P}_{15}(s)\left[\frac{1-\overline{S}_\phi(s)}{s}\right] \qquad \qquad \text{...(74)}$$

$$\overline{P}_{18}(s) = ph(t)\overline{P}_{11}(s)\left[\frac{1-\overline{S}_{u_1}(s)}{s}\right] \qquad \qquad \text{...(75)}$$

$$\overline{P}_{19}(s) = \frac{\lambda_2(s+l_1+\theta)}{(s+qh(y)+ph(y)+r_1)(s+l_1+\theta)+\eta(s+l_1)}\overline{P}_0(s) \qquad \text{...(76)}$$

$$\overline{P}_{20}(s) = \frac{\eta}{(s+l_1+\theta)}\overline{P}_{19}(s) \qquad \qquad \text{...(77)}$$

$$\overline{P}_{21}(s) = \frac{r_1(s+\theta+l_2)}{(s+\theta+l_2)(s+r_2)+\eta(s+l_2)}\overline{P}_{19}(s) + \frac{l_1\theta}{(s+\theta+l_2)(s+r_2)+\eta(s+l_2)}\overline{P}_{20}(s) \qquad \text{...(78)}$$

$$\overline{P}_{22}(s) = .\frac{l_1}{(s+\theta+l_2)}\overline{P}_{20}(s) + \frac{\eta}{(s+\theta+l_2)}\overline{P}_{21}(s) \qquad \qquad \text{...(79)}$$

$$\overline{P}_{23}(s) = \frac{r_2(s+\theta)}{(s+qh(y)+ph(y))(s+\theta)+\eta s}\overline{P}_{21}(s) + \frac{l_2\theta}{(s+qh(y)+ph(y))(s+\theta)+\eta s}\overline{P}_{22}(s) \qquad \text{...(80)}$$

$$\overline{P}_{24}(s) = \frac{\eta}{(s+\theta)}\overline{P}_{23}(s) + \frac{l_2}{(s+\theta)}\overline{P}_{22}(s) \qquad \qquad \text{...(81)}$$

$$\overline{P}_{25}(s) = ph(t)\overline{P}_{23}(s)\left[\frac{1-\overline{S}_\phi(s)}{s}\right] \qquad \qquad \text{...(82)}$$

$$\overline{P}_{26}(s) = ph(t)\overline{P}_{19}(s)\left[\frac{1-\overline{S}_{u_1}(s)}{s}\right] \qquad \qquad \text{...(83)}$$

$$\overline{P}_{27}(s) = \frac{\lambda_3}{(s+\lambda_1+\lambda_2)}\overline{P}_0(s) \qquad \qquad \text{...(84)}$$

$$\overline{P}_{28}(s) = \frac{\lambda_1}{(s+\lambda_2)}\overline{P}_{27}(s) \qquad \qquad \text{...(85)}$$

$$\overline{P}_{29}(s) = \frac{\lambda_2(s+\theta+l_1)}{(s+qh(y)+ph(y)+r_1)(s+\theta+l_1)+\eta(s+l_1)}\overline{P}_{28}(s) \qquad \text{...(86)}$$

$$\overline{P}_{30}(s) = \frac{\eta}{(s+\theta+l_1)}\overline{P}_{29}(s) \qquad \qquad \dots(87)$$

$$\overline{P}_{31}(s) = \frac{r_1(s+\theta+l_2)}{(s+r_2)(s+\theta+l_2)+\eta(s+l_2)}\overline{P}_{29}(s) + \frac{l_1\theta}{(s+r_2)(s+\theta+l_2)+\eta(s+l_2)}\overline{P}_{30}(s) \qquad \dots(88)$$

$$\overline{P}_{32}(s) = \frac{l_1}{(s+\theta+l_2)}\overline{P}_{30}(s) + \frac{\eta}{(s+\theta+l_2)}\overline{P}_{31}(s) \qquad \qquad \dots(89)$$

$$\overline{P}_{33}(s) = \frac{r_2(s+\theta)}{(s+qh(y)+ph(y))(s+\theta)+\eta s}\overline{P}_{31}(s) + \frac{l_2\theta}{(s+qh(y)+ph(y))(s+\theta)+\eta s}\overline{P}_{32}(s) \qquad \dots(90)$$

$$\overline{P}_{34}(s) = \frac{\eta}{(s+\theta)}\overline{P}_{33}(s) + \frac{l_2}{(s+\theta)}\overline{P}_{32}(s) \qquad \qquad \dots(91)$$

$$\overline{P}_{35}(s) = ph(t)\overline{P}_{29}(s)\left[\frac{1-\overline{S}_{u_1}(s)}{s}\right] \qquad \qquad \dots(92)$$

$$\overline{P}_{36}(s) = ph(t)\overline{P}_{33}(s)\left[\frac{1-\overline{S}_\phi(s)}{s}\right] \qquad \qquad \dots(93)$$

$$\overline{P}_{37}(s) = \frac{\lambda_2(s+\theta+l_1)}{(s+qh(y)+ph(y)+r_1)(s+\theta+l_1)+\eta(s+l_1)}\overline{P}_{27}(s) \qquad \dots(94)$$

$$\overline{P}_{38}(s) = \frac{\eta}{(s+\theta+l_1)}\overline{P}_{37}(s) \qquad \qquad \dots(95)$$

$$\overline{P}_{39}(s) = ph(t)\overline{P}_{37}(s)\left[\frac{1-\overline{S}_{u_1}(s)}{s}\right] \qquad \qquad \dots(96)$$

$$\overline{P}_{40}(s) = \frac{\eta}{(s+\theta+l_2)}\overline{P}_{41}(s) + \frac{l_1}{(s+\theta+l_2)}\overline{P}_{38}(s) \qquad \qquad \dots(97)$$

$$\overline{P}_{41}(s) = \frac{r_1(s+\theta+l_2)}{(s+r_2)(s+\theta+l_2)+\eta(s+l_2)}\overline{P}_{37}(s) + \frac{l_1\theta}{(s+r_2)(s+\theta+l_2)+\eta(s+l_2)}\overline{P}_{38}(s) \qquad \dots(98)$$

$$\overline{P}_{42}(s) = \frac{\eta}{(s+\theta)}.\overline{P}_{43}(s) + \frac{l_2}{(s+\theta)}.\overline{P}_{40}(s) \qquad \qquad \dots(99)$$

$$\overline{P}_{43}(s) = \frac{r_2(s+\theta)}{(s+qh(y)+ph(y))(s+\theta)+\eta s}\overline{P}_{41}(s) + \frac{l_2\theta}{(s+qh(y)+ph(y))(s+\theta)+\eta s}\overline{P}_{40}(s) \qquad \dots(100)$$

$$\overline{P}_{44}(s) = ph(t)\overline{P}_{43}(s)\left[\frac{1-\overline{S}_\phi(s)}{s}\right] \qquad \qquad \dots(101)$$

where

$$A(s) = \left[s + \lambda_1 + \lambda_2 + \lambda_3\right] - qh(y)B(s) - qh(y)C(s)D(s)B(s) - E(s)J(s)B(s) - F(s)$$

$$P(s)J(s)B(s) - K(s)D(s)B(s) - E(s)J(s)B(s) - qh(y)B(s)M(s) - qh(y)$$

$$C(s)D(s)M(s)B(s) - E(s)J(s)B(s)M(s) - F(s)K(s)D(s)M(s)B(s) - E(s)$$

$$J(s)M(s)B(s) + P(s)J(s)B(s)M(s) - qh(t)B(s)L(s)M(s) - qh(t)C(s)D(s)$$

$$B(s)L(s)M(s) - E(s)J(s)L(s)M(s)B(s) - F(s)P(s)J(s) - F(s)P(s)J(s)$$

$$B(s)L(s)M(s) - K(s)D(s)L(s)M(s)B(s) - E(s)J(s)B(s)L(s)M(s) - qh(y)$$

$$B(s)N(s)O(s) - qh(y)C(s)D(s)B(s)B(s)N(s)O(s) - E(s)J(s)N(s)O(s)B(s)$$

$$- F(s)P(s)J(s)B(s)N(s)O(s) - K(s)D(s)B(s)N(s)O(s) - E(s)J(s)N(s)O(s)$$

$$+ P(s)J(s)B(s)O(s) - ph(t)\overline{S}_u(s)B(s) - ph(y)C(s)D(s)B(s) - E(s)J(s)B(s)$$

$$- F(s)P(s)J(s)B(s) - K(s)D(s)B(s) - E(s)B(s)J(s)\overline{S}_\phi(s) - ph(y)B(s)M(s)$$

$$\overline{S}_{u_1}(x) - ph(y)C(s)D(s)B(s)M(s) - E(s)J(s)B(s)M(s) - F(s)K(s)D(s)B(s)$$

$$M(s) - E(s)J(s)M(s)B(s) + P(s)J(s)B(s)M(s)\overline{S}_\phi(s) - ph(y)B(s)L(s)M(s)$$

$$\overline{S}_{u_1}(s) - ph(y)C(s)D(s)B(s)L(s)M(s) + E(s)J(s)L(s)M(s)B(s) - F(s)P(s)$$

$$J(s)B(s)L(s)M(s) - K(s)D(s)B(s)L(s)M(s) - J(s)E(s)B(s)L(s)M(s)\overline{S}_\phi(s)$$

$$- ph(y)B(s)N(s)O(s)\overline{S}_{u_1}(s) - ph(y)C(s)D(s)B(s)N(s)O(s) + E(s)J(s)B(s)$$

$$N(s)O(s) + F(s)P(s)J(s)B(s)N(s)O(s) + K(s)D(s)N(s)O(s)B(s) + E(s)J(s)$$

$$B(s)N(s)O(s)\overline{S}_\phi(s) - ph(y)B(s)O(s)\overline{S}_{u_1}(s) - ph(y)C(s)D(s)B(s)O(s) + E(s)$$

$$J(s)B(s)O(s) + F(s)K(s)D(s)B(s)O(s) + E(s)J(s)O(s)B(s) + P(s)J(s)B(s)$$

$$O(s)\overline{S}_\phi(s) \qquad\qquad\qquad\qquad\qquad\qquad\qquad\qquad\qquad \dots(102)$$

Also up and down state probabilities of the system are given by

$$\overline{P}_{up}(s) = \overline{P}_0(s) + \overline{P}_1(s) + \overline{P}_2(s) + \overline{P}_3(s) + \overline{P}_4(s) + \overline{P}_6(s) + \overline{P}_7(s) + \overline{P}_{11}(s) + \overline{P}_{12}(s) + \overline{P}_{13}(s)$$

$$\overline{P}_{14}(s) + \overline{P}_{19}(s) + \overline{P}_{20}(s) + \overline{P}_{21}(s) + \overline{P}_{22}(s) + \overline{P}_{27}(s) + \overline{P}_{28}(s) + \overline{P}_{29}(s) + \overline{P}_{30}(s) + \overline{P}_{31}(s)$$

$$\overline{P}_{32}(s) + \overline{P}_{37}(s) + \overline{P}_{38}(s) + \overline{P}_{40}(s) + \overline{P}_{41}(s)$$

$$= \overline{P}_0(s) + M(s)\overline{P}_0(s) + L(s)M(s)\overline{P}_0(s) + M(s)\overline{P}_0(s)B(s) + J(s)M(s)\overline{P}_0(s)$$

$$B(s) + D(s)B(s)M(s)\overline{P}_0(s) + E(s)J(s)M(s)\overline{P}_0(s)B(s) + K(s)D(s)B(s) +$$

$$E(s)M(s)\overline{P}_0(s)B(s) + E(s)J(s)M(s)\overline{P}_0(s)B(s) + P(s)J(s)B(s)M(s)\overline{P}_0(s)$$

$$+ B(s)L(s)M(s)\overline{P}_0(s) + J(s)B(s)L(s)M(s)\overline{P}_0(s) + D(s)L(s)B(s)M(s)$$

$$\overline{P}_0(s) + E(s)J(s)L(s)M(s)\overline{P}_0(s)B(s) + P(s)J(s)B(s)L(s)M(s)\overline{P}_0(s) +$$

$$K(s)D(s)L(s)M(s)\overline{P}_0(s)B(s) + E(s)J(s)L(s)B(s)M(s)\overline{P}_0(s) + B(s)\overline{P}_0(s)$$

$$+ J(s)B(s)\overline{P}_0(s) + D(s)B(s)\overline{P}_0(s) + E(s)J(s)B(s)\overline{P}_0(s) + P(s)J(s)B(s)$$

$$\overline{P}_0(s) + K(s)D(s)B(s)\overline{P}_0(s) + E(s)J(s)B(s)\overline{P}_0(s) + O(s)\overline{P}_0(s) + N(s)O(s)$$

$$\overline{P}_0(s) + B(s)N(s)O(s)\overline{P}_0(s) + J(s)N(s)B(s)O(s)\overline{P}_0(s) + D(s)B(s)N(s)$$

$$O(s)\overline{P}_0(s) + E(s)J(s)B(s)N(s)O(s)\overline{P}_0(s) + P(s)J(s)B(s)N(s)O(s)\overline{P}_0(s)$$

$$+ K(s)D(s)B(s)N(s)O(s)\overline{P}_0(s) + E(s)J(s)N(s)O(s)\overline{P}_0(s)B(s) + O(s)$$

$$\overline{P}_0(s)B(s) + J(s)B(s)O(s)\overline{P}_0(s) + D(s)B(s)K(s)O(s)\overline{P}_0(s) + E(s)O(s)$$

$$\overline{P}_0(s) + P(s)J(s)O(s)\overline{P}_0(s)B(s) + D(s)B(s)O(s)\overline{P}_0(s) + E(s)J(s)O(s)$$

$$\overline{P}_0(s)B(s) \qquad\qquad\qquad\qquad\qquad \ldots(103)$$

$$\overline{P}_{down}(s) = \overline{P}_5(s) + \overline{P}_8(s) + \overline{P}_9(s) + \overline{P}_{10}(s) + \overline{P}_{15}(s) + \overline{P}_{16}(s) + \overline{P}_{17}(s) + \overline{P}_{18}(s) + \overline{P}_{23}(s)$$

$$+ \overline{P}_{24}(s) + \overline{P}_{25}(s) + \overline{P}_{26}(s) + \overline{P}_{33}(s) + \overline{P}_{34}(s) + \overline{P}_{35}(s) + \overline{P}_{36}(s) + \overline{P}_{39}(s) +$$

$$\overline{P}_{42}(s) + \overline{P}_{43}(s) + \overline{P}_{44}(s)$$

$$= ph(y)B(s)M(s)\overline{P}_0(s)\left[\frac{1-\overline{S}_{u_1}(s)}{s}\right] + C(s)D(s)B(s)M(s)\overline{P}_0(s) + E(s)J(s)$$

$$M(s)\overline{P}_0(s)B(s) + K(s)F(s)D(s)B(s)M(s)\overline{P}_0(s) + E(s)J(s)M(s)\overline{P}_0(s)$$

$$B(s) + P(s)J(s)B(s)M(s)\overline{P}_0(s) + Q(s)C(s)D(s)B(s)M(s)\overline{P}_0(s) + E(s)$$

$$J(s)B(s)M(s)\overline{P}_0(s) + F(s)K(s)D(s)B(s)M(s)\overline{P}_0(s) + E(s)J(s)M(s)$$

$$\overline{P}_0(s)B(s) + P(s)J(s)B(s)M(s)\overline{P}_0(s) + R(s)K(s)D(s)M(s)\overline{P}_0(s)B(s) +$$

$$E(s)J(s)B(s)M(s)\overline{P}_0(s) + P(s)J(s)B(s)M(s)\overline{P}_0(s) + ph(y)C(s)D(s)$$

$$B(s)M(s)\overline{P}_0(s) + E(s)J(s)B(s)M(s)\overline{P}_0(s) + F(s)K(s)D(s)B(s)M(s)$$

$$\overline{P}_0(s) + E(s)J(s)M(s)\overline{P}_0(s)B(s) + P(s)J(s)B(s)M(s)\overline{P}_0(s)\left[\frac{1-\overline{S}_\phi(s)}{s}\right] +$$

$$C(s)D(s)B(s)L(s)M(s)\overline{P}_0(s) + E(s)J(s)B(s)L(s)M(s)\overline{P}_0(s) + F(s)P(s)$$

$$J(s)B(s)L(s)M(s)\overline{P}_0(s) + K(s)D(s)B(s)L(s)M(s)\overline{P}_0(s) + E(s)J(s)L(s)$$

$$B(s)M(s)\overline{P}_0(s) + Q(s)C(s)D(s)B(s)L(s)M(s)\overline{P}_0(s) + E(s)J(s)L(s)B(s)$$

$$M(s)\overline{P}_0(s) + F(s)P(s)J(s)L(s)B(s)M(s)\overline{P}_0(s) + K(s)D(s)L(s)M(s)B(s)$$

$$\overline{P}_0(s) + E(s)J(s)L(s)M(s)\overline{P}_0(s)B(s) + R(s)P(s)J(s)B(s)L(s)M(s)\overline{P}_0(s)$$

$$+ K(s)D(s)B(s)L(s)M(s)\overline{P}_0(s) + E(s)J(s)L(s)M(s)\overline{P}_0(s)B(s)ph(y)C(s)$$

$$D(s)B(s)L(s)M(s)\overline{P}_0(s) + E(s)J(s)B(s)L(s)M(s)\overline{P}_0(s) + F(s)P(s)J(s)$$

$$B(s)L(s)M(s)\overline{P}_0(s) + K(s)D(s)L(s)B(s)M(s)\overline{P}_0(s) + E(s)J(s)L(s)B(s)$$

$$M(s)\overline{P}_0(s)\left[\frac{1-\overline{S}_\phi(s)}{s}\right] + ph(t)B(s)L(s)M(s)\overline{P}_0(s)\left[\frac{1-\overline{S}_{u_1}(s)}{s}\right] + C(s)D(s)$$

$$B(s)\overline{P}_0(s) + E(s)J(s)B(s)\overline{P}_0(s) + F(s)P(s)J(s)B(s)\overline{P}_0(s) + K(s)D(s)B(s)$$

$$\overline{P}_0(s) + E(s)B(s)J(s)\overline{P}_0(s) + Q(s)C(s)D(s)B(s)\overline{P}_0(s) + E(s)J(s)B(s)$$

$$\overline{P}_0(s) + F(s)P(s)J(s)B(s)\overline{P}_0(s) + K(s)D(s)B(s)\overline{P}_0(s) + E(s)B(s)\overline{P}_0(s)$$

$$J(s) + R(s)*B(s)\overline{P}_0(s)P(s)J(s) + K(s)D(s)B(s)\overline{P}_0(s) + E(s)J(s)B(s)$$

$$\overline{P}_0(s) + ph(t)C(s)D(s)B(s)\overline{P}_0(s) + E(s)J(s)B(s)\overline{P}_0(s) + F(s)P(s)J(s)$$

$$B(s)\overline{P}_0(s) + K(s)B(s)\overline{P}_0(s)D(s) + E(s)J(s)B(s)\overline{P}_0(s)\left[\frac{1-\overline{S}_\phi(s)}{s}\right] + ph(y)$$

$$B(s)\overline{P}_0(s)\left[\frac{1-\overline{S}_{u_1}(s)}{s}\right]C(s)D(s)B(s)N(s)O(s)\overline{P}_0(s) + E(s)J(s)B(s)N(s)$$

$$O(s)\overline{P}_0(s) + F(s)P(s)J(s)B(s)N(s)O(s)\overline{P}_0(s) + K(s)D(s)B(s)N(s)O(s)$$

$$\overline{P}_0(s) + E(s)J(s)B(s)N(s)O(s)\overline{P}_0(s) + Q(s)C(s)D(s)N(s)O(s)\overline{P}_0(s)B(s)$$

$$+ E(s)J(s)B(s)N(s)O(s)\overline{P}_0(s) + F(s)P(s)J(s)B(s)N(s)O(s)\overline{P}_0(s) + K(s)$$

$$D(s)B(s)N(s)O(s)\overline{P}_0(s) + E(s)J(s)B(s)N(s)O(s)\overline{P}_0(s) + R(s)B(s)P(s)$$

$$J(s)N(s)O(s)\overline{P}_0(s) + K(s)D(s)B(s)N(s)O(s)\overline{P}_0(s) + E(s)J(s)B(s)N(s)$$

$$O(s)*\overline{P}_0(s) + B(s)ph(y)N(s)O(s)\overline{P}_0(s)\left[\frac{1-\overline{S}_{u_1}(s)}{s}\right]ph(t)C(s)D(s)B(s)$$

$$N(s)O(s)\overline{P}_0(s) + E(s)J(s)B(s)N(s)O(s)\overline{P}_0(s) + F(s)P(s)J(s)B(s)N(s)$$

$$O(s)\overline{P}_0(s) + K(s)D(s)N(s)B(s)O(s)\overline{P}_0(s) + E(s)J(s)N(s)O(s)B(s)\overline{P}_0(s)$$

$$\left[\frac{1-\overline{S}_\phi(s)}{s}\right] + ph(t)B(s)O(s)\overline{P}_0(s)\left[\frac{1-\overline{S}_{u_1}(s)}{s}\right] + Q(s)C(s)D(s)B(s)O(s)$$

$$\overline{P}_0(s) + E(s)J(s)B(s)O(s)\overline{P}_0(s) + F(s)K(s)D(s)B(s)O(s)\overline{P}_0(s) + E(s)$$

$$J(s)O(s)B(s)\overline{P}_0(s) + P(s)J(s)B(s)O(s)\overline{P}_0(s) + R(s)K(s)O(s)\overline{P}_0(s) +$$

$$R(s)K(s)D(s)B(s)O(s)B(s)\overline{P}_0(s) + P(s)\overline{P}_0(s) + E(s)J(s)O(s)J(s)B(s)$$

$$O(s)\overline{P}_0(s) + C(s)D(s)B(s)O(s)\overline{P}_0(s) + E(s)J(s)B(s)O(s)\overline{P}_0(s) + F(s)D(s)$$

$$K(s)B(s)O(s)\overline{P}_0(s) + E(s)J(s)B(s)O(s)\overline{P}_0(s) + P(s)J(s)B(s)O(s)\overline{P}_0(s)$$

$$+ ph(t)C(s)D(s)B(s)O(s)\overline{P}_0(s) + E(s)J(s)O(s)B(s)\overline{P}_0(s) + F(s)K(s)$$

$$D(s)B(s)O(s)\overline{P}_0(s) + E(s)J(s)O(s)B(s)\overline{P}_0(s) + P(s)J(s)B(s)O(s)\overline{P}_0(s)$$

$$\left[\frac{1 - \overline{S}_\phi(s)}{s} \right] \qquad \qquad \dots(104)$$

Also it is noticeable that

$$\overline{P}_{up}(s) + \overline{P}_{down}(s) = 1/s$$

Where

$$\phi(x) = \exp[-\{(-\log u_1(x))^{\theta_1} + (-\log u_2(x))^{\theta_1}\}^{1/\theta_1}]$$

$$B(s) = \frac{\lambda_2(s + \theta + l_1)}{(s + qh(y) + ph(y) + r_1)(s + \theta + l_1) + \eta(s + l_1)}, J(s) = \frac{\eta}{(s + l_1 + \theta)}$$

$$C(s) = \frac{r_2(s + \theta)}{(s + qh(y) + ph(y))(s + \theta) + \eta s}, K(s) = \frac{\eta}{(s + \theta + l_2)}, L(s) = \frac{\lambda_3}{(s + \lambda_2)}$$

$$D(s) = \frac{r_1(s + \theta + l_2)}{(s + r_2)(s + \theta + l_2) + \eta(s + l_2)}, M(s) = \frac{\lambda_1}{(s + \lambda_2 + \lambda_3)}, N(s) = \frac{\lambda_1}{(s + \lambda_2)}$$

$$E(s) = \frac{l_1\theta}{(s + r_2)(s + \theta + l_2) + \eta(s + l_2)}, O(s) = \frac{\lambda_3}{(s + \lambda_1 + \lambda_2)}, P(s) = \frac{l_1}{(s + \theta + l_2)}$$

$$F(s) = \frac{l_2\theta}{(s + qh(y) + ph(y))(s + \theta) + \eta s}, Q(s) = \frac{\eta}{(s + \theta)}, R(s) = \frac{l_2}{(s + \theta)}$$

7. ASYMPTOTIC BEHAVIOUR

Using Able's lemma

$$\lim_{s \to 0} \{s\overline{F}(s)\} = \lim_{t \to \infty} F(t) = F(\text{say})$$

in equations (103) and (104), one can obtain the following time independent up and down state probabilities

$$P_{up} = \frac{1}{A(0)} + M(0)\frac{1}{A(0)} + L(0)M(0)\frac{1}{A(0)} + M(0)\frac{1}{A(0)}B(0) + J(0)M(0)\frac{1}{A(0)}$$

$$B(0) + D(0)B(0)M(0)\frac{1}{A(0)} + E(0)J(0)M(0)\frac{1}{A(0)}B(0) + K(0)D(0)B(0) +$$

$$E(0)M(0)\frac{1}{A(0)}B(0) + E(0)J(0)M(0)\frac{1}{A(0)}B(0) + P(0)J(0)B(0)M(0)$$

$$\frac{1}{A(0)} + B(0)L(0)M(0)\frac{1}{A(0)} + J(0)B(0)L(0)M(0)\frac{1}{A(0)} + D(0)L(0)B(0)$$

$$M(0)\frac{1}{A(0)} + E(0)J(0)L(0)M(0)\frac{1}{A(0)}B(0) + P(0)J(0)B(0)L(0)M(0)$$

$$\frac{1}{A(0)} + K(0)D(0)L(0)M(0)\frac{1}{A(0)}B(0) + E(0)J(0)L(0)B(0)M(0)\frac{1}{A(0)} +$$

$$B(0)\frac{1}{A(0)} + J(0)B(0)\frac{1}{A(0)} + D(0)B(s)\frac{1}{A(0)} + E(0)J(0)B(0)\frac{1}{A(0)} + P(0)$$

$$J(0)B(0)\frac{1}{A(0)} + K(0)D(0)B(0)\frac{1}{A(0)} + E(0)J(0)B(0)\frac{1}{A(0)} + O(0)\frac{1}{A(0)} +$$

$$N(0)O(0)\frac{1}{A(0)} + B(0)N(0)O(0)\frac{1}{A(0)} + J(0)N(0)B(0)O(0)\frac{1}{A(0)} + D(0)$$

$$B(0)N(0)O(0)\frac{1}{A(0)} + E(0)J(0)B(0)N(0)O(0)\frac{1}{A(0)} + P(0)J(0)B(0)N(0)$$

$$O(0)\frac{1}{A(0)} + K(0)D(0)B(0)N(0)O(0)\frac{1}{A(0)} + E(0)J(0)N(0)O(0)\frac{1}{A(0)}B(0)$$

$$+ O(0)\frac{1}{A(0)}B(0) + J(0)B(0)O(0)\frac{1}{A(0)} + D(0)B(0)K(0)O(0)\frac{1}{A(0)} + E(0)$$

$$O(0)\frac{1}{A(0)} + P(0)J(0)O(0)\frac{1}{A(0)}B(0) + D(0)B(0)O(0)\frac{1}{A(0)} + E(0)J(0)$$

$$O(0)\frac{1}{A(0)}B(0)$$

$$\dots(105)$$

$$P_{down} = ph(y)B(0)M(0)\frac{1}{A(0)}\overline{M}_{u_1} + C(0)D(0)B(0)M(0)\frac{1}{A(0)} + E(0)J(0)M(0)$$

$$\frac{1}{A(0)}B(0) + K(0)F(0)D(0)B(0)M(0)\frac{1}{A(0)} + E(0)J(0)M(0)\frac{1}{A(0)}B(0)$$

$$+ P(0)J(0)B(0)M(0)\frac{1}{A(0)} + Q(0)C(0)D(0)B(0)M(0)\frac{1}{A(0)} + E(0)J(0)$$

$$B(0)M(0)\frac{1}{A(0)} + F(0)K(0)D(0)B(0)M(0)\frac{1}{A(0)} + E(0)J(0)M(0)\frac{1}{A(0)}$$

$$B(0) + P(0)J(0)B(0)M(0)\frac{1}{A(0)} + R(0)K(0)D(0)M(0)\frac{1}{A(0)}B(0) + E(0)$$

$$J(0)B(0)M(0)\frac{1}{A(0)}+P(0)J(0)B(0)M(0)\frac{1}{A(0)}+ph(y)C(0)D(0)B(0)$$

$$M(0)\frac{1}{A(0)}+E(0)J(0)B(0)M(0)\frac{1}{A(0)}+F(0)K(0)D(0)B(0)M(0)\frac{1}{A(0)}$$

$$+E(0)J(0)M(0)\frac{1}{A(0)}B(0)+P(0)J(0)B(0)M(0)\frac{1}{A(0)}\overline{M}_\phi+C(0)D(0)$$

$$B(0)L(0)M(0)\frac{1}{A(0)}+E(0)J(0)B(0)L(0)M(0)\frac{1}{A(0)}+F(0)P(0)J(0)$$

$$B(0)L(0)M(0)\frac{1}{A(0)}+K(0)D(0)B(0)L(0)M(0)\frac{1}{A(0)}+E(0)J(0)L(0)$$

$$B(0)M(0)\frac{1}{A(0)}+Q(0)C(0)D(0)B(0)L(0)M(0)\frac{1}{A(0)}+E(0)J(0)L(0)B(0)$$

$$M(0)\frac{1}{A(0)}+F(0)P(0)J(0)L(0)B(0)M(0)\frac{1}{A(0)}+K(0)D(0)L(0)M(0)B(0)$$

$$\frac{1}{A(0)}+E(0)J(0)L(0)M(0)\frac{1}{A(0)}B(0)+R(0)P(0)J(0)B(0)L(0)M(0)\frac{1}{A(0)}$$

$$+K(0)D(0)B(0)L(0)M(0)\frac{1}{A(0)}+E(0)J(0)L(0)M(0)\frac{1}{A(0)}B(0)ph(y)C(0)$$

$$D(0)B(0)L(0)M(0)\frac{1}{A(0)}+E(0)J(0)B(0)L(0)M(0)\frac{1}{A(0)}+F(0)P(0)J(0)$$

$$B(0)L(0)M(0)\frac{1}{A(0)}+K(0)D(0)L(0)B(0)M(0)\frac{1}{A(0)}+E(0)J(0)L(0)B(0)$$

$$M(0)\frac{1}{A(0)}\overline{M}_\phi+ph(y)B(0)L(0)M(0)\frac{1}{A(0)}\overline{M}_{u_1}+C(0)D(0)B(0)\frac{1}{A(0)}+$$

$$E(0)J(0)B(0)\frac{1}{A(0)}+F(0)P(0)J(0)B(0)\frac{1}{A(0)}+K(0)D(0)B(0)\frac{1}{A(0)}+$$

$$E(0)B(0)J(0)\frac{1}{A(0)}+Q(0)C(0)D(0)B(0)\frac{1}{A(0)}+E(0)J(0)B(0)\frac{1}{A(0)}+$$

$$F(0)P(0)J(0)B(0)\frac{1}{A(0)}+K(0)D(0)B(0)\frac{1}{A(0)}+E(0)B(0)\frac{1}{A(0)}J(0)+$$

$$R(0)B(0)\frac{1}{A(0)}P(0)J(0)+K(0)D(0)B(0)\frac{1}{A(0)}+E(0)J(0)B(0)\frac{1}{A(0)}+$$

$$ph(y)C(0)D(0)B(0)\frac{1}{A(0)}+E(0)J(0)B(0)\frac{1}{A(0)}+F(0)P(0)J(0)B(0)$$

$$\frac{1}{A(0)} + K(0)B(0)\frac{1}{A(0)}D(0) + E(0)J(0)B(0)\frac{1}{A(0)}\overline{M}_{\phi} + ph(y)B(0)\frac{1}{A(0)}$$

$$\overline{M}_{u_1}C(0)D(0)B(0)N(0)O(0)\frac{1}{A(0)} + E(0)J(0)B(0)N(0)O(0)\frac{1}{A(0)} + F(0)$$

$$P(0)J(0)B(0)N(0)O(0)\frac{1}{A(0)} + K(0)D(0)B(0)N(0)O(0)\frac{1}{A(0)} + E(0)J(0)$$

$$B(0)N(0)O(0)\frac{1}{A(0)} + Q(0)C(0)D(0)N(0)O(0)\frac{1}{A(0)}B(0) + E(0)J(0)B(0)$$

$$N(0)O(0)\frac{1}{A(0)} + F(0)P(0)J(0)B(0)N(0)O(0)\frac{1}{A(0)} + K(0)D(0)B(0)N(0)$$

$$O(0)\frac{1}{A(0)} + E(0)J(0)B(0)N(0)O(0)\frac{1}{A(0)} + R(0)B(0)P(0)J(0)N(0)O(0)$$

$$\frac{1}{A(0)} + K(0)D(0)B(0)N(0)O(0)\frac{1}{A(0)} + E(0)J(0)B(0)N(0)O(0)\frac{1}{A(0)} +$$

$$B(0)ph(y)N(0)O(0)\frac{1}{A(0)}\overline{M}_{u_1}ph(y)C(0)D(0)B(0)N(0)O(0)\frac{1}{A(0)} + E(0)$$

$$J(0)B(0)N(0)O(0)\frac{1}{A(0)} + F(0)P(0)J(0)B(0)N(0)O(0)\frac{1}{A(0)} + K(0)D(0)$$

$$N(0)B(0)O(0)\frac{1}{A(0)} + E(0)J(0)N(0)O(0)B(0)\frac{1}{A(0)}\overline{M}_{\phi} + ph(y)B(0)O(0)$$

$$\frac{1}{A(0)}\overline{M}_{u_1} + Q(0)C(0)D(0)B(0)O(0)\frac{1}{A(0)} + E(0)J(0)B(0)O(0)\frac{1}{A(0)} +$$

$$F(0)K(0)D(0)B(0)O(0)\frac{1}{A(0)} + E(0)J(0)O(0)B(0)\frac{1}{A(0)} + P(0)J(0)B(0)$$

$$O(0)\frac{1}{A(0)} + R(0)K(0)O(0)\frac{1}{A(0)} + R(0)K(0)D(0)B(0)O(0)B(0)\frac{1}{A(0)} +$$

$$R(0)K(0)D(0)B(0)O(0)B(0)\frac{1}{A(0)} + P(0)\frac{1}{A(0)} + E(0)J(0)O(0)J(0)B(0)$$

$$O(0)\frac{1}{A(0)} + C(0)D(0)B(0)O(0)\frac{1}{A(0)} + E(0)J(0)B(0)O(0)\frac{1}{A(0)} + F(0)$$

$$D(0)K(0)B(0)O(0)\frac{1}{A(0)} + E(0)J(0)B(0)O(0)\frac{1}{A(0)} + P(0)J(0)B(0)O(0)$$

$$\frac{1}{A(0)} + ph(y)C(0)D(0)B(0)O(0)\frac{1}{A(0)} + E(0)J(0)O(0)B(0)\frac{1}{A(0)} + F(0)$$

$$K(0)D(0)B(0)O(0)\frac{1}{A(0)} + E(0)J(0)O(0)B(0)\frac{1}{A(0)} + P(0)J(0)B(0)O(0)$$

$$\frac{1}{A(0)}\overline{M}_\phi$$

where

$$A(0) = \lim_{s\to 0} A(s)$$

$$\overline{M}_{u_1} = \lim_{s\to 0}\left\{\frac{1-\overline{S}_{u_1}(s)}{s}\right\}, \overline{M}_\phi = \lim_{s\to 0}\left\{\frac{1-\overline{S}_\phi(s)}{s}\right\}$$

$$\phi(x) = \exp[-\{(-\log u_1(x))^{\theta_1} + (-\log u_2(x))^{\theta_1}\}^{1/\theta_1}], P(0) = \lim_{s\to 0} P(s) = \frac{l_1}{(\theta + l_2)}$$

$$B(0) = \lim_{s\to 0} B(s) = \frac{\lambda_2(\theta + l_1)}{(qh(y) + ph(y) + r_1)(\theta + l_1) + \eta(l_1)}, J(0) = \lim_{s\to 0} J(s) = \frac{\eta}{(l_1 + \theta)}$$

$$C(0) = \lim_{s\to 0} C(s) = \frac{r_2(\theta)}{(qh(y) + ph(y))(\theta) + \eta s}, K(0) = \lim_{s\to 0} K(s) = \frac{\eta}{(\theta + l_2)}$$

$$D(0) = \lim_{s\to 0} D(s) = \frac{r_1(\theta + l_2)}{(r_2)(\theta + l_2) + \eta(l_2)}, L(0) = \lim_{s\to 0} L(s) = \frac{\lambda_3}{(\lambda_2)}$$

$$E(0) = \lim_{s\to 0} E(s) = \frac{l_1\theta}{(r_2)(\theta + l_2) + \eta(l_2)}, M(0) = \lim_{s\to 0} M(s) = \frac{\lambda_1}{(\lambda_2 + \lambda_3)}$$

$$F(0) = \lim_{s\to 0} F(s) = \frac{l_2\theta}{(qh(y) + ph(y))(\theta) + \eta s}, N(0) = \lim_{s\to 0} N(s) = \frac{\lambda_1}{(\lambda_2)}$$

$$Q(0) = \lim_{s\to 0} Q(s) = \frac{\eta}{(\theta)}, R(0) = \lim_{s\to 0} R(s) = \frac{l_2}{(\theta)}, O(0) = \lim_{s\to 0} O(s) = \frac{\lambda_3}{(\lambda_1 + \lambda_2)}$$

8. PARTICULAR CASES

(i) When repair follows exponential distribution.

In this case the results can be derived by putting

$$\overline{S}_\phi(s) = \frac{\phi(x)}{s + \phi(x)}, \overline{S}_{u_1}(s) = \frac{u_1(x)}{s + u_1(x)} \qquad \qquad ...(107)$$

in equations (65) and (66), which yield

$$\overline{P}_{up}(s) = \frac{1}{A_1(s)} + M(s)\frac{1}{A_1(s)} + L(s)M(s)\frac{1}{A_1(s)} + M(s)\frac{1}{A_1(s)}B(s) + J(s)M(s)\frac{1}{A_1(s)}$$

$$B(s) + D(s)B(s)M(s)\frac{1}{A_1(s)} + E(s)J(s)M(s)\frac{1}{A_1(s)}B(s) + K(s)D(s)B(s)$$

$$+ E(s)M(s)\frac{1}{A_1(s)}B(s) + E(s)J(s)M(s)\frac{1}{A_1(s)}B(s) + P(s)J(s)B(s)M(s)$$

$$\frac{1}{A_1(s)} + B(s)L(s)M(s)\frac{1}{A_1(s)} + J(s)B(s)L(s)M(s)\frac{1}{A_1(s)} + D(s)L(s)B(s)$$

$$M(s)\frac{1}{A_1(s)} + E(s)J(s)L(s)M(s)\frac{1}{A_1(s)}B(s) + P(s)J(s)B(s)L(s)M(s)\frac{1}{A_1(s)}$$

$$+ K(s)D(s)L(s)M(s)\frac{1}{A_1(s)}B(s) + E(s)J(s)L(s)B(s)M(s)\frac{1}{A_1(s)} + B(s)$$

$$\frac{1}{A_1(s)} + J(s)B(s)\frac{1}{A_1(s)} + D(s)B(s)\frac{1}{A_1(s)} + E(s)J(s)B(s)\frac{1}{A_1(s)} + P(s)$$

$$J(s)B(s)\frac{1}{A_1(s)} + K(s)D(s)B(s)\frac{1}{A_1(s)} + E(s)J(s)B(s)\frac{1}{A_1(s)} + O(s)\frac{1}{A_1(s)} +$$

$$N(s)O(s)\frac{1}{A_1(s)} + B(s)N(s)O(s)\frac{1}{A_1(s)} + J(s)N(s)B(s)O(s)\frac{1}{A_1(s)} + D(s)$$

$$B(s)N(s)O(s)\frac{1}{A_1(s)} + E(s)J(s)B(s)N(s)O(s)\frac{1}{A_1(s)} + P(s)J(s)B(s)N(s)$$

$$O(s)\frac{1}{A_1(s)} + K(s)D(s)B(s)N(s)O(s)\frac{1}{A_1(s)} + E(s)J(s)N(s)O(s)\frac{1}{A_1(s)}B(s)$$

$$+ O(s)\frac{1}{A(s)}B(s) + J(s)B(s)O(s)\frac{1}{A_1(s)} + D(s)B(s)K(s)O(s)\frac{1}{A_1(s)} + E(s)$$

$$O(s)\frac{1}{A_1(s)} + P(s)J(s)O(s)\frac{1}{A_1(s)}B(s) + D(s)B(s)O(s)\frac{1}{A_1(s)} + E(s)J(s)$$

$$O(s)\frac{1}{A_1(s)}B(s) \qquad\qquad\qquad …(108)$$

$$\overline{P}_{down}(s) = ph(y)B(s)M(s)\frac{1}{A_1(s)}\frac{1}{s+u_1(x)} + C(s)D(s)B(s)M(s)\frac{1}{A_1(s)} + E(s)J(s)$$

$$M(s)\frac{1}{A_1(s)}B(s) + K(s)F(s)D(s)B(s)M(s)\frac{1}{A_1(s)} + E(s)J(s)M(s)\frac{1}{A_1(s)}$$

$$B(s) + P(s)J(s)B(s)M(s)\frac{1}{A_1(s)} + Q(s)C(s)D(s)B(s)M(s)\frac{1}{A_1(s)} + E(s)$$

$$J(s)B(s)M(s)\frac{1}{A_1(s)} + F(s)K(s)D(s)B(s)M(s)\frac{1}{A_1(s)} + E(s)J(s)M(s)$$

$$\frac{1}{A_1(s)}B(s) + P(s)J(s)B(s)M(s)\frac{1}{A_1(s)} + R(s)K(s)D(s)M(s)\frac{1}{A_1(s)}B(s) +$$

$$E(s)J(s)B(s)M(s)\frac{1}{A_1(s)} + P(s)J(s)B(s)M(s)\frac{1}{A_1(s)} + ph(y)C(s)D(s)$$

$$B(s)M(s)\frac{1}{A_1(s)} + E(s)J(s)B(s)M(s)\frac{1}{A_1(s)} + F(s)K(s)D(s)B(s)M(s)$$

$$\frac{1}{A_1(s)} + E(s)J(s)M(s)\frac{1}{A_1(s)}B(s) + P(s)J(s)B(s)M(s)\frac{1}{A_1(s)}\frac{1}{s+\phi(x)} +$$

$$C(s)D(s)B(s)L(s)M(s)\frac{1}{A_1(s)} + E(s)J(s)B(s)L(s)M(s)\frac{1}{A_1(s)} + F(s)P(s)$$

$$J(s)B(s)L(s)M(s)\frac{1}{A_1(s)} + K(s)D(s)B(s)L(s)M(s)\frac{1}{A_1(s)} + E(s)J(s)L(s)$$

$$B(s)M(s)\frac{1}{A_1(s)} + Q(s)C(s)D(s)B(s)L(s)M(s)\frac{1}{A_1(s)} + E(s)J(s)L(s)B(s)$$

$$M(s)\frac{1}{A_1(s)} + F(s)P(s)J(s)L(s)B(s)M(s)\frac{1}{A_1(s)} + K(s)D(s)L(s)M(s)B(s)$$

$$\frac{1}{A_1(s)} + E(s)J(s)L(s)M(s)\frac{1}{A_1(s)}B(s) + R(s)P(s)J(s)B(s)L(s)M(s)\frac{1}{A_1(s)}$$

$$+ K(s)D(s)B(s)L(s)M(s)\frac{1}{A_1(s)} + E(s)J(s)L(s)M(s)\frac{1}{A_1(s)}B(s)ph(y)C(s)$$

$$D(s)B(s)L(s)M(s)\frac{1}{A_1(s)} + E(s)J(s)B(s)L(s)M(s)\frac{1}{A_1(s)} + F(s)P(s)J(s)$$

$$B(s)L(s)M(s)\frac{1}{A_1(s)} + K(s)D(s)L(s)B(s)M(s)\frac{1}{A_1(s)} + E(s)J(s)L(s)B(s)$$

$$M(s)\frac{1}{A(s)}\frac{1}{s+\phi(x)} + ph(y)B(s)L(s)M(s)\frac{1}{A(s)}\frac{1}{s+u_1(x)} + C(s)D(s)B(s)$$

$$\frac{1}{A_1(s)} + E(s)J(s)B(s)\frac{1}{A_1(s)} + F(s)P(s)J(s)B(s)\frac{1}{A_1(s)} + K(s)D(s)B(s)$$

$$\frac{1}{A_1(s)} + E(s)B(s)J(s)\frac{1}{A_1(s)} + Q(s)C(s)D(s)B(s)\frac{1}{A_1(s)} + E(s)J(s)B(s)$$

$$\frac{1}{A_1(s)} + F(s)P(s)J(s)B(s)\frac{1}{A_1(s)} + K(s)D(s)B(s)\frac{1}{A_1(s)} + E(s)B(s)\frac{1}{A_1(s)}$$

$$J(s) + R(s)B(s)\frac{1}{A_1(s)}P(s)J(s) + K(s)D(s)B(s)\frac{1}{A_1(s)} + E(s)J(s)B(s)$$

$$\frac{1}{A_1(s)} + ph(y)C(s)D(s)B(s)\frac{1}{A_1(s)} + E(s)J(s)B(s)\frac{1}{A_1(s)} + F(s)P(s)J(s)$$

$$B(s)\frac{1}{A_1(s)} + K(s)B(s)\frac{1}{A_1(s)}D(s) + E(s)J(s)B(s)\frac{1}{A_1(s)}\frac{1}{s+\phi(x)} + ph(y)$$

$$B(s)\frac{1}{A_1(s)}\frac{1}{s+u_1(x)}C(s)D(s)B(s)N(s)O(s)\frac{1}{A_1(s)} + E(s)J(s)B(s)N(s)O(s)$$

$$\frac{1}{A_1(s)} + F(s)P(s)J(s)B(s)N(s)O(s)\frac{1}{A_1(s)} + K(s)D(s)B(s)N(s)O(s)\frac{1}{A_1(s)} +$$

$$E(s)J(s)B(s)N(s)O(s)\frac{1}{A_1(s)} + Q(s)C(s)D(s)N(s)O(s)\frac{1}{A_1(s)}B(s) + E(s)$$

$$J(s)B(s)N(s)O(s)\frac{1}{A_1(s)} + F(s)P(s)J(s)B(s)N(s)O(s)\frac{1}{A_1(s)} + K(s)D(s)$$

$$B(s)N(s)O(s)\frac{1}{A_1(s)} + E(s)J(s)B(s)N(s)O(s)\frac{1}{A_1(s)} + R(s)B(s)P(s)J(s)$$

$$N(s)O(s)\frac{1}{A_1(s)} + K(s)D(s)B(s)N(s)O(s)\frac{1}{A_1(s)} + E(s)J(s)B(s)N(s)O(s)$$

$$\frac{1}{A_1(s)} + B(s)ph(y)N(s)O(s)\frac{1}{A_1(s)}\frac{1}{s+u_1(x)}ph(y)C(s)D(s)B(s)N(s)O(s)$$

$$\frac{1}{A_1(s)} + E(s)J(s)B(s)N(s)O(s)\frac{1}{A_1(s)} + F(s)P(s)J(s)B(s)N(s)O(s)\frac{1}{A_1(s)} +$$

$$K(s)D(s)N(s)B(s)O(s)\frac{1}{A_1(s)} + E(s)J(s)N(s)O(s)B(s)\frac{1}{A_1(s)}\frac{1}{s+\phi(x)} +$$

$$ph(y)B(s)O(s)\frac{1}{A_1(s)}\frac{1}{s+u_1(x)} + Q(s)C(s)D(s)B(s)O(s)\frac{1}{A_1(s)} + E(s)J(s)$$

$$B(s)O(s)\frac{1}{A_1(s)} + F(s)K(s)D(s)B(s)O(s)\frac{1}{A_1(s)} + E(s)J(s)O(s)B(s)\frac{1}{A_1(s)}$$

$$+ P(s)J(s)B(s)O(s)\frac{1}{A_1(s)} + R(s)K(s)O(s)\frac{1}{A_1(s)} + R(s)K(s)D(s)B(s)O(s)$$

$$B(s)\frac{1}{A_1(s)} + R(s)K(s)D(s)B(s)O(s)B(s)\frac{1}{A_1(s)} + P(s)\frac{1}{A_1(s)} + E(s)J(s)O(s)$$

$$J(s)B(s)O(s)\frac{1}{A_1(s)} + C(s)D(s)B(s)O(s)\frac{1}{A_1(s)} + E(s)J(s)B(s)O(s)\frac{1}{A_1(s)} +$$

$$F(s)D(s)K(s)B(s)O(s)\frac{1}{A_1(s)} + E(s)J(s)B(s)O(s)\frac{1}{A_1(s)} + P(s)J(s)B(s)O(s)$$

$$\frac{1}{A_1(s)} + ph(y)C(s)D(s)B(s)O(s)\frac{1}{A_1(s)} + E(s)J(s)O(s)B(s)\frac{1}{A_1(s)} + F(s)K(s)$$

$$D(s)B(s)O(s)\frac{1}{A_1(s)} + E(s)J(s)O(s)B(s)\frac{1}{A_1(s)} + P(s)J(s)B(s)O(s)\frac{1}{A_1(s)}$$

$$\frac{1}{s+\phi(x)} \qquad\qquad\qquad \dots(109)$$

where

$$A_1(s) = \left[s + \lambda_1 + \lambda_2 + \lambda_3\right] - qh(y)B(s) - qh(y)C(s)D(s)B(s) - E(s)J(s)B(s) - F(s)$$

$$P(s)J(s)B(s) - K(s)D(s)B(s) - E(s)J(s)B(s) - qh(y)B(s)M(s) - qh(y)$$

$$C(s)D(s)M(s)B(s) - E(s)J(s)B(s)M(s) - F(s)K(s)D(s)M(s)B(s) - E(s)$$

$$J(s)M(s)B(s) + P(s)J(s)B(s)M(s) - qh(t)B(s)L(s)M(s) - qh(t)C(s)D(s)$$

$$B(s)L(s)M(s) - E(s)J(s)L(s)M(s)B(s) - F(s)P(s)J(s) - F(s)P(s)J(s)$$

$$B(s)L(s)M(s) - K(s)D(s)L(s)M(s)B(s) - E(s)J(s)B(s)L(s)M(s) - qh(y)$$

$$B(s)N(s)O(s) - qh(y)C(s)D(s)B(s)B(s)N(s)O(s) - E(s)J(s)N(s)O(s)B(s)$$

$$- F(s)P(s)J(s)B(s)N(s)O(s) - K(s)D(s)B(s)N(s)O(s) - E(s)J(s)N(s)O(s)$$

$$+ P(s)J(s)B(s)O(s) - ph(t)\frac{u_1(x)}{s+u_1(x)}B(s) - ph(y)C(s)D(s)B(s) - E(s)J(s)$$

$$B(s) - F(s)P(s)J(s)B(s) - K(s)D(s)B(s) - E(s)B(s)J(s)\frac{\phi(x)}{s+\phi(x)} - ph(y)$$

$$B(s)M(s)\frac{u_1(x)}{s+u_1(x)} - ph(y)C(s)D(s)B(s)M(s) - E(s)J(s)B(s)M(s) - F(s)$$

$$K(s)D(s)B(s)M(s) - E(s)J(s)M(s)B(s) + P(s)J(s)B(s)M(s)\frac{\phi(x)}{s+\phi(x)} -$$

$$ph(y)B(s)L(s)M(s)\frac{u_1(x)}{s+u_1(x)} - ph(y)C(s)D(s)B(s)L(s)M(s) + E(s)J(s)$$

$$L(s)M(s)B(s) - F(s)P(s)J(s)B(s)L(s)M(s) - K(s)D(s)B(s)L(s)M(s) -$$

$$J(s)E(s)B(s)L(s)M(s)\frac{\phi(x)}{s+\phi(x)} - ph(y)B(s)N(s)O(s)\frac{u_1(x)}{s+u_1(x)} - ph(y)C(s)$$

$$D(s)B(s)N(s)O(s) + E(s)J(s)B(s)N(s)O(s) + F(s)P(s)J(s)B(s)N(s)O(s) +$$

$$K(s)D(s)N(s)O(s)B(s) + E(s)J(s)B(s)N(s)O(s)\frac{\phi(x)}{s+\phi(x)} - ph(y)B(s)O(s)$$

$$\frac{u_1(x)}{s+u_1(x)} - ph(y)C(s)D(s)B(s)O(s) + E(s)J(s)B(s)O(s) + F(s)K(s)D(s)B(s)$$

$$O(s) + E(s)J(s)O(s)B(s) + P(s)J(s)B(s)O(s)\frac{\phi(x)}{s+\phi(x)} \qquad \dots(110)$$

Where

$$\phi(x) = \exp[-\{(-\log u_1(x))^{\theta_1} + (-\log u_2(x))^{\theta_1}\}^{1/\theta_1}]$$

$$B(s) = \frac{\lambda_2(s+\theta+l_1)}{(s+qh(y)+ph(y)+r_1)(s+\theta+l_1)+\eta(s+l_1)}, J(s) = \frac{\eta}{(s+l_1+\theta)}$$

$$C(s) = \frac{r_2(s+\theta)}{(s+qh(y)+ph(y))(s+\theta)+\eta s}, K(s) = \frac{\eta}{(s+\theta+l_2)}, L(s) = \frac{\lambda_3}{(s+\lambda_2)}$$

$$D(s) = \frac{r_1(s+\theta+l_2)}{(s+r_2)(s+\theta+l_2)+\eta(s+l_2)}, M(s) = \frac{\lambda_1}{(s+\lambda_2+\lambda_3)}, N(s) = \frac{\lambda_1}{(s+\lambda_2)}$$

$$E(s) = \frac{l_1\theta}{(s+r_2)(s+\theta+l_2)+\eta(s+l_2)}, O(s) = \frac{\lambda_3}{(s+\lambda_1+\lambda_2)}, P(s) = \frac{l_1}{(s+\theta+l_2)}$$

$$F(s) = \frac{l_2\theta}{(s+qh(y)+ph(y))(s+\theta)+\eta s}, Q(s) = \frac{\eta}{(s+\theta)}, R(s) = \frac{l_2}{(s+\theta)}$$

9. NUMERICAL COMPUTATION

9.1 Reliability Analysis

Reliability of the system is determined in following three situations. In the first section let us fix failure rates $\lambda_1 = 0.1$, $\lambda_2 = 0.2$, $\lambda_3 = 0.3$, $r_1 = 0.6$, $r_2 = 0.7$, $l_1 = 0.8$, $l_2 = 0.9$, repair rates $\Phi = u_1 = 0$, inspection rate h=0, $\theta_1 = 1$, $x = 1$, $y = 1$ and $\theta = 0.4$, $\eta = 0.5$. Also let the repair follows exponential distribution. Now by putting all these values in equation (103), using equation (107) and setting $t = 0, 1, 2, 3, 4, 5, 6, 7, 8, 9, 10$. In the second section we put $\theta = 2$ and in third section $\eta = 0$ while the other parameters are fixed as earlier and then following the same procedure as above, one can obtain Table 2 and Figure 2 which represent how reliability varies as the time increases corresponding to (i) θ (normal weather rate) $< \eta$ (abnormal weather rate) (ii) θ (normal weather rate) $> \eta$ (abnormal weather rate) and (iii) $\eta = 0$ *i.e.* system is performed under perfectly normal weather.

9.1.1 *Availability Analysis*

If we take $\lambda_1 = 0.1$, $\lambda_2 = 0.2$, $\lambda_3 = 0.3$, $r_1 = 0.6$, $r_2 = 0.7$, $l_1 = 0.8$, $l_2 = 0.9$, $\theta = 0.4$, $\eta = 0.5$, $p = 0.8$, $q = 0.2$, $\Phi = u_1 = 1$, inspection rate h = 1, $\theta_1 = 1$, $x = 1$ and $y = 1$. Also if the repair follows exponential distribution i. e. equation (107) holds, then putting all these values in equation (103) and taking inverse Laplace transformation, we get

$$P_{up}(t) = 0.2309541216\, e^{(-1.345506745t)} \sin(0.2760553762\, t) + 0.1095679093\, e^{(-1.345506745t)}$$

$$\cos(0.2760553762\ \ t) + 0.7185982883\ - 0.9935114620\ \ e^{(-0.6411182026\ t)}$$

$$- 0.1922672905\ \ e^{(-2.172832665\ t)} + 0.0389266182\ \ 2\ e^{-0.9369926106\ t)} +$$

$$0.2183766977\ \ e^{(-2.006986514\ t)} + 1.100309239\ \ e^{(-0.4510565187\ t)} \qquad\qquad \dots(111)$$

Again we set $\theta = 1.1$ and the other parameters as above and following the same procedure, we get

$$P_{up}(t) = 0.7986147888 + 0.4347645542 e^{(-0.6062890438t)} \sin(0.3280966861t)$$

$$0.09224766030\, e^{(-0.606289038t)} \cos(0.3280966861t) + 0.06702980106$$

$$e^{(-1.290921765t)} \sin(0.2297463596t) + 0.1527335982 e^{(-1.290921765t)}$$

$$\cos(0.2297463596t) - 0.06043232069 e^{(-2.698980240t)} - 0.04593364106$$

$$e^{(-1.976240331t)} + 0.06276991504 e^{(-2.5030357811t)} \qquad\qquad \dots(112)$$

Varying t from 0 to 10 in equation (111) and (112), we can obtain Table 3 and correspondingly Figure 3 representing the availability of the system when (i) $\theta < \eta$ (ii) $\theta > \eta$.

9.1.2 *M.T.T.F. Analysis*

Let us suppose that repair follows exponential distribution, then using equation (107) and from $M.T.T.F. = \lim_{s \to 0} \overline{P}_{up}(s)$, we have following four cases:

1. Setting $\lambda_1 = 0.1$, $\lambda_2 = 0.2$, $\lambda_3 = 0.3$, $r_1 = 0.6$, $r_2 = 0.7$, $l_1 = 0.8$, $l_2 = 0.9$, repair rates $\Phi = u_1 = 0$, inspection rate h = 0, $\theta_1 = 1$, $x = 1$, $y = 1$ and $\eta = 0.5$ and varying θ as 0.1, 0.2, 0.3, 0.4, 0.5, 0.6, 0.7, 0.8, 0.9, 1, one can obtain the variation of M.T.T.F. with respect to θ.
2. By taking $\lambda_1 = 0.1$, $\lambda_2 = 0.2$, $\lambda_3 = 0.3$, $r_1 = 0.6$, $r_2 = 0.7$, $l_1 = 0.8$, $l_2 = 0.9$, repair rates $\Phi = u_1 = 0$, inspection rate h = 0, $\theta_1 = 1$, $x = 1$, $y = 1$ and $\eta = 0.9$ and varying θ as 0.1, 0.2, 0.3, 0.4, 0.5, 0.6, 0.7, 0.8, 0.9, 1, one can observe how M.T.T.F. changes with the increase of abnormal weather rate.
3   Fixing $\lambda_1 = 0.1$, $\lambda_2 = 0.2$, $\lambda_3 = 0.3$, $r_1 = 0.8$, $r_2 = 0.7$, $l_1 = 0.8$, $l_2 = 0.9$, repair rates $\Phi = u_1 = 0$, inspection rate h = 0, $\theta_1 = 1$, $x = 1$, $y = 1$ and $\eta = 0.5$ and varying the value of θ as 0.1, 0.2, 0.3, 0.4, 0.5, 0.6, 0.7, 0.8, 0.9, 1, one can get the changes of M.T.T.F. with the increasing value of r_1.
4. Assuming $\lambda_1 = 0.1$, $\lambda_2 = 0.2$, $\lambda_3 = 0.3$, $r_1 = 0.6$, $r_2 = 0.7$, $l_1 = 0.8$, $l_2 = 0.9$, repair rates $\Phi = u_1 = 0$, inspection rate h = 0, $\theta_1 = 1$, $x = 1$, $y = 1$ and $\eta = 0.5$ and increasing the value of θ from 0.1 to 1.0, we obtain how M.T.T.F. varies with respect to r_2.
5. Letting $\lambda_1 = 0.1$, $\lambda_2 = 0.2$, $\lambda_3 = 0.3$, $r_1 = 0.6$, $r_2 = 0.7$, $l_1 = 1.1$, $l_2 = 0.9$, repair rates $\Phi = h = u_1 = 0$, $\theta_1 = 1$, $x = 1$, $y = 1$ and $\eta = 0.5$ and varying θ as 0.1, 0.2, 0.3, 0.4, 0.5, 0.6, 0.7, 0.8, 0.9, 1.0, we obtain the changes of M. T. T. F. with respect to l_1.
6. Taking the values $\lambda_1 = 0.1$, $\lambda_2 = 0.2$, $\lambda_3 = 0.3$, $r_1 = 0.6$, $r_2 = 0.7$, $l_1 = 0.8$, $l_2 = 1.2$, repair rates $\Phi = u_1 = 0$, inspection rate h = 0, $\theta_1 = 1$, $x = 1$, $y = 1$ and $\eta = 0.5$ and varying $\theta = 0.1$, 0.2, 0.3, 0.4, 0.5, 0.6, 0.7, 0.8, 0.9, 1.0, one can observe how the M. T. T. F. changes with respect to l_2. Variation of M.T.T.F with respect to above six cases have been given in the Table 4.

9.1.3 *Cost Analysis*

Let us assume that $\lambda_1 = 0.1$, $\lambda_2 = 0.2$, $\lambda_3 = 0.3$, $r_1 = 0.6$, $r_2 = 0.7$, $l_1 = 0.8$, $l_2 = 0.9$, $\theta = 0.4$, $\eta = 0.5$, p = 0.8, q = 0.2, $\Phi = u_1 = 1$, inspection rate h = 1, $\theta_1 = 1$, $x = 1$ and $y = 1$. Moreover, if the repair follows exponential distribution then using equations (107), we can obtain equation (111). If the service facility is always available, then expected profit during the interval (0, t] is given by

$$E_P(t) = K_1 \int_0^t P_{up}(t)dt - K_2 t$$

where K_1 and K_2 are the revenue per unit time and Service cost per unit time respectively, then

$$E_p(t) = K_1(-0.1119373745\, e^{-1.345506745 t}\cos(0.2760553762\, t) - 0.1486824264$$

$$e^{(-1.345506745\ t)}\sin(0.2760553762\ t) + 0.7185982883\ t + 1.549654117$$

$$e^{(0.6411182026\ t)} + 0.0884869293\ 4 e^{(-2.172832665\ t)} - 0.0415442104\ 7$$

$$e^{(-0.9369926106\ t)} - 0.1088082537\ e^{(-2.006986514\ t)} - 2.439404362\ e^{(0.4510565187\ t)}$$

$$+ 1.063553155\ ) - t\, K_2 \qquad\qquad ...(113)$$

When we increase normal weather rate to $\theta = 1.1$ and the other parameters are kept fixed, we can obtain equation (112), then expected profit is given by

$$E_p(t) = K_1(0.7986147888t - 0.4178438875e^{(-0.6062890438t)} \cos(0.3280966861t)$$

$$- 0.4909726845 \ e^{-0.6062890438 \ t)} \sin(0.3280966861 \ t) - 0.1236384741$$

$$e^{-1.290921765 \ t)} \cos(0.2297463596 \ t) - 0.0299199477 \ 2e^{(-1.290921765 \ t)}$$

$$\sin(0.2297463596 \ t) + 0.0223907977 \ 5e^{(-2.698980240 \ t)} + 0.0232429428 \ 4$$

$$e^{(-1.976240331t)}0.024806734475e^{-2.530357811t)} + 0.5206553558) - tK_2 \qquad \ldots(114)$$

Keeping $K_1 = 1$ and varying K_2 at 0.1, 0.2, 0.3, 0.4, 0.5 in equations (113) and (114), one can obtain Table 5 which is depicted by Figure 5.

9.1.4 *Sensitivity Analysis*

We perform a sensitivity analysis for changes in R(t) resulting from changes in system parameters r_1, r_2, l_1, l_2, θ and η.

Putting $\lambda_1 = 0.1$, $\lambda_2 = 0.2$, $\lambda_3 = 0.3$, $r_2 = 0.7$, $l_1 = 0.8$, $l_2 = 0.9$, $\theta = 0.4$, $\eta = 0.5$, $p = 0.8$, $q = 0.2$, $\Phi = u_1 = 0$, inspection rate h = 0, $\theta_1 = 1$, $x = 1$ and $y = 1$ in equation (103) and then differentiating with respect to r_1, we get $\dfrac{\partial R(s)}{\partial r_1}$. Taking inverse Laplace transformation gives

$\dfrac{\partial R(t)}{\partial r_1}$. Then varying time as t = 0, 1, 2, 3, 4, 5, 6, 7, 8, 9, 10 we get sensitivity for system reliability for r_1.

Using the same procedure described above, we can get sensitivity for system reliability for r_2, l_1, l_2, θ and η.

Now we perform a sensitivity analysis of changes in M.T.T.F. with respect to r_1, r_2, l_1 and l_2. Setting $\lambda_1 = 0.1$, $\lambda_2 = 0.2$, $\lambda_3 = 0.3$, $r_2 = 0.7$, $l_1 = 0.8$, $l_2 = 0.9$, $\theta = 0.4$, $\eta = 0.5$, $p = 0.8$, $q = 0.2$, $\Phi = u_1 = 0$, inspection rate h = 0, $\theta_1 = 1$, $x = 1$ and $y = 1$ in equation (103) then using equation (107), we get

$$M.T.T.F. = 3.676470588 \frac{2.064 \ r_1 + 1.1344}{.40 + 1.2r_1}$$

Differentiating it with respect to r_1, we have

$$\frac{\partial(M.T.T.F.)}{\partial r_1} = 7.588235294 \frac{1}{.40 + 1.2r_1} - \frac{4.411764706 \ (2.064 \ r_1 + 1.1344)}{(.40 + 1.2r_1)^2}$$

Using the same procedure,

$$\frac{\partial(M.T.T.F.)}{\partial r_2}, \frac{\partial(M.T.T.F.)}{\partial l_1}, \frac{\partial(M.T.T.F.)}{\partial l_2}, \quad \frac{\partial(M.T.T.F.)}{\partial \theta} \text{ and } \frac{\partial(M.T.T.F.)}{\partial \eta}$$

can be obtained. Numerical results of the sensitivity analysis for the system reliability and the M.T.T.F. are presented in Figures 6, 7 and 8 and Tables 10 and 11.

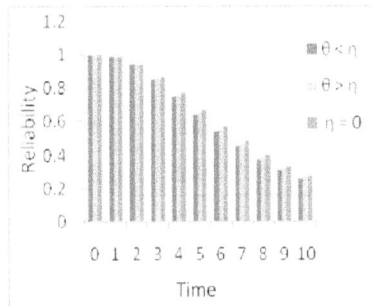

Fig. 2: Time vs. Reliability

Table 2: Time vs. Reliability

Time	Reliability ($\theta < \eta$)	Reliability ($\theta > \eta$)	Reliability ($\theta = 0$)
0	1	1	1
1	0.989534193	0.989749412	0.990298585
2	0.939730367	0.941957616	0.945325068
3	0.854358745	0.860471612	0.867853755
4	0.750955592	0.761089018	0.772022138
5	0.644533282	0.657543085	0.670765808
6	0.544304521	0.558695326	0.572864201
7	0.454716286	0.469204483	0.483240905
8	0.377170988	0.390873116	0.404053446
9	0.311392703	0.323807961	0.335726764
10	0.256308771	0.267226519	0.277716979

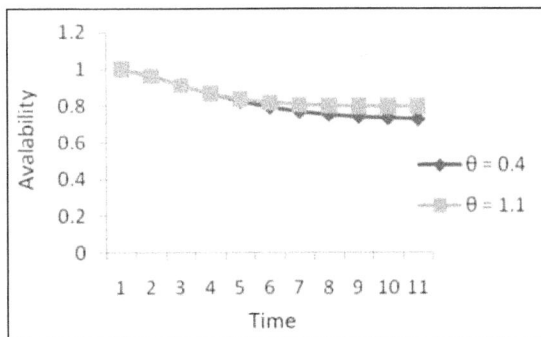

Fig. 3: Time vs. Availability

Table 3: Time vs. Availability

Time	Availability $(\theta < \eta)$	Availability $(\theta > \eta)$
0	1	1
1	0.962713949	0.962312151
2	0.911358034	0.911092032
3	0.864672675	0.868862923
4	0.825382205	0.838703524
5	0.794348801	0.819421975
6	0.771077035	0.808261632
7	0.754299066	0.802409414
8	0.742549602	0.799669145
9	0.734496325	0.798576054
10	0.729064034	0.798261434

Table 4: M.T.T.F. with Respect to Normal Weather Rate

θ	M.T.T.F.					
	$\lambda_1=.1, \lambda_2=.2,$ $\lambda_3=.3, \eta=.5,$ $r_1=.6, r_2=.7,$ $l_1=.8, l_2=.9$	$\lambda_1=.1, \lambda_2=.2,$ $\lambda_3=.3,$ $\eta=1.4,$ $r_1=.6, r_2=.7,$ $l_1=.8, l_2=.9$	$\lambda_1=.1, \lambda_2=.2,$ $\lambda_3=.3, \eta=.5,$ $r_1=.8, r_2=.7,$ $l_1=.8, l_2=.9$	$\lambda_1=.1, \lambda_2=.2,$ $\lambda_3=.3, \eta=.5,$ $r_1=.6, r_2=.9,$ $l_1=.8, l_2=.9$	$\lambda_1=.1, \lambda_2=.2,$ $\lambda_3=.3, \eta=.5,$ $r_1=.6, r_2=.7,$ $l_1=1.1, l_2=.9$	$\lambda_1=.1, \lambda_2=.2,$ $\lambda_3=.3, \eta=.5,$ $r_1=.6, r_2=.7,$ $l_1=.8, l_2=1.2$
0.1	7.719703979	7.541185145	7.492236027	7.600472814	7.567613831	7.540510074
0.2	7.74590164	7.562721965	7.5068306	7.61111111	7.597066436	7.575949365
0.3	7.768758226	7.582664528	7.519379844	7.620545075	7.623389662	7.607204118
0.4	7.788865545	7.60118213	7.530276816	7.628968254	7.647058828	7.634966777
0.5	7.806684839	7.618421053	7.539821292	7.636534839	7.668457371	7.659786004
0.6	7.822580645	7.634508347	7.548245618	7.643369177	7.687898089	7.682101976
0.7	7.836844685	7.649554871	7.555732488	7.64957265	7.705638699	7.702271823
0.8	7.849713059	7.663657769	7.562427411	7.655228757	7.721893491	7.720588235
0.9	7.861378803	7.676902537	7.568447636	7.660406887	7.736842103	7.737293324
1	7.872001216	7.689364691	7.57388862	7.665165162	7.750636287	7.752589041

Table 5: Time vs. Expected Profit

Time	$E_p(t)$			
	$\theta = 0.4 < \eta$		$\theta = 1.1 > \eta$	
	$K_2 = 0.2$	$K_2 = 0.4$	$K_2 = 0.2$	$K_2 = 0.4$
0	0	0	0	0
1	0.785144282	0.585144282	0.785030771	0.585030771
2	1.522007875	1.122007875	1.521328018	1.121328018
3	2.209483495	1.609483495	2.21033238	1.61033238
4	2.853835131	2.053835131	2.863119994	2.063119994
5	3.46301617	2.46301617	3.491381612	2.491381612
6	4.045128398	2.845128398	4.10467046	2.90467046
7	4.607336659	3.207336659	4.70966666	3.30966666
8	5.155400942	3.555400942	5.310518175	3.710518175
9	5.693664923	3.893664923	5.909547334	4.109547334
10	6.225264242	4.225264242	6.507925325	4.507925325

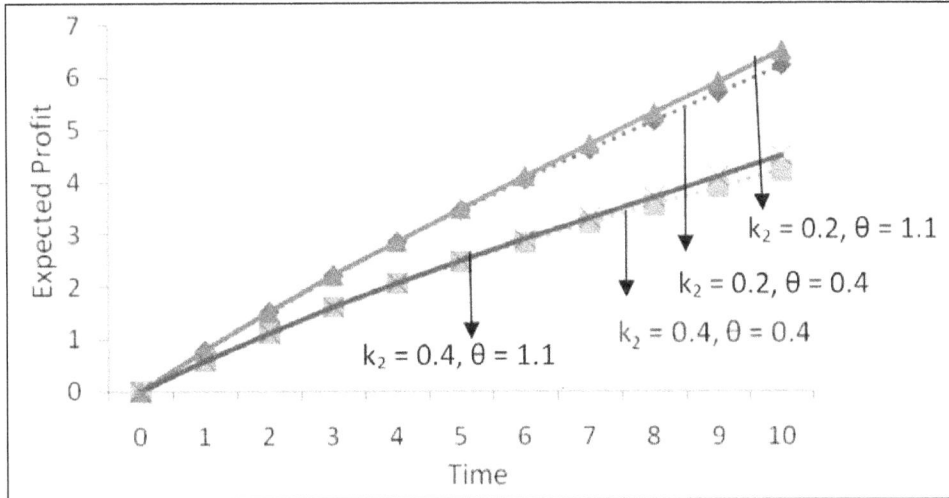

Fig. 5: Time vs. Expected Profit

Table 6: Sensitivity Analysis for System Reliability with Cases $r_1 = 0.6$ $r_2 = 0.7$, $l_1 = 0.8$, $l_2 = 0.9$

Time	$\dfrac{\partial R(t)}{\partial r_1}$	$\dfrac{\partial R(t)}{\partial r_2}$	$\dfrac{\partial R(t)}{\partial l_1}$	$\dfrac{\partial R(t)}{\partial l_2}$
0	0	0	0	0
1	−0.012612395	−0.009355376	−0.001630409	−0.002343675
2	−0.054928248	−0.037418525	−0.01186575	−0.015887907
3	−0.102366172	−0.066307314	−0.028417172	−0.03576278
4	−0.136289981	−0.085611783	−0.043982441	−0.052496175
5	−0.152241355	−0.093769296	−0.054244629	−0.061927088
6	−0.153237654	−0.093127316	−0.058460773	−0.064325774
7	−0.144339849	−0.086875424	−0.05779622	−0.061708383
8	−0.130100709	−0.077734755	−0.053935102	−0.056205258
9	−0.113806533	−0.067613831	−0.048376398	−0.049449688
10	−0.097520733	−0.057680066	−0.042208814	−0.042498625

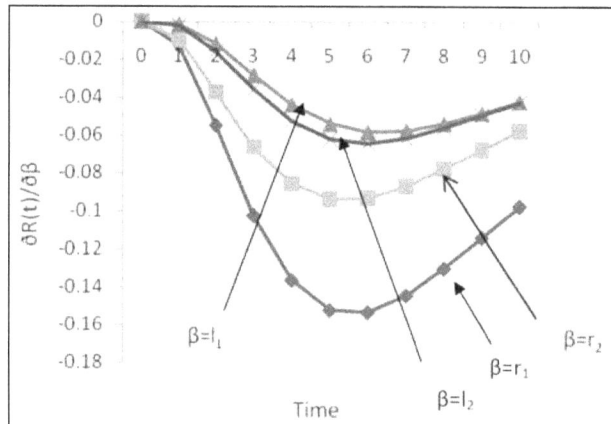

Fig. 6: Sensitivity Analysis for System Reliability with Cases $r_1 = 0.6$ $r_2 = 0.7$, $l_1 = 0.8$, $l_2 = 0.9$

Table 7: Sensitivity Analysis for System Reliability with Different Values of θ

Time	$\dfrac{\partial R(t)}{\partial \theta}$	
0	0	0
1	0.000158113	0.000176755
2	0.001788754	0.002170644
3	0.005153364	0.006626218
4	0.00877192	0.011730629
5	0.011426177	0.015676946
6	0.012738941	0.017760909
7	0.012878282	0.018127054
8	0.012203821	0.017267563
9	0.011066044	0.015696355
10	0.009731889	0.013814823

Table 8: Sensitivity Analysis for System Reliability with Different Values of η

Time	$\dfrac{\partial R(t)}{\partial \eta}$	
0	0	0
1	−0.001357609	−0.001186059
2	−0.008995712	−0.007109187
3	−0.019989491	−0.014688859
4	−0.029175294	−0.020337812
5	−0.03438224	−0.023058268
6	−0.03577841	−0.023312611
7	−0.034433949	−0.021953096
8	−0.031481689	−0.019740934
9	−0.027801816	−0.017214724
10	−0.023975535	−0.014704084

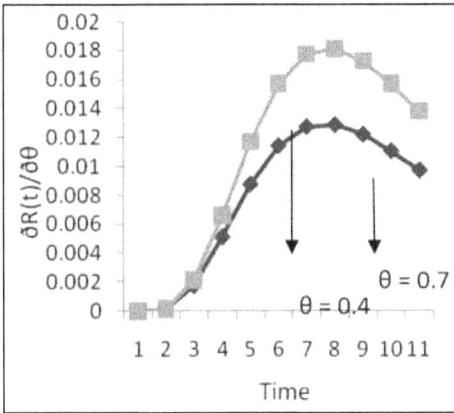

Fig. 7: Sensitivity Analysis for System Reliability with Different Values of θ

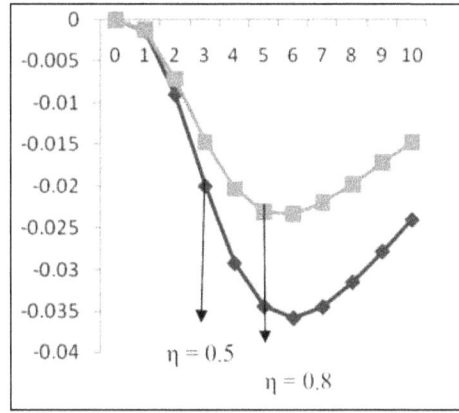

Fig. 8: Sensitivity Analysis for System Reliability with Different Values of η

Table 9: Sensitivity Analysis for the M.T.T.F. with Various Values of r_1, r_2, l_1 and l_2

r_1	$\dfrac{\partial (M.T.T.F.)}{\partial r_1}$	r_2	$\dfrac{\partial (M.T.T.F.)}{\partial r_2}$	l_1	$\dfrac{\partial (M.T.T.F.)}{\partial l_1}$	l_2	$\dfrac{\partial (M.T.T.F.)}{\partial l_2}$
0.1	−7.28332754	0.3	−2.49635568	0.6	−0.947712419	0.6	−0.588329083
0.3	−3.409646405	0.5	−1.455726091	0.8	−0.611963536	0.9	−0.611963536
0.6	−1.570003002	0.7	−0.952329459	1	−0.427515626	1.2	−0.625704497
0.9	−0.899110557	0.9	−0.671173818	1.2	−0.315436826	1.5	−0.634688973
1.2	−0.581702434	1.1	−0.498367069	1.4	−0.242282243	1.8	−0.641022292

Table 10: Sensitivity Analysis for the M.T.T.F. with Various Values of θ and η

θ	$\frac{\partial(M.T.T.F.)}{\partial\theta}$	η	$\frac{\partial(M.T.T.F.)}{\partial\eta}$
0.2	0.244154798	0.1	−0.797
0.4	0.188954257	0.3	−0.520758518
0.6	0.150364204	0.5	−0.364667918
0.8	0.122379628	0.7	−0.268567037
1	0.101468813	0.9	−0.205503973

10. CONCLUSIONS

The numerical results obtained indicate that reliability of the system decreases rapidly as the time increases w. r. t. all three cases (i) when the normal weather rate is less than abnormal weather rate (ii) when the normal weather rate is greater than the abnormal weather rate (iii) when the system is operated under perfect normal weather *i.e.* η = 0. One can observe from Figure 2 that the reliability of the system in different cases: (i) < (ii) < (iii). From Figure 3 one can easily conclude that the availability of the system decreases with the increment in time and later on it stabilizes at value 0.72. In this case θ < η but if we increases normal weather rate then availability of the system also decreases but the values obtained in this case is better than the values obtained in previous case.

It is clear from Table 4 that M.T.T.F. of the system goes on increasing with the increase of normal weather rate (θ) and it is greater when normal weather rate (θ) > abnormal weather rate (η) and lesser when θ < η. However there is a decline in the value of M.T.T.F when the failures rates r_1, r_2, l_1, l_2 and η are increased.

Fixing revenue cost per unit time at 1 and varying service cost at 0.2 and 0.4, one can obtain Table 5 and Figure 5 which represent (i) expected profit with respect to time when θ < η (ii) expected profit w. r. t. time when θ > η. It is very clear that the profit decreases as the service cost increases in both the cases but expected profit when θ > η is greater than expected profit when θ < η as time increases.

In Figure 6, along the time coordinate, we show the sensitivity of system reliability with respect to r_1 at 0.6, r_2 at 0.7, l_1 at 0.8 and l_2 at 0.9. It reveals that the sensitivity initially decreases and then tends to increase and attain a value −0.0975, −0.0576, −0.0422 and −0.0424 with respect to r_1, r_2, l_1 and l_2 respectively. It is clear from the graph that system reliability is more sensitive w. r. t. l_1. Figures 7 and 8 represent the sensitivity performance of the system reliability with respect to normal weather rate (θ) and abnormal weather rate (η). It is clear from the Figure 7 that when θ < η then system reliability is more sensitive and when θ > η then system reliability is less sensitive. From Figure 8 we can also see that when the abnormal weather rate is high system reliability is more sensitive. Moreover, Tables 9 and 10 show that sensitivity analysis for the M.T.T.F. with respect to r_1, r_2, l_1, l_2, η and θ. The signs of the values in these tables indicate an increase or decrease in the M.T.T.F. by changing the values of r_1, r_2, l_1, l_2, η and θ. We observe that sensitivity for the M.T.T.F. increases as r_1, r_2, l_1 and l_2 increases. It is clear from Table 10 that when we increase normal weather rate, the sensitivity for M.T.T.F.

decreases and it increases as we increase abnormal weather rate. Thus, the study reveals that the system performances can be made better if it is operated under normal weather.

REFERENCES

[1] Billinton, R., Wu, C. and Singh, G. (2002), "Extreme adverse weather modeling in transmission and distribution system reliability evaluation", in *14th PSCC, 24–28 June*, Sevilla, pp. 1–7.

[2] Goel, L.R. Gupta, R. and Rastogi, A. K. (1985), "Cost analysis of a system with partial failure mode and abnormal weather conditions", *Microelectronics and Reliability*, Vol. 25, No. 3, pp. 461–466.

[3] Gupta, R. and Goel, R. (1991), "Profit analysis of a two-unit cold standby system with abnormal weather condition", *Microelectronics Reliability*, Vol. 31, No. 1, pp. 1–5.

[4] Heidtmann, K.D. (1981), "A class of non-coherent systems and their reliability analysis", *Proceedings of 11th Annual Symposium Fault Tolerant Comput.* pp. 96–98.

[5] Jain, M. and Ghimire, R.P. (1997), "Reliability of k-r-out-of-N: G system subject to random and common cause failure", *Performance Evaluation an International journal*, Vol. 29, No. 3, pp. 213–218.

[6] Nelson, R.B. (2006), *An Introduction to Copulas, 2nd ed.,* Springer, New York, NY.

[7] Pham, H. and Upadhyaya, S. J. (1988), "The efficiency of computing the reliability of k-out-of-n system", *IEEE Transactions on Reliability,* Vol. 37, No. 5, pp. 521–523.

[8] Sen, P.K. (2003), "Copulas: concepts and novel applications", METRON-*International journal of statistics*, Vol. LXI, No. 3, pp. 323–353.

[9] Shao, J. and Lamberson, L.R. (1991), "Modelling a shared load k-out-of-n: G system", IEEE *Transactions on Reliability*, Vol. 40, No. 2, pp. 205–209.

[10] Upadhyaya, S. J. and Pham, H. (1993), "Analysis of non-coherent system and architecture for the computation of system reliability", *IEEE Transactions on Computers*, Vol. 42, No. 4, pp. 484–493.

Safety Critical Software of Software Reliability Growth Model Considering Log-Logistic Testing-Effort and Imperfect Debugging

S.P.V.N.D. Suneetha[1] and O. NagaRaju[2]

Acharya Nagarjuna University, Hyderabad–500020
E-mail: [1]suneetha.may23@gmail.com; [2]onrajunrt@gmail.com

ABSTRACT: The Log-logistic software reliability growth model that can capture the increasing/decreasing nature of the failure occurrence rate per fault Gokhale and Trivedi (1998). In this paper, we will first show that a Log-logistic Testing-Effort Function (TEF) can be expressed as a software development/testing-effort expenditure curve. We investigate how to incorporate the Log-logistic TEF into logistic software reliability growth models based on Non-Homogeneous Poisson Process (NHPP). The models parameters are estimated by Least Square Estimation (LSE) and Maximum Likelihood Estimation (MLE) methods.[1] The methods of data analysis and comparison criteria are presented. The experimental results from actual data applications show good fit. A comparative analysis to evaluate the effectiveness for the proposed model and other existing models are also performed.[2] Results show that the proposed models can give fairly better predictions. Therefore, the Log-logistic TEF is suitable for incorporating into logistic growth models. In addition, the proposed models are discussed under imperfect debugging environment.

Keywords: Software Reliability Growth Models, Testing-Effort Functions, Software Testing, Imperfect Debugging, Logistic Growth Model, Estimation Methods.

1. INTRODUCTION

The size and complexity of computer systems has grown significantly during the past decades. Computers are used in medical fields, businesses, chemical labs, air traffic control towers, ships, space ships, home appliances, communication, manufacture and many more. Software is a functioning element embedded in computers that plays vital role in the modern life. Errors are bound to happen as software is written by humans. Before, the focus was only on the design and reliability of the hardware. But, now increase in the demand of software has led to the study of the high quality reliable software development. Reliability is the most important aspect since it measures software failures during the process of software development. Software reliability is defined as the probability of failure free operation of a computer program for a specified time in a specified environment.[3] Many researches have been conducted over the past decades[4] and still going on, to study the software reliability. A common approach for measuring software reliability is by using an analytical model whose parameters are generally estimated from available data on software failures.[5,6]

A Software Reliability Growth Model (SRGM) is a mathematical expression of the software error occurrence and the removal process. In early 1970's, many Software Reliability Growth Models (SRGMs) have been proposed.[4,7,8] A Nonhomogeneous Poisson Process (NHPP) as the stochastic process has been widely used in SRGM. In the past years, several SRGMs based on NHPP which incorporates the Testing-Effort Functions (TEF) have been proposed by many authors.[2,6,9,10,11] The testing-effort can be represented as the number of CPU hours, the number of executed test cases, etc.[12, 13] Most of these works on SRGMs modified the exponential NHPP growth model[14] and incorporated the concept of testing-effort into an NHPP Model to describe the software fault detection phenomenon. Recently, Bokhari and Ahmad (2006)[1] also proposed a new SRGM with the Log-logistic testing-effort function to predict the behavior of failure and fault of software. However, the exponential NHPP growth model is some times insufficient and inaccurate to analyze real software failure data for reliability assessment. In this paper we show how to integrate a Log-logistic testing-effort function into logistic NHPP growth models[1, 15] to get a better description of the software fault detection phenomenon. The parameters of the model are estimated by Least Square Estimation (LSE) and Maximum Likelihood Estimation (MLE) methods. The statistical methods of data analysis are presented and the experiments are performed based on real data sets and the results are compared with other existing models. The experimental results show that the proposed SRGM with Log-logistic testing-effort function can estimate the number of initial faults better than that of other models and that the Log-logistic testing-effort functions is suitable for incorporating into logistic NHPP growth model. Further, the analyses of the proposed models under imperfect debugging environment are also discussed.

2. SRGM WITH TEF

A Software Reliability Growth Model (SRGM) explains the time dependent behavior of fault removal. The objective of software reliability testing is to determine probable problems with the software design and implementation as early as possible to assure that the system meets its reliability requirements. Numerous SRGMs have been developed during the last three decades and they can provide very useful information about how to improve reliability.[4,16] Among these models, exponential growth model and logistic growth model have been shown to be very useful in fitting software failure data. Many authors incorporated the concept of testing-effort into exponential type SRGM based on the NHPP to get a better description of the fault detection phenomenon. The testing-effort indicates how the errors are detected effectively in the software and can be modeled by different distributions[17,18] proposed the Log-logistic SRGM that can capture the increasing/decreasing nature of the failure occurrence rate per fault. Recently, Bokhari and Ahmad,[1] and Ahmad *et al.*[1,12,19] also presented how to use the Log logistic curve to describe the time-dependent behaviour of testing-effort consumptions during testing. The Cumulative testing-effort expenditure consumed in (0,t] is depicted in the following:

$$W(t) = \alpha(1 - e^{-\beta t^m})$$

... (1)

Therefore, the current testing-effort expenditure at testing *t* is given by:

$$w(t) = \alpha\beta \ t^{m-1} \ m e^{-\beta t^m}$$

... (2)

Where α is the total amount of testing-effort consumption required by software testing, β is the scale parameter, and δ is the shape parameter. The testing-effort $w(t)$ reaches its maximum value at time,

$$w(t) \frac{\alpha\left(\frac{t}{\lambda}\right)^{-\beta} \beta}{\left(1 + \left(\frac{t}{\lambda}\right)^{-\beta}\right)^2 t}$$

... (3)

The logistic NHPP software reliability growth model is known as one of the flexible SRGMs that can depict both exponential and S-shaped growth curves depending upon the parameter values. The model has been shown to be useful in fitting software failure data.[20] Ohba proposed that the fault removal rate increases with time and assumed the presences of two types of errors in the software. Later, modified the logistic model and incorporated the testing-effort in an NHPP model. Therefore, we show how to incorporate Log-logistic testing-effort function into logistic NHPP model. The extended logistics GM with Log-logistic testing-effort function is formulated on the following assumptions:

1. The software system is subject to failures at random times caused by errors remaining in the system.
2. Error removal phenomenon in software testing is modeled by NHPP.
3. The mean number of errors detected in the time interval $(t, t + \Delta t]$ by the current testing-effort expenditures is proportional to the mean number of detectable errors in the software.
4. The proportionality increases linearly with each additional error removal.
5. Testing-effort expenditures are described by the Log logistic TEF.
6. Each time a failure occurs, the error causing that failure is immediately removed and no new errors are introduced.
7. Errors present in the software are of two types: mutually independent and mutually dependent. The mutually independent errors lie on different execution paths, and mutually dependent errors lie on the same execution path. Thus, the second type of errors is detectable if and only if errors of the first type have been removed. According to these assumptions, if the error detection rate with respect to current testing-effort expenditures is proportional to the number of detectable errors in the software and the proportionality increases linearly with each additional error removal, we obtain the following differential equation:

$$\frac{dm(t)}{dt} \times \frac{1}{w(t)} = \emptyset(t)(n(t) - m(t))$$

... (4)

$$\emptyset(t) = b\left[r + (1 - r)\frac{m(t)}{n(t)}\right]$$

... (5)

And $n(0) = a$

r (>0) is the inflection rate and represents the proportion of independent errors present in the software, $m(t)$ be the Mean Value Function (MVF) of the expected number of errors detected in time $(0,t]$, $w(t)$ is the current testing effort expenditure at time t, a is the expected number of errors in the system, and b is the error detection rate per unit testing-effort at time t. Solving (3) with the initial condition $t = 0$, $W(t) = 0$, $m(t) = 0$, we obtain the MVF,

$$m(t) = \frac{a[1 - e^{-b(1-r) W(t)}]}{(1 - \gamma) + \left[(1 - r)/r\right]e^{-b(1-r) W(t)}}$$ ... (6)

If the inflection rate r = 1, the above NHPP model becomes equivalent to the exponential growth model.

The failure intensity at testing time t of the logistic NHPP model with testing-effort is given by,

$$\lambda(t) = \frac{ab[1 - r\gamma]w(t)e^{-b(1-r) W(t)}}{1 - \gamma\left[(r(1 - \gamma) + (1 - r)e^{-b(1-r) W(t)})\right]}$$ .... (7)

Furthermore, we describe a flexible SRGM with mean value function considering the Log-logistic testing-effort expenditure as,

$$m_{remaining}(t) = \frac{a(1 - r\gamma)e^{-b(1-r) W(t)}}{r(1 - \gamma) + (1 - r)e^{-b(1-r) W(t)}}$$ ... (8)

In addition, the expected number of errors to be detected eventually is,

$$m(\infty) = \frac{a\left[1 - e^{-b\alpha}\right]}{1 + ((1-r)/r)e^{-b\alpha}}$$ ... (9)

2.1 Imperfect-Software Debugging Models

An NHPP model is said to have perfect debugging assumption when $a(t)$ is constant, i.e., no new faults are introduced during the debugging process. An NHPP SRGM subject to imperfect debugging was introduced by the authors with the assumption that if detected faults are removed, then there is a possibility that new faults with a constant rate γ are introduced.[21] Let $n(t)$ be the number of errors to be eventually detected plus the number of new errors introduced to the system by time t, we obtain the following system of differential equations:

$$\frac{dm(t)}{dt} \times \frac{1}{w(t)} = \phi(t)(n(t) - m(t)) .$$

and $$\frac{dn(t)}{dt} = \gamma\frac{dm(t)}{dt} .$$

where $$\phi(t) = b\left[r + (1 - r)\frac{m(t)}{n(t)}\right]$$

and $$n(0) = a .$$

Solving the above differential equations under the boundary conditions $m(0) = 0$ and $W(0) = 0$, we can obtain the following MVF of logistic model with Log-logistic testing-effort under imperfect debugging.

3. ESTIMATION OF MODEL PARAMETERS

The parameters of the SRGM are estimated based upon the failure data collected by the MLE and LSE techniques.[1,20] The performance of the proposed model is then compared with other existing models. Experiments on three real software failure data are performed.

3.1 Least Square Method

The parameters α, β, and δ in the Log-logistic TEF can be estimated using the method of LSE. These parameters are determined for n observed data pairs in the form (t_k, W_k) ($k = 1, 2, ..., n; 0 < t_1 < ... < t_n$) where, W_k is the cumulative testing-effort consumed in time $(0, t_k]$. The estimators α, β and δ, which contribute the model with a greater fitting, can be obtained by minimizing:[22,23] ▣ Differentiating S with respect to α, β, and δ setting the partial derivatives to zero, the set of nonlinear equations are obtained respectively. Thus, the LSE of α is given by,

$$\hat{\alpha} = e^{\left[\sum_{k=1}^{n} l_k \ln W_k - \theta \sum_{k=1}^{n} l_k \ln\left(1 - e^{-\beta \cdot t_k^\delta}\right)\right] / \sum_{k=1}^{n} l_k}$$

$$\hat{\theta} = \frac{\sum_{k=1}^{n} l_k \ln W_k \ln\left(1 - e^{-\beta \cdot t_k^\delta}\right) - \ln\alpha \cdot \sum_{k=1}^{n} l_k \ln\left(1 - e^{-\beta \cdot t_k^\delta}\right)}{\sum_{k=1}^{n} l_k \left(\ln\left(1 - e^{-\beta \cdot t_k^\delta}\right)\right)^2}$$

The LSE of β and δ can be obtained numerically by solving the following equations:

$$\frac{\partial S}{\partial \beta} = 2\sum_{k=1}^{n} l_k \left\{\ln W_k - \ln\alpha - \theta \ln\left(1 - e^{-\beta \cdot t_k^\delta}\right)\right\} \cdot \left(-\frac{\theta \cdot t_k^\delta \cdot e^{-\beta \cdot t_k^\delta}}{1 - e^{-\beta \cdot t_k^\delta}}\right) = 0$$

3.2 Maximum Likelihood Method

Once the estimates of α, β, and δ are known, the parameters of the SRGM can be estimated through MLE method. The estimators of a, b, and r are determined for the n observed data pairs in the form (t_k, y_k) ($k = 1, 2, ..., n; 0 < t_1 < t_2 < ... < t_n$) where y_k is the cumulative number of software errors detected up to time t_k or $(0, t_k]$. Then the likelihood function for the unknown parameters a, b, and r in the NHPP model is given by,[25]

$$L'(a, b, r) = \prod_{k=1}^{n} \frac{\left[m(t_k) - m(t_{k-1})\right]^{(y_k - y_{k-1})}}{(y_k - y_{k-1})!} \cdot e^{-[m(t_k) - m(t_{k-1})]}$$

where, $t_0 \equiv 0$ and $y_0 \equiv 0$. The maximum likelihood estimates of SRGM parameters a, r and b can be obtained by solving the following three equations,

$$\hat{a} = \frac{y_n \left[1 + \lambda \phi_n\right]}{1 - \phi_n}$$

$$\frac{y_n}{r^2 \left[1 - \lambda\right]} + \frac{a\phi_n \left[1 - \phi_n\right]}{\left[r^2 \left[1 - \lambda \phi_n\right]\right]^2} = \sum_{k=1}^{n} \frac{(y_k - y_{k-1})\phi_k}{r^2 \left[1 - \lambda \phi_k\right]} - \sum_{k=1}^{n} \frac{(y_k - y_{k-1})\phi_{k-1}}{r^2 \left[1 - \lambda \phi_{k-1}\right]}$$

$$\sum_{k=1}^{n} \frac{(y_k - y_{k-1})\left[W(t_k)\phi_k - W(t_{k-1})\phi_{k-1}\right]}{\phi_{k-1} - \phi_k} + \sum_{k=1}^{n} \frac{(y_k - y_{k-1})\left[\lambda W(t_k)\phi_k\right]}{1 + \lambda \phi_k}$$

$$+\sum_{k=1}^{n}\frac{(y_k - y_{k-1})\left[\lambda W'(t_{k-1})\phi_{k-1}\right]}{1+\lambda\phi_{k-1}} = \frac{aW'(t_n)\phi_n\left[1-\lambda\right]}{\left[1+\lambda\phi_{n-1}\right]^2}$$

where $\quad \phi_k = e^{-bW(t_k)}, k = 1, 2, ..., n,$ and $\lambda = \dfrac{1-r}{r}$

The above can be solved by numerical methods to get the values of $a\hat{}$, $b\hat{}$ and $r\hat{}$.

4. DATA ANALYSIS AND EXPERIMENTS

4.1 Comparison Criteria

To check the performance of proposed SRGM with Log-logistic TEF, we use the following four criteria:

1. The Accuracy of Estimation (AE) is defined as,[26,27]

$$AE = \left|\frac{M_a - a}{M_a}\right|$$

   where, $a M$ is the actual cumulative number of detected errors after the test, and a is the estimated number of initial errors. For practical purposes, $a M$ is obtained from software error tracking after software testing.

2. The Mean of Squared Errors (MSE) (Long-termpredictions) is defined as,

$$MSE = \frac{1}{k}\sum_{j=1}^{k}\left[m(t_j) - m_i\right]^2$$

   where, $m(t_i)$ is the expected number of errors at time t_i estimated by a model, and m_i is the expected number of errors at time t_i. MSE gives a quantitative comparison for long-term predictions. A smaller MSE indicates a minimum fitting error and better performance.

3. The coefficient of multiple determinations is defined as,[28]

$$R^2 = \frac{S\left(\hat{\alpha}, 0.1\right) - S\left(\alpha, \hat{\beta}, \hat{\delta}\right)}{S\left(\hat{\alpha}, 0.1\right)}$$

   where, $\hat{\alpha}$ is the LSE of α for the model with only a constant term, that is, $\beta = 0$, and $\delta = 1$ in (12). It is

$$\ln \hat{\alpha} = \frac{1}{n}\sum_{k=1}^{n}\ln W_k$$

   Given by, Therefore, $R2$ measures the percentage of total variation about the mean accounted for by the fitted model and tells us how wella curve fits the data. It is frequently employed to compare models and assess which model provides the best fit to the data. The best model is the one which provides the higher R^2, that is, closer to 1.[29,30,31] To investigate whether a significant trend exists in the estimated testing-effort, one could test the hypotheses H_0: $\beta = 0$ and $\delta = 1$, against H_1: $\beta \neq 0$ or at least $\delta \neq 1$ using F-test by merely forming the ratio,

$$F = \frac{\left[S\left(\hat{\alpha},0.1\right) - S\left(\alpha, \hat{\beta},\hat{\delta}\right)\right]/2}{S\left(\hat{\alpha},0.1\right)/(n-3)}.$$

If the value of F is greater than $F_\alpha(2, n-3)$, which is the α percentile of the F distribution with degrees of freedom 2 and $n-3$, we can be $(1-\alpha)100$ percent confident that H_0 should be rejected, that is, there is a significant trend in the testing-effort curve.

4. The Predictive Validity is defined (Musa *et al.*, 1987; Musa, 1999) as the capability of the model to predict future failure behavior from present and past failure behaviour. Assume that, we have observed q failures by the end of test time t_q. We use the failure data up to time t_e $(\leq t_q)$ to determine the parameters of $m(t)$. The ratio,

$$\frac{\hat{m}(t_q) - q}{q}$$

is called the relative error. Values close to zero for relative error indicate more accurate prediction and hence, a better model. We can visually check the predictive validity by plotting the relative error for normalized test time t_e/t_q.

4.2 Numerical Examples

Data Set 1: The first set of actual data is from the study by Ohba.[31,32] The system is PL/1 data base application software, consisting of approximately 1,317,000 lines of code. During nineteen weeks of testing, 47.65 CPU hours were consumed and about 328 software errors were removed. Moreover, the total cumulative number of detected faults after a long time of testing was 358. The Estimated parameters α, β, and δ of the Log-logistic TE Fare:

$$\alpha = 1511.2265, \beta = 0.0022, \delta = 1$$

Figure 1, shows the fitting of the estimated testing-effort by using above estimates. The fitted curves and the actual software data are shown by solid and dotted lines, respectively. The estimated values of the parameters a, b, and r in (5) are:

$$a = 256.63, b = 0.0263, r = 1$$

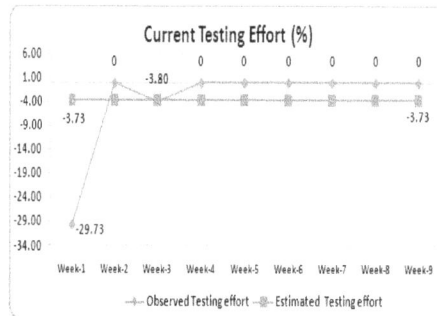

Fig. 1: Observed/Estimated Current Testing-Effort Function vs Time

Figure 2, illustrates a fitted curve of the estimated cumulative failure curve with the actual software data. The R^2 value for proposed Log-logistic TEF is 0.99574. Therefore, it can be said that the proposed curve is suitable for modeling the software reliability. Also, the calculated value F (= 4.9787) is greater than 0.05 F (2, 16). Therefore, it can be concluded that the proposed model is suitable for modeling the software reliability and the fitted testing effort curve is highly significant for this data set,

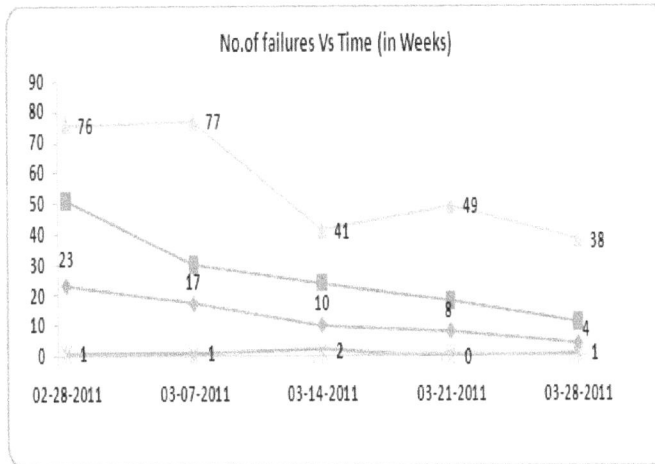

Fig. 2: Observed/Estimated Cumulative Number of Failures vs Time

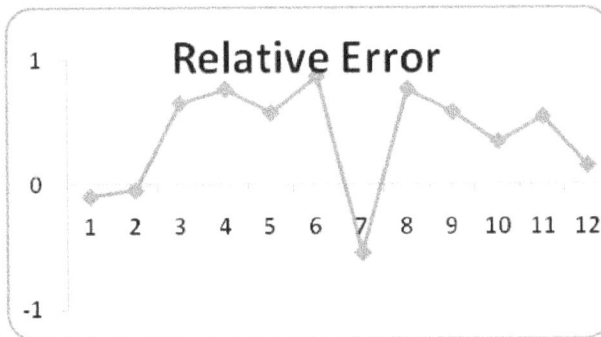

Fig. 3: Predictive Relative Error Curve

Table 1 lists the comparisons of proposed model with different SRGMs which reveal that the proposed model has better performance. Kolmogorov Smirnov goodness-of-fit test shows that the proposed SRGM fits pretty well at the 5 percent level of significance. Finally, the relative error in prediction of proposed model for this Data Set (DS) is calculated and illustrated by Figure 3. It is observed that relative error approaches zero as t_e approaches t_q and the error curve is usually within ±5 percent.[35]

Table 1: Comparison Results of Different SRGMs for DS1

Model	a (or) α	r (or) δ	b (or) β	AE%	MSE
Proposed model	256.63	1	0.0263	6.23	67.87
Bokhari Log-logistic model	565.73		0.0196	58.02	116.74
Yamada delayed S-shaped model	374.05		0.1976	4.48	168.67
Delayed S-Shaped with Logistic TEF	346.55		0.0936	3.20	147.61
Inflection S-shaped model	389.1	0.2	0.0935	8.69	133.53
G-O model	760.0		0.0323	112.29	139.82

Therefore, Figures 1 to 3 and Table 1 reveals that the proposed model has better performance than the other models. This model fits the observed data better, and predicts the future behavior well.

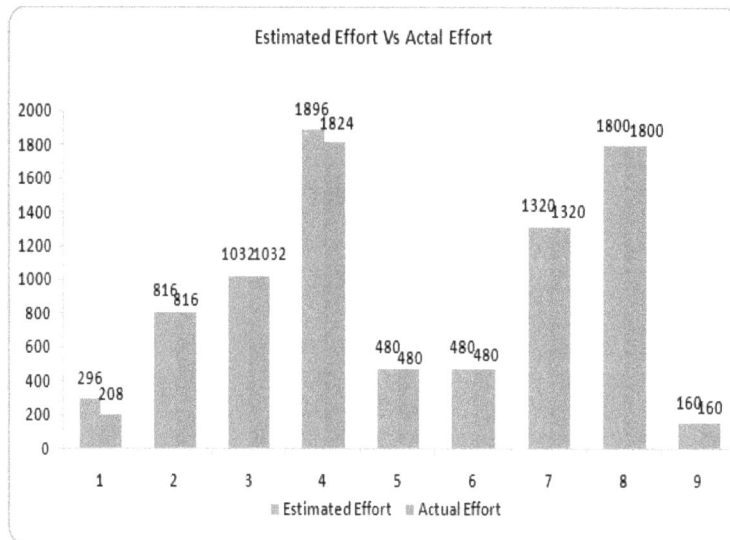

Fig. 4: Observed/Estimated Estimated Effort vs Actual Effort

Data Set2: The second set of actual data relates to the Release of Tandem Computer Project cited in Wood[36] from a subset of products for four separate software releases. In this research, only Release 1 is used for illustrations. There were 10000 CPU hours consumed, over the 20 week of testing and 100 software errors were removed. The estimated parameters of the TEF are obtained as:

$A = 10008.3743$, $\beta = 0.4835$, $\delta = 1$

Figure 4 illustrates the comparisons between the observed failure data and the estimated Log-logistic testing-effort data. Here, the fitted curves are shown as a solid and dotted line represents actual software data. The estimated parameters of the SRGM (5) are:

$A = 203.4875, b = 1.09 \times 10^{-4}, r = 1$

Table 2: Comparison Results of Different SRGMs for DS2

Model	a (or) α	r (or) δ	b (or) β	MSE
Proposed model	203.4875	1	1.09×10^{-4}	17.99
Bokhari Log-logistic model	135.91		0.000142	19.80
Yamada delayed S-shaped model	107.66		0.000266	22.76
Delayed S-haped with Logistic TEF	110.60		0.000226	39.69
Inflection S-shaped model	607.53	1.668	0.00854	87.44
G-O model	146.074		0.05156	52.36

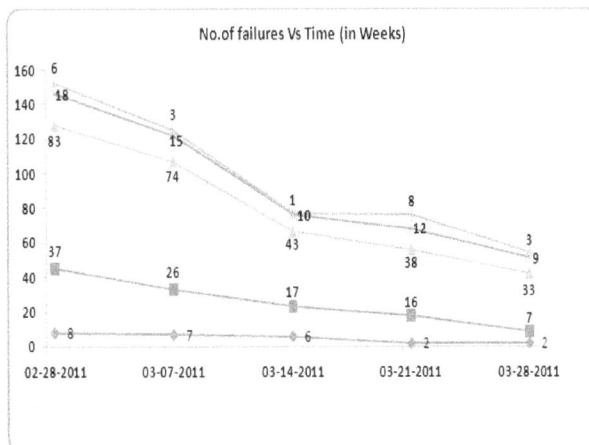

Fig. 5: Observed/Estimated Cumulative Number of Failures vs Time

Figure 5 illustrates a fitted curve of the estimated cumulative failure curve with the actual software data. The R^2 also known as the coefficient of determination depicts how well a curve fits the data.

A fit is more reliable when the value is closer to 1. The R_2 value for proposed Log-logistic TEF is 0.7860, which is very close to one. Moreover, the value of MSE is 18.04, which is very small compared to other SRGM. Table 2 lists the comparisons of proposed model with different SRGMs which reveal that the proposed model has better performance. It can therefore be observed that the Log-logistic TEF is suitable for modeling the proposed SRGM of this data set. Also the fitted testing-effort curve is significant since the calculated value F (= 7.3204) is greater than 0.05 F (2, 17). Kolmogorov Smirnov goodness-of-fit test shows that the proposed SRGM fits pretty well at the 5 percent level of significance.[37,38]

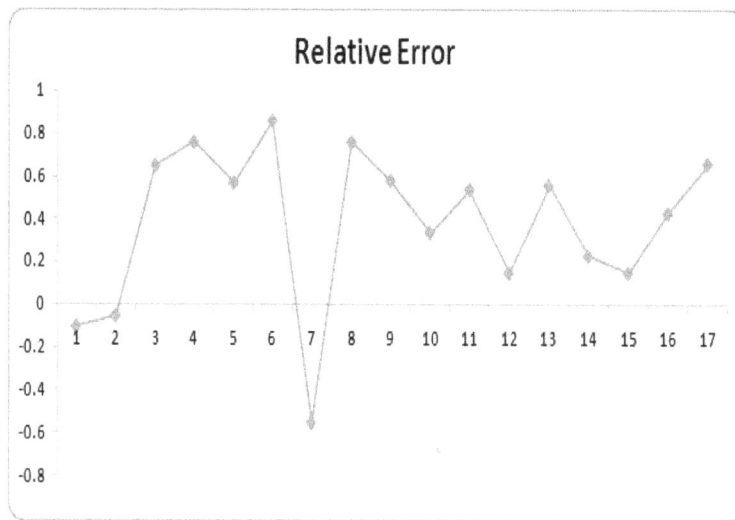

Fig. 1: Predictive Relative Error Curve

Following the work of Musa *et al.*,[39] the relative error in prediction for this data set is computed and the results are plotted in Figure 6. Figures 4–6 and Table 2 show that the proposed model has better performance and predicts the future behavior well.

4.3 Numerical Example on Imperfect Debugging

We consider the DS1 to discuss the issue of imperfect debugging for proposed SRGM. In order to validate the proposed SRGM under imperfect debugging, MSE is selected as the evaluation criterion. The parameters a, b, r in (8) can be solved by the method of MLE.

The estimated parameters with MSE of the proposed SRGM and some selected models for comparison under imperfect debugging. We observed that the value of MSE of the proposed SRGM with Log-logistic testing effort function is the lowest among all the models considered. Moreover, the estimated values $\delta = 1$ of all the models is close to but not equal to zero, thus the fault removal phenomenon may not be pure perfect debugging process. A fitted curve of the estimated cumulative number of failures with the actual software data and the RE curve for the proposed SRGM with Log-logistic testing-effort function with $\delta = 1$ under imperfect debugging.

5. CONCLUSION

In this paper, we have proposed a flexible SRGM based on NHPP model, which incorporates Log-logistic testing effort function with $\delta = 1$ into logistic model. The performance of the proposed SRGM is compared with other traditional SRGMs using different criteria. The results obtained show better fit and wider applicability of the proposed model on different types of real data applications. We conclude that the proposed flexible SRGM has better performance as compare to the other SRGMs and gives a reasonable predictive capability for the real failure data. We also conclude that the incorporated Log-logistic testing-effort function into logistic model is a flexible and can be used to describe the actual expenditure patterns more faithfully during software development. In addition, the proposed models under imperfect debugging environment are also discussed.

REFERENCES

[1] Ahmad, N., Bokhari, M.U., Quadri, S.M.K. and Khan, M.G.M. (2008), "The Exponen-tiated Weibull software reliability growth model with various testing-efforts and optimal release policy: a performance analysis", *International Journal of Quality and Reliability Management*, Vol. 25(2), pp. 211–235.

[2] Ahmad, N., Khan, M.G.M., Quadri, S.M.K. and Kumar, M. (2009), "Modeling and analysis of software reliability with Burr type X testing-effort and release-time determination", *Journal of Modelling in Management*, Vol. 4(1), pp. 28–54.

[3] Ahmad, N., Khan, M.G.M and Rafi, L.S. (2010), "A Study of Testing-Effort Dependent Inflection S-Shaped Software Reliability Growth Models with Imperfect Debugging", *International Journal of Quality and Reliability Management,* Vol. 27(1), 89–110.

[4] Ahmad, N., Khan, M.G.M. and Rafi, L.S. (2010a), "Software Reliability Modeling Incorporating Log-Logistic Testing-Effort with Imperfect Debugging", in *Proceedings of the International Conference on Modeling, Optimization and Computing (ICMOC–2010),* Durgapur, India, Published by American Institute of Physics, pp. 651–657.

[5] Bokhari, M.U. and Ahmad, N. (2006), "Analysis of a software reliability growth models: the case of log-logistic test-effort function", In: *Proceedings of the 17th IASTED International Conference on Modeling and Simulation (MS'2006),* Montreal, Canada, pp. 540–545.

[6] Bokhari, M.U. and Ahmad, N. (2007), "Software reliability growth modeling for Exponentiated Weibull functions with actual software failures data", *Advances in Computer Science and Engineering: Reports and Monographs, World Scientific Publishing Company,* Singapore, Vol. 2, pp. 390–396.

[7] Goel, A.L. and Okumoto, K. (1979), "Time dependent error detection rate model for software reliability and other performance measures", *IEEE Transactions on Reliability*, Vol. R-28, No. 3, pp. 206–211.

[8] Gokhale, S.S. and Trivedi, K.S. (1998), "Log-logistic software reliability growth model", In: *Proceedings of the 3rd IEEE International High Assurance Systems Engineering Symposium (HASE' 98),* Washington, USA, pp. 34–41.

[9] Huang, C.Y. (2005), "Performance analysis of software reliability growth models with testing-effort and change point", *Journal of Systems and Software*, Vol. 76, pp. 181–194.

[10] Huang, C.Y. and Kuo, S.Y. (2002), "Analysis of incorporating logistic testing-effort function into software reliability modeling", *IEEE Transactions on Reliability*, Vol. 51, No. 3, pp. 261–270.

[11] Huang, C.Y., Kuo, S.Y. and Chen, I.Y. (1997), "Analysis of software reliability growth model with logistic testing-effort function", In: *Proceeding of 8th International Symposium on IJCSNS International Journal of Computer Science and Network 170 Security,* Vol. 11, No. 1, January 2011 Software Reliability Engineering (ISSRE'1997), Albuquerque, New Mexico, pp. 378–388.

[12] Huang, C.Y., Kuo, S.Y. and Lyu, M.R. (2007), "An assessment of testing-effort dependent software reliability growth models", *IEEE Transactions on Reliability*, Vol. 56, No. 2, pp. 198–211.

[13] Kapur, P.K. and Garg, R.B. (1996), "Modeling an imperfect debugging phenomenon in software reliability", *Microelectronics and Reliability*, Vol. 36, pp. 645–650.

[14] Kapur, P.K., Garg, R.B. and Kumar, S. (1999), Contributions to Hardware and Software Reliability, World Scientific, Singapore.

[15] Kapur, P.K., Jha, P.C. and Bardhan, A.K. (2004), "Optimal Allocation of Testing Resource for a Modular Software", *Asia-Pacific Journal of Operational Research,* Vol. 21, No. 3, pp. 333–354.

[16] Kapur, P.K. and Younes, S. (1994), "Modeling an imperfect debugging phenomenon with testing effort", In: *Proceedings of 5th International Symposium on Software Reliability Engineering (ISSRE' 1994),* pp. 178–183.

[17] Khan, M.G.M., Ahmad, N. and Rafi, L.S. (2008), "Optimal Testing Resource Allocation for Modular Software Based on a Software Reliability Growth Model: A Dynamic Programming Approach", in: *Proceedings of the International Conference on Computer Science and Software Engineering (CSSE-2008),* Wuhan, China, IEEE Computer Society, pp. 759–762.

[18] Kumar, M., Ahmad, N. and Quadri, S.M.K. (2005), "Software reliability growth models and data analysis with a Pareto test-effort", *RAU Journal of Research*, Vol. 15 (1–2), pp. 124–128.

[19] Kuo, S.Y., Hung, C.Y. and Lyu, M.R. (2001), "Framework for modeling software reliability, using various testing efforts and fault detection rates", *IEEE Transactions on Reliability*, Vol. 50, No. 3, pp. 310–320.

[20] Lo, J.H. and Huang, C.Y. (2004), "Incorporating imperfect debugging into software fault correction process", In: *Proceeding of the IEEE Regional 10 Annual International Conference (TENCON'2004),* Chiang Mai, Thailand, pp. 326–329.

[21] Lyu, M.R. (1996), Handbook of Software Reliability Engineering, McGraw-Hill.

[22] Musa J.D. (1999), Software Reliability Engineering: More Reliable Software, Faster Development and Testing, McGraw-Hill.

[23] Musa, J.D., Iannino, A. and Okumoto, K. (1987), Software Reliability: Measurement, Prediction and Application, McGraw-Hill.

[24] Ohba, M. (1984), "Software reliability analysis models", *IBM Journal. Research Development,* Vol. 28, No. 4, pp. 428–443.

[25] Ohba, M. (1984a), "Inflection S-shaped software reliability growth models", Stochastic Models in Reliability Theory (Osaki, S. and Hatoyama, Y. Editors), pp. 144–162, Springer-Verlag Merlin.

[26] Pham, H. (2000), Software Reliability, Springer-Verlag, New York.

[27] Pham, H. (2007), "An imperfect-debugging fault-detection dependent para- meter software", *International Journal of Automation and Computing*, Vol. 4, No. 4, pp. 325–328.

[28] Pham, H., Nordmann L. and Zhang, X. (1999), "A general imperfect software debugging model with S-shaped fault detection rate", *IEEE Transactions on Reliability*, Vol. R–48, No. 2, pp. 169–175.

[29] Quadri, S.M.K., Ahmad, N. and Peer, M.A. (2008), "Software Optimal Release Policy and Reliability Growth Modeling", *In the Proceedings of 2nd National Conference on Computing for Nation Development*, INDIACom-2008, India, pp. 423–431.

[30] Quadri, S.M.K., Ahmad, N., Peer, M.A. and Kumar, M. (2006), "Nonhomogeneous Poisson process software reliability growth model with generalized exponential testing effort function", *RAU Journal of Research*, Vol. 16 (1–2), pp. 159–163.

[31] Shyur, H.J. (2003), "A stochastic software reliability model with imperfect and change point", *The Journal of Systems and Software*, Vol. 66, pp. 135–141.

[32] Wood, A. (1996), "Predicting software reliability", *IEEE Computers*, 11, pp. 69–77.

[33] Xie, M. (1991), Software Reliability Modeling, World Scientific Publication, Singapore.

[34] Xie, M. and Yang, B. (2003), "A study of the effect of imperfect debugging on software development cost", *IEEE Transaction on Software Engineering*, Vol. SE–29, No. 5, pp. 471–473.

[35] Yamada, S., Hishitani J. and Osaki, S. (1993), "Software reliability growth model with Weibull testing–effort: a model and application", *IEEE Transactions on Reliability*, Vol. R–42, pp. 100–105.

[36] Yamada, S. and Ohtera, H. (1990), "Software reliability growth models for testing effort control", *European Journal of Operational Research*, Vol. 46, 3, pp. 343–349.

[37] Yamada, S., Ohtera, H. and Norihisa, H. (1986), "Software reliability growth model with testing-effort", *IEEE Transactions on Reliability*, Vol. R-35, No. 1, pp. 19–23.

[38] Yamada, S., Ohtera, H. and Norihisa, H. (1987), "A testing effort dependent software reliability model and its application", *Microelectronics and Reliability*, Vol. 27, No. 3, pp. 507–522.

[39] Yamada, S. and Osaki, S. (1985), "Software reliability growth modeling: Models and applications", *IEEE Transaction on Software Engineering*, Vol. SE-11, No. 12, pp. 1431–1437.

[40] Yamada, S., Tokuno, K. and Osaki, S. (1992), "Imperfect debugging models with fault introduction rate for software reliability assessment", *International Journal of Systems Science*, Vol. 23, No. 12, pp. 2241–2252.

[41] Zhang, X., Teng, X. and Pham, H. (2003), "Considering fault removal efficiency in software reliability assessment", *IEEE Transaction on System, Man, and Cybernetics* - Part A, Vol. 33, No. 1, pp. 114–120.

Determining Optimum Software Release Time with Euler Distribution as a Prior for the Number of Undiscovered Bugs

Satadal Ghosh[1], Soumya Roy[2] and Ashis K. Chakraborty[3]

[1,2]Indian Institute of Science, Bangalore
[3]Indian Statistical Institute, Kolkata
E-mail: [1]satadal.ghosh@gmail.com; [2]roy.soumya.08@gmail.com;
[3]akchakraborty123@rediffmail.com

ABSTRACT: Optimum software release time under discrete set up has been studied by researchers recently. Bayesian decision making methodology using the actual observations from testing of software has been used by some researchers to arrive at an optimum release time for software. In such a framework, a prior distribution of the number of unidentified bugs in the software is assumed. The models then develop a maximum reward criterion to optimize the solution. In this paper, a specific prior distribution, namely, the Euler distribution is considered as the prior distribution for the number of unidentified bugs in the software and under this prior distribution, some interesting results for the optimum software release time are obtained. A sufficient condition for releasing the software considering an assumed reward criterion is derived under the afore-mentioned framework.

1. INTRODUCTION

In a typical software development project, static and dynamic testing includes different phases. Fixing a bug in the testing department of the software development organization itself is cheaper than doing the same in the client's premises. So testing at the developer's premises is quintessential though it involves enough cost-consuming processes. Also a fault in the developed software doesn't necessarily lead to a failure during operation by the client. So fixing all the bugs present in the software might be a tempting ideal condition, one can think of, but is practically infeasible and unachievable for the time and cost constraint in the competitive business scenario. Hence, there should be a judicial trade-off between the testing effort to be put in and the cost concerned with the testing. Therefore, finally the most sought after answer to the question 'When to stop testing and release the software' needs to be replied. This paper makes a contribution in this direction.

Musa et al.[1] discussed issues regarding the measurement and prediction of software reliability, while Pham[2] discussed about several practical problems in software reliability. Koch and Kubat[3] was one of the pioneers in the field of determination of optimum software release time. They developed a generic cost based decision procedure to determine the optimum software release time and gave a set of necessary and sufficient conditions for optimal software release time taking into account the cost benefit for the entire company. Musa and

Ackerman[4] introduced the basic concepts of software reliability measurement, e.g. Software failures and faults, operational profile, software failure occurrence, etc. They have discussed three models for software failure occurrence based on the potential of a fault to cause a failure in actual operation – static execution-time model, basic execution-time model and logarithmic Poisson execution-time model. Yang and Chao[5] defined two stopping criteria for debugging. In the first criteria, they considered whether the software had reached the desired reliability threshold; and in the other criteria, they considered whether testing costs could be justified by the growth in reliability. Based on these two criteria, they compared among various stopping rules. However, all these models were developed based on the change in software reliability in continuous time consideration. Chakraborty[6] analyzed the optimum software release time problem under discrete set up and developed a model assuming Euler Distribution[7,8] as the prior distribution for the number of unidentified bugs in the software and derived a one step posterior stopping rule, in a sufficient sense. However, in this model, Chakraborty[6] considered the tests in a case by case basis and assumed that a bug, once detected, gets repaired immediately, which is not the practice in most of the real life situations. Normally, in a typical software development scenario, debugging is carried out after testing the software either for a specified time or for a specified number of test cases. To overcome this drawback, later Das and Chakraborty[9] introduced the concept of test-phase, which they named as 'test bed'. A test phase was defined as a collection of several test cases. Furthermore, they assumed that the bugs, detected at any test bed, get repaired at the end of the corresponding test-phase.

In this paper, the testing process is depicted phase by phase, which is the most practical one. A model is developed assuming Euler Distribution as the prior distribution for the number of unidentified bugs in the software. An award or return function of testing is developed and the expected return is maximized in the form of a one test-bed posterior stopping rule in the sufficient sense. Also the necessary condition for continuation of testing is derived in terms the number of test-beds and total number of successes up to the corresponding test-bed in the form of maximum expected return.

2. MODEL DEVELOPMENT

2.1 Assumptions

While determining a strategy to stop testing, the following assumptions are considered:

1. In each phase of software testing, it is tested for a fixed number 'k' test cases. Each case will result in either a success (getting a bug) or a failure (not getting a bug). One can detect at most one bug by testing each case.

2. As and when the software fails, the faults are detected and the module containing the bug is identified and recorded. For a particular test phase (test bed), bugs are fixed after all the test cases in the test bed are run. Fixation of bug doesn't incur any additional cost.

3. When a fault is detected and fixed, a reward of 1 unit is achieved.

4. Cost (in the same unit in earlier assumption) incurred for testing each case is same and is 'c', such that c < 1.

2.2 Brief Note on Euler Distribution

Let $\pi(h,t,\lambda) = (\pi_0(h,t,\lambda), \pi_1(h,t,\lambda), \ldots)$ be the sequence of probabilities representing prior information about the number of undetected bugs at a time point 't' with a history 'h' regarding the detection of bugs and λ is a parameter of this prior distribution. Generalized Euler distribution comes as a convex combination of its posterior distributions as below:

$$\pi(h,t,\lambda) = \rho_1 \pi(hs, t+1, \lambda) + \rho_2 \pi(hf, t+1, \lambda) \qquad \ldots (1)$$

where, $\rho_1 + \rho_2 = 1$, $\rho_1 > 0$ & $\rho_2 \geq 0$ imply that,

$$\pi_n(h,t,\lambda) = \pi_0(h,t,\lambda)\tau^n[\prod_{r=0}^{n-1}(1-aq^r)]\,[\prod_{r=1}^{n}(1-q^r)]^{-1}, \tau = \frac{\lambda}{\rho_2}, a = \frac{\rho_2}{1-\lambda} \ldots (2)$$

$$\lambda = \sum_{i\geq 0} \pi_i(h,t,\lambda)(1-q^i) \qquad \ldots (3)$$

$\pi_n(hs, t+1, \lambda) = $ Probability that n bugs remain undetected in the software after a success.

$\pi_n(hf, t+1, \lambda) = $ Probability that n bugs remain undetected in the software after a failure.

Now, if $a = 0$ & $\rho_1 = 1$, the generalized Euler distribution reduces to the Euler distribution as below:

$$\pi_n(h,t,\lambda) = \pi_0(h,t,\lambda)\lambda^n[\prod_{r=1}^{n}(1-q^r)]^{-1} \qquad \ldots (4)$$

where, $\pi_0(h,t,\lambda) = \{S(\lambda,q)\}^{-1}$, where,

$$\{S(\lambda,q)\} = \sum_{i\geq 0} \frac{\lambda^i}{\prod_{r=0}^{i}(1-q^r)} \qquad \ldots (5)$$

and, $\pi_0(h,t,\lambda q) = \frac{\pi_0(h,t,\lambda)}{1-\lambda} \qquad \ldots (6)$

2.3 Euler Distribution as a Prior

Chakraborty[6] has shown that if Euler distribution i.e. $\pi(h,t,\lambda)$ is taken as a prior, then one step posterior is affected only by the failure, that is,

$$\pi(hs, t+1, \lambda) = \pi(h,t,\lambda) \qquad \ldots (7)$$

$$\pi(hf, t+1, \lambda) = \pi(h,t,\lambda q) \qquad \ldots (8)$$

Now, for a test bed consisting of 'k' test cases, the nature of the posterior distribution of undetected bugs after running all the test cases in the test bed will be derived. Let $\pi(h_x, t_x, \lambda)$ denote the prior distribution of the number of undiscovered bugs with history 'h_x' and 't_x' test cases used upto 'x'th test bed.

Lemma 1: Consider that in the 'x+1'th test bed, there are r $(0 \le r \le k)$ successes and $(k-r)$ failures. Then the one bed-posterior distribution of undetected bugs $\pi(h_x s^r f^{k-r}, t_{x+1}, \lambda)$ may be expressed as $\pi(h_x, t_x, \lambda q^{k-r})$, that is, the one bed-posterior distribution depends only on the number of failures in the test bed.

Proof: From (7) & (8), we have,

$$\pi(h_x s, t_x + 1, \lambda) = \pi(h_x, t_x, \lambda)$$

$$\pi(h_x f, t_x + 1, \lambda) = \pi(h_x, t_x, \lambda q).$$

We now consider, $\pi(h_x ss, t_x + 2, \lambda)$.

$$\pi_n(h_x ss, t_x + 2, \lambda) = \frac{\pi_{n+i}(h_x s, t_x + 1, \lambda) p_{n+i}}{\sum_{i \ge 0} \pi_i(h_x s, t_x + 1, \lambda) p_i}$$

$$= \frac{\pi_{n+i}(h_x, t_x, \lambda) p_{n+i}}{\sum_{i \ge 0} \pi_i(h_x, t_x, \lambda) p_i}$$

$$= \frac{\pi_0(h_x, t_x, \lambda) \lambda^{n-i} [\prod_{r=1}^{n+i}(1-q^r)]^{-i} p_{n+i}}{\lambda} \quad \text{(using (3) and from,}^{[8]}$$

$$p_i = 1 - q_i = 1 - q^i)$$

$$= \pi_n(h_x, t_x, \lambda) \text{ (using (4)).}$$

$$\therefore \pi(h_x ss, t_x + 2, \lambda) = \pi(h_x, t_x, \lambda) \qquad \qquad \dots (9)$$

Next, we consider, $\pi(h_x sf, t_x + 2, \lambda)$.

$$\pi_n(h_x sf, t_x + 2, \lambda) = \frac{\pi_n(h_x s, t_x + 1, \lambda) q_n}{\sum_{i \ge 0} \pi_i(h_x s, t_x + 1, \lambda) q_i}$$

$$= \frac{\pi_n(h_x, t_x, \lambda) q_n}{1-\lambda}$$

$$= \frac{\pi_0(h_x, t_x, \lambda)(\lambda q)^n [\prod_{r=1}^{n}(1-q^r)]^{-i}}{1-\lambda}$$

$$= \pi_n(h_x, t_x, \lambda q).$$

$$\therefore \pi(h_x sf, t_x + 2, \lambda) = \pi(h_x, t_x, \lambda q) \qquad \qquad \dots (10)$$

Next, we consider, $\pi(h_x fs, t_x + 2, \lambda)$.

$$\pi_n(h_x fs, t_x + 2, \lambda) = \frac{\pi_{n+i}(h_x f, t_x + 1, \lambda) p_{n+i}}{\sum_{i \ge 0} \pi_i(h_x f, t_x + 1, \lambda) p_i}$$

$$= \frac{\pi_{n+i}(h_x f, t_x + 1, \lambda) p_{n+i}}{\sum_{i \ge 0} \pi_i(h_x, t_x + 1, \lambda q) p_i}$$

$$= \frac{\pi_{n+i}(h_x, t_x + 1, \lambda q) p_{n+i}}{\lambda q} \quad \text{(using (8), (3) and noting that (3) holds}$$

$$\text{for any } \lambda)$$

$$= \frac{\pi_0(h_x, t_x, \lambda q)(\lambda q)^{n+1}[\prod_{r=1}^{n+1}(1-q^r)]^{-1}p_{n+1}}{\lambda q}$$

$$= \pi_0(h_x, t_x, \lambda q)(\lambda q)^n[\prod_{r=1}^n(1-q^r)]^{-1}$$

$$= \pi_n(h_x, t_x, \lambda q).$$

$$\therefore \pi(h_x fs, t_x + 2, \lambda) = \pi(h_x, t_x, \lambda q) \qquad \qquad \text{... (11)}$$

Next, consider, $\pi(h_x ff, t_x + 2, \lambda)$.

$$\pi_n(h_x ff, t_x + 2, \lambda) = \frac{\pi_n(h_x f, t_x + 1, \lambda)q_n}{\sum_{i \geq 0}\pi_i(h_x f, t_x + 1, \lambda)q_i}$$

$$= \frac{\pi_n(h_x, t_x + 1, \lambda q)q_n}{\sum_{i \geq 0}\pi_i(h_x, t_x + 1, \lambda q)q_i}$$

$$= \frac{\pi_0(h_x, t_x, \lambda q)(\lambda q)^n[\prod_{r=1}^n(1-q^r)]^{-1}q_n}{1-\lambda q}$$

$$= \pi_0(h_x, t_x, \lambda q^2)(\lambda q^2)^n[\prod_{r=1}^n(1-q^r)]^{-1} \quad \text{(Using (6) and}$$

noting that (6) holds for any λ).

$$= \pi_n(h_x, t_x, \lambda q^2).$$

$$\therefore \pi(h_x ff, t_x + 2, \lambda) = \pi(h_x, t_x, \lambda q^2) \qquad \qquad \text{... (12)}$$

Considering (9), (10), (11) & (12), we can infer that for any history 'h' & time 't' i.e. for any prior, one bed-posterior distribution depends on the number of failure in the test bed only. Therefore for the 'k-r' failures in the test bed we conclude that,

$$\pi(h_x s^r f^{k-r}, t_{x+1}, \lambda) = \pi(h_x, t_x, \lambda q^{k-r}) \qquad \qquad \text{... (13)}$$

This completes the proof of Lemma 1.

Lemma 2: Let the unconditional expected number of successes in 'i'th test bed be expressed as

$$\mu_{si}(\lambda) =$$

$$\sum_{r_1=0}^k \sum_{r_2=0}^k \cdots \sum_{r_i=0}^k \sigma_{k,r_1}\left(\pi(h_x, t_x, \lambda)\right)\sigma_{k,r_2}\left(\pi(h_x, t_x, \lambda q^{k-r_1})\right)\ldots\sigma_{k,r_i}\left(\pi\left(h_x, t_x, \lambda q^{k-\Sigma_{j=1}^{i}r_j}\right)\right) r$$

Then, $\mu_{si}(\lambda)$ is non-increasing in 'i' for all $i > 0$. Define $\mu_{s0}(\lambda) = 0$.

Proof:

$\mu_{s0}(\lambda)$ is defined as '0'.

$$\mu_{s1}(\lambda) = \sum_{r=0}^k r\sigma_{k,r}(\pi(h_x, t_x, \lambda)).$$

$$\mu_{s2}(\lambda) = \sum_{r_1=0}^k [\sum_{r_2=0}^k r_2\sigma_{k,r_2}(\pi(h_x, t_x, \lambda q^{k-r_1}))]\sigma_{k,r_1}(\pi(h_x, t_x, \lambda))$$

From Lemma 4.3 & 4.2 of Das and Chakraborty,[9] it is clear that $\pi(h_x, t_x, \lambda) >^{LR} \pi(h_x, t_x, \lambda q^{k-r})$ and hence, $\alpha\big(\pi(h_x, t_x, \lambda)\big) \geq \alpha\big(\pi(h_x, t_x, \lambda q^{k-r})\big)$.

$$\therefore \mu_{s2}(\lambda) = \sum_{r_1=0}^{k}\Big[\sum_{r_2=0}^{k} r_2 \alpha_{k r_2}\big(\pi(h_x, t_x, \lambda q^{k-r_2})\big)\Big]\alpha_{k r_1}\big(\pi(h_x, t_x, \lambda)\big)$$

$$\leq \sum_{r_1=0}^{k}\Big[\sum_{r=0}^{k} r\alpha_{k r}\big(\pi(h_x, t_x, \lambda)\big)\Big]\alpha_{k r_1}\big(\pi(h_x, t_x, \lambda)\big) = \sum_{r_1=0}^{k} \mu_{s1}(\lambda)\alpha_{k r_1}\big(\pi(h_x, t_x, \lambda)\big)$$

$$= \mu_{s1}(\lambda)\sum_{r_1=0}^{k} \alpha_{k r_1}\big(\pi(h_x, t_x, \lambda)\big) = \mu_{s1}(\lambda).$$

$$\therefore \mu_{s2}(\lambda) \leq \mu_{s1}(\lambda).$$

Now assume that for some 'l' $(l \geq 1)$, this holds true i.e. $\mu_{sl}(\lambda) \leq \mu_{s(l-1)}(\lambda) \leq \cdots \leq \mu_{s1}(\lambda)$.

Now,
$$\mu_{s(l+1)}(\lambda) =$$

$$\sum_{r_1=0}^{k}\sum_{r_2=0}^{k}\cdots\sum_{r_{l+1}=0}^{k} \alpha_{k r_1}\big(\pi(h_x, t_x, \lambda)\big)\alpha_{k r_2}\big(\pi(h_x, t_x, \lambda q^{k-r_2})\big)\cdots\alpha_{k r_{l+1}}\Big(\pi\big(h_x, t_x, \lambda q^{(l+1)k}\big)$$

$$=$$

$$\sum_{r_1=0}^{k}\sum_{r_2=0}^{k}\cdots\Big[\sum_{r_{l+1}=0}^{k} \alpha_{k r_{l+1}}\Big(\pi\big(h_x, t_x, \lambda q^{(l+1)k-\sum_{i=2}^{l+1}r_i}\big)\Big)\Big]r_{(l+1)}\Big]\alpha_{k r_2}\big(\pi(h_x, t_x, \lambda)\big)\alpha_{k r_2}\big(\pi(h_x, t_x, \lambda q^{k-r_2})\big)$$

$$\cdots\alpha_{k r_1}\Big(\pi\big(h_x, t_x, \lambda q^{lk-\sum_{i=2}^{l}r_i}\big)\Big)$$

$$\leq \sum_{r=2}^{k}\Big[\sum_{r_1=0}^{k}\cdots\sum_{r=0}^{k} r\alpha_{kn}\Big(\pi\big(h_x, t_x, \lambda q^{k-\sum_{i=1}^{l}r_i}\big)\Big)\alpha_{kr-1}\Big(\pi\big(h_x, t_x, \lambda q^{l-1 k-\sum_{i=1}^{l}r_i}\big)\Big)\cdots\alpha_{k r_1}\big(\pi(h_x, t_x, \lambda)\big)\Big]\alpha_{k r}\big(\pi(h_x, t_x, \lambda)\big)$$

$$= \sum_{r=0}^{k} \mu_{sl}(\lambda)\alpha_{k r}\big(\pi(h_x, t_x, \lambda)\big)$$

$$= \mu_{sl}(\lambda)\sum_{r=0}^{k}\alpha_{k r}\big(\pi(h_x, t_x, \lambda)\big)$$

$$= \mu_{sl}(\lambda).$$

Hence, by induction it is proved that for any 'l' $(l \geq 1)$,

$$\mu_{s(l+1)}(\lambda) \leq \mu_{sl}(\lambda) \leq \mu_{s(l-1)}(\lambda) \leq \cdots \leq \mu_{s1}(\lambda).$$

This completes the proof of lemma 2.

3. OPTIMUM DECISION RULE

3.1 Expression of the Maximum Expected Return after Testing for m Test Beds

Let $g_{m,k}\big(\pi(h_x,t_x)\big)$ be defined as the maximum expected return after testing for m test beds when the number of undiscovered bugs is represented by an initial prior distribution $\pi(h_x,t_x)$. Define, $g_{0,k}\big(\pi(h_x,t_x)\big) = 0$, and for $m \geq 1$, we have, from Das and Chakraborty,[9]

$$g_{m,k}\big(\pi(h_x,t_x)\big) = \max\Big\{ 0, -ck + \textstyle\sum_{r=0}^{k} \alpha_{k,r}\big(\pi(h_x,t_x)\big)\big[r + g_{m-1,k}\big(\pi(h_x s^r f^{k-r}, t_{x+1})\big)\big]\Big\} \cdots$$
$$\cdots (14)$$

Theorem 1: If $\pi(h_x,t_x) = \pi(h_x,t_x,\lambda)$ is an Euler distribution, then the solution to the stopping problem becomes one bed look-ahead rule and the expected return in that case can be expressed as,

$$g(\lambda) = -ck + \textstyle\sum_{r=0}^{k} \alpha_{k,r}\big(\pi(h_x,t_x,\lambda)\big)\big[r + g(\lambda q^{k-r})\big], \text{ for}$$
$$\textstyle\sum_{r=0}^{k} r\alpha_{k,r}\big(\pi(h_x,t_x,\lambda)\big) > ck$$
$$= 0 \text{ , for } \textstyle\sum_{r=0}^{k} r\alpha_{k,r}\big(\pi(h_x,t_x,\lambda)\big) \leq ck.$$

Proof: From (14),

$$g_{m,k}(\pi(h_x,t_x,\lambda)) = \max\left\{ 0, -ck + \sum_{r=0}^{k} \alpha_{k,r}\big(\pi(h_x,t_x,\lambda)\big)\Big[r + g_{m-1,k}\big(\pi(h_x s^r f^{k-r}, t_{x+1},\lambda)\big)\Big]\right\}$$

$$= \max\left\{ 0, -ck + \textstyle\sum_{r=0}^{k} \alpha_{k,r}\big(\pi(h_x,t_x,\lambda)\big)\Big[r + g_{m-1,k}\big(\pi(h_x,t_x,\lambda q^{k-r})\big)\Big]\right\}.$$

WLOG $g_{m,k}\big(\pi(h_x,t_x,\lambda)\big)$ can be written as $g_m(\lambda)$ and $g_{m,k}\big(\pi(h_x,t_x,\lambda q^{k-r})\big)$ can be written as $g_m(\lambda q^{k-r})$. Hence,

$$g_m(\lambda) = \max\Big\{ 0, -ck + \textstyle\sum_{r=0}^{k} \alpha_{k,r}\big(\pi(h_x,t_x,\lambda)\big)\big[r + g_{m-1}(\lambda q^{k-r})\big]\Big\} \quad \cdots (15)$$

As $m \to \infty$, we have,

$$g(\lambda) = \max\Big\{ 0, -ck + \textstyle\sum_{r=0}^{k} \alpha_{k,r}\big(\pi(h_x,t_x,\lambda)\big)\big[r + g(\lambda q^{k-r})\big]\Big\} \qquad \cdots (16)$$

From (15), it is clear that if $\mu_{\geq 1}(\lambda) \leq ck$, then $g_1(\lambda) = 0$ since $g_0(\lambda) = 0$. Also by lemma 2, it can be obtained that $\mu_{\geq 1}(\lambda) < ck$, for any $i > 0$ and hence, $g_i(\lambda) = 0$ for all i = 0, 1, 2…… and finally $g(\lambda) = 0$.

If $\mu_{\geq 1}(\lambda) > ck$, then only $g(\lambda) > 0$. This completes the proof of Theorem 1.

3.2 Sufficient Condition for 'Stopping Rule'

Now from (16),

$$-ck + \sum_{r_1=0}^{k} \alpha_{k,r_1} \left(\pi(h_x, t_x, \lambda) \left[r_1 + g(\lambda q^{k-r_1}) \right] \right)$$

$$= -ck + \sum_{r_1=0}^{k} r_1 \alpha_{k,r_1} \left(\pi(h_x, t_x, \lambda) \right) + \sum_{r_1=0}^{k} \alpha_{k,r_1} \left(\pi(h_x, t_x, \lambda) \right) g(\lambda q^{k-r_1})$$

$$= -ck + \sum_{r_1=0}^{k} r_1 \alpha_{k,r_1} \left(\pi(h_x, t_x, \lambda) \right)$$

$$+ \sum_{r_1=0}^{k} \alpha_{k,r_1} \left(\pi(h_x, t_x, \lambda) \right) \left[-ck + \sum_{r_2=0}^{k} r_2 \alpha_{k,r_2} \left(\pi(h_x, t_x, \lambda) \right) \right.$$

$$\left. + \sum_{r_2=0}^{k} \alpha_{k,r_2} \left(\pi(h_x, t_x, \lambda q^{k-r_1}) \right) g(\lambda q^{2k-r_1-r_2}) \right]$$

$$= -ck \left(1 + \sum_{r_1=0}^{k} \alpha_{k,r_1} \left(\pi(h_x, t_x, \lambda) \right) \right) + \left[\sum_{r_1=0}^{k} r_1 \alpha_{k,r_1} \left(\pi(h_x, t_x, \lambda) \right) \right.$$

$$+ \sum_{r_1=0}^{k} \alpha_{k,r_1} \left(\pi(h_x, t_x, \lambda) \right) \sum_{r_2=0}^{k} r_2 \alpha_{k,r_2} \left(\pi(h_x, t_x, \lambda q^{k-r_1}) \right) \right]$$

$$+ \sum_{r_1=0}^{k} \alpha_{k,r_1} \left(\pi(h_x, t_x, \lambda) \right) \sum_{r_2=0}^{k} \alpha_{k,r_2} \left(\pi(h_x, t_x, \lambda q^{k-r_1}) \right) g(\lambda q^{2k-r_1-r_2})$$

$$= -2ck + \left[\sum_{r_1=0}^{k} r_1 \alpha_{k,r_1} \left(\pi(h_x, t_x, \lambda) \right) + \sum_{r_1=0}^{k} \alpha_{k,r_1} \left(\pi(h_x, t_x, \lambda) \right) \sum_{r_2=0}^{k} r_2 \alpha_{k,r_2} \left(\pi(h_x, t_x, \lambda q^{k-r_1}) \right) \right]$$

$$+ \sum_{r_1=0}^{k} \alpha_{k,r_1} \left(\pi(h_x, t_x, \lambda) \right) \sum_{r_2=0}^{k} \alpha_{k,r_2} \left(\pi(h_x, t_x, \lambda q^{k-r_1}) \right) g \left(\lambda q^{2k-\sum_{i=1}^{2} r_i} \right)$$

$$= -2ck + \sum_{i=0}^{2} \mu_{si}(\lambda) + \sum_{r_1=0}^{k} \alpha_{k,r_1} \left(\pi(h_x, t_x, \lambda) \right) \sum_{r_2=0}^{k} \alpha_{k,r_2} \left(\pi(h_x, t_x, \lambda q^{k-r_1}) \right) g \left(\lambda q^{2k-\sum_{i=1}^{2} r_i} \right)$$

$$= -mck + \sum_{i=0}^{m} \mu_{si}(\lambda)$$

$$+ \sum_{r_1=0}^{k} \sum_{r_2=0}^{k} \cdots \sum_{r_m=0}^{k} \alpha_{k,r_1} \left(\pi(h_x, t_x, \lambda) \right) \alpha_{k,r_2} \left(\pi(h_x, t_x, \lambda q^{k-r_1}) \right) \cdots \alpha_{k,r_m} \left(\pi(h_x, t_x, \lambda q^{mk-\sum_{i=1}^{m-1} r_i}) \right) g \left(\lambda q^{mk-\sum_{i=1}^{m} r_i} \right)$$

Say, $\tau = \min\left[m: \mu_{sm}(\lambda) - ck \leq 0\right]$

Therefore, $\mu_{si}(\lambda) \leq ck$ for all $i \geq \tau$ by Lemma 2. Maximum expected return would be obtained at such τ.

Hence, the sufficient condition for 'stopping rule' may be simplified as below.

Stop testing at the 'τ'th test bed at which the unconditional expected number of successes would be just less than or equal to cost of testing of a test bed i.e.

$$\tau = \min\left[m: \mu_{sm}(\lambda) - ck \leq 0\right]$$

and, the maximum expected return,

$$g(\lambda) = \sum_{i=1}^{\tau-1}(-ck + \mu_{si}(\lambda)) = -(\tau - 1)ck + \sum_{i=1}^{\tau-1}\mu_{si}(\lambda).$$

3.3 Necessary Condition for 'Continuation'

Chakraborty[6] defined the notion of 'greater in likelihood ratio' between two priors, say $\pi(hs, t + 1)$ and $\pi(h, t)$, as $\pi(hs, t + 1) >^{LR} \pi(h, t)$ if the ratio $\frac{\pi_n(hs, t+1)}{\pi_n(h, t)}$ is non-decreasing in $n \geq 0$. From Remark 5.5.1 of Chakraborty,[6] we get if $\pi(hs, t + 1) >^{LR} \pi(h, t)$, then the continuation region is given by { $(k, r): k \geq 0, r > s(k)$}, where $s(k)$ is given as follows:

$$s(k) = \max\left\{r: g\left(\pi(hs^r f^{k-r}, t + k)\right) = 0\right\}, \text{ for each integer } k \geq 0 \text{ and}$$

$$s(k) = -1, \text{ if } g\left(\pi(hs^r f^{k-r}, t + k)\right) > 0, \text{ for all } r = 0, 1, 2, 3, \ldots, k.$$

Following this, Das and Chakraborty[9] has given the necessary condition for continuation of testing in test beds as below:

If $\pi(h_x s, t_x + 1) >^{LR} \pi(h_x, t_x)$, then the continuation region is given by,

$\{(m, \sum_{i=1}^{m} r_i): m > 0, \sum_{i=1}^{m} r_i > s(m)\}$, where $s(m)$ is given as follows:

$$s(m) = \max\left\{\sum_{i=1}^{m} r_i : g\left(\pi(h_x s^{\sum_{i=1}^{m} r_i} f^{\sum_{i=1}^{m}(k - r_i)}, t_{x+m})\right) = 0\right\}$$

$$s(m) = -1, \text{ if } g\left(\pi(h_x s^{\sum_{i=1}^{m} r_i} f^{\sum_{i=1}^{m}(k - r_i)}, t_{x+m})\right) > 0, \text{ for each } m > 0 \text{ and}$$

$r_i = 0, 1, 2, \ldots, k.$

For the Euler distribution with $\pi_n(h_x, t_x, \lambda)$ as a prior, since, $\pi(h_x, s, t_x + 1, \lambda) = \pi(h_x, t_x, \lambda)$, it can be said that $\pi(h_x, s, t_x + 1, \lambda) >^{LR} \pi(h_x, t_x, \lambda)$, that is, the assumption stated above is satisfied. Therefore, the continuation region is given by,

$$\{(m, \sum_{i=1}^{m} r_i) : m > 0, \sum_{i=1}^{m} r_i > s(m)\}, \text{ where } s(m) \text{ is given as follows:}$$

$$s(m) = \max \left\{ \sum_{i=1}^{m} r_i : g\left(\pi\left(h_x s^{\sum_{i=1}^{m} r_i} f^{\sum_{i=1}^{m}(k-r_i)}, t_{x+m}, \lambda\right)\right) = 0 \right\}$$

$$s(m) = -1, \text{ if } g\left(\pi\left(h_x s^{\sum_{i=1}^{m} r_i} f^{\sum_{i=1}^{m}(k-r_i)}, t_{x+m}, \lambda\right)\right) > 0, \text{ for each } m \geq 0 \text{ and }$$

$$r_i = 0, 1, 2, \dots, k.$$

But, here $\pi\left(h_x s^{\sum_{i=1}^{m} r_i} f^{\sum_{i=1}^{m}(k-r_i)}, t_{x+m}, \lambda\right) = \pi\left(h_x, t_x, \lambda q^{\sum_{i=1}^{m}(k-r_i)}\right)$.

Hence, the 'continuation region' can be given, in terms of the number of test-beds m and total number of successes upto m-th test-bed in the form of maximum expected return as below:

$$\{(m, \sum_{i=1}^{m} r_i) : m > 0, \sum_{i=1}^{m} r_i > s(m)\}, \text{ where } s(m) \text{ is given as follows:}$$

$$s(m) = \max \left\{ \sum_{i=1}^{m} r_i : g\left(\lambda q^{\sum_{i=1}^{m}(k-r_i)}\right) = 0 \right\}$$

$$s(m) = -1, \text{ if } g\left(\lambda q^{\sum_{i=1}^{m}(k-r_i)}\right) > 0, \text{ for each } m > 0 \text{ and } r_i = 0, 1, 2, \dots, k \text{ for all}$$

$$i = 0, 1, 2, \dots, m.$$

4. CONCLUSION

This paper has addressed the problem of finding optimum software release time based on the model developed under the assumption of Euler distribution as a prior for the number of undetected bugs present in the software. Tests were considered as phase by phase basis, where each test phase (test bed) consists of same number of test cases. A sufficient condition for optimum stopping rule has been developed in terms of unconditional expected number of successes and the maximum expected return at the optimum stopping time is derived. Also a necessary condition for continuation of testing the software was derived in terms of the number of test-beds and total number of successes up to the corresponding test-bed in the form of maximum expected return.

REFERENCES

[1] Musa, J.D., Iannino, A. and Okumoto, K. (1987), "Software Reliability Measurement, Prediction and Application", McGraw-Hill, New York.

[2] Pham, H. (2000), "Software Reliability", Springer-Verlag, Singapore.

[3] Koch, H. S. and Kubat, P. (1983), "Optimal Release Time of Computer Software", *IEEE Transactions on Software Engineering*, Vol. 9, No. 3, pp. 323–327.

[4] Musa, J.D. and Ackerman A.F. (1989), "Quantifying Software Validation: When to Stop Testing?", *IEEE Software*, Vol. 6, No. 3, pp. 19–27.

[5] Yang, M.C.K. and Chao, A. (1995), "Reliability-Estimation and Stopping-Rules for Software Testing, Based on Repeated Appearances of Bugs", *IEEE Transactions on Reliability*, Vol. 44, No. 2, pp. 315–321.

[6] Chakraborty, A.K. (1996), "Software Quality Testing and Remedies", PhD. Thesis.

[7] Benkherouf, L. and Bather, J.A. (1988), "Oil Exploration: Sequential Decisions in the Face of Uncertainty", *Journal of Applied Probability*, Vol. 25, pp. 529–543.

[8] Benkherouf, L. and Alzeid, A.A. (1993), "On the Generalized Euler Distribution", *Statistics and Probability Letters*, Vol. 18, pp. 323–326.

[9] Das, S. and Chakraborty, A.K. (2008), "Determination of Optimum Software Release Time Using a Discrete Set-Up".

Physics of Failure Based Reliability Prediction Method for a Two-Wheeler

C. Sasun[1] and Sushant Mohan Dewal[2]

TVS Motor Company Ltd, Harita, Hosur
E-mail: [1]c.sasun@tvsmotor.co.in; [2]Sushant.dewal@tvsmotor.co.in

ABSTRACT: Recent trends in Indian automotive industry have shown increasing importance of reliability and durability. Reliability perception of the customer is significantly impacting the brand loyalty. This paved way for a systematic 'Design for Reliability (DFR)' process during new product development stages. Reliability prediction methodology using 'Physics of Failure (POF)' is well-acknowledged and proven approach to reduce design iteration. While the POF methodology is well developed and applied with success in Electronics industry, the same is not the case with mechanical systems. Excellent attempts were made to develop reliability prediction models for generic mechanical components by Carderock Division of the Naval Surface Warfare Center (CDNSWC), USA. However, these models are very generic and needs to be validated before putting into use for specific application.

In this paper, an attempt is made to study these models outlined in the NSWC-07 handbook and their validity for automotive application. On successful verification of the models, applicable for the existing two-wheeler, the same are used to predict reliability for new products and optimize the designs. Later, these prediction results are compared with the actual field performance and found that the results have matched very well.

Keywords: Physics of Failure, Reliability Prediction, Case Study, Reliability Prediction of Two-Wheeler, Mechanical Reliability, NSWC-07.

1. INTRODUCTION

Automobile industries are increasingly facing a challenge to release reliable design within the prescribed time line. Brand loyalty and brand image are building upon the reliability of the product. Customers expect a new yet reliable product to be available in market in a short duration of time. In order to design a product that meets the reliability requirement, designers mostly rely on proven design codes and standards available in literature. However, design codes and standards may lead to over design by applying factor of safety or redundancy. Alternatively, designs may be traditionally freeze by test and fix method (Warburton *et al.*, 1998). A Design for Reliability program including reliability prediction in the early stage of a development program is suggested to support the design process (NSWC, 2007). Reliability predictions are mostly performed for a) Feasibility evaluation of the product b) Comparing competitive designs c) Identification of potential reliability problem and d) To provide reliability input to other R/M tasks (RIAC, 1995). The environment and operational conditions are deciding factors for the reliability and durability of components (in this paper the term "component" is used interchangeably for part, sub-system and product). Hence, reliability prediction techniques incorporating environmental and operational conditions with design characteristics of a product are gaining importance.

Various standards for carrying out reliability prediction are available in literature such as: Mil-Std 217F, RIAC 217Plus, SR-332 (Bellcore), IEC62380, EPRD, NPRD, etc. The basic of reliability prediction method is the Mil-Std 217F (DOD of USA, 1991), developed for electronics components. The development of this document was made possible because of the standardization and mass production of electronic parts. The part stress analysis described in the handbook incorporated the stress factors for predicting reliability of the components. Reliability prediction models and techniques as in NPRD are available for mechanical components. In contrast to electronic components, mechanical components Failure Modes (FM) are largely dependent on time and stress. Even for electronic components the reliability prediction approach based on historical database may produce highly variable assessments (Pecht, 1996). This is mainly due to a) the database are not completely updated for all components b) most of the database have equipment removal from the field as the failure data which may not be equal to the part failure and c) reliability prediction models are based on industry specific value of failure rate which are neither vendor nor device specific. Use of reliability assessment techniques which are based on root-cause analysis of failure mechanism, FMs and failure causing stress have found to be effective in reliability prediction and design improvements of products. Physics of Failure (PoF) approach to reliability uses knowledge of a product's expected life-cycle usage, environmental stress load profiles, knowledge of material strength, degradation properties of the architecture and technologies the product is based upon. Reliability assessments based on physics-of-failure methods move reliability growth up into the design process by using scientific basis for estimating product life under actual operating conditions.

Mechanical components fail mostly because of fatigue, creep, impact, wear, corrosion, etc. All these FMs are highly dependent on the material properties, basic geometry of the part and the usage condition. Mechanical components are used in vast applications, which may change the nature of failure of the component. Warburton (Warburton, 1998) has worked towards development of procedure and techniques for assessing the reliability of components from a fundamental understanding of the degradation, failure process and their relation to the underlying operational, environment, material and design variables. However, statistical model of failure rate or reliability prediction have not been developed in his work. Brissaud *et al.* (Brissaud *et al.*, 2008) have developed a PoF based reliability evaluation technique focused on Safety Instrumented Systems.

NSWC-07 (NSWC, 2007) contains PoF based reliability prediction models for mechanical equipment. PoF models are very sensitive to the design, operational and environmental factors. A little or no feedback makes it difficult to justify the use of these methods in reliability prediction of a two-wheeler. Hence, this generic model cannot be used until it is proven for the type of components for specific application, as the usage without understanding the implications can lead to catastrophic results. The traditional methods seldom provide correct and accurate results on the reliability prediction of mechanical components hence; PoF approach is preferred for reliability prediction of mechanical components. In this paper, PoF models available in NSWC are explored and their applicability is studied for two-wheelers. The PoF methods available in NSWC are applied on critical components in a two-wheeler with smaller capacity engines (less than 200cc engine capacity). Field failure rate for the specific component is used to validate the PoF models on two-wheeler components.

Section 2 of this paper explains the PoF approach for reliability prediction. Section 3 includes the exploration procedure for the PoF based reliability prediction models available in literature. Section 4 provides case studies conducted on critical components of two-wheeler. Exploration of appropriate PoF model and comparison with field failure rate is provided to validate the applicability of PoF models outlined in NSWC-07 on two-wheeler. In Section 5 the interpretation of results obtained in case studies are discussed under the fundamental assumption of PoF based reliability prediction models, followed by the conclusion.

2. PHYSICS OF FAILURE APPROACH

Life of a component is dependent on some influencing factors like: temperature, humidity, load, vibration, contamination, etc. Considering these influencing factors, several well-proven methods have been developed. One such model is Arrhenius model, which considers thermal stress for predicting reliability. The Arrhenius model is defined as in Eqn 1 (ReliaSoft Corporation, 2007). The activation energy used in this model is often very different for similar components (Eusgeld *et al.*, 2008). Hence, this model is dependent on test data and not much on the underlying physical model. Similar to Arrhenius model, some of generic models based on the influencing factors available are Eyring, Inverse Power Law, Temperature-Humidity and Temperature-Non-thermal. These generic models can be applied on components to predict the reliability of the component given the change in stress. As discussed, these models depend on the testing to predict the reliability of the components. Hence, these models are more supportive towards the test and fix methodology for design.

$$R(t) = Ae - \frac{E_A}{K/T} \qquad \qquad \dots (1)$$

where,

      E_A = Activation Energy
      R   = Speed of reaction
      A   = Unknown non-thermal constant
      K   = Boltzman's constant
      T   = Absolute temperature

Designers are interested in reliability prediction methods, which states life of component as a function of design (geometry), usage (cycles/time), environment (stresses i.e. load, temperature, vibration, humidity, etc.) and material characteristics. It is a known fact that the life of a component is proportional to the stress experienced by the component. The stress acting upon a component can be expressed as a function of the design and environment. Life of a component can be predicted by taking the aspect of material properties, usage and the stresses acting on the component. Cruse *et al.* (Cruse *et al*, 1997) by an example of spring has explained the procedure for calculating the failure rate of a component. First the ratio of stress level to material tensile strength is found and then the fatigue relationship is applied to incorporate material property with the stress level changes to predict the life (in cycles) of the component.

The steps are represented pictorially in Figure 2. In PoF based reliability prediction model the failure rate is generally of the form as in equation 2,

$$\lambda_{comp} = \lambda_{comp,B} \prod_{i=1}^{n} C_i \qquad \qquad \dots (2)$$

where,

λ_{Comp} = Failure rate of the component
$\lambda_{Comp,B}$ = Base failure rate
C_i = Correction factor

This base failure rate is the failure rate at the reference material and stress level, i.e. at reference material properties, design and environment conditions. The base failure rate, $\lambda_{comp,B}$ can be obtained from the historical database or from lab test data-for a set of known parameter values. Alternatively, these failure rates can be 'expressed' in the form of an equation that is derived from engineering design fundamentals, pertaining to the FM-mechanism combination. In practice, a combination of both `historical data' and 'stress-life' relationships is adopted with good success. The correction factor here is the ratio of stresses and material properties. If a stress factor does not change then the corresponding correction factor is unity. As explained above, the ratio of stresses can be calculated and then clubbed with the effect of material change to derive the correction factors. The Correction factors, C_i, are empirically arrived at relationships between the 'factor' and the failure rate. Typically, it would include, factors like: Temperature, Fluids, Contamination, Stress levels, Size, Surface finish. etc. Once we have the correction factor and the base failure rate the failure rate of the component can be obtained by using eqn 2.

The method described above computes the life of the component against the FM addressed. A component may fail as a result of multiple FM-Mechanisms. These FMs are assumed to be in series for practical purposes, as the occurrence of any FM will result in failure of the component. The life of a component due to multiple FMs can be computed by using equation 3,

$$\lambda_{comp} = \sum_{i=1}^{n} \left(\lambda_{b,i} \prod_{j=1}^{f,i} C_{j,i} \right) \qquad \qquad \dots (3)$$

where,

n denotes number of FMs for the component
(f, i) denotes the number of correction factors for i^{th} FM
i denotes the FM
j denote number of correction factors.
$\lambda_{b,i}$ denotes base failure rate for i^{th} FM and
$C_{j,i}$ denotes the jth correction factor for i^{th} FM .

To predict the life of the component a relation of cycle to required life variable is required. Using this relationship the life of the component can be predicted. Hence, basic of PoF approach is to model the life of the component to their design, usage, environment and material characteristics. As an alternative way of looking at PoF, Lu (Lu *et al.*, 2009) has started with a relation between the effect of environment and material properties. Later, an equation was developed taking the aspects of the relation developed in earlier step and the design parameters to predict the life (in cycle) of the component. The basics of PoF approach is as depicted in the Figure 1.

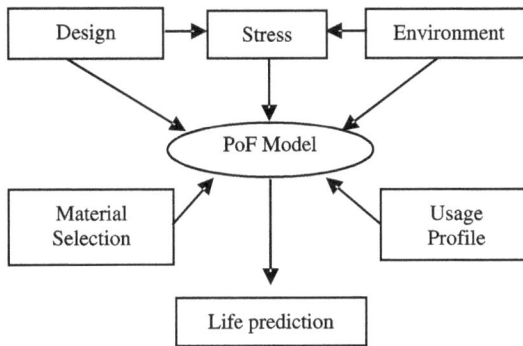

Fig. 1: PoF Approach

Considering the component to be in series, the failure rate for a system can be computed using equation 4,

$$\lambda_{\text{system}} = \sum_{i=1}^{n} \left(\lambda_{b,i} \prod_{j=1}^{f,i} C_{j,i} \right) \qquad \qquad \dots (4)$$

where,

n	denotes number of parts in the system
(f,i)	denote number of correction factors that affect the failure rate for i^{th} FM
$\lambda_{b,i}$	denotes base failure rate for i^{th} part in the system and
$C_{j,i}$	denotes the correction factor for j^{th} FM in i^{th} part.

PoF approach works by identifying the dominant cause of failure of the equipment and then modeling the dominant failure mechanism to predict the reliability of the equipments.

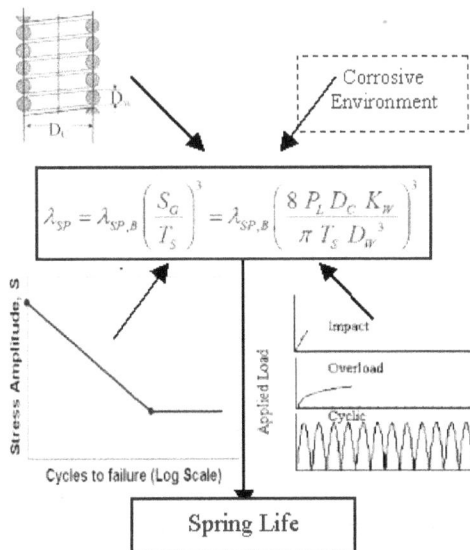

Fig. 2: PoF Model for Reliability Prediction of Spring

3. RELIABILITY PREDICTION FOR TWO-WHEELER

Excellent work has been done on the PoF based reliability prediction procedures for mechanical components in NSWC-07. PoF based models are built for mechanical equipment and correction factors based on the classification of stresses have been calculated and listed in the handbook. However, the applicability of the methods to two-wheeler industry is still not known. Moreover, the base failure rate proposed in the handbook may not be applicable to Indian automotive industry – due to vast differences in material processing and component manu-facturing technologies. This is due to no or a very little feedback on PoF methods for the specific use of mechanical equipment in the two-wheeler industry. In this work, the model as outlined in NSWC-07 are explored for the two-wheeler components based on the knowledge of usage, design, environment and material.

Critical parts in a two-wheeler are selected for predicting reliability and studying the applicability of reliability prediction procedures as outlined in NSWC-07. Relevant data pertaining to the design and material properties are collected/computed from the design and material reports/ manuals of the component. The parameters related to the environment like temperature, loads, etc. are collected from field and nominal values have been assigned to them. Usage pattern is arrived at based on experience and data collection from field. Using this information and engineering judgment the correction factors as mentioned in the manual are selected. Failure rate for the component are calculated based on selected correction factor and the equation provided in the manual. Reliability prediction from this PoF model is compared with the field failure rate data at different intervals in time. The proposed model developed is also verified on different types of two-wheelers. The procedure for reliability prediction of two-wheeler is given in Figure 3.

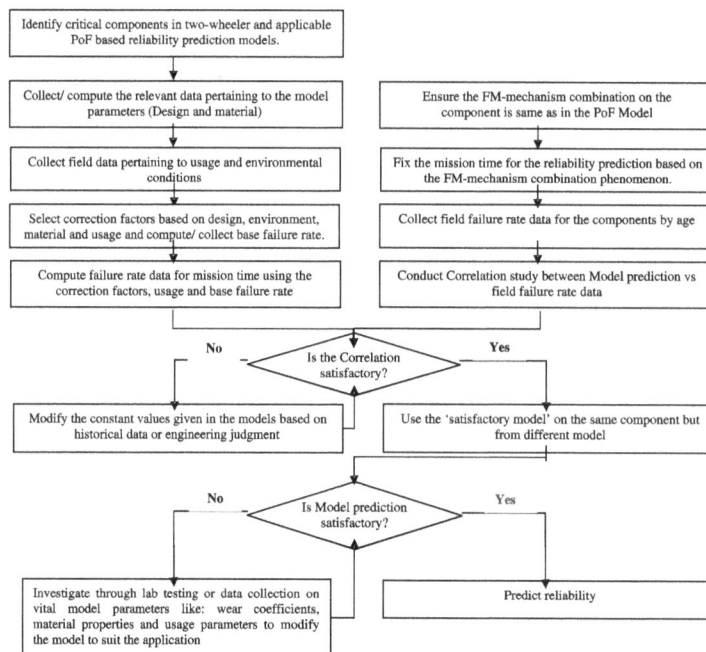

Fig. 3: Flowchart of the Reliability Prediction for Two-Wheeler

4. CASE STUDY

On critical components of a two-wheeler the procedure as outlined in Section 3 is followed. The case studies are provided for the following components.

(a) Disc Pad Brake

(b) V-Belt

(c) Poppet Valve

(d) Clutch Plates.

Case studies on these critical components are discussed in Sections 4.1–4.4.

4.1 Disc Pad-Brake

Brakes are among the most critical system from safety point of view in a two-wheeler. Failure of brakes may lead to customer dissatisfaction and even catastrophic results. These parameters are recorded and fed into the model. For application in two-wheeler, a disc pad brake system is considered to have the following schematic. The brake pads as in Figure 5, are of two types a) moving pad and b) stationary pad. The resin binder used in pad for automotive applications specifically for two-wheeler are mostly oil modified phenolic based for their resistance to high heat and performance balance between noise and wear life. The disc rotor as in Figure 4 is made of iron with highly machined surfaces where the brake pads contact it. Just as the brake pads wear out over time, the rotor also undergoes some wear, usually in the form of ridges and groves where the brake pad rubs against it. After a wear limit is reached the life of disc rotor is over and should be replaced with a new set. Slotted annulus disc type is used in the two-wheeler studied in this paper. The caliper is hydraulic controlled via a master cylinder, which regulates the pressure based on the force applied on the front disc lever or the rear brake pedal. A brake caliper is shown in Figure 6.

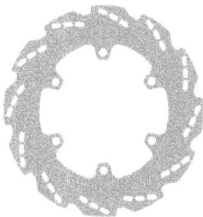

Fig. 4: Disc Rotor   **Fig. 5:** Brake Pad   **Fig. 6:** Brake Caliper

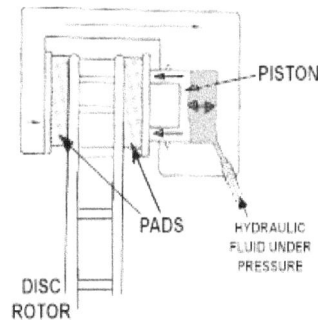

The reliability model of a brake sub-system as outlined in NSWC is based on the wear rate modeling. It is observed in field also that the major contributor to the failure rate of the brake is the wear of the pad/disc. As the FM of the component and the design, environment, material and usage characteristics are well within the scope of the PoF model built in NSWC-07, it is applied.

Applicability of PoF model is studied on front and rear disc-pad brakes. As mentioned in NSWC-07 the main factors influencing reliability of brake sub-system are:

1. Brake type
2. Dust Contamination
3. Operating temperature.

The failure rate of the brake system in presence of influencing factors is given by eqn 5. The C_{BT} is related to design parameter; C_{RD} is related to material properties and C_t is related to environment parameter.

$$\lambda_{FR} = \lambda_{FR, B} \times C_{BT} \times C_{RD} \times C_t \qquad \qquad \ldots (5)$$

where,

$\qquad C_{BT}$ = Correction factor for Brake Type
$\qquad C_{RD}$ = Correction factor for Resin Binder
$\qquad Ct$ = Correction factor for temperature
$\qquad \lambda_{FR,B}$ = Base Failure rate

$$\lambda_{FR, B} = (K_o \times p \times V_s \times t_a)/d \qquad \qquad \ldots (6)$$

where,

$\qquad K_o$ = Wear Rate coefficient
$\qquad V_s$ = Sliding velocity
$\qquad t_a$ = Brake actuation time
$\qquad p$ = Nominal pressure between Pad and Disc
$\qquad d$ = Pad Thickness

4.1.1 *Front Disc-Pad*

Using eqn. 6 the base failure rate for the front disc-pad brake is calculated. The failure rate for front brake is calculated in accordance with eqn. 5, using correction factors as in this case. The operating temperature of nominal value equal to 70°C is recorded for the front disc-pad and Liner type used is Resin Asbestos (Light duty). The failure rate obtained using 5 with the usage is compared to the field failure rate. The comparison plot is given in Figure 7. The reliability prediction is within acceptable range. Hence, the PoF model on the front disc brake in a two-wheeler is applicable.

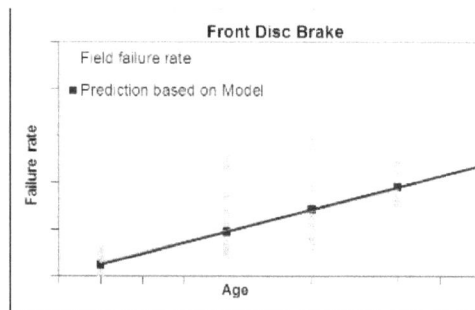

Fig. 7: Reliability Prediction of Front Disc Brake

4.1.2 *Rear Disc-Pad*

In same line, as for front-brake the base failure rate and failure rate for rear brake is computed using eqns. 5 and 6. Relevant correction factors using the design, environment, material and usage are selected for making the reliability prediction. The operating temperature of nominal value equal to 90°C is recorded for the rear disc-pad and Liner type used is Resin Asbestos (Light duty). The comparison plot of reliability prediction with field failure rate is given in Figure 8. It is found that the reliability prediction is within acceptable range and is on pessimistic side as the conditions considered are more on severe side. Hence, the PoF model outlined in NSWC–07 on the rear disc brake in a two-wheeler is applicable.

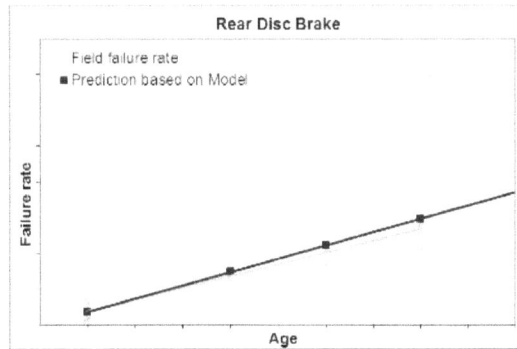

Fig. 8: Reliability Prediction of Rear Disc Brake

4.2 V-Belt

V-Belt is a critical part in the transmission system of the vehicle, as the failure of V-Belt can lead to immobility of a vehicle. This will lead to customer satisfaction and hence, ensuring reliability of v-belt is crucial in a two-wheeler. V-belt considered is as shown in Figure 9. In this case study, the V-Belt is between two pulleys, which have variable ratio in accordance to the speed of the vehicle. Hence, the V-Belt experiences variable torque and speeds. Here, the V-Belt experience operating vs. design load in a ratio of 1:1.3. Nominal value for the pulley diameter is 95 mm. The nominal temperature recorded is 65°C. The belt type factor corresponds to SPA type as mentioned in NSWC-07 and the belt is intermittently used with a non-uniform torque and light shock. Dominant FMs observed in V-Belt are cracks and cuts in belt. These FMs are shown in Figure 10. The design, environment, material and usage, FMs fall within the scope of the PoF model outlined in NSWC-07. Hence, the model is taken up to check for the validity of prediction on a two-wheeler.

Fig. 9: V-Belt with CVT sub-System

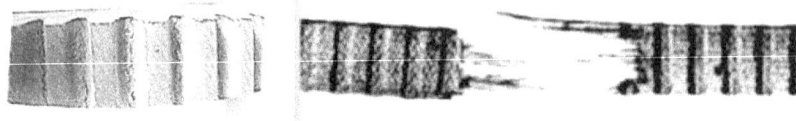

Fig. 10: V-Belt Crack and Belt Broken

The base failure rate is taken from the historical database. Using eqn 7 the failure rate is calculated using the usage as in field and is compared with the field failure rate data collected to verify the model,

$$\lambda_{BD} = \lambda_{BD,B} \times C_{BL} \times C_{PD} \times C_t \times C_{BT} \times C_{BV} \times C_{SV} \quad \ldots (7)$$

where,

$\lambda_{BD,B}$ = Base failure rate
C_{BL} = Correction factor for Belt Loading
C_{PD} = Correction factor for Pulley diameter
C_t = Correction factor for Operating temperature
C_{BT} = Correction factor for Belt type
C_{BV} = Correction factor for Drive operating service
C_{SV} = Correction factor for Shock environment service

The comparison plot of reliability prediction is given in Figure 11. The developed model is applied to predict the reliability of a new product. For the new product the operation vs design load is in a ratio of 1:1.3. Nominal value of the pulley diameter is 87 mm and the nominal temperature recorded is 90°C. The belt type is same as the previous product. Failure rate are calculated using the developed model. After the product matures in the field the field failure rate is estimated and a good correlation between model prediction and the field failure rate is observed. The comparison plot for new product is given in Figure 12.

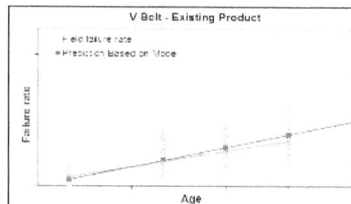

Fig. 11: Reliability Prediction of V-Belt (existing product)

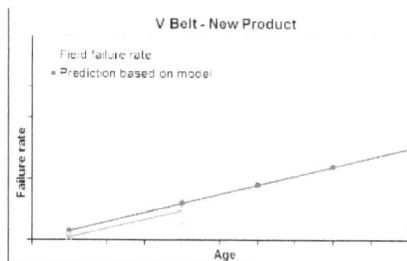

Fig. 12: Reliability Prediction of V-Belt (new product)

4.3 Poppet Valve-Intake

Engine Intake valve is a critical item in a two-wheeler. For a two-wheeler the degradation of engine valve directly affects the performance of the engine and may lead to FMs, which requires replacement of the engine valves and affected parts. The intake valve is an important component that affects the engine reliability; hence ensuring reliability of an engine intake valve (poppet) is required. Engine valve for a two-wheeler is shown in Figure 13. The main FMs of engine valve in field are the Fatigue fracture of the spring, wear and carbon deposition on the valve seat area as presented in Figure 14. The design, environment, material and usage parameters are found to be well within the spec of manual. The failure rate of a Poppet valve is given by eqn 8.

Fig. 13: Engine Valve (Valve Seat, Valve Spring and Valve)

Fig. 14: FM Engine Valve (Wear Seat Area, Carbon Deposition, Spring Fatigue Breakage)

$$\lambda_{PO} = \lambda_{PO,B} \times C_P \times C_Q \times C_F \times C_v \times C_N \times C_S \times C_{DT} \times C_{Sw} \times C_W \qquad \ldots (8)$$

where,

$\lambda_{PO,B}$	= Base failure rate
C_P	= Fluid pressure factor
C_Q	= Allowable leak factor
C_F	= Surface finish factor
C_v	= Fluid viscosity factor
C_N	= Effect of contamination factor
C_S	= Seat stress factor
C_{DT}	= Seat base diameter factor
C_{Sw}	= Seat Land Width Factor
C_W	= Effect of Fluid flow rate factor

The base failure rate is taken from the historical database. A two-wheeler having a small- IC engine is considered. The capacity of engine is nearly 100 CC. Valves meets the required surface finish. Applying usage rate and other design parameters to the model relevant correction factors are selected. The reliability prediction is made based on the usage and eqn 8. The reliability prediction result from the model is compared with field failure rate data and comparison plot is given in Figure 15.

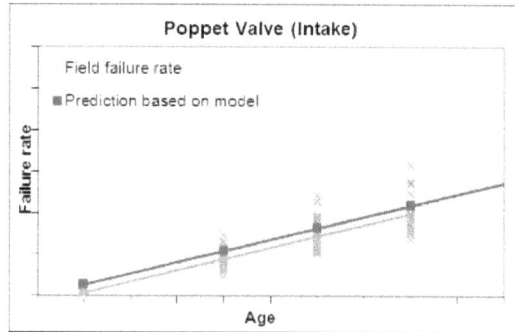

Fig. 15: Reliability Prediction of Valves

4.4 Clutch Plate

The engine life is directly related to the clutch plate life. Poor reliability of clutch plates lowers the overall reliability of an engine. Ideally, a clutch plate should have a life equivalent to the life of other engine parts. Hence, ensuring reliability of clutch plates becomes important in order to gain customer satisfaction. The clutch considered in this study is multi-plate type and a clutch plate is as shown in Figure 16.

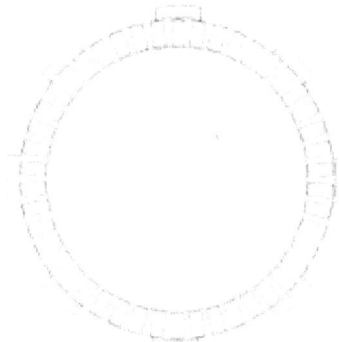

Fig. 16: A Clutch Plate

The basic function of a clutch plate is to transfer torque to the mating surface by the means of friction. This results in wear of the plates. Reliability prediction model of clutch plates in NSWC-07 is based on the wear phenomenon, the same is observed in a two-wheeler. The Dominant Failure Modes in Clutch plates are the wear of liner and formation of cracking on the liner as shown in Figure 17. Also, the design, environment, material and usage are falling in the range considered in the manual. Based on this understanding the model from NSWC-07 is explored and failure rates are compared with the field failure rate. The Number of drive plates in a clutch assembly is four. Liner material is of light duty resin asbestos type and other parameters as recorded are taken for building the model. The failure rates are found to have a good correlation with field failure rate, verifying the models applicability. The failure rate of a clutch plate is given by eqn 9 and the base failure rate is given by eqn 10. The comparison plot for failure rate is given in Figure 18,

$$\lambda_{CF} = \lambda_{CF,B} \times C_{NP} \times C_t \qquad \qquad \dots (9)$$

where,

$\lambda_{CF,B}$ = Base failure rate
C_{NP} = Correction factor for number of plates
C_t = Correction factor for operating temperature

$$\lambda_{CF,B} = (K_o \times p \times V_s \times t_a \, d) \qquad \qquad \dots (10)$$

where,

K_o = Wear rate coefficient
p = Nominal pressure between drive and driven plate
V_s = Sliding velocity
t_a = Clutch slip time
d = Clutch liner thickness

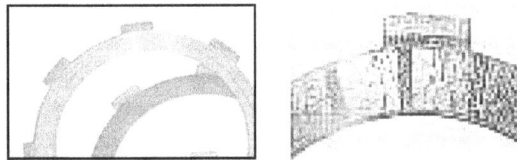

Fig. 17: FMs of Clutch Plate Wear and Cracking Of Liner

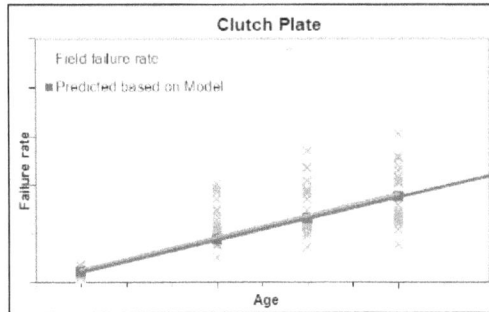

Fig. 18: Reliability Prediction of Clutch Plate

5. DISCUSSION

PoF approach based reliability predictions are generally more accurate than traditional reliability prediction techniques (Pecht, 1996). As the PoF models rely on design, material, geometry and stress profiles, a deep understanding of the component's design, environment, material and usage is necessary. In actual field usage, the failure inducing parameters like: temperature, loads, speeds, etc. are random variables and hence the 'median values' to be considered for the parameters, should be supported by data collection through instrumented vehicles before using these models for reliability prediction.

The PoF approach as detailed in NSWC-07 is limited by the failure rate prediction for a specified interval of usage. This interval of usage is mostly bounded by the FMs i.e. the FM should not change in this interval. Most of the PoF models in NSWC-07 are based on the principle of capacity depletion. In other words, the handbook takes basic life deciding factor of the part (lining thickness, fatigue endurance limit, etc.) into account to arrive at a failure rate considering the design parameters and stress levels.

The PoF models are highly sensitive to the design parameters and the operational and environmental factors. If applied without a proper understanding of underlying failure mechanism and design parameters the prediction results can be wrong and may lead to catastrophic results. The failure mechanisms are heavily dependent on time for mechanical components. Depending on the time elapsed the failure mechanism can change and hence, selection of mission time is very critical for performing reliability prediction based on PoF. The mission time considered in this paper is based on the knowledge of FM-Mechanism combination for the component. Applicability of NSWC-07 reliability prediction methods on two-wheelers is justified for only the mission period considered. Deep investigations supported by engineering judgments are required to apply the methods for extending the mission period considered.

A critical point to take into consideration while developing a PoF model is that the variables used in developing the models are independent (Cruse *et al.*, 1997). In the case studies the design parameters are not dependent on each other under the project feasibility boundary, i.e. given that the project feasibility in terms of cost, tooling capability, performance parameters, etc. is met the design parameters are statistically independent to each other. Also the variability in design, usage, environment and the material properties are not taken into consideration and hence the solution lies around the nominal values selected for the input to model. It is because of this reason that the model has to be validated on a specific application before being implemented on the components.

6. CONCLUSION

The PoF based reliability prediction techniques are found to have good correlation to the field failure rate. Hence, PoF based techniques can be used to predict reliability. The applicability of PoF models outlined in NSWC-07 on two-wheelers is verified in this paper.

The correctness of the reliability prediction methods depends on the quality of input data fed to the PoF model. The environment, usage data collected from field should be adequate and representative of field conditions for correct reliability estimates. All case studies in this paper are presented with good quality data input and the prediction result of models matches closely with the field failure rate.

The PoF based reliability prediction requires information on design, environment, material and usage. If these parameters can be approximated then the PoF model can be applied on new products for various applications. Reliability prediction was performed for a new product in the case study for V-Belt, using the available information. This prediction is compared with the field failure rate. The result of comparison validates the reliability prediction model. Hence, the PoF based model can be applied on two-wheeler during New Product Development.

Acknowledgement

The authors would like to express their gratitude to Mr. Vinay Harne, President NPD and Mr. N Jayaram, Vice President R&D., for their continued support to the present study. The authors also express their gratitude towards the management of TVS Motor Co. Ltd., for allowing them to publish this work.

REFERENCES

[1] Brissaud, F., Charpentier, D., Fouladirad, M., Barros, A. and Bérenguer, C. (2008), "Safety Instrumented System reliability evaluation with Influencing Factors", European Safety and Reliability Conference, ESREL, Valencia, Spain.

[2] Cruse, T.A., Harris, D.O., Kowal, M.T., Mahadevan, S. and Stracener, J.T. (1997), Reliability-based mechanical design, Marcel Dekker, New York, NY.

[3] Department of Defence of USA [DoD of USA] (1991), MILHDBK- 127F, Reliability Prediction of Electronic Equipment, DoD of USA, Philadelphia.

[4] Eusgeld, I., Freiling, F. and Reussner, R. (2008), Dependability metrics: advanced lectures, Springer-Verlag, Berling, Heidelberg, Germany.

[5] Lu, H., Bailey, C. and Yin, C. (2009), "Design for reliability of power electronics modules", Microelectronics Reliability, Vol. 49, Issue 9–11, pp. 1250–1255.

[6] Naval Surface Warfare Center [NSWC] (2007), Handbook of Reliability Prediction Procedures for Mechanical Equipment, NSWC, Maryland

[7] Pecht, M. (1996), "Why the traditional reliability prediction models do not work – is there an alternative?", Electronics Cooling, Vol. 2, pp. 10–12.

[8] Reliability Information Analysis Center [RiAC], (1995), Nonelectronic Parts Reliability Data, RiAC, Utica, NY.

[9] ReliaSoft Corporation (2007), "Accelerated Life Testing Reference", ReliaSoft Publishing, Tucson, AZ.

[10] Warburton, D., Strutt, J.E. and Allsopp, K. (1998), "Reliability prediction procedures for mechanical components at the design stage", *Journal of Process Mechanical Engineering,* Issue 4, pp. 213–224.

[11] Wyrwas, E., Condra, L. and Hawa, A. (2011), "Accurate Quantitative Physics-of-Failure Approach to Integrated Circuit Reliability", IPC APEX Expo, Las Vegas, Nevada.

An Extension of General Class of Change Point and Change Curve Modeling for Life Time Data Considering More than One Change Point Present in the Data

Ishita Basak[1] and Ashis K. Chakraborty[2]

Indian Statistical Institute, Kolkata

E-mail: [1]ishita.basak@gmail.com; [2]akchakraborty123@rediffmail.com

ABSTRACT: Change point hazard rate models arise in many life time data analysis, as it is often reasonable to assume that early failures occur at one rate and later on at different rates. The points at which the change of hazard rate takes place in the data are called change points. In this article, we propose a model with two change points which practically extends the works of Patra and Dey [2002]. We propose a Bayesian approach and provide inference conditional upon the observed data. The proposed Bayesian models are fitted using Markov Chain Monte Carlo method. We illustrate the proposed methodology with an application to modeling life times of the printed circuit board.

Keywords: Change Point, Hazard Function, Survival Fraction, Kaplan Meier Plot, Bayesian Inference, Metropolis-Hastings Algorithm.

1. INTRODUCTION

In life time data analysis, for example, in clinical trials, undesirable side effects may cause a different failure rate after a threshold time or in cancer prevention trials, it is often reasonable to assume that any effect of treatment is not immediate but affects the risk of failure only after a lag.

The classical change point problem arises from the observation of a sequence of random variables $X_1, X_2, ..., X_n$ such that $X_1, X_2, ..., X_\eta$ have a common distribution F and $X_{\eta+1}, X_{\eta+2}, ..., X_n$, $(\eta \leq n)$ have the distribution G with $F \neq G$. The index η, called the change point, is usually unknown and has to be estimated from the data. The above is the situation when a single change point η is present in the data. But in reality this may not be the case. More than one change points may be present in the data.

Patra and Dey (2002) proposed a general class of models which incorporates the change point model for hazard function and in general for curves which are functions of survival times. But they considered only one change point present in the data.

Here we try to propose a general change point model for hazard function considering more than one change points present in the data. We use Bayesian methodologies [Ghosh,

Delampady, Samanta (2006)] to infer about the model parameters where the model for the hazard function is mainly governed by the empirical plot, for example, Kaplan Meier Plot. For some data, we may get a rough idea from the plot about the fact that more than one change point may be present in the data.

2. GENERAL FORMULATION

Let T be a random variable denoting the lifetime distribution with density function $f_T(t)$ and survival function $S_T(t)$. A suitable change point hazard function $h_T(t) = f_T(t)/S_T(t)$ considering single change point η present in the data is defined as,

$$h_T(t) = h_1(t)I(0 \leq t \leq \eta) + h_2(t)I(t < \eta)$$

Where, $I(A) = 1$ if $x \varepsilon A$ and 0, otherwise; and $\eta \varepsilon R^+$ is the possible unknown threshold parameter and $h_1(t)$ and $h_2(t)$ are the hazard functions in $[0, \eta]$ and $[\eta, \infty]$ respectively.

Now, suppose that there are two change points present in the data viz. η_1 and η_2, then the change point hazard function is given by,

$$h_T(t) = h_1(t)I(0 < t \leq \eta_1) + h_2(t)I(\eta_1 < t \leq \eta_2) + h_3(t)I(t > \eta_2)$$

where $I(A) = 1$ if $x \varepsilon A$ and 0, otherwise, and $\eta_1, \eta_2 \varepsilon R^+$ are the possible unknown threshold parameters and $h_1(t), h_2(t)$ and $h_3(t)$ are the hazard functions in $[0, \eta_1]$, $[\eta_1, \eta_2]$ and $[\eta_2, \infty]$ respectively.

The cumulative hazard function is given by, $H_T(t) = \int_0^t h_T(u)du$. Now, there can be three cases in calculating cumulative hazard.

Case 1: $t \leq \eta_1$,

$$\int_0^t h_T(u)du = \int_0^t h_1(u)du = H_1(t).$$

Case 2: $\eta_1 < t \leq \eta_2$,

$$\int_0^t h_T(u)du = \int_0^{\eta_1} h_1(u)du + \int_{\eta_1}^t h_2(u)du = H_1(\eta_1) + H_2(t) - H_2(\eta_1).$$

Case 3: $t > \eta_2$,

$$\int_0^t h_T(u)du = \int_0^{\eta_1} h_1(u)du + \int_{\eta_1}^{\eta_2} h_2(u)du + \int_{\eta_2}^t h_3(u)du$$

$$= H_1(\eta_1) + H_2(\eta_2) - H_2(\eta_1) + H_3(t) - H_3(\eta_2)$$

Therefore, $H_T(t) = H_1(t)I(t < \eta_1) + [H_1(\eta_1) + \{H_2(t) - H_2(\eta_1)\}]I[\eta_1 < t < \eta_2]$

$$+[H_1(\eta_1) + \{H_2(\eta_2) - H_2(\eta_1)\} + \{H_3(t) - H_3(\eta_2)\}]I[t > \eta_2]$$

Hence the density function, denoted by $g(t)$, is given by,

$$g(t) = h_T(t)exp\left[-\int_0^t h_T(u)du\right] = h_T(t)exp[-H_T(t)]$$

$$= [h_1(t)I(0 < t \leq \eta_1) + h_2(t)I(\eta_1 < t \leq \eta_2) + h_3(t)I(t > \eta_2)] \times$$

$$exp[-H_1(t)I(t < \eta_1) - \{H_1(\eta_1) + H_2(t) - H_2(\eta_1)\}I(\eta_1 < t < \eta_2)$$
$$- \{H_1(\eta_1) + H_2(\eta_2) - H_2(\eta_1) + H_3(t) - H_3(\eta_2)\}I(t > \eta_2)]$$
$$= h_1(t)exp[-H_1(t)]I(t < \eta_1)+$$
$$h_2(t)exp[-\{H_1(\eta_1) + H_2(t) - H_2(\eta_1)\}]I(\eta_1 < t < \eta_2)+$$
$$h_3(t)exp[-\{H_1(\eta_1) + H_2(\eta_2) - H_2(\eta_1) + H_3(t) - H_3(\eta_2)\}]I(t > \eta_2)]$$

Now let $p_1 = \int_0^{\eta_1} h_1(u)exp(-H_1(u))du$

$$= \int_0^{\eta_1} e \, xp(-H_1(u))d(H_1(u))$$
$$= [-exp(-H_1(u))]_0^{\eta_1}$$
$$= 1 - exp(-H_1(\eta_1))$$

and $p_2 = \int_{\eta_2}^{\infty} h_3(u)exp[-\{H_1(\eta_1) + H_2(\eta_2) - H_2(\eta_1) + H_3(u) - H_3(\eta_2)\}]du$

$$= exp[-H_1(\eta_1) - H_2(\eta_2) + H_2(\eta_1) + H_3(\eta_2)] \int_{\eta_2}^{\infty} h_3(u)exp(-H_3(u))du$$
$$= exp[-H_1(\eta_1) - H_2(\eta_2) + H_2(\eta_1) + H_3(\eta_2)] \int_{\eta_2}^{\infty} e \, xp(-H_3(u))d(H_3(u))$$
$$= exp[-H_1(\eta_1) - H_2(\eta_2) + H_2(\eta_1) + H_3(\eta_2)][-exp(-H_3(u))]_{\eta_2}^{\infty}$$
$$= exp[-H_1(\eta_1) - H_2(\eta_2) + H_2(\eta_1) + H_3(\eta_2)] \times [exp(-H_3(\eta_2))]$$
$$= exp[-H_1(\eta_1) - H_2(\eta_2) + H_2(\eta_1)]$$

and therefore, $1 - p_1 - p_2$

$$= 1 - [1 - exp(-H_1(\eta_1))] - [exp[-H_1(\eta_1) - H_2(\eta_2) + H_2(\eta_1)]]$$
$$= exp(-H_1(\eta_1)) - exp[-H_1(\eta_1) - H_2(\eta_2) + H_2(\eta_1)]$$

Also, $\int_{\eta_1}^{\eta_2} h_2(u)exp[-\{H_1(\eta_1) + H_2(u) - H_2(\eta_1)\}]du$

$$= exp[-H_1(\eta_1) + H_2(\eta_1)] \int_{\eta_1}^{\eta_2} h_2(u)exp(-H_2(u))du$$
$$= exp[-H_1(\eta_1) + H_2(\eta_1)] \int_{\eta_1}^{\eta_2} e \, xp(-H_2(u))d(H_2(u))$$
$$= exp[-H_1(\eta_1) + H_2(\eta_1)][-exp(-H_2(u))]_{\eta_1}^{\eta_2}$$
$$= exp[-H_1(\eta_1) + H_2(\eta_1)][exp[-H_2(\eta_1) - exp(-H_2(\eta_2))]$$
$$= exp(-H_1(\eta_1)) - exp[-H_1(\eta_1) - H_2(\eta_2) + H_2(\eta_1)]$$
$$= 1 - p_1 - p_2$$

Hence, $g(t)$ can be written as,

$$g(t) = p_1 . f_1(t) + (1 - p_1 - p_2) . f_2(t) + p_2 . f_3(t)$$

where, $f_1(t) = \frac{h_1(t)exp(-H_1(t))}{1 - exp(-H_1(\eta_1))} I(t < \eta_1)$ which corresponds to a right truncated distribution.

and $f_2(t) = \frac{h_2(t)exp(-H_1(\eta_1) - H_2(t) + H_2(\eta_1))}{exp(-H_1(\eta_1)) - exp[-H_1(\eta_1) - H_2(\eta_2) + H_2(\eta_1)]} I(\eta_1 < t \le \eta_2)$

and $f_3(t) = \frac{h_3(t)exp(-H_1(\eta_1)-H_2(\eta_2)+H_2(\eta_1)-H_3(t)+H_3(\eta_2))}{exp(-H_1(\eta_1)-H_2(\eta_2)+H_2(\eta_1))}I(t > \eta_2)]$ which corresponds to a left truncated distribution.

It can be easily shown that $f_i(t)$'s are density functions. Therefore, $g(t)$ can be written as a mixture of three density functions.

Next we provide the explicit form of the survival function and the hazard function corresponding to different components of the mixture distribution.

Survival function of the first component,

$$S_1^*(t) = \overline{F_1}(t) = 1 - F_1(t)$$

$$= 1 - \int_0^t f_1(u)du$$

$$= 1 - \int_0^t \frac{h_1(u)exp(-H_1(u))}{1 - exp(-H_1(\eta_1))}du$$

$$= 1 - \int_0^t \frac{exp(-H_1(u))}{1 - exp(-H_1(\eta_1))}d(H_1(u))$$

$$= 1 - \frac{[-exp(-H_1(u))]_0^t}{1 - exp(-H_1(\eta_1))}$$

$$= 1 - \frac{1 - exp(-H_1(t))}{1 - exp(-H_1(\eta_1))}$$

$$= \frac{exp(-H_1(t)) - exp(-H_1(\eta_1))}{1 - exp(-H_1(\eta_1))}I(t < \eta_1)$$

Hazard function of the first component,

$$h_1^*(t) = \frac{f_1(t)}{S_1^*(t)} = \frac{h_1(t)exp(-H_1(t))}{exp(-H_1(t)) - exp(-H_1(\eta_1))}I(t < \eta_1)$$

Survival function of the second component,

$$S_2^*(t) = \overline{F_2}(t) = 1 - F_2(t)$$

Survival function of the second component,

$$S_2^*(t) = \overline{F_2}(t) = 1 - F_2(t)$$

$$= 1 - \int_{\eta_1}^t f_2(u)du$$

$$= 1 - \int_{\eta_1}^t \frac{h_2(u)exp(-H_2(u) + H_2(\eta_1))}{1 - exp(H_2(\eta_1) - H_2(\eta_2))}du$$

$$= 1 - \frac{exp(H_2(\eta_1))(exp(-H_2(\eta_1)) - exp(H_2(t)))}{1 - exp(H_2(\eta_1) - H_2(\eta_2))}$$

$$= 1 - \frac{1 - (exp(H_2(\eta_1)) - H_2(t))}{1 - exp(H_2(\eta_1) - H_2(\eta_2))}$$

$$= \frac{1 - exp(H_2(\eta_1) - H_2(\eta_2)) - 1 + (exp(-H_2(\eta_1)) - H_2(t))}{1 - exp(H_2(\eta_1) - H_2(\eta_2))}$$

$$= \frac{exp(H_2(\eta_1))[-(exp(-H_2(\eta_2)) + exp(-H_2(t)))]}{1 - exp(H_2(\eta_1) - H_2(\eta_2))} I(\eta_1 < t < \eta_2)$$

Hazard function of the second component,

$$h_2^*(t) = \frac{f_2(t)}{S_2^*(t)} = \frac{h_2(t)exp(-H_2(t) + H_2(\eta_1))}{exp(H_2(\eta_1))[-exp(-H_2(\eta_2)) + exp(-H_2(t))]} I(\eta_1 < t < \eta_2)$$

Survival function of the third component,

$$S_3^*(t) = \overline{F_3}(t) = 1 - F_3(t)$$

$$= 1 - \int_{\eta_2}^{t} f_3(u)du$$

$$= 1 - \int_{\eta_2}^{t} h_3(u)exp(-H_3(u) + H_3(\eta_2))du$$

$$= 1 - exp(H_3(\eta_2))[exp(-H_3(\eta_2)) - exp(-H_3(t))]$$

$$= exp(-H_3(t) + H_3(\eta_2))I(t > \eta_2)$$

Hazard function of the third component,

$$h_3^*(t) = \frac{f_3(t)}{S_3^*(t)} = h_3(t)I(t > \eta_2)$$

3. SURVIVAL FRACTION

A population of a life time distribution is said to have a survival fraction if the corresponding survival function does not tend to zero but to a strictly positive fraction as time approaches infinity [Patra and Dey, 2002]. Naturally this type of survival function is not proper and hence, it will be, instead, referred to as an improper survival function. An example of this situation is in accelerated life time experiments where a certain fraction of population never fails. The survival function in our case is given by,

$$S_T(t) = exp(-H_T(t))$$

$$= exp[-\{H_1(t)I(t < \eta_1) + [H_1(\eta_1) + \{H_2(t) - H_2(\eta_1)\}]I[\eta_1 < t < \eta_2]$$

$$+ [H_1(\eta_1) + \{H_2(\eta_2) - H_2(\eta_1)\} + \{H_3(t) - H_3(\eta_2)\}]I[t > \eta_2]\}]$$

Hence, $lim_{t \to \infty} S_T(t) = exp(-lim_{t \to \infty} H_T(t))$

$$L_i = lim_{t \to \infty} H_i(t) = \int_0^\infty h_i(u)du \text{ for } i = 1, 2, 3.$$

Therefore, $lim_{t \to \infty} H_T(t) = H_1(\eta_1) + (H_2(\eta_2) - H_2(\eta_1)) + (L_3 - H_3(\eta_2))$

So an improper or a proper survival function will be obtained depending on the finiteness of L_3. The value of the improper survival function at infinity will be called the survival fraction. So modeling survival fraction is solely based on the choice of $h_3(.)$. Note that for a simple hazard function without a change point we simply plug in $\eta_1 = \eta_2 = 0$ and construct the survival function accordingly.

4. INCORPORATION OF BASELINE HAZARD

Suppose we incorporate baseline hazard in the above models and replace $h_i(t)$ by $h_0(t)h_i(t)$, for $i = 1, 2, 3$, as in Patra and Dey [2002].

Consequently the overall hazard function is given by,

$$h_T(t) = h_0(t)h_1(t)I(0 < t \le \eta_1) + h_0(t)h_2(t)I(\eta_1 < t \le \eta_2) + h_0(t)h_3(t)I(t > \eta_2)$$

Let us define, $M_{0i}(t) = \int_0^t h_0(u)h_i(u)du$ for $i = 1, 2, 3$.

Then the cumulative hazard function is expressed as,

$$H_T(t) = M_{01}(t)I(t < \eta_1) + [M_{01}(\eta_1) + \{M_{02}(t) - M_{02}(\eta_1)\}]I[\eta_1 < t < \eta_2]$$
$$+ [M_{01}(\eta_1) + \{M_{02}(\eta_2) - M_{02}(\eta_1)\} + \{M_{03}(t) - M_{03}(\eta_2)\}]I[t > \eta_2]$$

Again the limit $lim_{t \to \infty} M_{0i}(t) = \int_0^\infty h_0(u)h_i(u)du$ can be finite or infinite depending upon the choice of both $h_0(u)$ and $h_i(u)$'s.

Here, we only need to concentrate on the limit corresponding to $M_{03}(t)$ and if it is finite then survival fraction will be obtained.

The density can be expressed as,

$$g(t) = p_1.f_1(t) + (1 - p_1 - p_2).f_2(t) + p_2.f_3(t)$$

Where,

$$p_1 = 1 - exp(-M_{01}(\eta_1))$$
$$p_2 = exp[-M_{01}(\eta_1) - M_{02}(\eta_2) + M_{02}(\eta_1)]$$
$$1 - p_1 - p_2 = exp(-M_{01}(\eta_1)) - exp[-M_{01}(\eta_1) - M_{02}(\eta_2) + M_{02}(\eta_1)]$$

and

$$f_1(t) = \frac{h_0(t)h_1(t)exp(-M_{01}(t))}{1 - exp(-M_{01}(\eta_1))}I(t < \eta_1)$$

$$f_2(t) = \frac{h_0(t)h_2(t)exp(-M_{01}(\eta_1) - M_{02}(t) + M_{02}(\eta_1))}{exp(-M_{01}(\eta_1)) - exp[-M_{01}(\eta_1) - M_{02}(\eta_2) + M_{02}(\eta_1)]}I(\eta_1 < t \le \eta_2)$$

$$f_3(t) = \frac{h_0(t)h_3(t)exp(-M_{01}(\eta_1) - M_{02}(\eta_2) + M_{02}(\eta_1) - M_{03}(t) + M_{03}(\eta_2)}{exp(-M_{01}(\eta_1) - M_{02}(\eta_2) + M_{02}(\eta_1))}I(t > \eta_2)]$$

5. AN EXAMPLE

Meeker and Lu Valle (1995) discussed a dataset on lifetimes of printed circuit boards tested using Relative humidity (RH) as the accelerating stress. In particular, there are 72 circuit boards tested at each of four stress levels, 49.5% RH, 62.8% RH, 75.4% RH and 82.4% RH. The authors noted that due to problems with test equipment there were several circuit boards that did not yield useful information. Therefore the resulting dataset consists of 70 boards at stress levels of 49.5% RH, 75.4% RH and 82.4% RH and 68 boards at 62.8% RH level. The boards are monitored periodically, thus the data were interval censored. For each unit only the interval of consecutive inspection times containing the failure time was recorded. In addition to interval censoring, there are several circuit boards that did not fail. Therefore, data were also subject to right censoring. In particular there were 48 censored observations at 4078 hours at the 49.5% RH and 11 at 3067 hours at the 62.8% RH. All of the circuit boards did fail at 75.4% RH and 82.4% RH. This dataset was extensively investigated by Sinha *et al.* (1999) at all four humidity levels. They assumed that failure to a unit is caused by several latent competing risks of manufacturing defects. In practice this is an appropriate assumption for many manufacturing processes. For circuit boards salt residues may be left on the surface of the circuit board due to faulty manufacturing process. Under humidity, these salt residues ionize over time to create conductive filaments between the two poles (cathode and anode) to cause short circuits. Now humidity level affects this process in two ways – it may cause any salt particle to ionize sooner and it may increase the number of latent risk factors (number of bridges which may potentially ionize to become conductive filaments) of the board. Even minor faults (small traces of salt bridges), which may not cause failure (ionize) under normal stress level, are at risk of becoming fatal (at the risk of ionizing) under accelerated stress (humidity).

The original data is given in the following Table 1:

Table 1: Circuit Board Accelerated Life Test Data with Relative Humidity 82.4%

		82.4% Relative Humidity								
Observations 01–09	Lower	14	18	30	42	46	50	58	62	66
	Upper	18	22	34	46	50	54	62	66	70
	Count	1	1	8	3	3	3	4	1	5

Observations 10–18	Lower	70	74	78	82	86	90	94	98	106
	Upper	74	78	82	86	90	94	98	102	110
	Count	2	2	1	4	8	4	5	5	1

Observations 18–25	Lower	110	114	122	130	134	138	142
	Upper	114	118	126	134	138	142	146
	Count	2	1	1	1	1	1	2

The data in Table 1 represents the number of failures observed in a four-hourly interval. If some interval is missing in the table, it indicates that there were no failures observed in that four-hour interval. For example, the intervals 22–26 hours and 26–30 hours are missing from the table, indicating that there were no failures during these two intervals. "Lower" and "Upper" denotes the lower and upper limits of interval (time in hours) during which at least one failure is observed and "Count" denotes the number of failures occurred in that interval. For example, for observation 1, "Lower", "Upper" and "Count" are 14, 18 and 1 respectively implying that there is one failure in the time interval of 14 to 18 hours. Here exact time at which the failure has occurred is not recorded, only the interval in which failure has occurred is recorded.

For illustrative purpose we are only concentrating on the fourth humidity level of the original data where we do not have any survival fraction of the circuit board population. However we can build up the model incorporating the survival fraction or the so called LFP (Limited Failure Population) for the lower humidity levels as well by the technique discussed in previous sections.

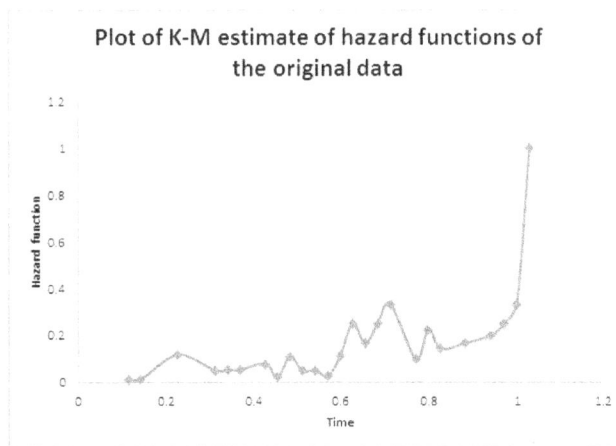

Fig. 1: Plot of the Kaplan Meier Estimates

Figure 1 gives the Kaplan Meier plot of the hazard function for the humidity level of 82.4% RH. Actual time points (hourly) were scaled by 140. The hazard plot is based on the minimum of several unknown life times of the salt particle because a circuit board would fail if at least one salt particle gets ionized. Also the number of such salt particles is random.

Let x_{ij} represent the time when the jth salt particle of the ith board ionize, $j = 1, ..., M_i$ and $i = 1, ..., n$. Clearly, the circuit boards fails at $T_i = min(x_{i_1}, ..., x_{i_{M_i}})$. Thus, T_i is the observed time to failure of the ith board. In addition, we assume that, x_{ij} are iid with CDF $F^*(.)$ for $j = 1, ..., M_i$ and $i = 1, ..., n$. and M_i are iid Poisson with parameter θ for $i = 1, ..., n$.

The smoothed Kaplan Meier plot given in Figure 2 clearly shows that more than one change point may be present in the data. An appropriate non-parametric test suggested by Pettit [1978] is carried out to find the number of change points in the data. An initial estimate of η_1 and η_2, the two possible change points, is taken as 0.59 and 0.89 respectively based on the non-parametric test.

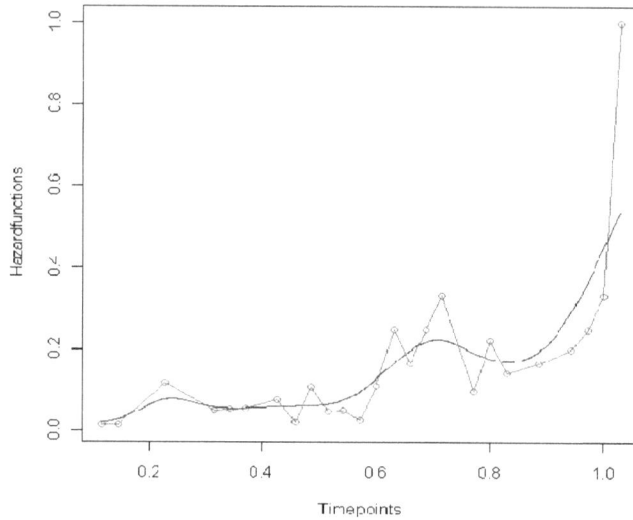

Fig. 2: The Smoothed Curve for Hazard Functions of the Data

Figure 2 gives the smoothed estimated hazard for T and will be used to get a functional form of $h_T(t)$. The first component of hazard before the change at η_1 occurred seems to be linear in time. The second component after the change at η_1 has ocurred and before the change at η_2 occurred is assumed to be quadratic in time and the third component after the change at η_2 has occurred is taken as a power function in time. Thus,

$$h_0(t) = \theta, \; h_1(t) = \alpha + \beta t, \; h_2(t) = \lambda + \mu t + \rho t^2, \; h_3(t) = \delta\gamma t^{\gamma-1}$$

where $\theta, \alpha, \beta, \lambda, \mu, \rho, \delta, \gamma$ and η_1, η_2 are positive valued parameters.

$$M_{01}(t) = \int_0^t h_0(u)h_1(u)du$$
$$= \int_0^t \theta\,(\alpha + \beta u)du$$
$$= [\alpha\theta u + \theta\beta u^2]_0^t$$
$$= \theta(\alpha t + \beta t^2/2)$$

$$M_{02}(t) = \int_0^t h_0(u)h_2(u)du$$
$$= \int_0^t \theta\,(\lambda + \mu u + \rho u^2)du$$
$$= \theta[\lambda u + \mu u^2/2 + \rho u^3/3]_0^t$$
$$= \theta(\lambda t + \mu t^2/2 + \rho t^3/3)$$

$$M_{03}(t) = \int_0^t h_0(u)h_3(u)du$$
$$= \int_0^t \theta\,(\delta\gamma u^{\gamma-1})du$$
$$= \theta[\delta\gamma u^\gamma]_0^t$$
$$= \theta(\delta\gamma t^\gamma)$$

Therefore,

$$f_1(t) = \frac{\theta(\alpha + \beta t)exp(-\theta(\alpha t + \beta t^2/2)}{1 - exp(-\theta(\alpha \eta_1 + \beta \eta_1^2/2)} I(t < \eta_1)$$

$$f_2(t) = \frac{\theta(\lambda + \mu t + \rho t^2)exp[-\theta\{\lambda(t - \eta_1) + \frac{\mu}{2}(t^2 - \eta_1^2) + \frac{\rho}{3}(t^3 - \eta_1^3)\}]}{1 - exp[-\theta\{\lambda(\eta_2 - \eta_1) + \frac{\mu}{2}(\eta_2^2 - \eta_1^2) + \frac{\rho}{3}(\eta_2^3 - \eta_1^3)\}]} I(\eta_1 < t \le \eta_2)$$

$$f_3(t) = \theta\delta\gamma t^{\gamma - 1}exp(-\theta\delta(t^\gamma - \eta_2^\gamma))I(t > \eta_2)]$$

and

$$p_1 = 1 - exp(-\theta(\alpha \eta_1 + \beta \eta_1^2/2))$$

$$p_2 = exp[-\theta\{(\alpha \eta_1 + \beta \eta_1^2/2) - (\lambda(\eta_2 - \eta_1) + \mu(\eta_2^2 - \eta_1^2)/2 + \rho(\eta_2^3 - \eta_1^3)/3)\}]$$

Hence,

$$g(t) = p_1. f_1(t) + (1 - p_1 - p_2). f_2(t) + p_2. f_3(t)$$

We do not have any censoring for the dataset, so the likelihood becomes,

$$L(\text{parameters}/t) = \prod_{i=1}^{n} g(t_i)$$

To perform the Bayesian analysis [Bolstad (2010)] we use the following priors for various parameters,

$\theta \sim$ Gamma (θ_1, θ_2), $\alpha \sim$ Gamma (α_1, α_2), $\beta \sim$ Gamma (β_1, β_2),
$\gamma \sim$ Gamma (γ_1, γ_2),

$\delta \sim$ Gamma(δ_1, δ_2), $\lambda \sim$ Gamma(λ_1, λ_2), $\mu \sim$ Gamma(μ_1, μ_2), $\rho \sim$ Gamma(ρ_1, ρ_2),

$\eta_1 \sim$ Uniform$(t_{(1)}, t_{(n)})$, $\eta_2 \sim$ Uniform$(\eta_1, t_{(n)})$.

where Gamma(a, b) denotes the probability density of a gamma distribution given by,

$$\frac{b^a exp(-bx)x^{a-1}}{\Gamma(a)} I(x > 0).$$

The Gamma prior for θ is chosen because of the conjugate nature of Gamma distribution for Poisson parameter. The choice of gamma prior is traditional for Weibull parameters γ and δ as well. The remaining parameters can be real valued but we choose Gamma priors to make them positive valued as encouraged by emperical observations. The hyperparameters are selected carefully to make the priors diffuse. The results are not sensitive to the choice of the hyper parameters as long as the priors are sufficiently diffused. Based on empirical observations we restrict the first change point η_1 between $t_{(1)}$ and $t_{(n)}$, the minimum and maximum failure time and the second change point between η_1 and $t_{(n)}$.

Full Conditional Posterior Distribution of the Model Parameters
[Gelfand and Smith (1990)]

If a random variable X takes n values $x_1, x_2, \ldots\ldots x_n$ and has the density $f(x/p_1, p_2, \ldots\ldots p_n)$, where $p_1, p_2\ldots\ldots p_n$ are the parameters of the distribution and

$p_i \sim \pi(p_i)$, the prior distribution of the parameter p_i, then the full conditional posterior distribution of parameter p_i given the values of the rest of the parameters, is as follows:

$$\pi(p_i/rest) \propto \prod_{i=1}^{n} f\left(x_i/p_1, p_2, \ldots \ldots p_n\right)\pi(p_i)$$

Hence the full conditional posterior distribution for parameter $\theta, \alpha, \beta, \gamma, \delta, \lambda, \mu, \rho, \eta_1, \eta_2$ are given respectively as:

$$\pi(\theta/rest) \propto \left(\prod_{t_i=1}^{t(\eta_1)} f_1\left(t_i\right)\right)\left(\prod_{t_i=t(\eta_1)+1}^{t(\eta_2)} f_2\left(t_i\right)\right)\left(\prod_{t_i=t(\eta_2)+1}^{n} f_3\left(t_i\right)\right)\pi(\theta)$$

$$\propto \frac{\theta^{n+\theta_1-1}exp(-\theta\left[\sum_{i=1}^{t(\eta_1)}(\alpha t_i + \beta t_i^2/2) + \lambda\sum_{i=t(\eta_1)+1}^{t(\eta_2)}(t_i - \eta_1) + \frac{\mu}{2}\sum_{i=t(\eta_1)+1}^{t(\eta_2)}(t_i^2 - \eta_1^2)\right]}{[1 - exp(-\theta(\alpha\eta_1 + \beta\eta_1^2/2))]^{\eta_1}\left[1 - exp(-\theta\lambda(\eta_2 - \eta_1) - \theta\frac{\mu}{2}(\eta_2^2 - \eta_1^2) - \theta\frac{\rho}{3}(\eta_2^3 - \eta_1^3)\right]^{\eta_2-\eta_1}}$$

$$\times exp(-\theta\left[+\frac{\rho}{3}\sum_{i=t(\eta_1)+1}^{t(\eta_2)}(t_i^3 - \eta_1^3) + \delta\sum_{i=t(\eta_2)+1}^{n}(t_i^\gamma - \eta_2^\gamma) + \theta_2\right]$$

$$\pi(\alpha/rest) \propto \left(\prod_{t_i=1}^{t(\eta_1)} f_1\left(t_i\right)\right)\pi(\alpha)$$

$$\propto \left(\frac{\prod_{t_i=1}^{t(\eta_1)}(\alpha+\beta t_i)\alpha^{\alpha_1-1}exp(-\theta(\sum_{t_i=1}^{t(\eta_1)} t_i + \alpha_2)\alpha)}{[1-exp(-\theta(\alpha\eta_1 + \beta\eta_1^2/2))]^{\eta_1}}\right)$$

$$\pi(\beta/rest) \propto \left(\prod_{t_i=1}^{t(\eta_1)} f_1\left(t_i\right)\right)\pi(\beta)$$

$$\propto \left(\frac{\prod_{t_i=1}^{t(\eta_1)}(\alpha+\beta t_i)\beta^{\beta_1-1}exp(-\theta(\sum_{t_i=1}^{t(\eta_1)} t_i^2/2 + \beta_2)\beta)}{[1-exp(-\theta(\alpha\eta_1 + \beta\eta_1^2/2))]^{\eta_1}}\right)$$

$$\pi(\gamma/rest) \propto \left(\prod_{t_i=t(\eta_2)+1}^{n} f_3\left(t_i\right)\right)\pi(\gamma)$$

$$\propto \prod_{t_i=t(\eta_2)+1}^{n} t_i^{\gamma-1} \, exp(-(\sum_{t_i=t(\eta_2)+1}^{n} \theta\,\delta(t_i^\gamma - \eta_2^\gamma) + \gamma_2\gamma))\gamma^{n-n_2+\gamma_1-1}$$

$$\pi(\delta/rest) \propto \left(\prod_{t_i=t(\eta_2)+1}^{n} f_3\left(t_i\right)\right)\pi(\delta)$$

$$\propto Gamma(n - n_2 + \delta_1 - 1, \sum_{t_i=t(\eta_2)+1}^{n} \theta\left(t_i^\gamma - \eta_2^\gamma\right) + \delta_2)$$

$$\pi(\lambda/rest) \propto \left(\prod_{t_i=t(\eta_1)+1}^{t(\eta_2)} f_2\left(t_i\right)\right)\pi(\lambda)$$

$$\propto \frac{\prod_{t_i=t(\eta_1)+1}^{t(\eta_2)}(\lambda+\mu t_i+\rho t_i^2)\lambda^{\lambda_1-1}exp(-(\theta\sum_{i=t(\eta_1)+1}^{t(\eta_2)}(t_i-\eta_1)+\lambda_2)\lambda)}{\left[1-exp(-\theta\lambda(\eta_2-\eta_1)-\theta c\mu 2(\eta_2^2-\eta_1^2)-\theta\frac{\rho}{3}(\eta_2^3-\eta_1^3)\right]^{\eta_2-\eta_1}}$$

$$\pi(\mu/rest) \propto \left(\prod_{t_i=t(\eta_1)+1}^{t(\eta_2)} f_2\left(t_i\right)\right)\pi(\mu)$$

$$\propto \frac{\prod_{t_i=t(\eta_1)+1}^{t(\eta_2)}(\lambda+\mu t_i+\rho t_i^2)\mu^{\mu_1-1}exp(-(\theta\sum_{i=t(\eta_1)+1}^{t(\eta_2)}(t_i^2-\eta_1^2)+\mu_2)\mu)}{\left[1-exp(-\theta\lambda(\eta_2-\eta_1)-\theta\frac{\mu}{2}(\eta_2^2-\eta_1^2)-\theta\frac{\rho}{3}(\eta_2^3-\eta_1^3)\right]^{\eta_2-\eta_1}}$$

$$\pi(\rho/rest) \propto \left(\prod_{t_i=t(\eta_1)+1}^{t(\eta_2)} f_2\left(t_i\right)\right)\pi(\rho)$$

$$\propto \frac{\prod_{t_i=t(\eta_1)+1}^{t(\eta_2)}(\lambda+\mu t_i+\rho t_i^2)\rho^{\rho_1-1}exp(-(\theta\sum_{i=t(\eta_1)+1}^{t(\eta_2)}(t_i^3-\eta_1^3)+\rho_2)\rho)}{\left[1-exp(-\theta\lambda(\eta_2-\eta_1)-\theta\frac{u}{2}(\eta_2^2-\eta_1^2)-\theta\frac{\rho}{3}(\eta_2^3-\eta_1^3)\right]^{\eta_2-\eta_1}}$$

$$\pi(\eta_1/rest) \propto \left(\prod_{t_i=1}^{t(\eta_1)} f_1\left(t_i\right)\right)\left(\prod_{t_i=t(\eta_1)+1}^{t(\eta_2)} f_2\left(t_i\right)\right)\pi(\eta_1)$$

$$\propto \frac{exp(-\theta(\lambda\sum_{i=t(\eta_1)+1}^{t(\eta_2)}(t_i-\eta_1)+\frac{\mu}{2}\sum_{i=t(\eta_1)+1}^{t(\eta_2)}(t_i^2-\eta_1^2)+\frac{\rho}{3}\sum_{i=t(\eta_1)+1}^{t(\eta_2)}(t_i^3-\eta_1^3))}{[1-exp(-\theta(pha\eta_1+\beta\eta_1^2/2))]^{\eta_1}\left[1-exp(-\theta\lambda(\eta_2-\eta_1)-\theta\frac{\mu}{2}(\eta_2^2-\eta_1^2)-\theta\frac{\rho}{3}(\eta_2^3-\eta_1^3))\right]^{\eta_2-\eta_1}}$$

$$\pi(\eta_2/rest) \propto \left(\prod_{t_i=t(\eta_1)+1}^{t(\eta_2)}f_2(t_i)\right)\left(\prod_{t_i=t(\eta_2)+1}^{n}f_3(t_i)\right)\pi(\eta_2)$$

$$\propto \frac{exp(-\theta\delta(\sum_{i=t(\eta_2)+1}^{n}(t_i^\gamma-\eta_2^\gamma)))}{\left[1-exp(-\theta\lambda(\eta_2-\eta_1)-\theta\frac{\mu}{2}(\eta_2^2-\eta_1^2)-\theta\frac{\rho}{3}(\eta_2^3-\eta_1^3)\right]^{\eta_2-eta_1}}$$

Estimates of the Model Parameters

We use the Metropolis Hastings algorithm [Chib and Greenberg (1995)] to estimate the model parameters. The estimates along with their respective standard deviation and their 95% high probability density credible intervals are given in the following table:

Table 2: Estimates of the Model Parameters

Parameters	Posterior Mean	Standard Deviation	95% HPD Credible Interval
θ	5.00000000000	0.00494	(4.90716,5.0989)
α	0.00098500449	3.2×10^{-5}	(0.00098,0.001012)
β	0.01908990000	1.12×10^{-4}	(0.0189523,0.0191235)
γ	6.87265968000	0.0064184	(6.8702365,6.8812563)
δ	0.02309933167	4.6×10^{-4}	(0.0230124,0.2312568)
λ	0.00818608970	6.61×10^{-5}	(0.008023564,0.0082356)
μ	0.02325830500	7.83×10^{-5}	(0.02312564,0.23275698)
ρ	0.00603826800	4.16×10^{-5}	(0.00592365,0.00605428)
η_1	0.60324325699	7.43×10^{-5}	(0.5986213,0.60342658)
η_2	0.87651912546	1.52×10^{-4}	(0.8712564,0.8845697)

Hence the change point estimates are at time points $\widehat{\eta_1} = 0.570743$ and $\widehat{\eta_2} = 0.771523$ respectively and the hazard function is given by,

$$h_T(t) = h_0(t)h_1(t)I(0 < t \le \eta_1) + h_0(t)h_2(t)I(\eta_1 < t \le \eta_2) + h_0(t)h_3(t)I(t > \eta_2)$$

where

$$h_0(t) = 5.0$$
$$h_1(t) = 0.00098500449 + 0.0190899t$$
$$h_2(t) = 0.0081860897 + 0.023258305t + 0.006038268t^2$$
$$h_3(t) = 0.02309933167 \times 6.87265968t^{6.87265968-1}$$

In Figure 3 we compare the hazard function estimated using two change point model to a single change point model [Patra and Dey, 2002] against the smoothed Kaplan Meier plot of the given data.

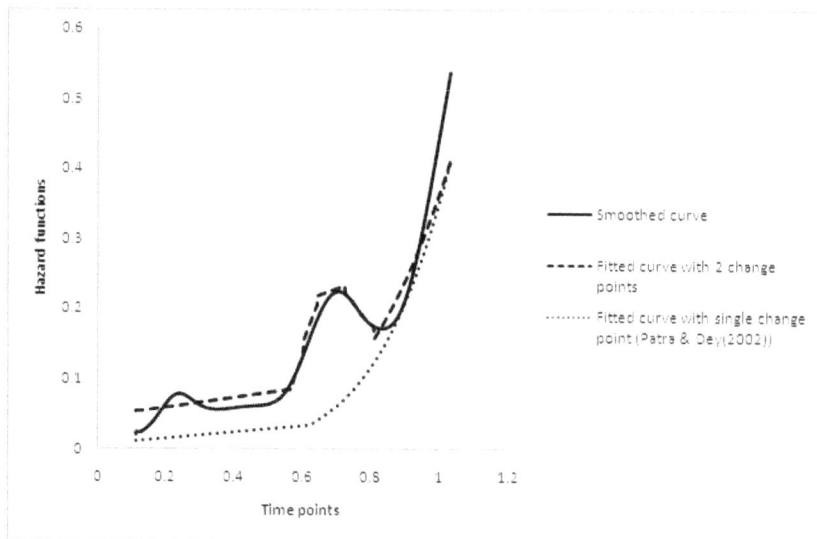

Fig. 3: Comparison between the Proposed Two Change Point Model and the Single Change Point Model Proposed by Patra and Dey [2002] over Smoothed Hazard Function.

Figure 3 shows that two change point model proposed by us gives a better fit compared to single change point model proposed by Patra and Dey (2002) for the given data.

6. PREDICTIVE VALIDATION

In order to validate our model, we fit the model to the above data excluding the the last three data points and then predict the excluded data points using the model. The result is shown in the Figure 4 below.

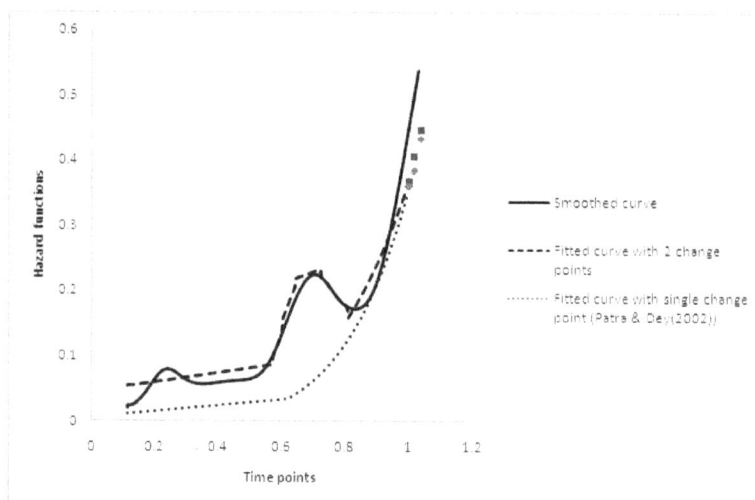

Fig. 4: Predictive Validation

Figure 4 depicts that the prediction error of the proposed two change point model is less compared to that of the single change point model of Patra and Dey [2002].

7. CONCLUSION

There may be several cases where the failure rate curve takes turn due to various reasons. Some of the reasons may be explained if the process parameters and their inter-relationships are well established. Many a times such explanations are not available, but the graph itself indicates that the failure rates might have changed due to some reasons or other at various points in the observed lifetime of the component. Authors try to explain this phenomenon as a change point model. Most of these authors have tried out assuming that only at one point change of failure rate has taken place. But there are cases, for example, the data of printed circuit board given by Meeker and Lu Valle (1995), where the failure rate graph indicates that more than one change points can explain the data better. We provide here a model with multiple change points and observed that the data may be better explained by using a multiple change point model rather than a single change point model. The predictive validity also becomes better with multiple change points in the printed circuit board example.

REFERENCES

[1] Patra, K. and Dey, D. (2002), "A general class of change point and changed curve modeling for life time data", *Annals of Institute of Statistical Mathematics.*, Vol. 54, No. 3, pp. 517–530.

[2] Meekar, W.Q. and LuValle, M.J. (1995), "An accelerated life tests model based on reliability kinetics", *Technometrics,* Vol. 37, No. 2, pp. 133–146.

[3] Chib, S. and Greenberg, E. (1995), "Understanding Metropolis Hastings Algorithm", *The American Statistician*, Vol. 49, pp. 327–335.

[4] Gelfand, A.E. and Smith A.F.M. (1990), "Sampling based approaches to calculating marginal densities", *Journal of American Statistical Association*, Vol. 85, pp. 398–409.

[5] Bolstad, W.M. (2010), *Understanding Computational Bayesian Statistics*, Wiley, New York.

[6] Ghosh, J.K. Delampady, M. and Samanta, T. (2006), *An Introduction to Bayesian Analysis Theory and Methods*, Springer, New York.

Performance Analysis of Industrial System under Corrective and Preventive Maintenance

Manwinder Kaur[1], Arvind Kumar Lal[2],
Satvinder Singh Bhatia[3] and Akepati Sivarami Reddy[4]

[1,2,3]School of Mathematics and Computer Applications,
Thapar University Patiala–147004 (Punjab), India
[4]Department of Biotechnology and Environmental Sciences,
Thapar University Patiala 147004 (Punjab), India
E-mail: [1]Manwinder.Kaur@live.com; [2]aklal2002in@yahoo.co.in;
[3]ssbhatia63@yahoo.com; [4]siva19899@gmail.com

ABSTRACT: This paper presents performance analysis of a stochastic model representing an industrial system with five units using supplementary variable technique. Four units are working under preventive as well as corrective maintenance while the fifth one undergoes for replacement. It is assumed that the subsystem undergoing preventive maintenance follows constant failure and variable repair rates, and variable failure and repair rates while undergoing corrective maintenance. In the case of replacement, both failure and repair rates are considered to be constant. The analysis has been carried out to compare the effect for various possible combinations of failure and repair rates in terms of reliability, availability and cost of the industrial system.

Keywords: *Reliability, Availability, Preventive Maintenance, Supplementary Variable, Cost, Maintenance Sheet.*

1. INTRODUCTION

Quantitative and qualitative are the two kinds of reliability analysis which are used to study the performance of system. Qualitative reliability analysis is intended to identify the various modes and failure causes that contribute to the unreliability of the device while the quantitative reliability analysis utilizes pertinent failures data in conjunction with a reliability model to produce numerical estimate of the device reliability (Martz 2002). Markov models are one of the quantitative techniques and considered to be appropriate throughout literature since 1930 to till date (Garg *et al.,* 2010; Sharma and Kumar 2010; Wang and Chen 2009). Markov models are non-deterministic mathematical models defined using the concept of stochastic process. The stochastic process is a collection of time dependent random or stochastic variables when characterized by Markov property. This property states that the distribution of future state depends only on the current state, not on the whole history of the random variable.

Markov modeling is the most comprehensive technique used today as well difficult to handle large systems (Knegtering and Brombacher 1999). However some of the complex manufacturing plants such as Plastic Manufacturing Plant (Gupta *et al.,* 2007), Butter Oil Manufacturing

Plant(Gupta *et al.*, 2005), Thermal Power Plant(Gupta and Tewari 2009), Urea Fertilizer(Kumar and Pandey 1993) and Hot Stand by Industrial system (Rizwan *et al.*, 2010) are studied and analyzed on the basis of these Models. Further, various other techniques such as Genetic Algorithm(Kumar *et al.*, 2010), GABLT (Sharma and Kumar 2010), and Stochastic Reward petri nets (Sachdeva *et al.*, 2009) have also been used as an extension in the analyses of the process to find the optimal solutions. These extensions were actually used to find the feasible values of failure and repair rates of the process industries.

Preventive and Corrective maintenance (Nakagawa 2005) as well as cost optimality (Kuo 2001) with respect to reliability, to improve the design of product has been extensively studied in literature. Both are wide interest of researchers as an independent area of reliability engineering. One can find numerous papers and techniques for the same. Without going into the depths of these fields, we try to work with the simple model markov approach for a complex industrial system in the operation fields. This model can be changed to different one as per the requirement of industries, and discussed in the form of subcases. Even though if one like to go on for detailed research work for maintenance under cost consideration and some markovian approaches the recent published work of Nakagawa (Nakagawa 2005) will be helpful as it holds a good bibliography of literature as well. In case of cost optimality, one can find the works of (Kuo 2001) as a good help with numerous references within the text.

In the present work, an attempt has been made to develop a quantitative analysis model with respect to cost optimization with its application on an industrial system. The quantitative analysis is based on Markov Chain defining system states having five units, where four units are working under preventive as well as corrective maintenance while the fifth one undergoes for replacement. It is assumed that the subsystem undergoing preventive maintenance follows constant failure and variable repair rates, and variable failure and repair rates while undergoing corrective maintenance. In the case of replacement, both failure and repair rates are considered to be constant. The analysis has been carried out to compare the effect for various possible combinations of failure and repair rates in terms of reliability, availability and cost of the industrial system.

2. SYSTEM MODEL

Let us suppose a system having five units (marked as ABCDE), where four units (ABCE) are working under preventive as well as corrective maintenance while the fifth one (marked as D) undergoes for replacement. To study the performance of the system following are the assumption stated while modeling.

2.1 Model Assumptions

1. At the initial point all units are working as good as new and termed this state as operating state of the system.
2. After a specified time of operation out of five, four units of the system intended for Preventive Maintenance (PM) with elapsed time of continuance to get back in operating state under the assumption that all units have fixed different time for preventive maintenance. That is, if unit A works for x_A hours only then the preventive maintenance takes places,

if it shows failure before this time then this leads to corrective maintenance action for the system unit A with elapsed failure time y_A and elapsed repair time x_A.

3. Elapsed failure time, y_i follow exponential distribution. This distribution function given as:

$$G(y_i;\theta) = \begin{cases} e^{\theta y_i}, & \text{if } y_i \geq 0 \\ 0, & \text{otherwise} \end{cases} \qquad \text{... (1)}$$

Similarly, elapsed repair times follow exponential distribution function.

4. After completion of each PM, system goes back to state zero with elapsed repair time x, having variable transition rate $(1-b)\beta_1(x)$. Here, b is an indicator variable, and defined as

$$b = \begin{cases} 0 & \text{for idealized Maintenance} \\ 1 & \text{for faulty Maintenance} \end{cases} \qquad \text{... (2)}$$

5. It is assumed that the subsystem undergoing preventive maintenance follows constant failure and variable repair rates, and while corrective maintenance same units follows variable failure and repair rates. In the case of unit 'D' which undergoes for replacement, follows constant failure and repair rates.

6. Replacement policy T based on the working age of unit D is used. Under this policy, a unit in the system is replaced at some point of time by a new and identical one with replacement rate approximately zero or negligible. To decide on the replacement policy one can follow the work of Nakagawa (Nakagawa 2005).

2.2 Model Analysis

Now under above assumptions and notations, stochastic process is defined as the family or collection of random or stochastic variable as $\{N(t), t \in T\}$ as,

$$N(t) = f(w,t) = \begin{cases} 0 & w \in A^{\circ}B^{\circ}C^{\circ}D^{\circ}E^{\circ} \\ 1 & w \in A^{m}B^{g}C^{g}D^{g}E^{g} \\ 2 & w \in A^{g}B^{m}C^{g}D^{g}E^{g} \\ 3 & w \in A^{g}B^{g}C^{m}D^{g}E^{g} \\ 4 & w \in A^{g}B^{g}C^{g}D^{g}E^{m} \\ 5 & w \in A^{g}B^{g}C^{g}D^{re}E^{g} \\ 6 & w \in A^{cm}B^{g}C^{g}D^{g}E^{g} \\ 7 & w \in A^{g}B^{cm}C^{g}D^{g}E^{g} \\ 8 & w \in A^{g}B^{g}C^{cm}D^{g}E^{g} \\ 9 & w \in A^{g}B^{g}C^{g}D^{g}E^{cm} \end{cases}, w \in S, t \in T \qquad \text{... (3)}$$

Thus, the state space of the stochastic process $S = \{0,1,2,3,4,5,6,7,8,9\}$ is discrete having time index set T, as a collection of continuous time epochs being its elements. Further, we can divide the state space set into three sets as $S_1 = \{0\}$ representing operating state subset, $S_2 = \{1,2,3,4\}$ representing preventive maintenance state space subset and $S_3 = \{6,7,8,9\}$ as corrective maintenance state space subset. The resulting stochastic process is markovian under the assumption that probabilities of the states of the state space subsets S_2 and S_3 are conditional probability with respect to their supplementary variables introduced for elapsed repair and elapsed failure time as discussed in the works of Cox (Cox 1955). The probability consideration of various states can be computed using supplementary variable technique (Gaver 1963; Cox 1955; Gupta, 2005) using transition diagram shown in Figure 1 as:

$$P_o(t) = e^{-(A_0+A_1)}[1 + \int (A_3(t) + \beta_5 P_5(t))e^{(A_0+A_1)}dt] \qquad \ldots (4)$$

$$P_5(t) = e^{-\beta_5 t}[\alpha_5 \int P_0(t)e^{\beta_5 t}dt] \qquad \ldots (5)$$

$$P_i(x_i,t) = e^{-\int \beta(x_i)dx_i}[\alpha_i P_o(t-x_i) + \int \alpha_i P_0(t)e^{-\int \beta(x_i)dx_i}dx_i], \ i = 1 to 4 \qquad \ldots (6)$$

$$P_{i+5}(y_i,x_i,t) = e^{-\int \eta(x_i)dx_i}[\gamma(y_i)P_o(t-x_i) + \int(b\beta_i(x_i)P_7(x_i,t) + \gamma_i(y_i)P_o(t))e^{\int \eta(x_i)dx_i}dx_i] \quad i = 1 to 4 \qquad \ldots (7)$$

where,

$$A_0 = \sum_{i=1}^{5}\alpha_i; \ A_1 = \sum_{i=1}^{5}\gamma_i(y_i); \ A_3(t) = \sum_{i=1}^{4}\int(1-b)\beta_i(x_i)P_i(x_i,t)dx + \sum_{i=1}^{4}\int(\eta_i(x_i)P_{i+5}(y_i,x_i,t)dx\,dy$$

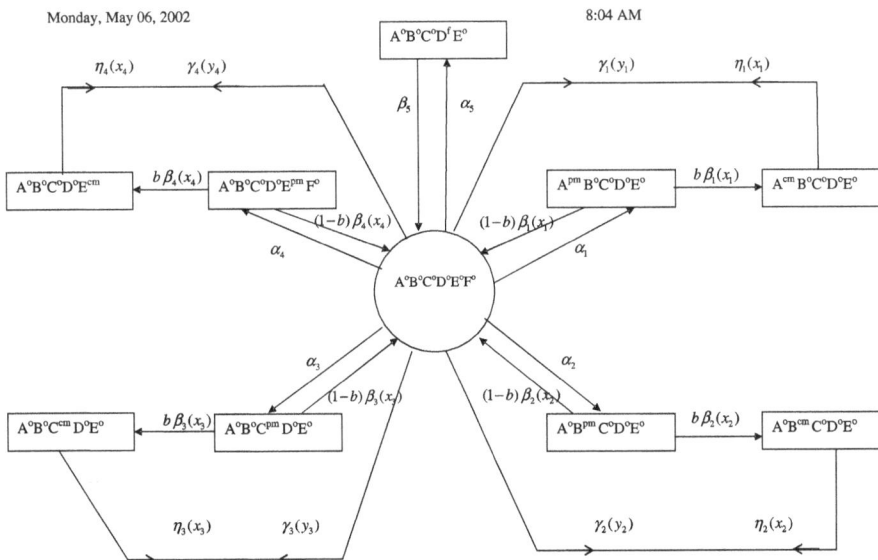

Fig. 1: Transition Diagram of System

2.3 Performance Indices

Taking the S_3 set as absorbing state space, reliability of the system defined as,

$$R(t) = 1 - \sum_{i=1}^{4} P_{i+5}(t)$$

$$\ldots (8)$$

Long run Availability of the system can be computed as,

$$A_\infty = \lim_{t \to +\infty} A(t) = 1 - \lim_{t \to +\infty} \sum_{i=1}^{5} P_i(t) - \lim_{t \to +\infty} \sum_{i=1}^{4} P_{i+5}(t))$$

$$\ldots (9)$$

In addition, one may replace the system based on less reliability given by system. Now, we can define overall problem as simple LPP (Kuo 2001) as,

$$Max \ T_p = C_r R(t) - CSSF$$

subject to

$$CSSF \leq c(t); \ R(T) \leq r(t); \ \sum_{j=0}^{9} P_j(t) = 1 \ ;$$

$$for \ all \ 0 \leq P_j(t) \leq 1; C_{min} \leq C_j \leq C_{max}, \forall \ j = 0 \ to \ 9$$

$$\ldots (10)$$

where $c(t)$ is z percent of revenue cost (C_r) obtained from the system working output and CSSF as overall *Cost Spent on System Functioning* and can be calculated as:

$$CSSF = \sum_{j=0}^{9} C_j P_j - \text{fixed cost}$$

$$\ldots (11)$$

Here, we check if the *CSSF* cost is less than or equal to z percent of the overall Total profit per unit of time (T_p) then there will be no loss in using the present system with ongoing maintenance schedule. Else, either a reschedule for maintenance is necessary as per the solution obtained from the constraints or the system need to be replaced completely if the *CSSF* cost exceeds fifty percent of overall profit of industry. The value of z to be decided by industries themselves.

3. NUMERICAL EXAMPLE: AN INDUSTRIAL SYSTEM APPLICATION

This model can be applicable from small to large scale industrial system. A textile industrial based mixture system has been selected to analyze the performance in terms of reliability, availability and cost parameters. This mixture system consists of cooker gasket, pressure gauge, safety valve, motor, and mixture filter and involves inspection and greasing to some components including checking of nut and bolts tightness. To fit the model, we take Cooker

Gasket as A, Pressure gauge as B, Safety valve as C and Mixture filter as D and Motor as E. To define the failure and repair parameters value, one needs to note down the data as per Maintenance sheet given in Table 1. Thus, Reliability analysis can be carried out following equation (8) as,

$$R(t) = 1 - \sum_{i=1}^{4} (e^{-\int \eta(x_i)dx_i}[\gamma(y_i)P_o(t-x_i) + \int (b\beta_i(x_i)P_7(x_i,t) + \gamma_i(y_i)P_o(t))e^{\int \eta(x_i)dx_i} dx_i])$$

$$\ldots (12)$$

Under perfect preventive maintenance, Reliability can be analyzed as,

$$R(t) = 1 - \sum_{i=1}^{4} (e^{-\int \eta(x_i)dx_i} \gamma(y_i)P_o(t-x_i))$$

$$\ldots (13)$$

System performance in terms of steady state availability can be analyzed by taking limits as $t \to \infty$ for the equations (4–7) by taking constant rate of transitions in equation (9) under the normalizing condition, that is, sum of all probabilities is one (Gupta *et al.,* 2005; Gupta *et al.,* 2005, 2007). Hence, the steady state availability calculated as,

$$A_\infty = \cfrac{1}{1 + \sum_{i=1}^{5} \cfrac{\alpha_i}{\beta_i} + \sum_{i=1}^{4} \cfrac{\gamma_i + b\alpha_i}{\eta_i}}$$

$$\ldots (14)$$

And, under perfect preventive maintenance, Steady state availability can be investigated as,

$$A_\infty = \cfrac{1}{1 + \sum_{i=1}^{5} \cfrac{\alpha_i}{\beta_i} + \sum_{i=1}^{4} \cfrac{\gamma_i}{\eta_i}}$$

$$\ldots (15)$$

To analyze the performance of system from a complex model, time independent analysis is appropriate substitute for transient solution of complex markov models. Thus, in absence of transient solution of complex model which results to the computations of reliability and time dependent availability as well, steady state solution can be used as an effective alternative for performance analysis of the system and so can be replaced in the objective function and one of the constraints. The resulting LPP will be,

$$Max \ T_p = C_r \ A_\infty - CSSF \quad \text{subject to}$$

$$CSSF \le c(t); \ A_\infty \le r; \ \sum_{j=0}^{9} P_j = 1 ;$$

$$\text{for all } 0 \le P_j \le 1; C_{min} \le C_j \le C_{max}, \forall j = 0 \text{ to } 9 \qquad \ldots (16)$$

Table 1: Maintenance Sheet for Industry Persons

Industry name and Unit number: xyz					
Maintenance manager Name: xyz					
Name of repairman, handling this complete system : xyz					
Unit Representations:					
A	Cooker Gasket	C	Safety valve	E	Motor
B	Pressure gauge	D	Mixture filter		

System start operating first time on:	
System complete wear out:	

Suggested figure for maintenance are given below as

Unit	PM carried after 's' hours	Time taken by repair man to carry out PM of unit in hours	System shows sudden breakdown after 't' hours due to	Time taken by system to repair the system back in operation after correcting sudden fault/failure
A	150	0.5	7200	5 hours
B	10	0.08	4800	3.5 hours
C	20	0.1667	2400	0.666 hours
D	100	0.25	–	–
E	40	1	600	5 hours

For record Maintenance					
PM rounds record, maintained by repairman under supervision of maintenance managers					
Round	For 'A' unit	For 'B' unit	For 'C' unit	For 'D' unit	For 'E' unit
1	145 / 0.5*	10/0.5	100/20	100/0.25	40/1 (#1)
On Date:	9 Oct 2011*	3 Oct 2011	6 Oct 2011	6 Oct 2011	3 Oct 2011
2					
On date					
3					
On date					
4					
…					
Corrective Maintenance record, maintained by repairman under supervision of maintenance managers					
Round	For 'A' unit	For 'B' unit	For 'C' unit	For 'D' unit	For 'E' unit
1	7200 hrs/3 hrs	None	None	None	None
On Date:	6 March 2011	–	–	–	–
2.					
On date					
3.					
…					
Remarks**					

#1 Motor shows early signs of failure than expecting, keep on check after every 5 to 10 hrs. of working of system.
* After working of x hours / PM carried by repairman within t hours on some date given.
** Mark "#" to corresponding failure to time ratio (for example see Unit E data for PM tasks) to give any remark for suspecting something which can damage or any information to keep the system running.
After complete damage of system, like wear out time: Main authorities are responsible to submit it dully signed copies to industrial main unit.

Performance of the system optimized using the availability analysis under the cost considerations using equations (16). The resulting analysis under prefect preventive maintenance for the industrial system discussed in the following section with respect to maintenance sheet, shown in Table 1.

4. RESULTS AND DISCUSSIONS

It was found that reliability of system goes up and down depending on the system corrective maintenance states and time spent by system under repair and can be referred from Figure 2. Reliability of system using equation (13) computed by generating data for repair and failure rate variation of system unit over time using equation (1). It was also found that if elapsed failure time of system is greater in comparison to previous elapsed failure times, system shows improved reliability and keeps on decreasing at variable rates.

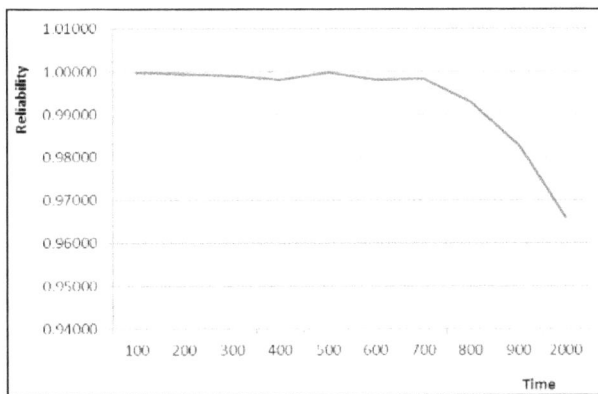

Fig. 2: Reliability of the Industrial System

Analysis of sensitive unit figured out on the basis of steady state availability of the system, which shows that motor is sensitive to operation of system in comparison to other system unit failure, which restore on time as well. The least on which effects the system operation is Mixture filter represented as D unit (Figure 3). Rest of the system's unit failure affects system equally. So,

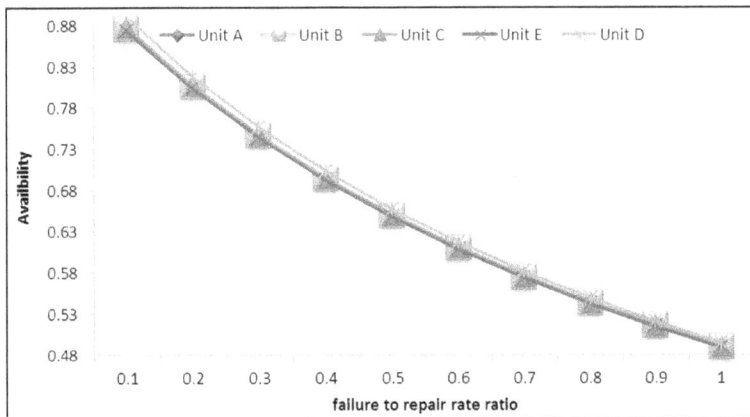

Fig. 3: Availability of the Industrial System

it would be suggested that industry needs to keep the failure to repair rate ratio close to 0.1 for each system unit to achieve the maximum possible availability. The current ongoing system availability is 0.9523 with failure to repair rate ratio 0.0033, 0.0056, 0.0083, 0.0025 and 0.025 for alpha to beta ratio corresponding to unit A, B, C, D and E while 0.00069, 0.00069, 0.00028, 1 for gamma to neta ratio corresponding to unit A, B,C and E .

System unit cost considerations in respect to this system to overall industrial system are within the cost constraints of the industry as per the industry correspondence, thus the cost analyses carried out taking some average value arbitrary.

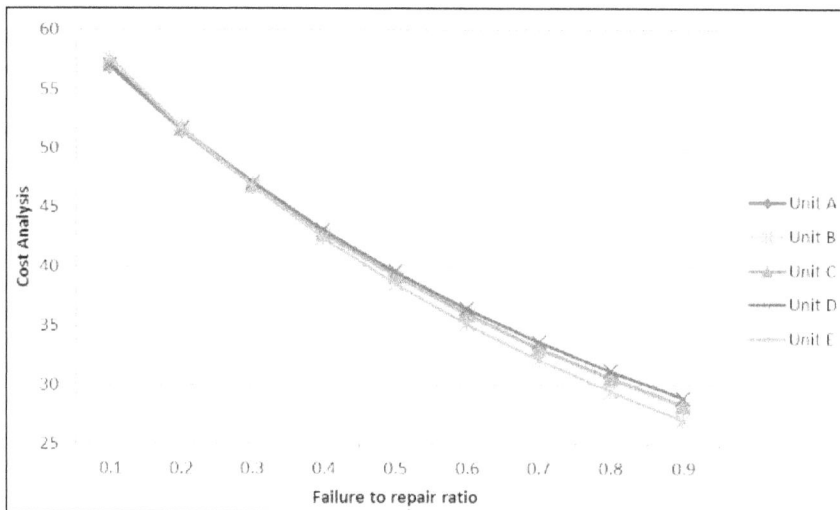

Fig. 4: Cost Analysis of the Industrial System

From Figure 4, it was found that maintenance of Motor (Unit E) which highly affects the profit obtained from this system while the least affected in terms of profit found to be Filter (Unit D). It was also observed that if the failure to repair ratio of each unit of the system is optimized to be less than 0.1 than this industrial system works at maximum availability as well as profit. For more details on Cost analysis, one can refer the works of various authors (Kumar and Pandey 1993; Kuo 2001; Tuteja and Malik 1992; Tuteja and Taneja 1992; Wang and Chen 2009).

In the present work we developed a theoretical model of quantitative analysis based on Markov Chain defining the system having five units, where four units are working under preventive as well as corrective maintenance while the fifth one undergoes for replacement using a supplementary variable technique. We also analyzed how an industrial system can be analyzed based on this model for required parameters as per need of the industry.

ACKNOWLEDGMENT

Authors thankfully acknowledge the University Grant Commission, New Delhi (Grant No. F. 2-3/2008(Policy/SR) for providing the financial support.

NOTATION

re	Unit under replacement
o	Unit under Operating state
g	Good state of unit but not operating
cm	Unit undergoes for corrective maintenance
pm	Unit undergoes for corrective maintenance
α_i	Constant failure rates of ith system unit, $i = 1\,to\,5$
$\beta_i(x_i), B(x_i;\theta)$	Variable repair rates and distribution function of PM time of the ith system unit, $i = 1\,to\,4$
$\eta_i(x_i), N(x_i;\theta)$	Variable repair rates and distribution function of CM time of the ith system unit, $i = 1\,to\,4$
$\gamma_i(y_i), G(y_i;\theta)$	failure rate and distribution function of the system's unit failed from normal working conditions to ith state, $i = 6\,to\,9$ in the interval $(y, y+\Delta)$ onditioned that it is has not failed up to time y.
$P_0(t)$	Probability that system is working in full operating condition at time t.
$P_i(t)$	Probability that system is in ith state at time t, $i = 1, 2$.
$P_i(x_i, t)$	Probability that system is in state i at time t and has elapsed repair time x_Z due to unit Z where $i = 1,...4$ and $Z = A, B, C, E$.
$P_{i+5}(y_i, x_i, t)$	Probability that system is in state i at time t and has an elapsed failure time y and elapsed repair time x_Z due to unit Z where $i = 1,...,4$ $Z = A, B, C, E$.
C_0	is the maximum revenue obtained by the system functioning being in operating states
C_i	for the states $i = 1\,to\,4$ is the preventive maintenance cost(PMC) per day and for $i = 5\,to\,9$ is corrective maintenance cost per day as defined in (Dhillon 1999).
$CSSF$	cost spent on system functioning

REFERENCES

[1] Cox, D.R. (1955). "The analysis of non-Markovian stochastic processes by the inclusion of supplementary variables." Cambridge Univ Press.

[2] Dhillon, B.S. (1999). Engineering maintainability: how to design for reliability and easy maintenance: Gulf Professional Publishing.

[3] Ebeling, C.E. (2000). An introduction to reliability and maintainability engineering: Tata McGraw-Hill Education.

[4] Garg, S., Singh, J. and Singh, D.V. (2010). "Availability and maintenance scheduling of a repairable block-board manufacturing system." International Journal of Reliability and Safety 4(1):104–118.

[5]   Gaver, D.P. (1963). "Time to failure and availability of paralleled systems with repair." Reliability, IEEE Transactions on 12(2):30–38.

[6]   Gupta, P., Lal, A.K., Sharma, R.K. and Singh, J. (2005). "Numerical analysis of reliability and availability of the serial processes in butter-oil processing plant." International Journal of Quality & Reliability Management 22(3):303–316.

[7]   Gupta, P., Lal, A.K., Sharma, R.K. and Singh, J. (2007). "Analysis of reliability and availability of serial processes of plastic-pipe manufacturing plant: A case study." International Journal of Quality and Reliability Management 24(4):404–419.

[8]   Gupta, P., Singh, J. and Singh, I.P. (2005). "Mission Reliability and Availability Prediction of Flexible Polymer Powder Production System." Opsearch-New Delhi- 42(2):152.

[9]   Gupta, S. and Tewari, P.C. (2009). "Simulation modeling and analysis of a complex system of a thermal power plant." *Journal of Industrial Engineering and Management* 2(2):387–406.

[10] Knegtering, B. and Brombacher, A.C. (1999). "Application of micro Markov models for quantitative safety assessment to determine safety integrity levels as defined by the IEC 61508 standard for functional safety." Reliability engineering and system safety 66(2):171–175.

[11] Kumar, D. and Pandey, P.C. (1993). "Maintenance planning and resource allocation in a urea fertilizer plant." Quality and reliability engineering international 9(5):411–423.

[12] Kumar, S., Tewari, P.C., Kuma, S. and Gupta, M. (2010). "Availability optimization of CO-Shift conversion system of a fertilizer plant using genetic algorithm technique." Bangladesh Journal of Scientific and Industrial Research 45(2):133–140.

[13] Kuo, W. (2001). Optimal reliability design: fundamentals and applications: Cambridge Univ Pr.

[14] Martz, H.F. (2002). "Reliability Theory." Encyclopedia of physical science and technology 14:143–159.

[15] Nakagawa, T. (2005). Maintenance theory of reliability: Springer Verlag.

[16] Rizwan, S.M., Khurana, V. and Taneja, G. (2010). "Reliability Analysis of a Hot Standby Industrial System." *International Journal of Modelling and Simulation* 2010 30(5).

[17] Sachdeva, A., Kumar, P. and Kumar, D. (2009). "Behavioral and performance analysis of feeding system using stochastic reward nets." The International Journal of Advanced Manufacturing Technology 45(1):156–169.

[18] Sharma, S.P. and Kumar, D. (2010). "Stochastic behaviour and performance analysis of an industrial system using GABLT technique." International Journal of Industrial and Systems Engineering 6(1):1–23.

[19] Tuteja, R.K. and Malik, S.C. (1992). "Reliability and profit analysis of two single-unit models with three modes and different repair policies of repairmen who appear and disappear randomly." Microelectronics and reliability 32(3):351–356.

[20] Tuteja, R.K. and Taneja, G. (1992). "Cost-benefit analysis of a two-server, two-unit, warm standby system with different types of failure." Microelectronics and reliability 32(10):1353–1359.

[21] Wang, K.H. and Chen, Y.J. (2009). "Comparative analysis of availability between three systems with general repair times, reboot delay and switching failures." Applied Mathematics and Computation 215(1):384–394.

PART–III

Six Sigma

Six Sigma, Quality Function Deployment and TRIZ—An Amalgamation

M. Shanmugaraja[1], M. Nataraj[2] and N. Gunasekaran[3.]

[1,2]Government College of Technology, Coimbatore, Tamil Nadu, India
[3]Angel College of Engineering and Technology, Tirupur, Tamil Nadu, , India
E-mail: [1]raja8011@yahoo.co.in; [2]m_natanuragct.yahoo.com; [3]guna_kct_cbe_tn_in.yahoo.com

ABSTRACT: Today organisations have a sound pursuit and passionate to understand the ins and outs of performance excellence for creating and delivering products or service in an increasingly better, faster and on cheaper basis. Performance excellence is holistic phenomenon that can be achieved through an integrated framework of strategy, improvement and innovation. Every organization seekssuch a framework to achieve the performance excellence goal by entwining the activities of strategy, improvement and innovation. This paperproposes a performance excellence framework which progresses in a lockstep, as the output of one piece become the inputs for the next, driven by change oriented strategic and organisational improvement. A grand unification of methodologies like Six Sigma, Quality Function Deployment and TRIZ is used in this framework to optimize the interplay of strategy, improvement and innovation. Theproposed conceptual model serves asa gate-way to amplify a high value concept of business to offer maximum outcome within the minimal constraints of the business.

Keywords: *Organization, Performance Excellence, Strategy, Improvement, Innovation, Six Sigma, QFD, TRIZ.*

1. BEGINNING

A study on influences of business over the last 50 years and longer revealed a list of names like Shewart, Ford, Deming, Drucker, Ono, Gates, Hamel, Womack, Harry, Welch, Walton and many others (Silverstein, *et al.*, 2008). Each one holds a key to some important aspect of business performance improvement-some aspect of how an organization creates and delivers value on an increasingly better, faster, and cheaper basis. And some hold the keys to how a business innovate itself periodically to bring on new and different value propositions, which are then relentlessly improved until the cycle needs to repeat itself again (Slocum and Domb, 2004). But today's global scenario routed the business improvement through performance excellence which needs holistic thinking and holistic action. It is known that to achieve performance excellence, very best companies have an integrated framework within which they entwine the activities of strategy, improvement, and innovation on an ongoing basis (Slocum, 2003). In other words, if the goal of an organization is to do what it does better, faster, and cheaper, they can achieve that goal by optimizing the interplay of strategy, improvement, and innovation (Gavetti *et al.*, 2005). Exactly how an organization compiling doesn't really matter, as long as it keeps the three major parts of performance excellence in good relations, and as long as it keeps each of the key elements within them in good relations, too (Kermani, 2004). This is where we look at combining the elements of performance excellence into a holistic model by integrating all parts as tightly and consistently as possible.

This article proposes a Performance Excellence Model (PEM) that progresses in a lockstep, as the outputs of one piece become the inputs for the next piece. This model is underpinned with performance improvement methods like Six Sigma, Quality Function Deployment and TRIZ. Among the pool of methodologies practiced today, the reason to choose the stated initiatives are; Six Sigma is a proven process improvement program which can minimize the defect rate and thereby reduces the cost, assures quality at cheaper (Park, 2003); QFD is known tool for lining the end users with the manufacturers so that the dynamic characteristics of customer demand can be incorporated in the process to assure the fastest response to customers; TRIZ is a Russian innovative program aims to improve the present performance to an elevated level through innovation to assure better product or service quality (Dourson, 2004). The big deal is in tying all the proven elements of performance success into a coherent system in which all resources and tools are brought to bear on the continuous lifecycle of business evolution (Hipple, 2005). With all the elements of the PEM working in coordinated rhythm, an organization can advance its current and future products, processes, transactions, and services in a way that increases the probability of greater long-term profitability and financial growth.

2. UNDERPINNING OF PEM MODEL

PEM is developed by adopting the grand unification approach to enable organizations to plan and execute it in a synergistic fashion. The elements of PEM build progressively, balanced, coordinated, and integrated with effective change leadership based on strategic priorities. PEMis conceptualized in such way that among the domains of performance excellence, strategy and innovation have a lot to learn from the domain of improvement, particularly when it comes to making strategy and innovation more methodological, deployable, measurable, predictable, and controllable. The groundwork closely follows classical Axiomatic Design (see Figure 1) but with a change.

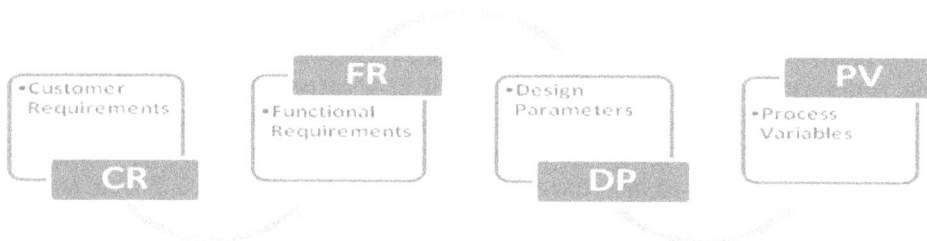

Fig. 1: Classic Axiomatic Design Domains

The classical axiomatic design model doesn't really address the strategic components of performance excellence, because it's deliberately focused on operations (Zhao, 2005). An additional domain "Societal Need" (SN) is added to the existing axiomatic design domains to develop PEM approach as shown in Figure 2. The reason is that societal need is the absolute root of business, as all corporations exist to meet the one penultimate objective of fulfilling human needs. Therefore, a company doesn't make watches; it enables people to know the time.

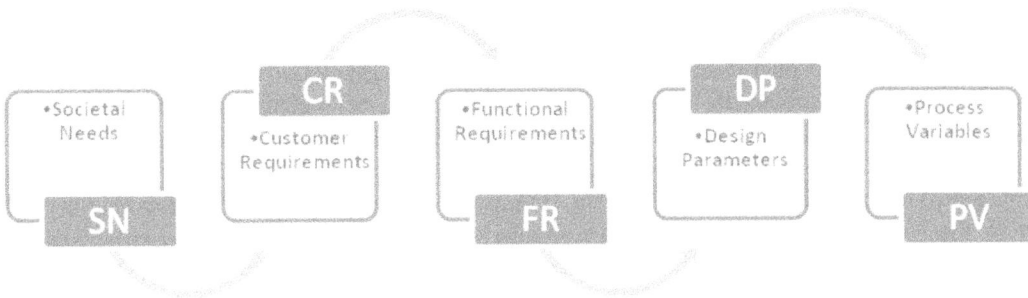

Fig. 2: Performance Excellence Model

Although, in classical Axiomatic model, the tools of improvement have blended well with each other, the tools of strategy and innovation have not. Further all three big domains of strategy, improvement and innovation have not connected themselves very well. By the introduction of additional domain to the classical frame work, the resultant PEM model becomes operable. In this new model a very logical progression of how to get from a customer requirement through societal need to a critical process variable, the performance of which has to be optimized and reliable over time if there is any chance of performance excellence is presented. As shown in figure III, TRIZ approach is designed to filter the Societal Requirements (SN) from which Customer Requirements (CR) are identified using Quality Function Deployment (QFD). On the customer requirement end, Six Sigma covers the gamut of all other domains. Function Modelling from TRIZ can also employ in defining functional requirements (Hipple, 2003).Contradiction matrix of TRIZ technique can be used at the final domains of this model to resolve contradictions if anything presents (Domb *et al.*, 1998). So each of these domains enables the next in progressive fashion, and this gives glimpse into a more sequential approach to the networked reality of performance excellence variables.

SN	CR	FR	DP	PV
		Six Sigma	Six Sigma	Six Sigma
	QFD	QFD		
TRIZTRIZTRIZTRIZ				

Fig. 3: Applications of Six Sigma, QFD and TRIZ to the Various Domains of PEM

3. PEM EXCELLENCES

Performance Excellence Model development (See Figure 4) starts with strategic thinking through TRIZ and entails envisioning the fuzzy edge of latent societal and customer needs. With a balanced picture of how the organization can preserve its current position, while also evolving into something different, it can derive great value from Strategic Planning methods with QFD, which connect strategy and execution as inseparable parts of a seamless whole. Key outcomes of strategic thinking and strategic planning are focused vision, meaningful missions, prioritized goals, aligned objectives, rational metrics, optimized tactics, and systematic review cycles.

This is the robust framework within which an organization can synergize otherwise disparate attempts to improve operations and innovate new value propositions. With priorities and metrics defined and rationally deployed, the high-performing organization can use Process Management to define and operate its key, subordinate, and enabling processes, all of which are aligned with strategic direction. With sound process architecture in place, Six Sigma method is engaged to optimize performance from the standpoint of time and quality. In the language of PEM, these elements serve to preserve a company's position and success, because they enable an organization to optimize its current offerings, capabilities, technologies, and processes. Such optimization and constant improvement are baseline requirements for keeping pace with the progressive nature of business. Key outcomes of Six Sigma are operational stability, measurement integrity, reduced waste, minimized inventory, reduced variation, higher margins, increased capacity, reduced cycle time, lower costs, and other benefits, not the least of which is a culture of people relentlessly committed to improvement.

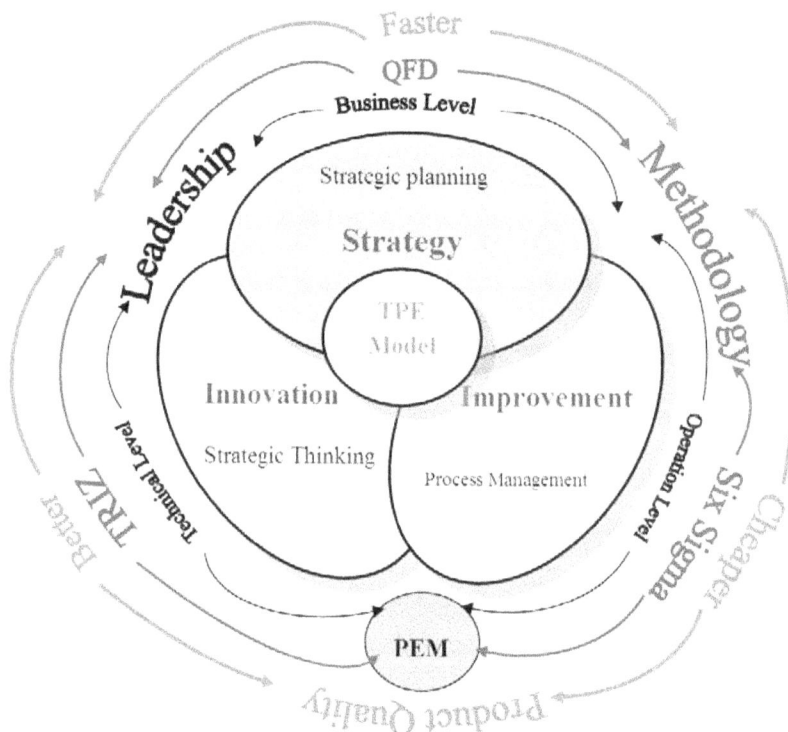

Fig. 4: Performance Excellence Model

4. PEM CONSTITUENTS

Practically, all of business is the drive to accomplish the mission of the organization better, faster, and cheaper. To make products better, faster, and cheaper; To deliver services better, faster, cheaper. And to operate processes better, faster, and cheaper. From a colloquial perspective, that's what all the various management tools do: They play some role in improving quality (better), redesigning processes (better still), spawning reinvention and growth (a lot better), minimizing waste and inventory (faster), reconfiguring operations (faster again), and, of course, cutting costs (cheaper) as a by-product of all this and a direct result of such targeted initiatives as corporate reorganizations, mergers, and technology implementations. In the PEM funnel (See Figure 5), the business mission is realized by mixing the strategic initiatives Six Sigma, QFD and TRIZ in an appropriate proportion to assure better quality product or process at cheaper in a fastest time frame.

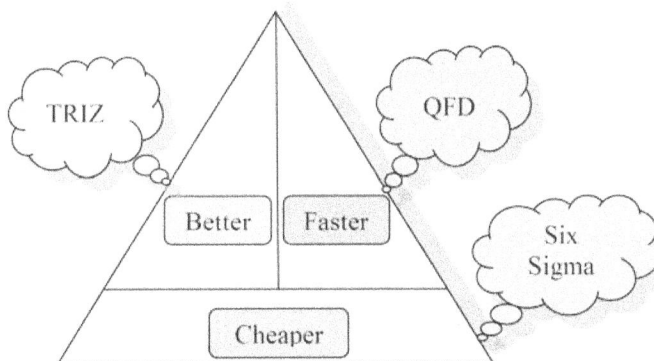

Fig. 5: PEM Constituents

4.1 Six Sigma

4.1.1 *Definition*

From a statistical perspective, Six Sigma is a metric of process measurement symbolized by the Greek letter σ that represents the amount of variation with a normal data distribution (Averboukh, 2006). Fundamentally, Six Sigma quality level relates to 3.4 Defects Per Million Opportunities (DPMO). The focus of Six Sigma is not on counting the defects in processes, but the number of opportunities within a process that could result in defects so that causes of quality problems can be eliminated before they are transformed into defects (De Foe and Bar-El, 2002). From a business perspective, Six Sigma could be described as a process that allows companies to drastically focus on continuous and breakthrough improvements in everyday business activities to increase customer satisfaction (Park, 2003; Yeung, 2007).

4.1.2 *Methodology*

There are two major improvement methodologies in Six Sigma (Schroeder *et al.*, 2008). The first methodology, DMAIC, is used to improve already existing processes and can be divided into five phases; define, measure, analyze, improve and control. Several studies have shown successful cases of DMAIC application in a variety of contexts such as healthcare, thermal power plants, retailing, financial services and manufacturing process (Kumar *et al.,* 2008; Roger*et al.*, 2008). In contrast, the second methodology, design for Six Sigma (DFSS), is used for new processes or when the existing processes are unable to achieve business objectives such as customer satisfaction (Nonthaleerak and Hendry, 2006). DFSS methodology can also be divided into five phases (DMADV); define, measure, analyze, design and verify (McAdam, 2005). Antony (2008) refers to DFSS as a powerful approach to design products andprocesses in a cost effective and simple manner. Applications of DFSS are also varied from high-tech manufacturing to designing new housing (Linderman *et al.*, 2006).

4.1.3 *Benefits*

When Six Sigma is implemented successfully, it will offer a disciplined approach for improving effectiveness and efficiency in a broad range of businesses. Six Sigma benefits are related to various areas such as reduction in process variability, reduction in in-process defect levels, reduction in maintenance inspection time, improving capacity cycle time, improving inventory on-time delivery, increasing savings in capital expenditures, increase in profitability, reduction of operational costs, reduction in the Cost of Poor Quality (COPQ), increase in productivity, reduction of cycle time, reduction of customer complaints, improved sales and reduced inspection (Kumar *et al.*, 2008; Kwakand Anbari, 2006). Six Sigma benefits for service organizations may involve improved accuracy of resources allocation, improving accuracy of reporting, reduced documentary defects, improving timely and accurate claims reimbursement, streamlining the process of service delivery, reduced inventory of equipment, reduced service preparation times, improved customer satisfaction, reduced defect rate in service processes, reduced variability of key service processes, transformation of organizational culture from fire-fighting mode to fire-prevention mode with the attitude of continuous improvement of service process performance, reduced process cycle time and hence, achieve faster service delivery, reduced service operational costs, increased market share, improved cross-functional teamwork across the entire organization, increased employee morale, reduced number of non-value added steps in critical business processes through systematic elimination, leading to faster delivery of service, reduced Cost of Poor Quality (COPQ) (costs associated with late delivery, customer complaints, costs associated with misdirected problem solving, etc.), increased awareness of various problem solving tools and techniques, leading to greater job satisfaction for employees, improved consistency level of service through systematic reduction of process variability and effective management decisions due to reliance on data and facts rather than assumptions and gut-feelings (Smith, 2001; Chakrabarty and Tan, 2007).

4.1.4 *Tools and Techniques*

Six Sigma tool has a specific role and is often narrow in focus, where as Six Sigma technique has a wider application and requires specific skills, creativity and training (Antony, 2008). Examples of Six Sigma tools include Pareto analysis, root cause analysis, process mapping or process flow chart, Gantt chart, affinity diagrams, run charts, histograms, Quality Function Deployment (QFD), Kanomodel, brainstorming, etc. Examples of Six Sigma techniques include Statistical Process Control (SPC), process capability analysis, Suppliers-Input-Process-Output-Customer (SIPOC), SERVQUAL, benchmarking, etc. Moreover, a Six Sigma technique can utilize various tools. For example, Statistical Process Control (SPC) is a technique that utilizes various tools such as control charts, histograms, root cause analysis, etc. satisfaction (Nonthaleerak and Hendry, 2006).

4.2 Quality Function Deployment [QFD]

QFD is a powerful tool that addresses strategic and operational decisions in businesses (Mazur, 2000). It provides a means of translating customer requirements into appropriate technical requirements for each stage of product development and production. QFD is all of the following (Chan *et al.*, 2002):

- Understanding customer requirements;
- Systematic thinking about quality;
- Adding value through quality maximization and customer satisfaction;
- Designing a comprehensive quality system for customer satisfaction; and
- Developing strategies that can put a company ahead of its competition.

A typical approach to QFD is the four-phase process by the American Supplier Institute (ASI) in the USA, which is admired and widespread (Slocum and Domb, 2004). Four phases to be considered can be briefly described as below:

Phase 1: Qualitative customer requirements are translated into design-independent, measurable, and quality characteristics of the product.

Phase 2: Examination of the relationship between the quality characteristics and the various components or parts of the design. The result of phase II is a prioritization of the component parts of the design in terms of their ability to meet the desired quality characteristic performance level.

Phase 3: Prioritization of manufacturing processes and specifications for key process parameters that are deployed to the fourth and final phase.

Phase 4: The key manufacturing processes and associated parameters are translated into work instructions, control and reaction plans, and training requirements necessary to ensure that the quality of key parts and processes is maintained.

4.3 TRIZ

Theory of Inventive Problem Solving (TRIZ) is a problem-solving, analysis and forecasting tool derived from the study of patterns of invention in the global patent literature (Altshuller, 1988).

The important principles and methods in TRIZ are (Grierson *et al.*, 2003);

- 40 Principles of Invention
- 76 Standard Inventive Solutions
- Trends of Evaluation of Technical Systems
- Alshuller's Contradiction Matrix
- ARIZ [Algorithm for Inventive Problem Solving].

By using the above said principles and methods, TRIZ provides a dialectic thinking of understanding the problem as a system to image the ideal solution first and then to solve contradictions

(Chang-qing *et al.*, 2006). To develop inventive solutions to problems, TRIZ methodology utilizes the following thinking tools (Silverstein, *et al.*, 2008).

1. Inventive Principles for business and management problems
2. Separation Principles for resolving organizational contradictions and conflicts
3. Substance-Field analysis for visualizing highly complex systems
4. Anticipatory Failure Recognition for prediction and evaluation of risks
5. Utilizing system resources for effective cost- saving decisions
6. Pattern of evaluation of technical systems to support systematic and multidimensional thinking
7. Innovation Situation Questionnaire for problem identification and CTQ identification
8. Contradiction Matrix to resolve physical and technical conflicts
9. Standard Solutions to inventive problems
10. ARIZ algorithm to eliminate contradictions which causes the problem.

Besides the above said tools, TRIZ also employs certain more tools like Level of Innovation, Laws of System Evolution, Ideal Final Result, Ideality degree, Functional analysis, Mind Mapping, Physical, Chemical and geographical effect to develop or produce innovative products with high quality and profit with an Anticipatory future failure protection to increase market share of the organization. Because of proven innovativeness, this world-class practice is used by some of the world's most innovative corporations includes Proctor and Gamble, Boeing, Siemens, 3M, Hewlett-Packard, Eli Lilly, Honeywell, NASA, Toyota, Intel, Johnson and Johnson, Motorola, and many others [Mann, 2002; Belski *et al.*, 2003; David Silverstein *et al.*, 2008].

5. IMPLEMENTING PEM

Maybe organizations should be training PEM champions and practitioners in addition to training black belts, Kaizen event facilitators, TRIZ experts, risk managers, and so on. Yes, as known, the initial resistance is, "Oh no, not more people to train!" But the reality is that corporate skill sets have to—must—keep up with an evolving world. And in this evolving world, the force of leadership will have to make tangible steps toward performance excellence integration. Therefore, it's wise for an organization to implement the parts of PEM that it can, with the vision of the whole in mind. Often, when an organization implements an initiative like PEM, the force of the initiative is strong enough to dislodge inertia, and the whole structure of performance begins to shift. Major operational improvements result in pressure on R&D to innovate the next product, which it does, which puts pressure on the strategy makers to think smarter about making a better world.

6. PRACTICALLY SPEAKING

The adoption of a Performance Excellence Model (PEM) can be quite daunting, as it involves the entire organization. The organization typically must assess all of its current initiatives, and it engages in prioritization sessions to decommission activities that do not support the strategic plan—or activities that could interfere with the PEM deployment. Of course the input to these sessions is the strategic plan. If this body or work is missing or incomplete, it must be produced, and it must contain a vision, a service mission, an economic mission, strategies, goals and objectives, a scorecard, and an operational review cycle. The other input is the presence and participation of all senior leaders from all functional areas. This is mandatory work, even in the midst of an environment in which organizational resources are at full capacity.

By simple laws of physics, it takes a certain force to redirect the motion of a body, and the force required at the outset is always much greater than the force required to sustain. We are just making the simple proposition that, if possible, an organization should consider tying all its inevitable initiatives into one strong, cohesive cord and using one initial force to get them all in action. By lifting that mass off the ground with one lever, the working parts (initiatives) are free to combine organically, as well as in an actively managed way. So the projects and efforts of one initiative dovetail with the efforts of another in a more seamless and fluid way than they would have had one initiative been launched at a separate time in a separate way by a different group of people. The inherent consequences of variation teach us that implementing a PEM model is better than implementing each contained initiative by itself in a sequenced manner. Once again, there is nothing wrong with bringing the pieces of a PEM model together over time in a sequential, coordinated way—especially if each is introduced as part of a bigger picture from the outset. But even if not, there is nothing wrong with focusing on innovation by means of TRIZ for one cycle, then changing the focus to operational excellence by means of Six Sigma, then adding additional capabilities during further cycles.

Simply put, taking a PEM approach is more efficient and effective and, if done right, will yield a more impressive ROI over a certain defined midterm time frame, such as five years. The net return is more sizeable in all the important areas of corporate performance, not just certain ones. The idea of an organization creates an event in the marketplace some might call "blazing growth" when all the performance indicators are buzzing. At its essence, this growth is a function of bringing new forms and solutions to customers, who buy them and keep buying them because the competition can keep up on only one of the two critical fronts of business success—preservation or evolution—but not both. In other words, no company grows and sustains its growth if it is good at only innovation or improvement. This is what a leadership team should have in mind when it configures its PEM deployment and decides how to load and allocate its resources. If more initiatives have been decommissioned and many resources made available, then the PEM deployment can proceed rapidly across a wide range of applications. Otherwise the pace and initial scope are determined on a case-by-case basis.

7. CONCLUSION AND REMARKS

If all the gears and wheels of performance success are defined into an overarching model, an organization will know better how to sequence and direct its portfolio of interventions over time. As one initiative gains momentum and makes an impact, others are planned, because the effect of one program will dwindle as it bumps up against challenges it's not equipped to handle. Of course running a successful enterprise is a much larger task than any one program or initiative can handle. There has been enormous movement toward holism and consolidation in managerial know-how over the past 100 years, and this will continue by sheer force of evolution. It seems inevitable that the domains of strategy, improvement, and innovation will come together even more over time as their methods connect and blend. And in the context of Performance

Excellence Model, the core force of your vision, missions, and goals determines the impact of the many methods you implement to achieve them. TRIZ can fuel the development of robust strategic thinking. Then, QFD determines adequate strategic plan to accomplish customer needs. The rest of performance excellence is taken care by Six Sigma in some rational configuration, magnify that intent into reality.

REFERENCES

[1] Altshuller, G. (1988), *Creativity as an Exact Science*, Gordon and Breach, New York.

[2] Antony, J. (2008), 'Can Six Sigma be effectively implemented in SMEs?', *International Journal of Productivity and Performance Management*, Vol. 57 No. 5, pp. 420–3.

[3] Averboukh, E.A. (2006), 'Six Sigma trends: Six Sigma in financial services', *TRIZ Journal*, April.

[4] Belski, I., Kaplan, L., Shapiro, V., Vaner, L., Wong PengWai (2003), "SARS and 40 Principles for Eliminating Technical Contradictions: Creative Singapore." *The TRIZ Journal*, June.

[5] Chakrabarty, A. and Tan, K. (2007), "The current state of Six Sigma application in services", *Managing Service Quality*, Vol. 17, No. 2, pp. 194–208.

[6] Chan, L.-K. and Wu, M.-L. (2002), "Quality function deployment: a literature review", *European Journal of Operational Research*, Vol. 143, pp. 463–97.

[7] Chang-qing, G., Ke-zheng, H. and Yong, Z. (2006), "The Contrast and Application of the Problem Resolving Tools of TRIZ", *Machine Design and Research*, Vol. 22, No. 1, pp. 13–19.

[8] De Foe, J.A. and Bar-El, Z. (2002), 'Creating strategic change more efficiently with a new design for Six Sigma process', *Journal of Change Management*, Vol. 3, No. 1, pp. 60–80.

[9] Domb, E., Miller, J., MacGran, E. and Slocum, M. (1998), "The 39 Features of Altshullers Contradiction Matrix", *The TRIZ Journal*, November.

[10] Dourson, S. (2004), "The 40 Inventive Principles of TRIZ Applied to Finance." *The TRIZ Journal*, October.

[11] Gavetti, Giovanni and Rivkin, J. (2005), "How Strategists Really Think: Tapping the Power of Analogy," *Harvard Business Review*, April, p. 54.

[12] Grierson, B., Fraser, I., Morrison, A., Niven, S. and Chisholm, G. (2003), "40 Principles—Chemical Illustrations" *The TRIZ Journal*, July.

[13] Hipple, J. (2003), 'The integration of TRIZ problem solving techniques with other problem solving and assessment tools', *Proceedings of TRIZCON2003, 5th Annual Altshuller Institute for TRIZ Studies International Conference*, Philadelphia, PA, USA.

[14] Hipple, J. (2005), 'The integration of TRIZ with other ideation tools and processes as well as with psychological assessment tools', *Creativity and Innovation Management*, Vol. 14, No. 1, pp. 22–33.

[15] Kermani, A.H.M. (2004), 'Algorithm of inventive quality improvement integration of ARIZ and Six Sigma methodology', *Proceedings of TRIZCON2004, 6th Annual Altshuller Institute for TRIZ Studies International Conference*, Seattle, WA, USA.

[16] Kumar, U.D., Nowicki, D., Ramirez-Marquez, J.E. and Verma, D. (2008), "On the optimal selection of process alternatives in a six sigma implementation", *International Journal of Production Economics*, Vol. 11, No. 1, pp. 456–467.

[17] Kwak, Y.H. and Anbari, F.T. (2006), "Benefits, obstacles and future of Six Sigma approach", *Technovation*, Vol. 26, pp. 708–15.

[18] Linderman, K., Schroeder, R. and Choo, A. (2006), "Six Sigma: the role of goals in improvement teams", *Journal of Operations Management*, Vol. 24, pp. 779–90.

[19] Mazur, G. (2000), 'QFD 2000: integrating supporting methodologies into quality function deployment', *Proceedings of the 12th Symposium on Quality Function Deployment*, Novi, MI, USA: QFD Institute.

[20] Mann, D. (2002), "Disruptive Advertising: TRIZ and the Advertisement." *The TRIZ Journal,* October.

[21] McAdam, R., Hazlett, S. and Henderson, J. (2005), "A critical review of Six Sigma: exploring the dichotomies", *International Journal of Organizational Analysis*, Vol. 3, No. 2, pp. 51–174.

[22] Nonthaleerak, P. and Hendry, L. (2006), "Six Sigma: literature review and key future research areas", *International Journal of Six Sigma and Competitive Advantage*, Vol. 2 No. 2, pp. 105–61.

[23] Park, S.H. (2003), 'Six Sigma for quality and productivity promotion', *Asian Productivity Organization*, New Delhi.

[24] Roger, G.S., Linderman, K., Liedtke, C. and Choo, A.S. (2008), "Six Sigma: Definition and underlying theory", *Journal of Operations Management*, Vol. 26, pp. 536–554.

[25] Schroeder, R., Linderman, K., Liedtke, C. and Choo, A. (2008), "Six Sigma definition and underlying theory", *Journal of Operations Management*, Vol. 26, No. 4, pp. 536–54.

[26] Silverstein, D., DeCarlo, N. and Slocum, M. (2008), *INsourcing Innovation How to Achieve Competitive Excellence Using TRIZ*, Auerbach Publications, Tylor and Francis Group, New York London.

[27] Slocum, M.S. (2003), 'Total product/process development system where Six Sigma meets TRIZ and QFD', *Proceeding of ETRIA World Conference: TRIZ Future 2003*, Aachen, Germany.

[28] Slocum, M.S. and Domb, E. (2004), 'The integration of comprehensive QFD, TRIZ, and Six Sigma in an axiomatically driven total product/process development system', *Proceedings of TRIZCON2004, 6th Annual Altshuller Institute for TRIZ Studies International Conference*, Seattle, WA, USA.

[29] Smith, L.R. (2001), 'Six Sigma and the evolution of quality in product development', *Six Sigma Forum Magazine*, Vol. 1, No. 1, pp. 28–35.

[30] Yeung, V. (2007), "Six Sigma paradigm shift", *International Journal of Six Sigma and Competitive Advantage*, Vol. 3, No. 4, pp. 317–32.

[31] Zhao, X. (2005), 'Integrated TRIZ and Six Sigma theories for service/process innovation', *Proceedings of ICSSSM '05, 2005 International Conference on Services Systems and Services Management*, Vol. 1, Nos. 13–15, pp. 529–532.

The Use of Design of Experiments in Design for Six Sigma for Improving Cable Toughness

Arup Ranjan Mukhopadhyay

SQC & OR Unit, Indian Statistical Institute, Kolkata, India
E-mail: armukherjee@yahoo.co.in

ABSTRACT: This study is centered on the application of designed experiments (L_{18} OA) for improving two responses—tensile strength and shore hardness—of a rubber compound used for sheathing of power cables. Separate analysis has been done for these two responses. However, the set of significant process parameters for tensile strength and that for shore hardness are found to be different. The study demonstrates the procedure of finding a unique optimum operating condition for maximizing both the responses through a combined weighted response. Confirmatory trials substantiate the finding. It demonstrates the importance of statistically designed experiments for implementation of Six Sigma – especially in the context of Design for Six Sigma (DFSS).

Keywords: Tensile Strength, Shore Hardness, Combined Response, Orthogonal Array (OA), Main Effect, Interaction, ANOVA, Confidence Interval (CI), Predicted Value.

1. INTRODUCTION

The power cable at times needs to demonstrate its ability to withstand heavy drags in the form of too much of friction caused by rocky and quite rough earth surface during laying operation. In order to withstand these heavy drags, the cable needs to have adequate toughness. The tensile strength and shore hardness are two measures of this toughness. Since the outer surface of the cable is formed by the sheathing operation, the quality problem turns out to be developing the sheathing compound in such a manner that makes the outer surface of the cable tough or robust enough to overcome the frictional damage. The tensile strength and shore hardness are two different measures of the toughness of this rubber compound.

In the cable making process, insulation and sheathing are two vital stages. Rubber compounds are used to make insulation and sheath materials for cables. Among the various sheath materials, elastomeric sheath, which is popularly termed as SE4, is the crucially important material. To prepare this SE4 rubber compound, Chlorinated Polyethylene (CPE) is used.

2. BACKGROUND INFORMATION

The research and development cell of the organization had tried out trials, which failed to achieve the minimum requirements of tensile strength (9 N/mm^2) and shore hardness (70 shore A scale) even though other characteristics – elongation at break and tear strength – not only met their respective minimum requirements of 250% and 5 N/mm, they were found to be much higher than their minimum requirements. While elongation at break was more than 600% on most of the occasions, tear strength ranged from 7 to 10.

3. OBJECTIVE

Therefore, the principal focus of this study is to develop a sheathing compound made of rubber that has enhanced tensile strength and shore hardness to conform to the specified requirement and side by side that satisfies other physical properties like elongation at break, tear strength etc. through determining the corresponding unique optimum operating condition.

Broadly speaking, this study focuses on application of statistically designed experiment with the solution of a product and process design problem. It is the purpose of this study to demonstrate that if statistically designed experiments and Six Sigma are combined then it would be quite powerful for quality improvement effort especially in the context of DFSS.

4. THE PROCESS

The receipt of an order from the customers' end leads to assigning the responsibility of formulation of new insulation and sheathing material to the Research and Development (R&D) department. The R&D formulates the new product in the manner described in Figure 1.

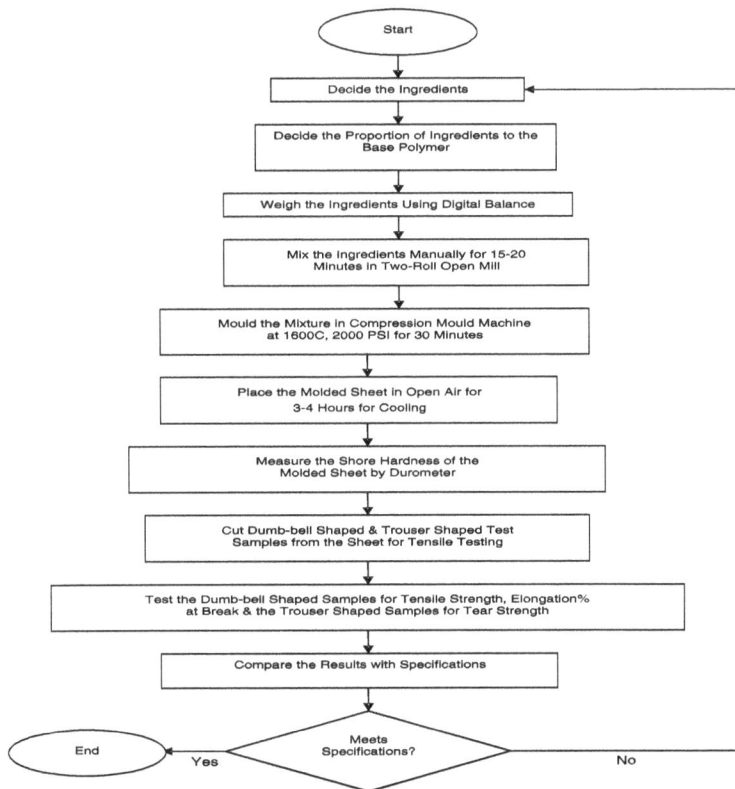

Fig. 1: Flow Chart for Formulating the Sheathing Compound by R&D

The process of mixing has been further elaborated in Figure 2.

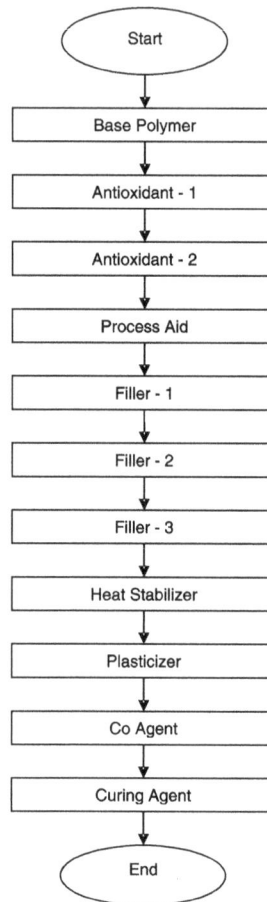

Fig. 2: Flow Chart Describing the Sequence of Mixing

Note that major degradation of rubber takes place due to oxygen and ozone attack. Antioxidants prevent this attack.

Fillers are non-reinforcing in nature and these are added to reduce cost.

A plasticizer is a chemical added to rubber to make it softer and more flexible.

The co-agent enhances the rate of curing and contributes to the development of desirable properties in the cured material.

The curing agent causes cross-linking to occur. Through cross-linking chemical bonds are set up between molecular chains (Babbit, 1978).

Discussion with the concerned technicians and literature survey (Ghosh, 1990) result in identifying the mixing ingredients and their possible ranges as mentioned in Table 1.

Table 1: Ingredients and Their Possible Ranges

Name of the Ingredient	Range in gm
Base Polymer	100 (fixed)
Heat Stabilizer	10 (fixed)
Additive	1 (fixed)
Process Aid	2 (fixed)
Antioxidant-1	0.5–1.5
Antioxidant-2	1–2
Filler-1	25–35
Filler-2	30–40
Filler-3	5–15
Plasticizer	5–10
Curing Agent	7–8
Co Agent	1–2

5. FACTORS AND LEVELS

Based on the findings of Table 1 it can be figured out that the number of variable factors is eight. Out of these eight factors, for one factor, namely co-agent, two levels have been considered. For the remaining seven factors, three equidistant levels have been considered for experimentation. The factors and the corresponding levels are furnished in Table 2.

Table 2: Factors and Levels Chosen for Experimentation

Factors	Level-1 (gm)	Level-2 (gm)	Level-3 (gm)
Co Agent	1.0	2.0	–
Curing Agent	7.0	7.5	8.0
Antioxidant-1	0.5	1.0	1.5
Antioxidant-2	1.0	1.5	2.0
Plasticizer	5.0	7.5	10.0
Filler-1	25.0	30.0	35.0
Filler-2	30.0	35.0	40.0
Filler-3	5.0	10.0	15.0

6. CHOICE OF THE OA

The choice of which OA to use depends on two things. They are; (a) the number of factors and interactions of interest and (b) the number of levels for the factors of interest. In this experiment, the number of factors is eight of which seven factors have three levels and one factor has two levels. Based on the process knowledge it has been anticipated that there might be an interaction between curing agent and co agent.

The degree of freedom for each factor is the number of levels minus one. The degree of freedom for an interaction is the product of the degrees of freedom of the interacting factors. In this case the required total degree of freedom is $7 \times (3–1) + 1 \times (2–1) + 1 \times (3–1) \times (2–1) = 17$.

The total degree of freedom available in an OA (v_{LN}) is equal to the number of trials minus one.

$v_{LN} = N{-}1$

When a particular OA is chosen for an experiment, the following inequality must be satisfied (Ross, 1989).

$v_{LN} \geq v_{\text{required for factors and interactions}}$ ... (1)

The nearest array, which satisfies the inequality (1) is L_{18}. Hence, L_{18} OA has been chosen for carrying out the experiment.

7. EXPERIMENTAL LAYOUT

The factors co-agent and curing agent are respectively assigned to column 1 and column 2 of the L_{18} array. The remaining factors are assigned to the remaining columns as mentioned in Table 3.

Table 3: Experimental Layout of L_{18} Array

Trial No.	Column No.							
	1 Co-agent	2 Curing agent	3 Antioxidant1	4 Antioxidant2	5 Plasticizer	6 Filler3	7 Filler1	8 Filler2
1	1	1	1	1	1	1	1	1
2	1	1	2	2	2	2	2	2
3	1	1	3	3	3	3	3	3
4	1	2	1	1	2	2	3	3
5	1	2	2	2	3	3	1	1
6	1	2	3	3	1	1	2	2
7	1	3	1	2	1	3	2	3
8	1	3	2	3	2	1	3	1
9	1	3	3	1	3	2	1	2
10	2	1	1	3	3	2	2	1
11	2	1	2	1	1	3	3	2
12	2	1	3	2	2	1	1	3
13	2	2	1	2	3	1	3	2
14	2	2	2	3	1	2	1	3
15	2	2	3	1	2	3	2	1
16	2	3	1	3	2	3	1	2
17	2	3	2	1	3	1	2	3
18	2	3	3	2	1	2	3	1

Since, all the columns of the L_{18} array have been used, to estimate the error variance two replicate measurements of tensile strength and shore hardness have been taken per trial. The replicates represent two different mixing and molding operations for a trial.

8. ANALYSIS

Analysis of variance has been carried out separately for tensile strength and shore hardness. Before doing analysis of variance, Bartlett's test has been carried out for both the responses to see the existence of homoscedasticity (Montgomery, 2004). The χ_0^2 values are found to be 17.67 and 22.47 for tensile strength and shore hardness respectively. The corresponding tabulated value of χ^2 at 5% level of significance for 17 degrees of freedom is 27.59. Hence, the assumption of homoscedasticity is valid for doing ANOVA. The ANOVA tables are given in Table 4 and Table 5 respectively for tensile strength and shore hardness. It can be seen that for tensile strength the significant factor is curing agent. On the other hand, for shore hardness the significant factors are co agent, plasticizer, filler 1 and filler 3.

Table 4: ANOVA for Tensile Strength

SOV	DF	SS	MS	F_0	P-value
Co Agent	1	0.1275	0.1275	0.19	0.667
Curing Agent	2	9.8397	4.9198	7.37	0.005
Co Agent × Curing Agent	2	0.4929	0.2465	0.37	0.696
Antioxidant-1	2	0.1590	0.0795	0.12	0.888
Antioxidant-2	2	0.7503	0.3752	0.56	0.580
Plasticizer	2	1.8341	0.9171	1.37	0.278
Filler-1	2	4.2179	2.1090	3.16	0.067
Filler-2	2	3.8433	1.9216	2.88	0.082
Filler-3	2	1.2243	0.6121	0.92	0.417
Replication	18	12.0078	0.6671		
Total	35	34.4968			

Table 5: ANOVA for Shore Hardness

SOV	DF	SS	MS	F_0	P-value
Co Agent	1	48.3952	48.3952	26.99	0.000
Curing Agent	2	12.5808	6.2904	3.51	0.052
Co Agent × Curing Agent	2	3.0067	1.5034	0.84	0.449
Antioxidant-1	2	4.1017	2.0509	1.14	0.341
Antioxidant-2	2	6.0858	3.0429	1.70	0.211
Plasticizer	2	66.5308	33.2654	18.55	0.000
Filler-1	2	32.2371	16.1186	8.99	0.002
Filler-2	2	7.5975	3.7988	2.12	0.149
Filler-3	2	152.9121	76.4561	42.63	0.000
Replication	18	32.2814	1.7934		
Total	35	365.7291			

Since, two different sets of significant factors are found, the optimum operating level for one set may not become the optimum for the other set. For example, filler-1 is found to be significant (P-value 0.002) for maximizing shore hardness. The corresponding optimum level is found to be 3. However, level 3 of filler-1 is not the optimum level for maximizing tensile strength with the corresponding P-value as 0.067. The optimum level is 2. It can be seen from the ANOVA Table 6 that while doing analysis with the combined response consisting of both tensile strength and shore hardness, the factor filler-1 has turned out to be significant with P-value as 0.044. The optimum level of filler-1 for the combined response, is found to be 2 instead of 3 for shore hardness. In order to obtain the combined response comprising both tensile strength and shore hardness, weighted values of the responses have been added. The weights have been found as follows.

9. DETERMINATION OF WEIGHTS

Replication error for tensile strength = 0.6671 and the replication error for shore hardness = 1.7934. Reciprocal of the replication error for tensile strength $= \dfrac{1}{0.6671} = 1.4990.$ Similarly, reciprocal of the replication error for shore hardness $= \dfrac{1}{1.7934} = 0.5576.$ Therefore, the weightage for tensile strength $= \dfrac{1.4990}{1.4990 + 0.5576} = 0.73$ and the weightage for shore hardness $= \dfrac{0.5576}{1.4990 + 0.5576} = 0.27.$ Hence, combined response = 0.73 × tensile strength + 0.27 × shore hardness. Note that more weightage has been given to tensile strength since its measurements are more precise and less weightage has been given to shore hardness whose measurements are less precise.

10. RESULTS OF THE COMBINED RESPONSE

The results of the analysis of the combined weighted response and allied pictorial displays are given in the following.

Table 6: ANOVA for the Combined Response (0.73 × TS + 0.27 × SH)

SOV	DF	SS	MS	F_0	P-value
Co Agent	1	2.6168	2.6168	6.40	0.021
Curing Agent	2	10.5468	5.2734	12.90	0.000
Co Agent × Curing Agent	2	0.1283	0.0642	0.16	0.856
Antioxidant-1	2	0.6241	0.3120	0.76	0.481
Antioxidant-2	2	1.3660	0.6830	1.67	0.216
Plasticizer	2	6.7728	3.3864	8.28	0.003
Filler-1	2	3.0432	1.5216	3.72	0.044
Filler-2	2	2.4221	1.2110	2.96	0.077
Filler-3	2	16.9896	8.4948	20.78	0.000
Replication	18	7.3581	0.4088		
Total	35	51.8679			

The effects of significant factors for the combined response are shown in Figure 3 to Figure 7. It can be observed from these figures that to maximize both tensile strength and shore hardness, the second level of co agent, third level of curing agent, first level of plasticizer, second level of filler-1 and third level of filler-3 are to be maintained.

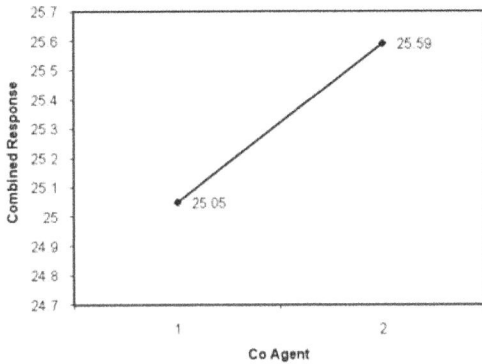

Fig. 3: Effect of Co-agent on the Combined Response

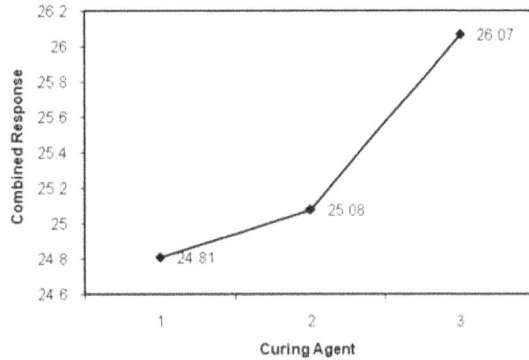

Fig. 4: Effect of Curing Agent on the Combined Response

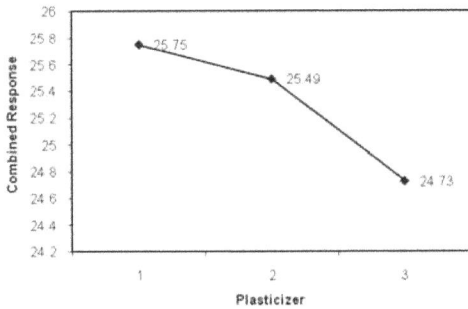

Fig. 5: Effect of Plasticizer on the Combined Response

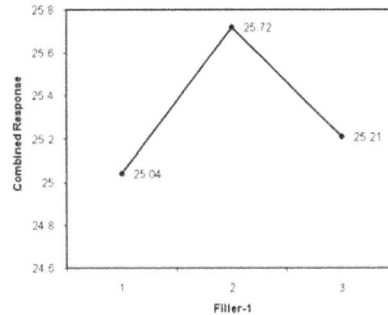

Fig. 6: Effect of Filler-1 on the Combined Response

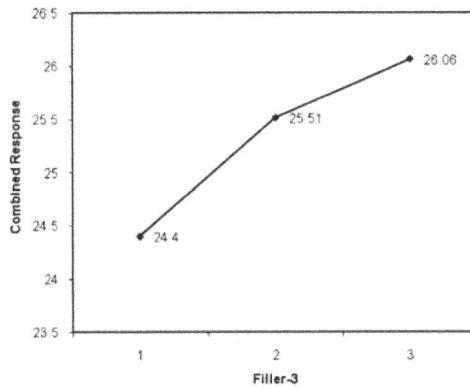

Fig. 7: Effect of Filler-3 on the Combined Response

Other factors that have been found to be insignificant from the ANOVA of the combined response (Table 6), can be maintained at any one of the three levels based on the cost or other criteria such as convenience of operation.

11. CONCLUSION

The optimum levels of significant factors are furnished in Table 7.

Table 7: Optimum Factor Level to Maximize the Combined Response
$(0.73 \times TS + 0.27 \times SH)$

Significant Factor	Optimum Level	Value of the Level (gm)
Co-agent	2	2.0
Curing Agent	3	8.0
Plasticizer	1	5.0
Filler-1	2	30.0
Filler-3	3	15.0

11.1 Estimating the Mean

The mean for the combined response has been estimated as 27.90 and the means for the tensile strength and shore hardness have been estimated as 10.44 and 75.12 respectively. This has been found based on the additivity of the factorial effects (Ross, 1989).

$$\hat{\mu} = \bar{A}_2 + \bar{B}_3 + \bar{E}_1 + \bar{F}_2 + \bar{H}_3 - 4\bar{T}$$

where $\bar{T}$ is the average of the entire experimental results for the concerned response.

Recall that the minimum requirement for tensile strength is 9 N/mm^2 and the minimum requirement of shore hardness is 70. From this the minimum requirement for the combined response has been worked out to be 25.47.

11.2 Confidence Interval around the Estimated Mean

The 95% confidence interval has been worked out to be ± 0.71 for the combined response, ± 0.90 for tensile strength and ± 1.48 for shore hardness. The formula for CI (Taguchi, 1986) is given in the following.

$$CI = \sqrt{\frac{F_{0.05,1,18}MS_{Error}}{\eta_{eff}}}$$

$$\eta_{eff} = \frac{36}{1 + \left[\text{total degrees of freedom associated with items used in } \hat{\mu} \text{ estimate}\right]}$$

$$= \frac{36}{1+9} = 3.6$$

12. CONFIRMATORY TRIAL

A confirmatory trial has been carried out to verify the conclusions from the previous round of experimentation. The optimum condition is set for the significant factors and levels and six mixings and moldings are made. Subsequently six tests have been conducted for a characteristic. It can be observed that the 95% confidence intervals as well as the minimum requirement of the tensile strength and shore hardness (Table 8) and the combined response (Table 9) are complied with.

Table 8: Results from Confirmatory Trial for the Individual Responses

Response	Observed Values	Average Value	Predicted Value	95% Lower Confidence Interval	95% Upper Confidence Interval	Minimum Requirement
Tensile Strength (N/mm^2)	10.77, 10.15, 9.98, 9.88, 9.89, 10.10	10.13	10.44	9.53	11.34	9
Shore Hardness	74, 75, 74, 75, 76, 75	74.83	75.12	73.63	76.60	70
Elongation at Break (%)	600, 580, 600, 620, 600, 640	606.67		–	–	250
Tear Strength (N/mm)	8.41, 9.64, 8.32, 9.15, 8.87, 8.26	8.78		–	–	5

Table 9: Results from Confirmatory Trial for the Combined Response ($0.73 \times$ TS + $0.27 \times$ SH)

Observed Values for Tensile Strength	Observed Values for Shore Hardness	Combined Response					
		Values	Average	Predicted Value	95% Lower Confidence Limit	95% Upper Confidence Limit	Minimum Requirement
10.77	74	27.84					
10.15	75	27.66					
9.98	74	27.27	27.60	27.90	27.19	28.61	25.47
9.88	75	27.46					
9.89	76	27.74					
10.10	75	27.62					

13. IMPACT ON SIX SIGMA

This study shows the importance of robust statistically designed experiments for implementation of Six Sigma—especially in the context of Design for Six Sigma (DFSS) (Park and Antony, 2008). The demonstrated approach to seek the solution of the product and process design problem may be of interest to Six Sigma professionals and other quality improvement specialists.

REFERENCES

[1] Babbit, R.O. (1978), *The Vanderbilt Rubber Handbook*, R. T. Vanderbilt Company, Inc. New York.

[2] Ghosh, P. (1990), *Polymer Science and Technology of Plastics and Rubbers*, Tata McGraw-Hill Publishing Company Ltd. New Delhi.

[3] Montgomery, D.C. (2004), *Design and Analysis of Experiments*, 6th ed. New York: John Wiley and Sons.

[4] Park, S.H. and Antony, J. (2008), *Robust Design for Quality Engineering and Six Sigma*, World Scientific, Singapore.

[5] Ross, P.J. (1989), *Taguchi Techniques for Quality Engineering*, New York: McGraw-Hill.

[6] Taguchi, G. (1986), *Introduction to Quality Engineering*. Tokyo: Asian Productivity Organization.

Pend Volumes Reduction: A Case Study

Joyson Peter Coelho

Hinduja Global Solutions Ltd., Bangalore, India
E-mail: joyson3714@teamhgs.com

ABSTRACT: Six Sigma provides an effective mechanism to focus on customer requirements through improvement of process quality. The current paper describes a Six Sigma project which was initiated with an objective of reducing in-process inventory against a less-predictable incoming volume and associated overhead expenses. This project included improvement of Schedule compliance, Quality compliance, Cycle time reduction and Error minimization. Turnaround time was considered as the Critical to Quality metric in addition to resource optimization. Different Statistical tools were used to analyze the critical factors along with their relative importance on cycle time minimization. The Improve phase included the formulation of action plan for the identified issues in accordance to the priority. On their successful implementation, 'After Improvement Values' were evaluated in terms of tangible and intangible benefits. DPMO analysis had also been carried out in the control phase to ensure sustained improvement.

Keywords: Cycle Time, Incoming Volume, After Improvement values, Critical to Quality metric, Pend Volumes, Pend Aging.

1. INTRODUCTION

The Six Sigma methodology is implemented in organizations through a structured project based approach popularly known as DMAIC approach or design, measure, analyze, improvement and control. Six Sigma's bottom line results normally flow from successful completion Six Sigma projects. The quality of Six Sigma projects differentiates successful Six Sigma campaigns from average ones (Bertels and Patterson, 2003). Six-Sigma, at many organizations, simply means a measure of quality that strives for near perfection. But the statistical implications of a Six Sigma program go well beyond the qualitative eradication of customer-perceptible defects. It's a methodology that is well rooted in mathematics and statistics. The basic objective of the Six Sigma project is to reduce the output variation so that customer's aggregate experience with the process over time will result in not more than a specified limit of defects per million opportunities. The Methodologies have been applied in several industries such as Manufacturing, Service Industry, Insurance, etc. to successfully improve the productivity and minimizing the errors employing resource allocation. The current projects aim at applying the concept of six-sigma in an Activity based Control Model to reduce the overhead cost.

MTV Claims is an adjudication process for settlement of claims submitted by the Providers or members. Claims are received either in paper or electronic format and the system is programmed to pay or deny a claim. From above, it is quite evident that claims will either be processed if all the necessary information is available or else it would be kept on hold. Such claims are called as Pends. MTV claims is an Activity Based Costing model from a financial perspective.

Claims which are completely processed first time within the scope of MTV claims are eligible for getting paid. Working on pends don't get counted for any receipts of payment to Hinduja Global Solutions. Over time, it was observed that the volumes of pends is going out of control and extra resources are required to manage the pends. To reduce this overhead cost, project on Reduction of MTV Claims pends was undertaken.

1.1 Identification of the Project

This Project Selection Criteria tool concentrated on Benefits and Efforts on this project. It calculated the Benefits in terms of expense reduction, customer satisfaction and efforts were calculated in terms of project cycle time, investment required and team members required. For the current project, the benefits were more to Hinduja Global Solutions and a satisfactory level of customer satisfaction to the clients. The problem statement was defined as "High Volumes of pend claims between the range of 6,000–7,000 per day contributing 28% towards the overall Inventory for the Month of Nov 2010. The scope of the project was restricted to all MTV claims pend scenarios which were current and which would be encountered in future. Having said this, there were always pend claims which fell as client exclusions or any new scenarios which the client themselves had not encountered.

2. RESEARCH METHODOLOGY

2.1 Define Phase

The basic purpose of the charter is to authorize a project manager to start an approved project and use resources to accomplish the goals of the project. Using a project charter approved by the stakeholders, problem statement, goal statement, scope, team members and milestones were defined. Also, a strong business case to convince the stakeholders explaining them the need of the project was given.

Six Sigma Project Charter

Project Name	MTV Claims Pend & TAT Reduction			Project number	BLR-GRDN-201010-036
Process Name	Garden			Sponsor	Prabhakar/Richard
Sub Process Name	MTV Claims			Project Lead	Joyson Coelho
Date Chartered	05/11/2011	Revision		Project Mentor	Ameet/Nanda Kumar

Business Case	Problem Statement	Purpose
Currently on average 6 500 claims are pended which requires daily monitoring follow ups and resolution to avoid delayed payments. It is of utmost importance to streamline the pend management process while coordinating with various client departments to reduce the overall pend volume and thereby reducing the TAT on Pend Inventory. Cost of handling & maintaining pends is borne by the process and this do not contribute towards revenue. Reduction in Pend TAT will help avoid Interest and Penalty Payouts by Client	• High Volumes of claims between the range of 6 000- 7 000 against internal bench mark of 4000 leading to excess use of resources in non revenue generating processes. (Data considered for Nov 2010)	**Goal Statement:** • Reduce Overall Pend Volumes from average 6367/day to </=3000/day by March 2011 **Potential Benefits:** • Total Estimated cost savings/year $20 456 • Eradicate claims falling into the Interest & Penalty purview due to aging • Meeting Client Expectation • Reduction in Rework • TAT improvement

Project scope Includes	Team Members
All MTV Claims Pend Scenarios-Current & New	Srividya Srikumar Sathish Saravanan Mahesh R Reddy Prasad
Project scope Excludes	
Client Exclusions. Un-encountered case scenarios Pends with Higher TAT requirements	

Project Milestones	Target Date	Actual Date	Tollgate Review Date
Start 01/11/2010		End 29/04/2011	
Define	05/11/2010	05/11/2010	10/11/2010
Measure	30/11/2010	26/11/2010	01/12/2010
Analyze	10/12/2010	22/12/2010	08/01/2011
Improve	31/03/2011	31/03/2011	06/04/2011
Control	29/04/2011	03/05/2011	11/05/2011

Project CTB (Critical to Business) Details: Process CTBs/CTQs/CTDs	Unit	Baseline	Current	Target
Pends	Pends/day	6367	6367	</= 3000

Project Charter Sign-off with Date	
Sponsor	Prabhakar/Richard
Project Lead	Joyson Coelho
Mentor	Ameet/Nanda Kumar

Fig. 1: Project Charter

A SIPOC diagram helps to identify the process outputs and the customers of those outputs so that the voice of the customer can be captured. For the Insurance Claims adjudication process SIPOC was designed to handle high overall pend volumes.

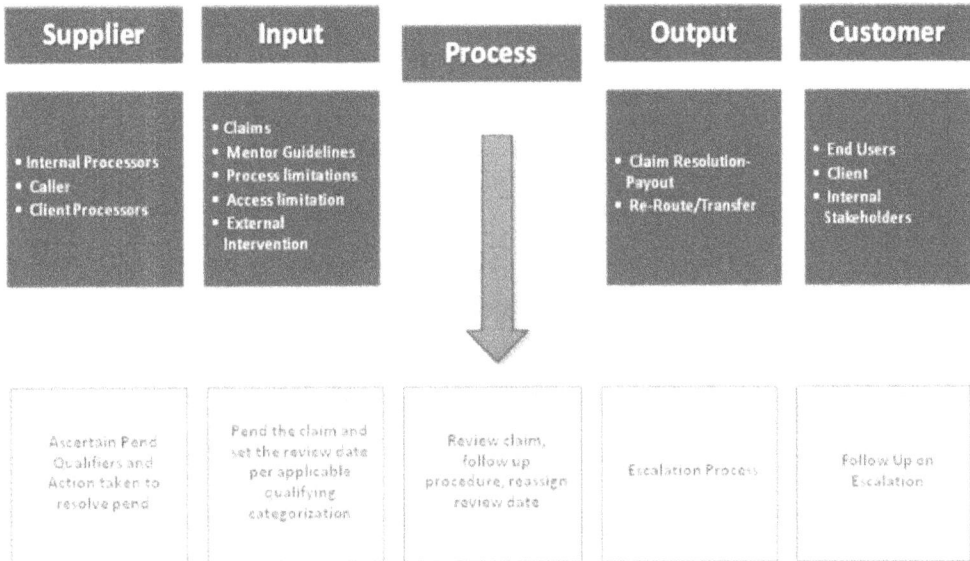

Fig. 2: SIPOC

As an initiative a workflow for Pends resolutions was drawn.

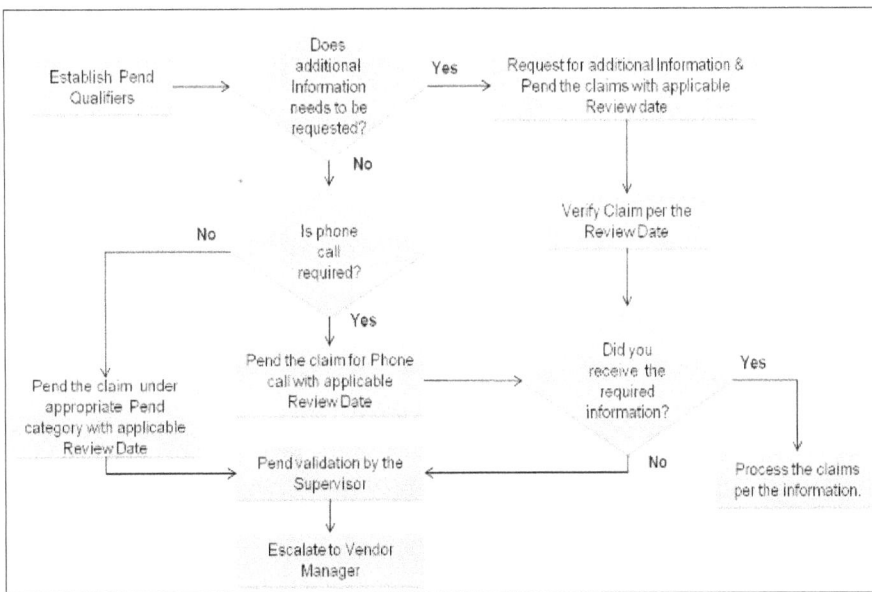

Fig. 3: Process Map

2.2 Measure Phase

Base Line

Sample data on which base lining was done is shown below.

Table 1: Sample Data

Date	Total Number of Claims/Day	Defect
1–Aug	3,129	1

To measure the performance of the process DPMO and Process Sigma is calculated.

Target: Achieve Business Goal of 3000 Pend Volumes per Day

Opportunities: There is one opportunity for each unit to be defective.

Defect: Any unit that does not meet the specification of a CTQ is a defect. In this case, a unit can have only one defect. So definition of defect and a unit being defective is the same. If the Pend Volumes is more than 3000/day we say the unit is defective.

Calculation for the sample data

Defect = Pend Volumes > 3000/day

Total Number of Units (Days) = 61

Opportunities/Unit = 1

Total Number of Opportunities = 61

Total Number of Defects = 61 (from Baseline Data)

Proportion of Defects = Total Number of Defects/ Total Number of Opportunities

Proportion of defects = 61/61 = 1

Defects Per Million Opportunities, $\text{DPMO} = \dfrac{\text{Total Number of Defects}}{\text{Total Number of Opportunities}} \times 1000000$

$\quad$ DPMO = 1 × 1000000 = 1000000

$\quad$ Yield = 1- Proportion of Defects = 1 − 1 = 0 ==> Process Sigma = 0

Therefore,

$\quad$ DPMO = 1000000 and Process Sigma = 0

DPMO and Process Sigma showed poor performance of the process. Hence, actions need to be taken to find the potential causes and remove them. Brain Storming was done to obtain a First level Fishbone Diagram to identify the potential causes for High Pend Volumes.

Machine

Man

System Malfunction

No Proper check on pends & Follow-up

Access limitation

Providers/Members submitting insufficient info

Claims Bounce Back

System failure Temporary access denial

Inadequate Training/Knowledge

High Pend Volumes

Mentor Instructing to Pend

Additional information unavailable in mentor guidelines

TPA not responded with the information

Escalation Reponses not received in time from POCs

Lack of Repricing info EOB Member information

Misinterpretation of the claim

Mentor (Guideline) discrepancies

Claim incorrectly pended

Communication gap within various dept

Continuous Alerts/ Updates

Clarification responses not received from SMEs

Claims waiting Mentor Updation

Material

Method

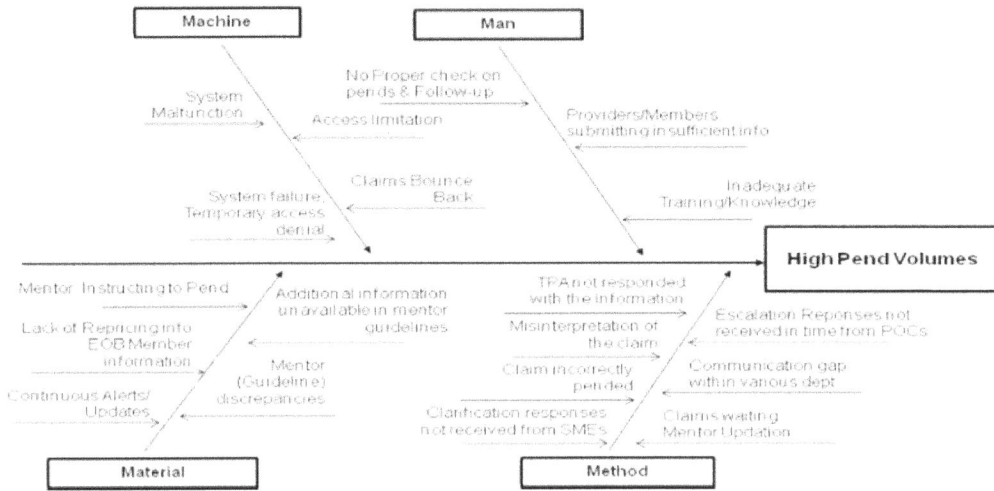

Fig. 4: Cause and Effect Diagram

The above Fishbone diagram shows various reasons that can lead to high pend volumes. Based on management's understanding of the process these causes were again given ranking according to their priority with the help of a cause and effect matrix.

Table 2: Priority Ranking Matrix

Y= Overall Pend Volume	Y0 (0–10 days)	Y1 (11–39 days)	Y2 (40+ days)	← Aging bucket volumes	
Contribution/ Weightage →	0.23	0.52	0.25		
Potential Causes	Relationship Ratings			Sum of Products	Rank
Inadequate Training/Knowledge	9	1	0	2.59	4
Providers/Members submitting insufficient info	3	1	0	1.21	
No Proper check on pends and Follow-up	3	1	0	1.21	
Additional information unavailable in mentor guidelines	1	0	0	0.23	
Mentor (Guideline) discrepancies	1	9	3	5.66	2
Mentor Instructing to Pend	3	0	0	0.69	
Lack of Re-pricing info, EOB, Member information	3	1	1	1.46	
Access limitation	3	3	3	3.00	3
Continuous Alerts/Updates	1	0	0	0.23	

System Malfunction	1	0	0	0.23	
Claims Bounce Back	1	3	3	2.54	5
System failure, Temporary access denial	1	0	0	0.23	
TPA not responded with the Information	3	9	3	6.12	1
Misinterpretation of the claim	3	0	0	0.69	
Claim incorrectly pended	3	1	0	1.21	
Clarification responses not received from SMEs	3	1	0	1.21	
Escalation Reponses not received in time from POCs	1	9	3	5.66	2
Communication gap within various dept.	1	1	0	0.75	
Claims waiting Mentor Updation	0	1	3	1.27	

2.3 Analyze Phase

Since Pend Volume did not follow Normal Distribution, Kruskal-Wallis non-parametric test was performed to find out if different types of Pends had different effect on overall pend volumes.

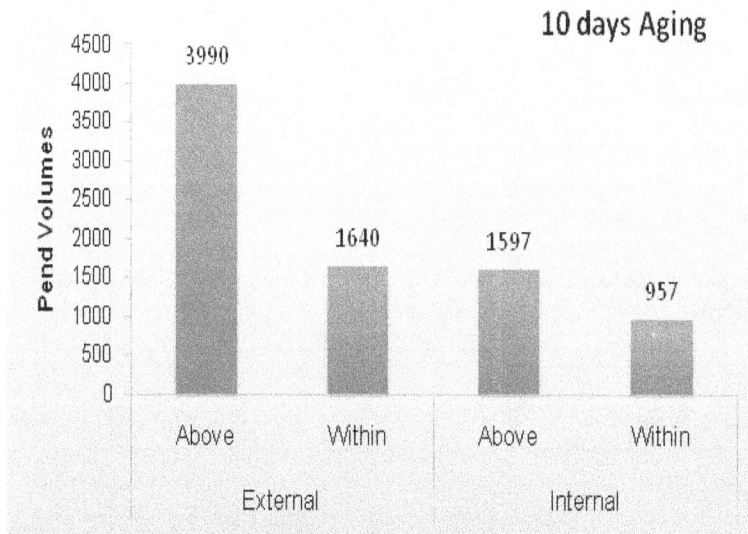

Fig. 5: Pend Type vs. Pend Volume

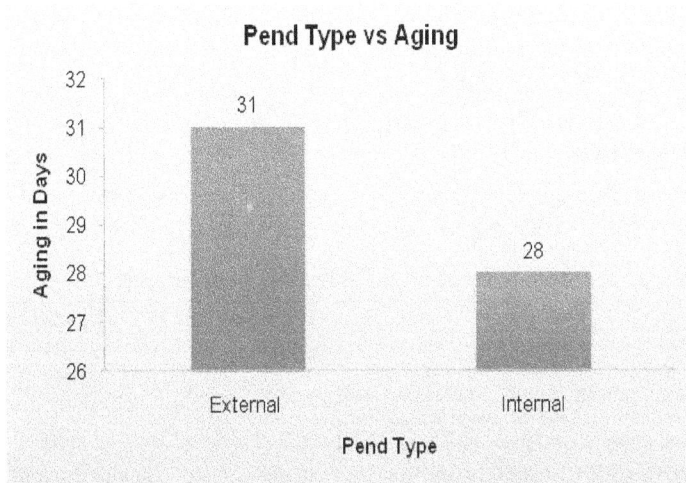

Fig. 6: Pend Type vs. Pend Aging

Kruskal-Wallis test for equality of median infer that Pend Volumes are significantly different for External and Internal Pends. Similarly, Pend aging values are significantly different for External and Internal Pends. The same analysis has been carried out on Pend categories versus Aging (above 10 days) and it has been found that Pend categories are significantly different.

Major contributors for High Pend Volumes for Aging beyond 10 days were found out using Pareto Analysis.

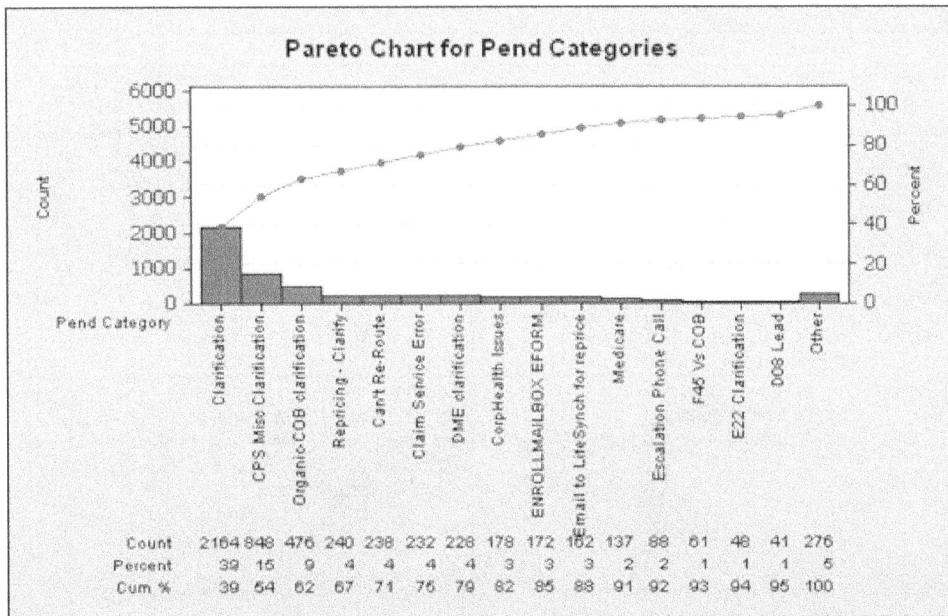

Fig. 7: Pareto Chart for Pend Categories

Second level of Fishbone diagram for categories was done. The root cause analysis brought out some action plans which could be done directly from Hinduja team perspective and some which needed management involvement.

2.4 Improvement Phase

Solutions to the Causes

Table 3: Root Cause Summary and Ranking

Cause	Root Cause	Dependability	Possible Solution	Rank
Client Clarification	All the claims (duplicate scenarios) with Multiple modifiers were pended under Client clarification due to disconnect in the mentor guidelines, which is taking greater time to get fixed.	External	Mentor Streamlining and Escalation matrix	2
CPS Clarification	New work (CPS Queue) which was transitioned recently, claims are pended 1. Due to the discrepancies in the mentor guidelines. 2. Awaiting clarification responses from Client POC's	External	Mentor Streamlining	2
Organic COB Clarification	1. Ambiguity in the mentor guidelines. 2. Claims were getting pended with secondary edit EXD91 post resolving the primary ex code which was not a trained EX code for HGSL and all the claims were bouncing back to macess pend box.	Internal	Mentor Streamlining and Escalation matrix Training/Coaching on the new Ex codes	2 4
Claim Service Error	1. Limited Security Access 2. Bottle Neck for ASO Run Access 3. No Access to Client application	Internal	Request for additional access	3
Email to Lifesynch for Reprice	Claims were pended awaiting response from Lifesynch area.	External	Design Pend Management Framework with TAT specifications in consensus with all related dept.	1
Duplicate Claims	Due to the ambiguity in the mentor guidelines (re-inventory claims EX4D1and EX4D4).	External	Mentor Streamlining	2
Medicaid Claims/ E22 Clarification	HGSL was not authorized to process Medicaid Florida claims. We were informed to route the claims by placing EX 2CP, but all the claims were bounced back to adjuster pend box.	External	Mentor Streamlining and Escalation matrix	2
Code Logic E-Form/ CIS BSS E-Form/ Corphealth Issue/ Leased Network email	1. No timely responses 2. Multiple requests 3. Communication Gap between Dept.	External Communication gap could be attributed to Internal reasons	Design Pend Management Framework with TAT specifications in consensus with all related dept. Training on/standardizing email scripting	1 4
Duplicate Claims	Due to the ambiguity in the mentor guidelines (re-inventory claims EX4D1and EX4D4).	External	Mentor Streamlining	2

Improvement action plans were accordingly implemented.

Table 4: Action Plans

Action Plan	Method/Approach	Responsibility
Pend Management Framework	1. Outline Procedures 2. TAT for all categories and Dept. 3. Follow Up Procedure 4. Escalation Levels	Process Manager Vendor Manager
Security Access	Requesting for HO and ASO Run Out Access.	Process Manager Vendor Manager
Mentor (Guideline) Streamlining	1. Establish need for Mentor Updation with Client SME's 2. Log Mentor Feedback on Duplicate, Enrollment 3. Follow CPO Procedures/Discussion	SME Process Leads
Refresher Training	1. Customized training on COB, Medicaid and Manual Deals. 2. Training scores cut off 97%	Process Trainer Process Leads
Pends Alignment and Review per the Framework	1. Align all outstanding pends to new Framework 2. Escalation and follow up on all pends beyond TAT 3. Monitor escalation timelines. 4. Monitor responses and resolutions.	POC's-Internal Process Leads Process Manager
Develop Customized Measuring system	1. Automatic tracking and Internal Escalation of Claims falling beyond TAT 2. System to track the pend History of the claim	Process Manager Process Improvement Manager Software Development Team

2.5 Control Phase

Defect = Pend Volumes > 3000/day, Total Number of Unit (Days) = 45, Opportunities/Unit = 1.

Total Number of Opportunities = 45, Total Number of Defects = 2, Proportion of Defects = Total Number of Defects/ Total Number of Opportunities, Proportion of defects = 2/45 = 0.044444

$$\text{Defects Per Million Opportunities, DPMO} = \frac{\text{Total Number of Defects}}{\text{Total Number of Opportunities}} \times 1000000$$

DPMO = 0.044444 × 1000000 = 44444.44

And Yield = 1- Proportion of Defects = 1 − 0.044444 = 0.955556 ==> Process Sigma = 1.701288

Therefore, DPMO = 44444.44 and Process Sigma = 1.701288

Benefits in terms of tangible aspects can be found out from DPMO chart improvement. The graph and the table below shows the extent to which DPMO and the time taken to resolve the claims reduced after the management had implemented the suggestions.

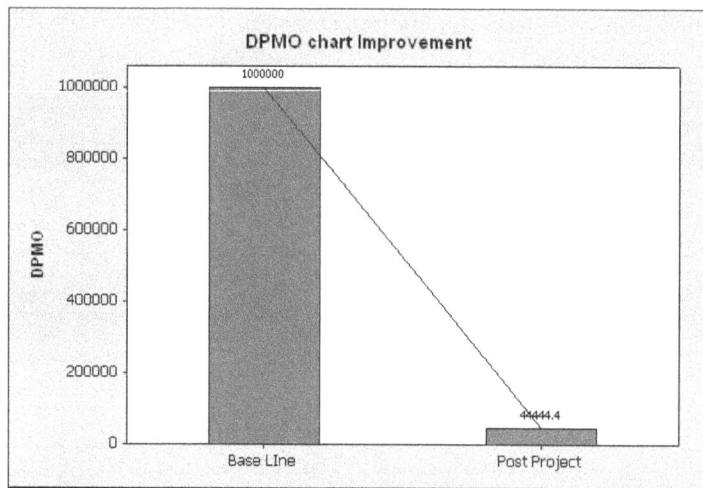
Fig. 8: DPMO Improvement

3. BENEFITS

Table 5: Benefit from the Project

Benefit from Project	Start of the Project	After the Project
Average Time Deployed on Pend Validation/Processors/Day (mins)	30	15
No of Processors	100	99
Total average time/Day (mins)	3000	1485
Average Time Deployed for Pend Review/Day (mins)	1500	540

Intangible Benefits

- Interest and Penalty paid to the client reduced
- Rework cost to Hinduja was avoided to an extent
- Timely Payment resulting in End Customer Satisfaction
- Reduced Escalation to client in Payment.

4. CONCLUSION

The problem statement was defined as "High Volumes of pend claims between the range of 6,000–7,000 per day contributing 28% towards the overall Inventory". During the Measure Stage data pertaining to volumes of pends, various reasons for pends which could be classified as external and internal were collected. Within these collected data contribution of various factors to the pend volumes were observed. Data about the aging of pends were also considered for analysis. First level Fishbone diagram was done to root out the factors causing high number of pends. To understand the performance in terms of measurable units DPMO calculation was done. Process sigma value showed that the performance was not good. The potential causes for high pend volumes were diagnosed by various statistical analysis. Remedial plans were suggested based on the analysis. Subject Matter Experts worked upon providing feedback for improving the process documentation to the client side team which did the necessary changes. A new workflow was worked up on to resolve pends on the first attempt. Action plans were devised to sustain the improvements. The scope of the project was restricted to all Insurance claims pend scenarios which were current and which would be encountered in future.

REFERENCES

[1] Huang, C., Chen, K.S. and Chang, T. (2010). An application of DMADV Methodology for increasing the Yield Rate of Surveillance Cameras, Microelectronics Reliability, 50, 266–272.

[2] Jain, R. and Lyons, A.C. (2009). The Implementation of Lean Manufacturing in the UK Food and Drink Industry, *International Journal of Services and Operations Management,* 5(4), 548–573.

[3] Krishna, R., Dangayach, G.S., Motwani, J. and Akbulut, A.Y. (2008). Implementation of Six Sigma in a Multinational Automotive Parts Manufacturer in India: A Case Study, *International Journal of Services and Operations Management,* 4(2), 246–276.

[4] Schon, K. (2006). Implementing Six Sigma in a Non-American Culture, *International Journal of Six Sigma and Competitive Advantage,* 2 (4), 404–428.

[5] Sokovic, M., Pavletic, D. and Fakin, S. (2005), Application of Six Sigma Methodology for Process Design, *Journal of Materials Processing Technology,* 162–163, 777–783.

Application of DFSS Framework in Adaptive Radiation Therapy Planning Proof of Concept

Prashant Kumar* and Yogish Mallya

Philips Electronics India Ltd., Bangalore, India
*E-mail: prashant.gupta@philips.com

ABSTRACT: Research/exploratory projects are often trapped in the complexity of problem solving and it becomes difficult to keep an end-to-end customer centric focus. Also, it is common to follow trial-error approach to formulate solutions and the perceived improvements may not always be statistically significant. The objective of this paper is to share our experiences on the effectiveness of Six Sigma methodology applied to Proof- Of-Concept generation. The project attempts to generate a deformable image registration based solution which can reduce the workload issues associated with Adaptive Radiation Therapy (RT) planning. A staged approach: Define, Identify, Design, Optimize, Verify and Monitor is followed to achieve Six Sigma process capability. Objective categorization, prioritization and statistical evaluation of process parameters at each stage greatly help in realizing the goal.

Keywords: DFSS, Adaptive Radiation Therapy, RT Planning, Deformable Image Registration.

1. INTRODUCTION

The role of external beam radiation therapy in cancer treatment is to deliver high energy radiation to irradiate the tumor. The goal is to maximize the dose of radiation to the tumor while minimizing the dose to the surrounding normal tissue. To achieve this goal, Radio therapy (RT) planning is done using the CT (Computed Tomography) image of the patient. RT planning is a complex and time consuming process. The clinician first needs to delineate\contour the tumor and other critical structures in the CT image slices of the patient. This is followed by setting up beam configuration for dose delivery. The best dose delivery settings are obtained by optimizing the beam configuration and other parameters in such a way that the dose delivered to various critical structures and tumor is as per the clinical guidelines. The dose delivery plan is clinically validated and transferred to the dose delivery machine for treating the patient, refer to Figure 1. Manual slice wise contouring of critical structures is a tedious and time consuming operation. It consumes several hours of radiation oncologist who is a scarce resource in a hospital set up. Radiation is delivered in a fractionated manner i.e. small quantities of dose over a period of four to eight weeks. During this period patient's anatomy might change due to several factors such as weight loss, shrinking of tumor and organ filling. As a result the use of initial dose delivery plan for the current state of the patient would result in inaccurate dose delivery. Adaptive radiation therapy i.e. re-planning for dose delivery using the latest CT image of the patient addresses this issue but it also increases the workload of the clinician.

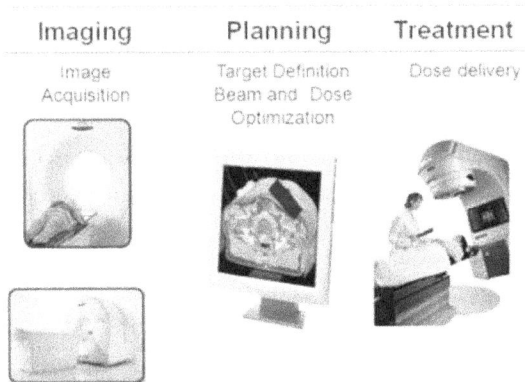

Fig. 1: RT Planning and Delivery

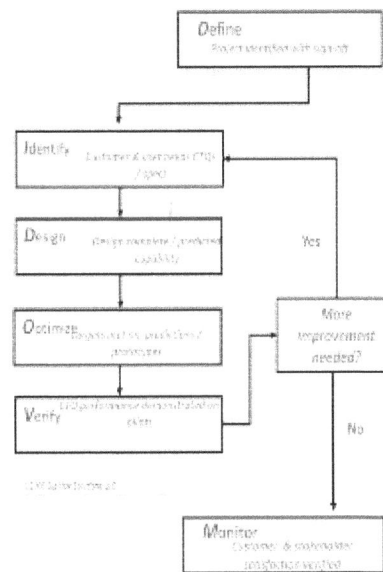

Fig. 2: Staged DFSS Approach

Our project attempts to come up with a proof of concept which can reduce the workload of the clinician in an adaptive RT planning setup. Towards this we focus on bringing down the manual effort required in manual contouring of patient's secondary images i.e. the new CT images which are acquired with respect to the latest state of the patient. This is achieved by computing a point-to-point mapping between the older and latest image of the patient using a deformable image registration algorithm and using this mapping to transfer contours from the old image to the latest image of the patient. DFSS framework is applied to execute this project. A staged six step approach-Define, Identify, Design, Optimize, Verify and Monitor is followed. Refer to Figure 2. The objective of this paper is to highlight the benefits achieved through DFSS.

2. DFSS METHODOLOGY

2.1 Define

In the define phase we formally define the project scope and goal. Through benchmarking of existing algorithms and solutions we arrive at the baseline performance and the improvement that we want to achieve through this project. Based on benchmarking it is established that we need to achieve at least 50% reduction in the execution time without compromising the accuracy of the algorithm. The use of SIPOC—Supplier, Input, Process, Output, and Customer helped us gain a complete understanding of the providers, process, stakeholders and consumers. We exercised Operational Profile to get an insight into the customer and user groups, actual usage environment of our solution, modes in which it operates, operations that it should support and the way end user is actually going to use it. This exercise greatly helped in understanding the various other demands on our solution instead of just focusing on algorithm development.

2.2 Identify

In this phase, Critical to Quality (CTQ) requirements are defined and respective measurable targets are set against those requirements to compute six sigma capabilities. The CTQs are further subdivided into sub-CTQs to further breakdown the problem into individual measurable components. Kano classification (refer to Figure 3) of customer and business needs is done to divide the various solution features into Dissatisfiers-must do, Satsisfiers-good to add value, Delighters-wow factors for the users though not a must do. Kano classification helps in prioritization of tasks and keeping an eye on the progress of the project against the customer and business needs. Figure 3 depicts the Kano classification exercise.

Deformable image registration algorithm, contour warping and image warping workflow are the must do items. Providing validation tools adds value as the user doesn't need to depend on external tools for validating the output of our algorithm. The items in the —delighters category are the other features apart from the basic workflow which would help make our solution attractive. The implementation of these items has the lowest priority but it is good to keep account of the items which are popped up as a result of SIPOC and Operation Profile.

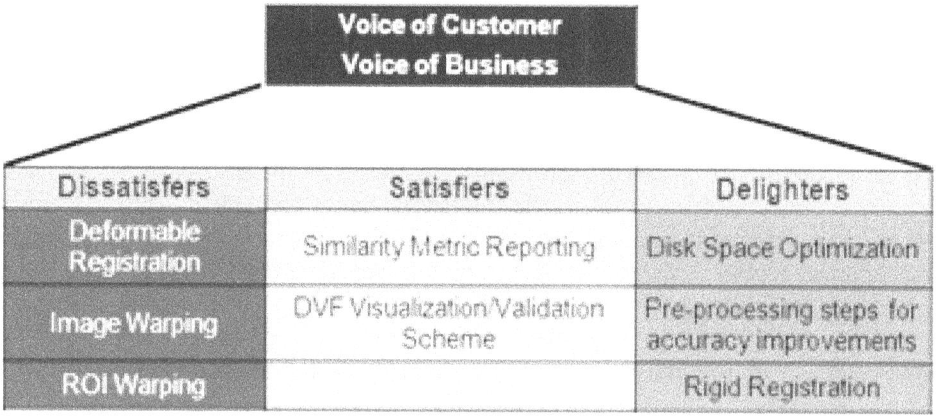

Fig. 3: Kano Classification

2.3 Design

Transfer function of the process is established in the Design phase. Categorization of input parameters into controlled, constant and noise variables is crucial to understand their effect on the output of the process. Input-output relation is mapped using a transfer function; Design of Experiments is an effective tool to generate a transfer function. However, in our case we could establish the transfer function based on our domain knowledge and it happens to be a simple linear relationship between the input variable, see Figure 4. The total time for the adaptive planning workflow is a linear summation of the time consumed by various features. We use X? (Unknown Xs) to denote the delighter features which we plan to address based on the availability of resources and time. Design options available for implementation of our algorithm are evaluated using weighted Pugh Matrix. Figure 5 depicts the various parameters against which the three design options were evaluated.

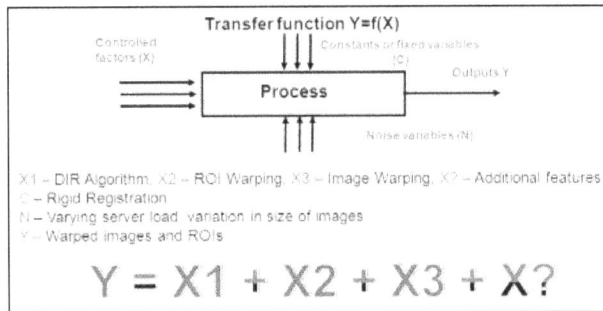

Fig. 4: Transfer Function

Though the default researcher psychology pushed us towards in house development of the algorithm, feasibility studies and a weighted approach based on various parameters suggested that the first option to extend an existing library is more appropriate.

Design Option	ITK Based Demons Optimization	In-house development of DIR	GPU Based speed up
Selection Criterion			
Effort Required	4	1	3
Expertise	5	3	4
Additional H/W resources	5	4	1
Project Schedule	5	2	1
Risk Involved	5	3	3
Speed up gain	3	3	5
Total Score	**27**	**16**	**17**

Fig. 5: Weighted Pugh Matrix

2.4 Optimize and Verify

Various algorithmic optimizations like multi-resolution framework for processing of image, limiting the extent of the image being passed to the computation routines, establishing a convergence criterion for the iterative algorithm, etc. are done to achieve the desired time performance. Prioritization of improvement items is done by looking at the current performance and Kano classification. Performance improvement against these optimizations is done statistically. Often we get trapped in false positives due to random evaluation instead of following a statistical framework. A sample population of more than 30 samples is used to make sure that normality tests can be applied successfully. Hypothesis testing is done to establish normality of input sample distribution. Once normality of input samples is established, hypothesis testing using Two-sample T-test is done to evaluate the improvement in performance against the attempted optimizations. Figures 6, 7 and 8 depict the statistical measurements for normality test, box plot comparison and capability analysis respectively. All evaluations are done against a confidence interval of 95%. Six sigma goals are achieved as a result of various optimization tricks. Failure modes are prone to modifications in the process, so Failure Mode Effect Analysis (FMEA) is done periodically to ensure that the safety critical aspects are never ignored. FMEA is helpful in making sure that the integrity of the process/solution is not busted due to component modifications, also the safety concerns in a healthcare product are captured and addressed appropriately.

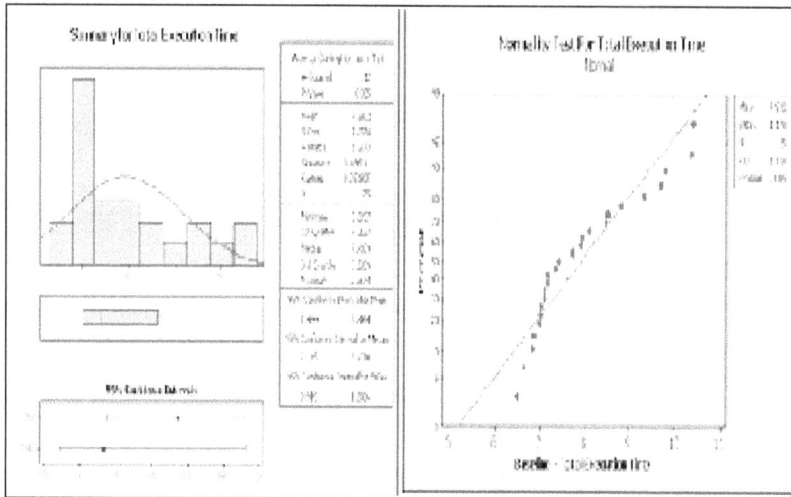

Fig. 6: Normality Test on Sample Population

2.5 Monitor

The performance of our algorithm at various clinical evaluation sites is continuously monitored. The proof of concept is transferred to engineering and we are participating in their design and implementation discussions to ensure that the same performance is achieved in the end product.

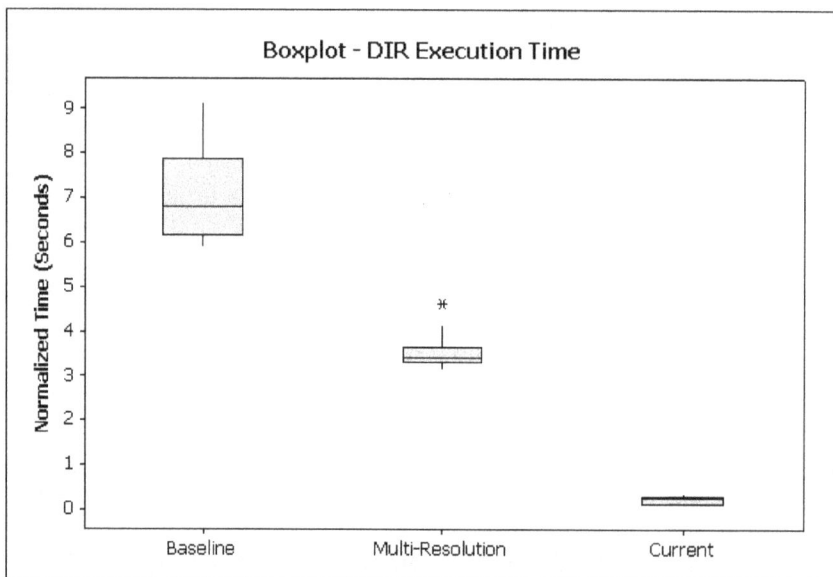

Fig. 7: Box plot to Evaluate Performance Improvement

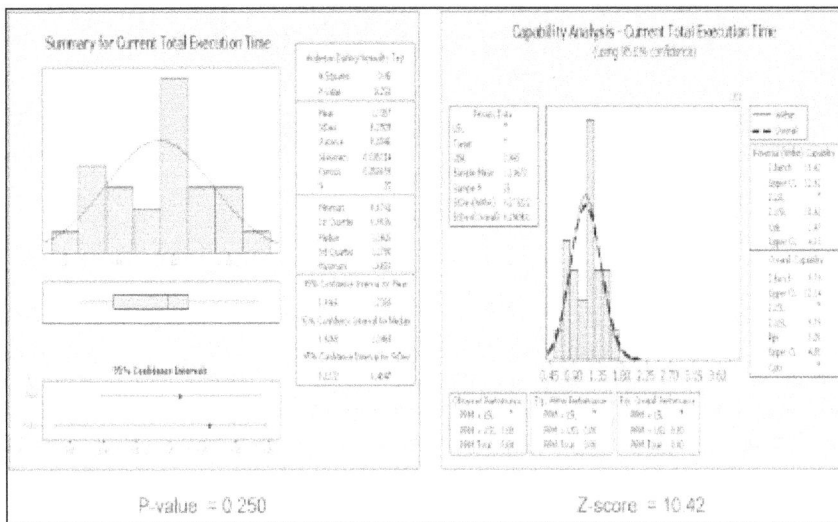

Fig. 8: Capability Analysis—Current Performance against Spec Limits

3. CONCLUSION

DFSS framework helped us keep focus on various user and business needs while maintaining a quantitative approach to decision making and problem solving. Accounting for the usage environment, clinical user needs and statistical evaluation of results ensured that the performance evaluation across different sites did not produce and unexpected results. The proof of concept is accepted by the engineering team at the first go without any major modifications. This is because the operation profile helped us understand the engineering requirement and we implemented a modular design for reuse. Six sigma process capability is achieved inclusive of all features put under the delighter category.

ACKNOWLEDGEMENT

The authors thankfully acknowledge the help and guidance of Mr. Ajit Ashok Shenvi in completing this project. He is the Senior Program Manager for Quality and Regulatory at Philips Innovation Campus, Bangalore and holds a Black Belt in Six Sigma methodology.

Error Reduction: A Case Study

A.M. Romesh Kumar Corera

Hinduja Global Solutions, Bangalore, India
E-mail: romeshkumarc@teamhgs.com

ABSTRACT: Six Sigma is a structured data-driven methodology that can be used to achieve near-perfection in process performance. The current project applied this methodology to improve precision rate in output quality of an international health insurance claims adjudication process. Initial conventional Corrective and Preventive action approaches were proved to be ineffective and the team opted for Six Sigma methodology to achieve sustained business results. Monthly data aggregation was done and a statistical model was developed at a micro-transactional level to identify the top contributors of errors. This scientific approach helped in breaking a conventional business notion for error generation. Interventions for improvement included formation of quality task force, development of advanced training program and interactive knowledge sessions. The successful application of Six Sigma techniques resulted in reduction of error followed by consistency in sustenance and introduction of *Development Action Plans* for processor performance monitoring.

Keywords: *Adjudication Process, Six Sigma Methodologies, Development Action Plans, DMAIC Methodology, Perception Analysis, Error Categories.*

1. INTRODUCTION

The purpose of process improvement is to eliminate the root causes of performance deficiencies in processes that already exist in the organization. These performance deficiencies may be causing real problems for the organization, or may be preventing it from working as efficiently and effectively as it could. Sometimes simply improving existing processes is not enough, and, therefore, new processes will need to be designed, or existing processes will need to be re-designed. This paper focuses on this aspect to improve the process.

HMO is one of the two main adjudication processes where performances are rigorously measured and monitored by client. QAP is one of the audit methods where samples are randomly picked and audited. Under QAP audit program 1% of a processor's production per day and 100% claims where the payout is greater than $5k were audited.

To improve the precision rate, client also adopted a new audit method called stratified audit where the entire production population was stratified based on the dollar value of payouts and then random samples from each stratum were picked for audit. If SLA's were not met, then "Performance Credits" would be evoked.

2. METHODOLOGY

DMAIC methodology is applied to minimize the error. The subsequent paragraph describes the methods in detail.

2.1 Define

Although the required SLA has been achieved on QAP random audit method, the performance started to dip and posed a threat of not meeting SLAs. Also, the performance in Stratified audit was inconsistent. The project charter clearly outlines the scope of the project and the target was to reduce errors to meet SLAs, both in QAP Random and Stratified Audits.

Define

Project Title	Error Reduction in HMO - QAP [Quality Audit Program]
Problem Definition	Last 3 months have seen an increasing trend in financial and statistical errors, thereby posing a threat to achieve KPMs Average for Dec. 08, Jan & Feb. 09 = 106 errors Financial : 70 errors & Statistical : 36 errors
Scope	HMO Claims Adjudication process
Goal Statement	To reduce Financial & Statistical errors from 106 to 30 errors by Aug 09 as an improvement plan for QAP results.
Tangible Benefits	Expected savings : $ 10,000 per annum [Opportunity Cost]
Target Beneficiary	Client & HGSL
Expected cust. Benefits	First Claim Resolution & Rework Reduction
Monitorable parameters	1. Total Claim Accuracy 2. Financial Accuracy
Project Number	BGL/POLO/200903/02

Metrics

Base line (Current) Measures of Y:

Month	Statistical	Financial	Total
Dec'08	39	74	113
Jan'09	37	58	95
Feb'09	41	70	111

Project Start	18 Mar 2009
Project End	20 Oct 2009
Duration	7 months

Critical to Business / Customer

Big Y's	Company's Goal: To reduce QAP errors
Small y's	Concentrate on Statistical & Financial errors
Business Implication	If the performance is not improved, then we would be paying an estimated penalty of $10,000 per year and also potentially lose significant amount of business (un-estimated)

The Team

Project Champion	Rajaram Natarajan - VP
Black Belt	Ravindra MG
Process Head	Romesh
Team Members	Gowri Sunil Shilpa Neetha
Support Required	Availability of SME's for Data collection, Brainstorming, initiating & implementation of actions
Domain Expert	HMO claim processing - SME's

Fig. 1: Project Charter

Also, the critical to Quality Tree has been defined in order to identify the critical factors impacting the customer from the business perspective.

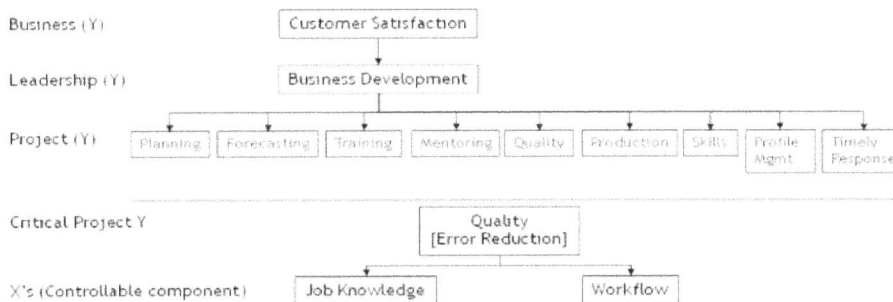

Fig. 2: CTQ Tree

The process map to get a brief overview of the process in also highlighted below.

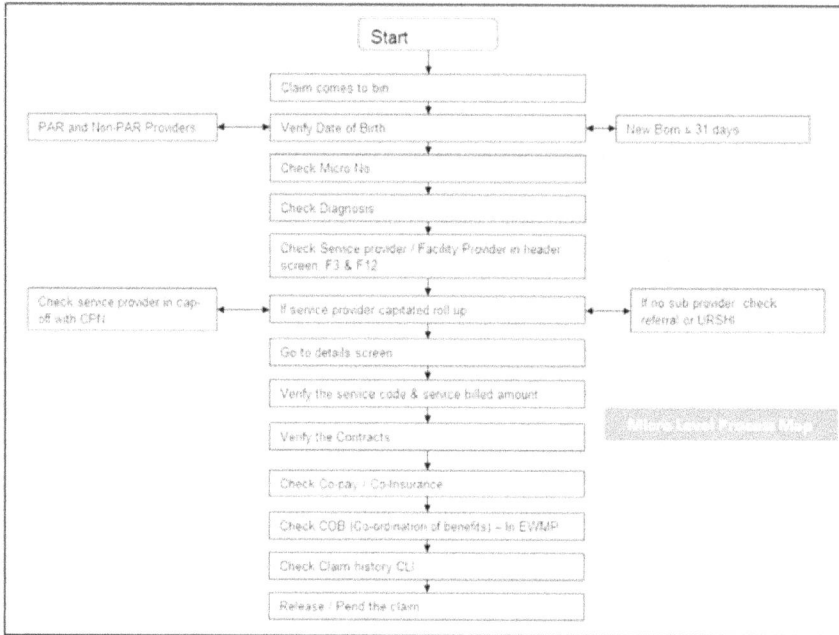

Fig. 3: Process Map

2.2 Measure Phase

The Quality scores from the months of Dec'08, Jan'09 and Feb'09 were considered for baseline. Data was collected from Client audits. MSA was not required as audit was performed at the client end.

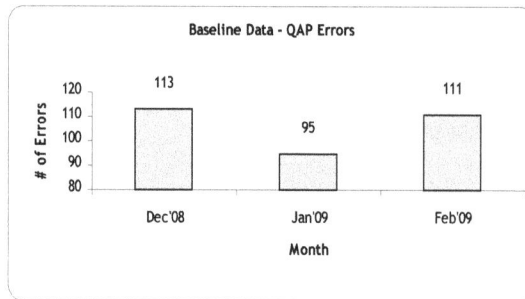

Fig. 4: QAP Errors

Z – Score has been calculated based on the following data.

Number 1 of Errors (D) = 319, Number of Claims Audited (N) = 36024, Number of Opportunities (O) = 1. Hence, DPMO = 8855 and Sigma Level Ω 3.87.

2.3 Analyze

The audit data collected was analyzed to arrive at the top error contributors by using the Pareto diagram. A brain storming session consisting of the process owner, SMEs and Team leads was conducted to arrive at the potential root causes for the top error categories identified.

The causes were then classified into 4 main drivers: a) Education – 14%, b) People – 65%, c) Procedure – 09% and d) System – 12%. An analysis was carried out on the major drivers: i) Team wise ii) Tenure wise and iii) Processor wise. The top teams and the processors that need attention were identified.

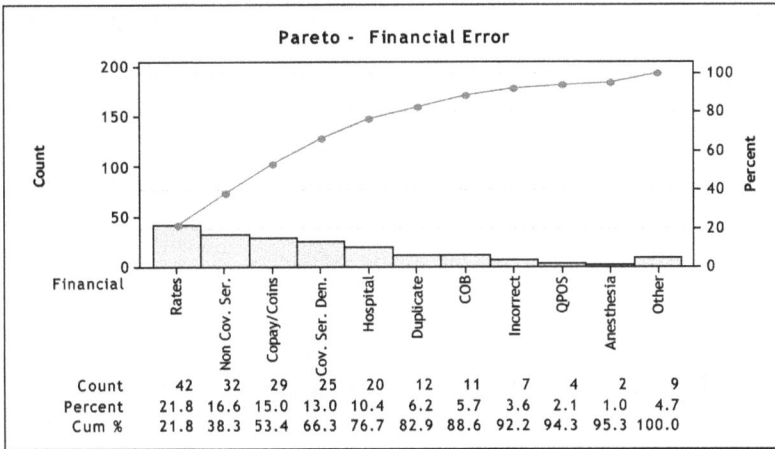

Fig. 5: Pareto Chart for Financial Error Contributors

The identified top error trends were: Rates Calculation, Non-Covered Services Denied, Copay/Coins, Covered Services Denied, Hospital Claims.

A Pareto chart of category wise error contribution was also developed.

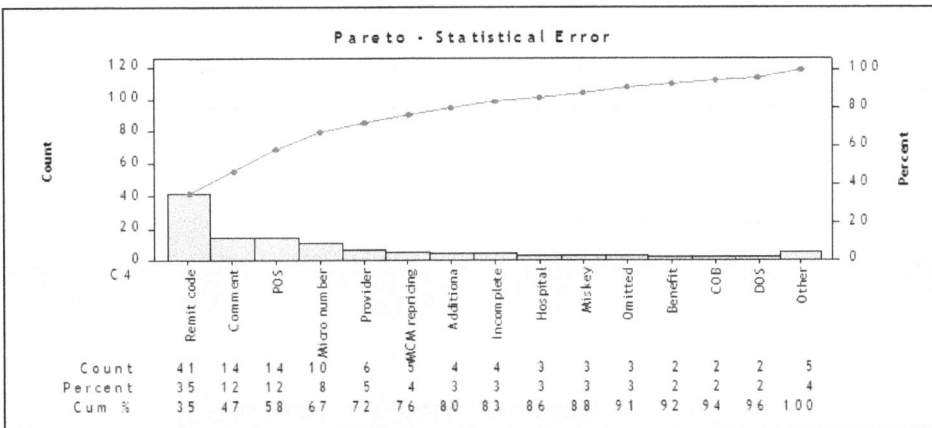

Fig. 6: Pareto Chart for Statistical Error Contributors

Considering major contributors the following error categories were publicized as "Zero Tolerance Errors": Omitted comments, Non Covered Services Paid, Rates.

A cause and effect diagram was also prepared through brainstorming considering the top Errors.

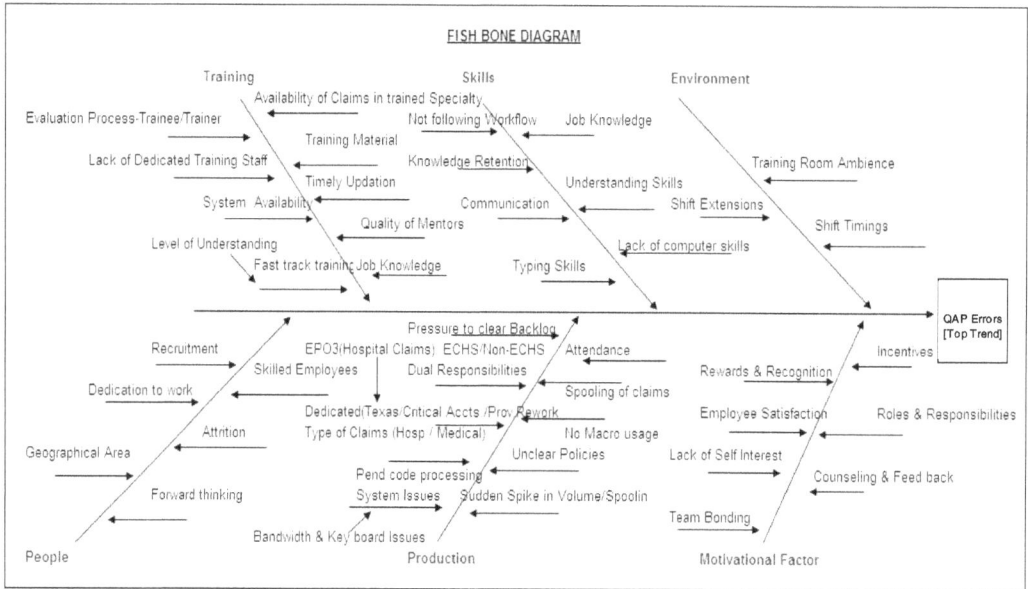

Fig. 7: Fish bone Diagram

Based on the Brain storming sessions on the Top Error Categories, with the team leads and process SMEs, the errors were broadly categorized under the 4 drivers:

(a) Educational b) People c) System and d) Procedure.

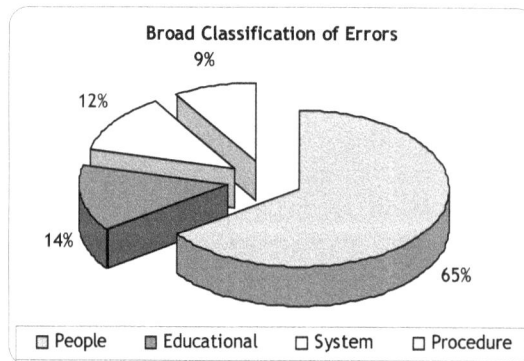

Fig. 8: Classification of Errors (RCA Approach)

It was found that 65% of problem was driven by people's "Skill and Will" to process.

The major findings are summarized below:

Table 1: Summary of Analysis

Analysis By	Findings	Remarks
Team	Teams/Individuals that top the list – RBCO, New Albany, Bismarck	Both Financial and Statistical errors were considered for rating
Batch	Batch 2, batch 20 and 13 contributes to high errors	Batch 20, Batch 2 and Batch 13 comes under Fin. and Stat.
Processors	35% of Processors contribute to 80% of error [51 processors out of 144]	17 Processors reflect under Fin. and Stat.

Perception Analysis: During brainstorming sessions and counseling sessions with the erring processors, most of them had a opinion that majority of errors occur due to Spool-in hours. It was also perceived that more errors occured on a Saturday which is not a regular working day, when the monitoring and coaching systems were not available in full strength. Therefore, a study was conducted to analyze the perceptions that were deeply rooted on the floor.

Inference: Based on the analysis we found the perception is only a "MYTH".

2.4 Improve

Alternate solutions were developed for the root causes identified. Pilot runs were made and tests were conducted to measure the impact of actions implemented. Sigma level improved from 3.87 to 4.14, successfully meeting the targeted numbers.

Table 2: Ideas and Solutions

People [Contribution - 65%]	Educational [Contribution - 14%]	System [Contribution - 12%]	Procedure [Contribution - 14%]

• Forming dedicated teams based on sites
•Formation of "Quality Circles/Task Force" for each segment
•*Visual Control* -
  - workflows
  - Posters
•*Contest:* Launched activity of month, Contest planned to ensure learning the "Fun-way"
 	
Month / Events
May' 09 — Quiz
Jun' 09 — Query & Answer Week
Jul' 09 — Poster Competition
Aug' 09 — Policy Hunt
Sep' 09 — Scramble
Oct' 09 — Word Building
Nov' 09 — Collage
	• *Training Programs:* Planned and executed 'Weekend Training Programs' on top 3 error categories - Rates Calculation, Co-pay / coins, Covered Services Denied
	• *DAP [Development Action Plan]:*
	  - Gemba / PEP talk session
	  - 'Teach to Learn' sessions by identified processors
	• *Knowledge Bank:* Unclear Policies & procedures - clarification from client
	• Daily Quick Tips
		• Developed power tools that would help processors by fetching accurate information and at a quicker time -	• Created Hotspots [Error Proof]
			• "Walk the floor" - SMEs made easily accessible to processors

2.4.1 *Inference*

It was evident from the "Box plot" that there is a significant improvement after implementation of Error reduction action Plans. A "2 Proportion T-Test" was also conducted and found that there is a significant improvement after implementation of solutions.

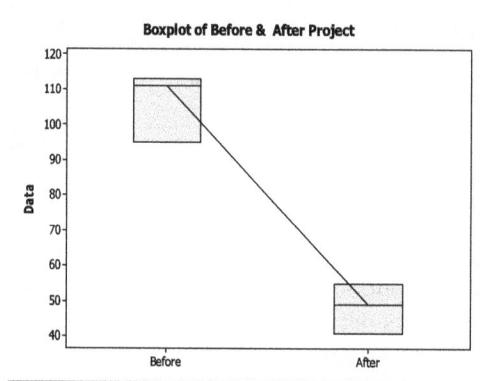

Fig. 9: Box-Plot of Pre and Post Project

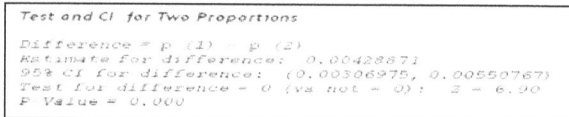

Fig. 10: Results of two-proportion t-test

2.4.2 *Benefits Achieved*

Formation of Quality Circles, introduction of DAP process and Events of the month has been a great success.

QAP errors saw consistent reduction from 111 in Feb'09 to 41 in Oct'09.

This has also helped us in reducing the total number of errors in stratified audit and thus meeting SLA's. The estimated cost savings is projected at $10,000 per annum.

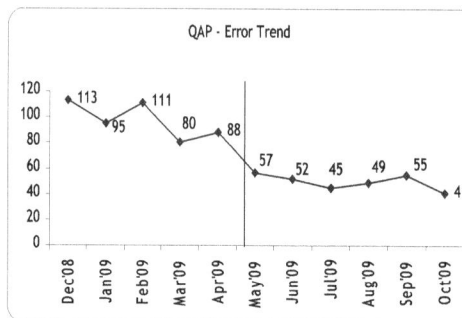

Fig. 11: QAP Error Trend after Implementation

2.5 Control

DAP, Grading methodology and FIT are institutionalized to sustain, improve the achieved results.

DAP [Developmental Action Plan]–Dashboards have been created and the performances of the processors were shared through Progress card on a daily basis. Based on grading, DAP team is formed every month.

Grading Methodology–Based on performance of processors in Internal and External [QAP] audits processors are classified under different grades – "A, B, C and D".

FIT [Fast Interactive Training]–It's a 5 day training program to improve the skill levels of targeted audience. Using the FIT tracker tool, we observed that 70% of the audiences were consistently error free. This obligated us to innovate and come out with the Grading methodology which also helped to widen our focus on resource utilization.

To test the efficacy of the FIT programs we conducted a paired t-test and it was statistically evident that knowledge levels had improved.

3. CONCLUSION

Followings are the major findings from the project:

- Variance in claim handling management – Lower the value of the claim, lesser the significance assigned.
- Approach adopted earlier: Concentration on 'vital few' and negligent towards 'trivial many'
- Importance of Stratification.
- Performance Management Technique-Grading Methodology.

REFERENCES

[1] Al-Araidah, O., Momani, A., Khasawneh, M. and Momani, M. (2010), Lead-Time Reduction Utilizing Lean Tools Applied to Healthcare: The Inpatient Pharmacy at a Local Hospital, Journal for Healthcare Quality, 32(1), 59–66.

[2] Ali, M. (2004). Six-sigma Design through Process Optimization using Robust Design Method, Master Thesis at Concordia University, Montreal, Canada.

[3] Bisgaard, S. and Does, R. (2009). Quality Quandaries: Health Care Quality-Reducing the Length of Stay at a Hopital, Quality Engineering, 21, 117–131.

[4] Cheng, Y.H. (2005). The Improvement of Assembly Efficiency of Military Product by Six-Sigma. NCUT Thesis Archive, Taiwan.

[5] Cournoyer, M.E., Renner, C.M., Lee, M.B., Kleinsteuber, J.F., Trujillo, C.M., Krieger, E.W. and Kowalczyk, C.L. (2010), Lean Six Sigma tools, Part III: Input metrics for a Glovebox Glove Integrity Program, *Journal of Chemical Health and Safety*, Article in press, 412, 1–10.

[6] Edgardo, J., Escalante, V. and Ricardo, A.D.P. (2006), An application of Six Sigma methodology to the manufacture of coal products, World Class Aplications of Six Sigma, 98–124.

[7] Hook, M. and Stehn, L. (2008), Lean Principles in Industrialized Housing Production: The Need for a Cultural Change, Lean Construction Journal, 20–33.

[8] Huang, C., Chen, K.S. and Chang, T. (2010), An application of DMADV Methodology for increasing the Yield Rate of Surveillance Cameras, Microelectronics Reliability, 50, 266–272.

[9] Jain, R. and Lyons, A.C. (2009), The Implementation of Lean Manufacturing in the UK Food and Drink Industry, *International Journal of Services and Operations Management,* 5(4), 548–573.

[10] Krishna, R., Dangayach, G.S., Motwani, J. and Akbulut, A.Y. (2008), Implementation of Six Sigma in a Multinational Automotive Parts Manufacturer in India: A Case Study, *International Journal of Services and Operations Management,* 4(2), 246–276.

[11] Schon, K. (2006), Implementing Six Sigma in a Non-American Culture, *International Journal of Six Sigma and Competitive Advantage*, 2 (4), 404–428.

[12] Sokovic, M., Pavletic, D. and Fakin, S. (2005), Application of Six Sigma Methodology for Process Design, *Journal of Materials Processing Technology,* 162–163, 777–783.

Improved Up Selling Using FMEA

Gurupreet Singh Khanuja

Firstsource Solutions Limited
E-mail: gurpreet.khanuja@firstsource.com

ABSTRACT: One of our key client's (No. 1 private insurance company in India) indicated satisfaction with overall performance of Firstsource Solutions Limited (FSL) on key metrics. During the recession (2009–10) clients' business objective was focused on Upselling services to remain as a market leader.

The FSL team aligned their goals in line with clients' expectations and a Six Sigma project was initiated after analyzing huge scope to improve upsell percentage, with the objective of improving revenues for clients and increasing conversion opportunities at sales end of clients.

Various factors impacting the up selling were identified through brain storming and were filtered through Fishbone diagram. A detailed FMEA was done to prioritize and increase efficiency on up selling front. This was done to ensure preventive actions are put in place against each of the failure modes which would translate into higher up selling numbers and reduced Post Tele-Call Drop (PTD)[*] percentage. Once the improved solutions were implemented, FMEA analysis indicated significant reduction of RPN for the identified failure modes/potential causes. Significant improvement observed on all the worked failure modes/causes.

Through continuous review of FMEA document and implementation of solutions, up sell percentage improved (increased by around 100%). Clients appreciated the efforts and were highly pleased with the outcome of the project.

Keywords: FMEA, Call Centre, Upselling, Post Tele-call Drop, Insurance.

1. INTRODUCTION

The client is a No. 1 private insurance company in India with a policy base of around 5 million and average call volume per month of 1,00,000. FSL manages customer support through inbound calls, emails, transaction processing, webchat and level 2 escalations. Clients were highly satisfied with overall performance of FSL on key metrics and applauded the team's efforts during Quarterly meetings. FSL team was appreciative of client feedback but was always keen to add more value, thus wanted to identify key improvement areas which can bring considerable delight and value to the client.

---

[*] Post tele call drop – Once an appointment is fixed with customer for a sales executive visit, the details are shared with the sales team. Field sales executive calls the customer before visiting the customer to ensure availability and confirm the appointment. If customer rejects for meeting stating it was planned with his/her consent, the transaction is considered as post tele-call drop.

The recent recession hitting across the industries, impacted clients business equally. Clients' business objective for the FY 2009–2010 was focused on Upselling to recover from the situation and remain a leader. Clients were expecting FSL team's expertise to support them in the situation. The FSL team aligned their goals in line with clients' expectations and since the call centre happens to be a touch point frequently used by customers to reach business, the scope of improvement was significant and very evident.

The FSL group was keen on establishing themselves in the niche market of service providers to insurance clients. This was taken as an opportunity to move towards that direction and gain expertise of the domain leading to great amount of focus from the management team towards the improvement initiative.

FSL did an internal study on some of the data points and came out with the observation that improvement in upsell percentage will immensely increase the revenue through new business and reduction in Post Tele-Call Drops (PTD) will greatly increase the conversion opportunities for sales team helping finalize the investment deals. The team decided to address this issue and use Six Sigma methodology to improve the current performance level.

Firstsource's Six Sigma Program is central to its Process Improvement philosophy. Six Sigma is a key tool in the overall deployment framework based on Process Management, Quality Assurance and Process Improvement. This model is governed and supported by the Business Quality Council (BQC) at the Apex level which ensures the prioritization, elimination of bottlenecks and the successful delivery of identified initiatives.

Firstsource has a strong focus on building a quality culture which is reflected by the Black Belt to Employees span of 1:650 and an Organization Quality DNA of 71% Yellow Belts and 61% Green Belts. This is further strengthened with Organization-wide Six Sigma awards initiative that rewards and recognizes exemplary performance. We believe that the deployment of our process excellence framework has improved our service delivery levels and as a result enhanced value- add to our customers.

A Six Sigma project was therefore initiated to improve upsell percentage, with the objective of improving revenues for clients and increasing conversion opportunities at sales end of clients. Clients were delighted and assured all support to this project.

Through the use of six sigma methodology, specifically FMEA tool, the project team approached the potential causes in scientific way at each step. The results were achieved by the project team just not fulfilled the expectations of the project six sigma project, but delivered significant gains through increased bottom line for FSL and clients were profited through the improved revenue through premium from new policies. The team won internal award for successfully contributing to business objective for both FSL and clients.

2. METHOD

Team used the DMAIC approach for problem identification and implementing solutions.

A project team was formed which included team members from the process across both the locations serving clients and across different domains (quality/operations). Client indicated keen interest to be involved and provided support at different stages of the project. The project lead then collected the baseline data for the period Feb '09–April '09 and established the problem statement for the period of Feb, March, April'09, the up sell % is 9.47% which is very low and PTD % is 43.89% which is very high against client expectation. The less % age of up sell and high % age of PTD results in client dissatisfaction and revenue loss. To create an impact FSL wanted to target more than 100% improvement in upsell percentage and 20% reduction in PTD. Clients were pleased to know the improvement targets as that would significantly translate into revenues.

2.1 CTQ

The team prepared the CTQ tree to clearly understand the broad focus areas. This showed that the highest contribution to client's business comes from genuine leads generated in customer queues. Care queue formed 65% of overall volumes and minimum premium of an HNI (High Net worth Individuals) customer is $2500. The team decided to keep other customer/non-customer queues (Share (Insurance advisors) and Prospect (Prospective customers)) were out of scope of the project due to no potential upselling possibilities.

2.2 SIPOC

High Level Process Map (SIPOC) was then mapped to study the various Suppliers, Inputs, Process, Outputs and Customer.

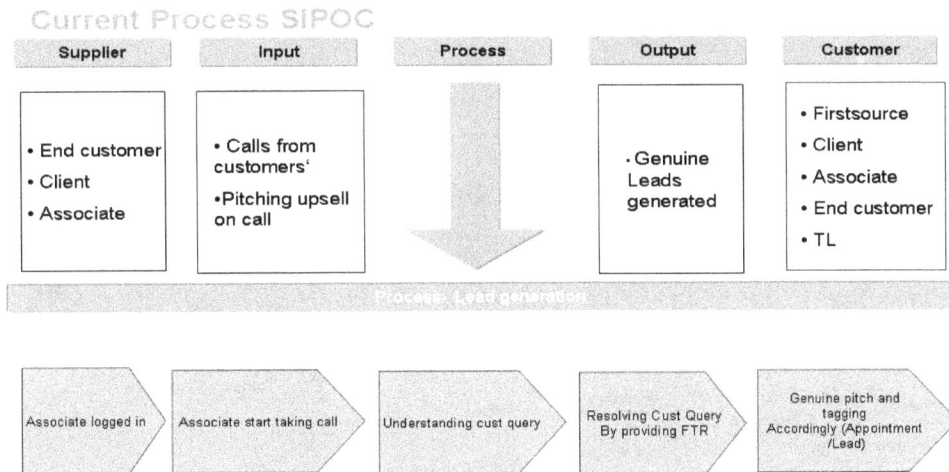

2.3 CTQ Characteristics

Before the data was collected, it was imperative to understand the operational definition of up sell and PTD. The definition provided by the clients was 'Up Sell—Any customer confirming to meet the Financial Service Consultant (FSC) will be considered successful lead on correct tagging of details. PTD A lead for which customer denies confirmation about appointment when FSC tries to meet them for discussing the investment plans.'

2.4 Process Capability

Data was collected for the baseline period of 3 months. The process capability was measured and the as-is process was found to be operating at 911377 DPMO and 0.15 Zst for Up Sell, 454464 DPMO and 1.61 Zst for PTD.

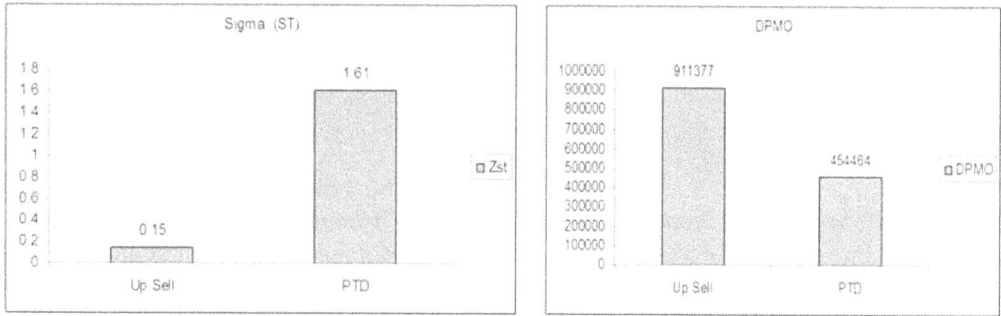

2.5 Brainstorming

In order to find different factors potentially impacting the out put metric Y (Up sell and PTD), brainstorming session was conducted using Fishbone diagram which included client team members, project team and associates.

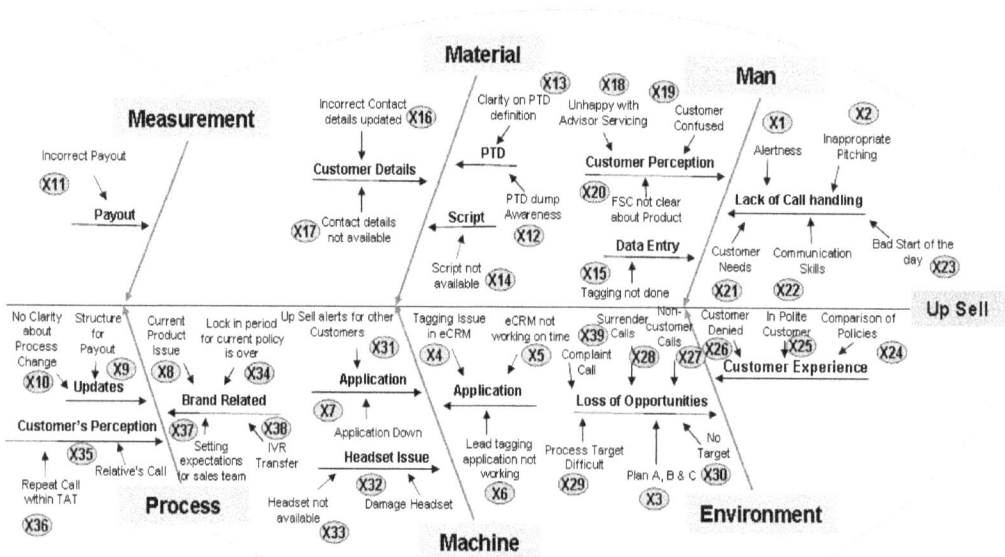

2.6 FMEA

Through the brainstorming activity, varied factors surfaced for the team which were under-process, measurement, material, man, machine, Mother Nature. Considering the following challenges, team looked at FMEA tool for improving the process:

- Need of a well designed reliable process that is documented—In order to sustain the improvement, a well documented improvement process was required for the team.
- Innovation for people related problems—The management used different techniques/methods to improve the performance earlier. To come over issues specifically related to people, a toll was needed that helps innovate.
- Importance of involvement of people for this people driven metric—Based on the local culture, involvement of front line staff resulted into mellow down the resistance to great extent that was experienced in the past. FMEA gives that opportunity of involving people from various fronts to contribute and give the feeling of being involved.
- Lack of risk quantification in other available tools—The associates risk for identified causes (majorly being related to people) could not be quantified as easily as through FMEA.

A detailed FMEA was planned to identify the potential failure modes, understand the associated risks and prioritize them based on their severity, occurrence and detection. This was to ensure preventive actions are put in place against each of the failure modes which would translate into higher up sell numbers and reduced PTD percentage. Following step by step activities were done to find the risk priority number for each of the failure modes:

I. **Creation of Team:** A team involving people from domains of critical importance impacting delivery of the metric were involved:

1. Subject Matter Experts (SME's),
2. Associates (front line staff),
3. TL's and QA's (performance owners – responsible for coaching and feedback)
4. Managers – program owners
5. Clients – considering the improvement ideas from clients can be of great value and that will also help netrulaize bias (if any) during the activity.

II. Road Map: Road map decided to carry out the required activities to ensure clarity of objective across the team and time bound it. The frequency of meetings, meeting organizer, role of team members, guidelines to resolve disconnect, all such aspects were decided before starting the activity.

III. Brief out done with the team sharing objective, road map, details of the tool (flow of failure mode, causes, rating scale, RPN calculation and significance).

IV. Methodology

1. *Failure mode identification:* The failure modes were identified through the brainstorming done earlier.
2. *Potential failure effects:* Through the team discussion, team arrived to common potential failure effects for each of the identified failure modes.

3. *Severity number:*[1] Following scale was used to assign rating to the severity for a potential failure effect:

(a) Low – 1

(b) Medium – 3

(c) High – 9

4. *Potential causes identification:* For all the identified potential failure effects, potential causes were listed down.

5. *Occurrence rating:* After arriving to the list of potential causes, their occurrence was identified. This was done in two stages, for the causes where the data was available; rating was given according to that. For others, rating was arrived through discussion using the following scale:

(a) Low – 1

(b) Medium – 3

(c) High – 9

6. *Current controls:* For each of the identified causes, a quick check was done for available controls and same were listed down against appropriate causes.

7. *Detect ability:* Rating based on the robustness of current controls already deployed in the system were arrived on following scale:

(a) Low – 9

(b) Medium – 3

(c) High – 1

8. *Risk Priority Number:* Using these severity, occurrence and detect ability ratings, RPN was derived.

V. RPN Identification: Potential failure causes with highest RPN (729) were picked up to be focused on for improvement. The responsibilities were distributed across the team (including client members) against agreed timelines. Following is a snap shot of the prioritized potential failure causes:

Process Step	Potential Failure Modes	Potential Failure Effects	Sev	Potential Causes	OCC	Current Controls	DET	RPN
Up Sell and PTD	Payout	Low Upsell High PTD	9	Low Motivation	9	Payout SOP	9	729
		Low Upsell High PTD	9	Incorrect Payout	9	Payout SOP	9	729
	Improper pitching	Low Upsell High PTD	9	Skill issue	9	PTD Dump	9	729
	PTD	Low Upsell High PTD	9	Application not working on time	9	PTD Dump	9	729

[1]Ratings for instances were data was not available were arrived through team consensus. Individual ratings are collected from all the team members and in case of non-unanimous rating; a discussion (reasoning to rating) is done till consensus is reached.

Process Step	Potential Failure Modes	Potential Failure Effects	Sev	Potential Causes	OCC	Current Controls	DET	RPN
	Call handling skills	Low Upsell High PTD	9	Implementation of Plan A, B and C as per volumes	3	None	9	243
					9		3	243
	Application	Low Upsell High PTD		alertness	3	None	9	243
		Low Upsell High PTD	9	Script not available	3	None	9	243
	Updates	Low Upsell High PTD	9	Incorrect/incomplete resolution	9	Quality monitoring	3	243
	Improper pitching	Low Upsell High PTD	9	Floor awareness of daily/weekly performance	9	Daily dashboard	3	243
	Data entry	Low Upsell High PTD	9	incorrect/no tagging	3	PTD Dump	9	243

FMEA - Pre (Upsell)

VI. Solution Scoping: Through the prioritized potential failure modes and its causes, brainstorming was conducted to identify solutions to mistake proof to the extent possible. The improvement started with identifying in-scope factors. The project team then used Control Impact matrix to classify the root causes into high, medium in control and out of Control.

		Impact		
		High	Medium	Low
Control	In Control	• Incorrect/incomplete resolution • Low motivation • Plan A, B and C • High PTD • Structure for Payout • Incorrect Payout • Script not available • Inappropriate pitching • Incorrect contact details • Skill issue • Awareness about performance	• Alertness • Lead not tagged/Inappropriate tagging • Comparison of policies	• Customer confused • Impolite customer • Complaint call • Surrender call
	Out Of Control	• Customer needs • POP up for other customer account • Tagging issue in SPAARC	• High Call volume • Unhappy with advisor servicing • Repeat call within TAT • Transfer calls (IVR/other queues)	• SPAARC Not working • S2S form not working • Bad start of the day • Relative call

VII. Solution Implementation: The high impacting and in-control issues were picked up for solutioning. Client support, as expected came in handy during the solution implementation stage. As part of the solution design, clients made changes to the application, provided clarity about the PTD's, modified the upsell scripts to be customer friendly as per suggestions.

Up Sell Pop Up Introduced

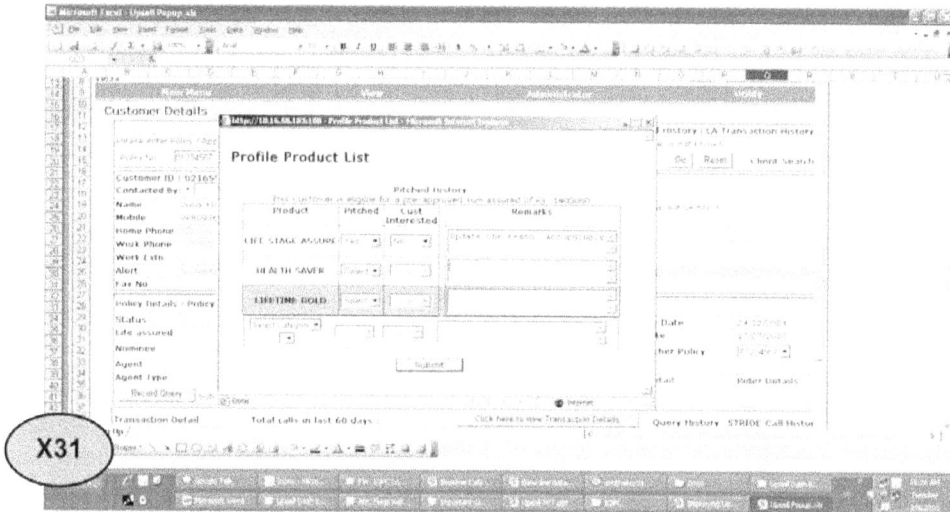

Tagging issue - Upsell not tagged properly
Action Taken: Upsell pop up introduced where in associate can tag the lead for that pop up

Script Changes Done

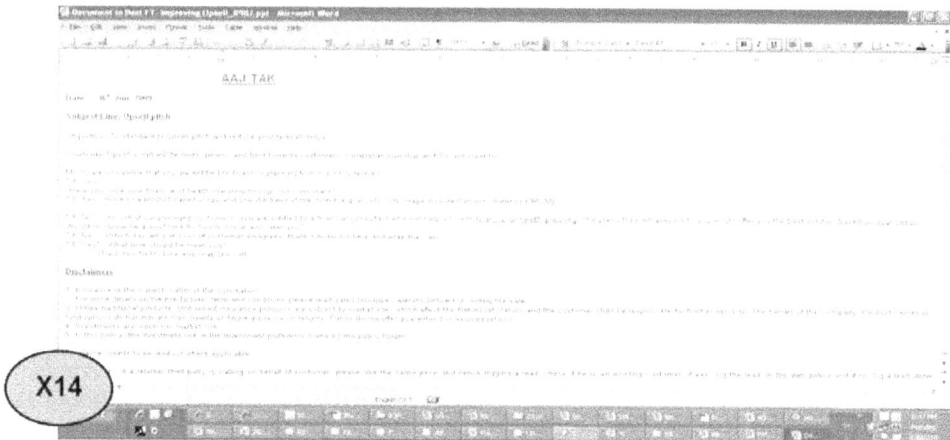

Proper script for pitching up sell is not available - Inconsistent scripts being used across the floor
Action Taken: Upsell FSC based Script is in place

Colored Flags put on the floor to avoid leakage of opportunities due to changes in plan.

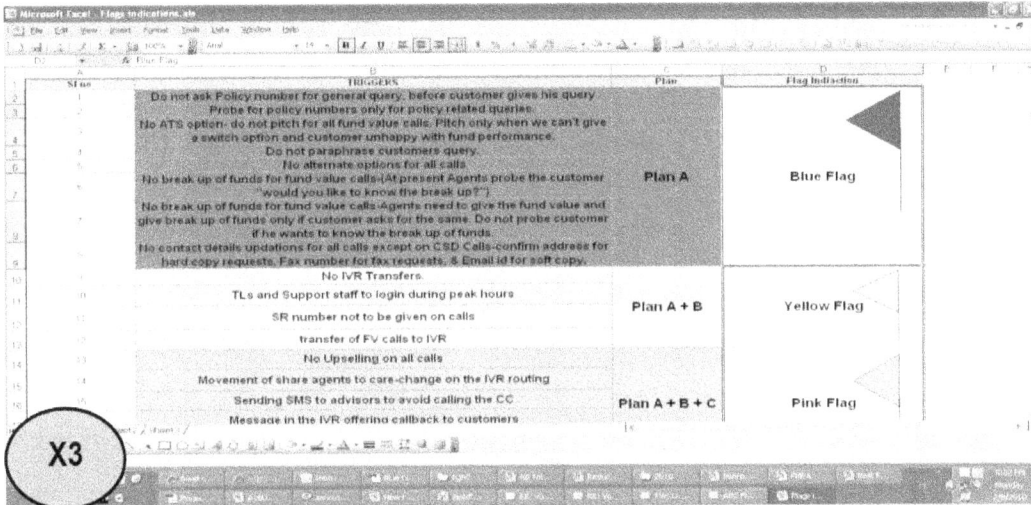

X3

Lost Upsell opportunity due to different plans through the day - Associate has to start upselling when he is in Plan A & B & stop when on Plan C.
Action Taken: Introduce Flags for awareness of active plan

RnR introduced to encourage and motivate people, involved clients as well.

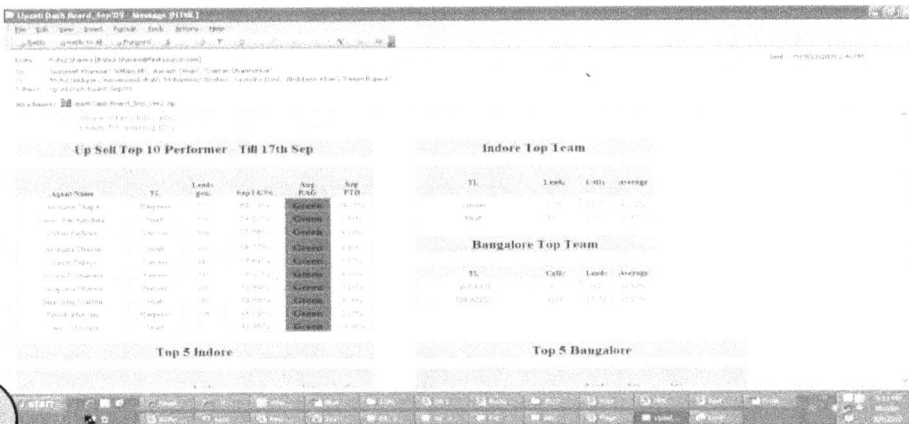

X39

Low motivation - No communication about payout
Action Taken:

• Pay out timelines Sop created for reference and clear understanding
• Up sell Performance included in RnR
• Clients involvement in RnRs

Payout SOP created, communicated to every one, made readily accessible at common drive. Final payout is shared only after confirmation from clients to avoid confusion.

X9 Payout structure issue **X11** Incorrect payouts done – Changes in Payout structure, Incorrect payouts given

Action Taken:

- Pay out Grid Has been made
- Now the performance is measured and shared, pay out is only discussed once finalized with clients

Dashboard with exhaustive details created to effectively track and control the metric.

Communication to people about changes/enhancements related to Upsell and PTD.

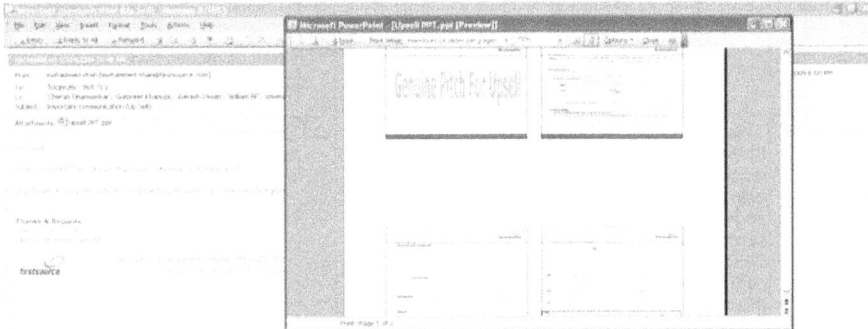

Inappropriate Pitching   Skill issue

Action Taken: New Presentation created with the help of Best performing associates. All bottom performers were trained on this

VIII. RPN Validation: Once the improvement solutions were implemented, FMEA analysis was done again to validate the improvement seen on RPN for the identified failure modes/ potential causes. Significant improvement observed on all the worked failure modes/causes as indicated below.

Process Step	Potential Failure Modes	RPN	Recommended Action	Responsibility	Status	Post Sev	Post Occ	Post Det	Rpn	Remarks	Status
Up Sell and PTD	Payout	729	- Pay out timelines Sop created for reference and clear understanding - Up sell Performance included in RnR	Gurpreet	Closed	9	1	1	9	Publishing the pay out details in common drive. The daily dashboard shares the performance details, the pay out details are shared only when it is finalized with clients	Ongoing
		729	- Associates are getting the pay out on time as per grid - Now the performance is measured and shared, pay out is only discussed once finalized with clients	Gurpreet	Closed	9	1	1	9	Process is rigorously followed and exceptions are being reported	Closed

Process Step	Potential Failure Modes	RPN	Recommended Action	Responsibility	Status	Post Sev	Post Occ	Post Det	Rpn	Remarks	Status
	Improper pitching	729	- New Presentation created with the help of Best performing associates. - All bottom performers were trained on this	Gurpreet	Closed	9	3	9	243	Sharing in dashboard on daily basis where associates is standing	Ongoing
	PTD	729	- Script allows (genuine Pitch to be check with quality to measure) application pop up form from quality parameters	Gurpreet	Closed	9	3	9	243	FSC based Script is on place and PTD % is publishing on daily basis	Ongoing
	Call handling skills	243	- Introduce Flags for awareness of active plan	Shafi/ Gurpreet.	Ongoing	9	1	3	27	Closed	
		243	- List of bottom performers to be published on a daily basis	Shafi/ Gurpreet.	Ongoing		5	3	135	Ongoing	
	Application	243	- Upsell to be reiterated during all team briefings	Shafi/ Gurpreet.	Ongoing	9	1	3	27	Ongoing	
		243	- Upsell FSC based Script is in place	Clients/Ops	Closed	9	1	3	27	Closed	
	Updates	243	- TL"S to be part of 2 roll outs and process related con calls in a month…and Sq needs to be part of 2 GK rolls of in month	Shafi/ Gurpreet.	Ongoing	9	3	1	27	Con calls MOM, TL wise weekly calendar and Briefing tracker in place	Ongoing
	Improper pitching	243	- Performance management through PTD and Upsell dash Board (frequency - daily)	Gurpreet	Closed	9	3	3	81	Closed	 FMEA - Post (Upsell)
	Data entry	243	- Upsell pop up introduced where in associate can tag the lead for that pop up	Clients/Ops	Closed	9	3	3	81	- FSC based Script is on place and PTD % is publishing on daily basis - The complete population is covered through the a session to clarify understanding	Ongoing

3. RESULTS

Team statistically established the improved performance using one of the strongest soft tools available.

The up sell percentage improved to 23% and more and PTD came down below 35%. Fisher exact test used to establish significance of improvement in output of upsell and PTD.

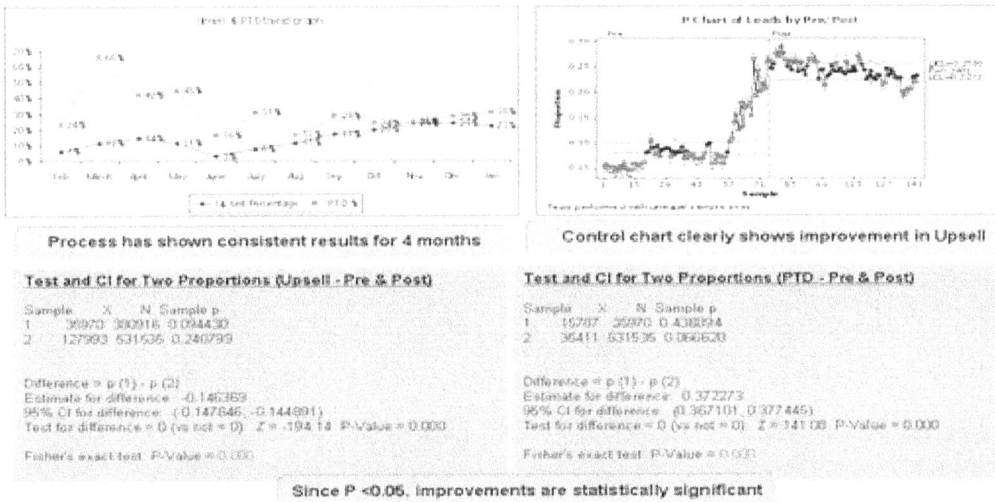

Process has shown consistent results for 4 months	Control chart clearly shows improvement in Upsell

Since P <0.05, improvements are statistically significant

4. CONTROL PLAN

Team prepared the control plan based on the solutions implemented which gave improvement of 300% in Zst for upsell and 25% improvement for PTD. Key control measures were:

- Periodic review of FMEA document
- Regular RnR
- Timely publishing of payout
- Daily dashboard and review
- Focus on associates in RED category
- Experts' comments section to share best practices through dashboard among associates.

Client's mail read "After all the motivating and driving in the month of November, we generated a whooping 38 thousand plus leads across Care and HNI............While these numbers may show some highs and lows and a long way forward, it does not stop us from highlighting some achievers:

- Shafi for consistent improvement and superior performance in the HNI queue.
- Sameer for lead generation as high as 39% and 45%.
- Amazing RAG turnaround by Padmini for the month of Dec.
- Continuous drive by Ibrahim and Zeenal for Bangalore center to move agents from Amber to Green and by Sameer for Indore.

Plan	Owner	Frequency
Regular Rewards & Recognition - Certificates for top 3 LG% associates across the center - Best FL for top LG%	Operations	Daily/Weekly/Monthly
Pay out publishing on common folder - Associates can directly access the common folder and they can predict approx. pay out	Operations	Monthly
Meeting red associates(less than 10% LG%) and fixing the target week on week	Managers	Daily
Publishing dashboard on daily basis -Creating awareness where they are currently standing	WFM	Daily
Expert comments in dash board -Associates comments (sharing best practices) publishing in dash board	Operations	Daily

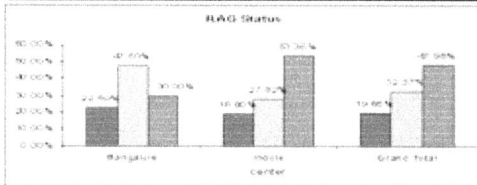

Red category associates reduced from 43% to 19.65% overall
Green category associates increased from 28% to 47.98% overall

Zst improved by more than 300% for Upsell and more than 25% for PTD

Organization results

- Clients appreciated the efforts and were highly pleased with the outcome of the project.

Great job everyone!!! Let's work together to reach 30% for the month of Dec."

4.1 Quantification of Benefits

Earnings from partnering the improvement for clients and Firstsource:
- Potential annualized signed off QNI for Firstsource is $94,851.
- The potential earnings contributes to 2–4% of gross margins in bottom line for Firstsource.
- Realized QNI savings of $46,608.21 (9 operational months in between the Up Selling was stopped due to operational reasons).

4.2 Replication

A module on upsell was added to the training curriculum to set expectation right from the training period. The clients have taken the solution kit and shared it with other channels e.g.: Webchat.

The learning's were shared with other programs within FSL. Project findings and solutions has been shared with team handling another leader in telecom industry as best practice sharing.

Reduction of Rework Percentage for Traditional Claims Process: A Case Study

Shereena Mody

Hinduja Global Solutions, Bangalore, India
E-mail: shereena12607@teamhgs.com

ABSTRACT: Six Sigma is an effective method to improve process performance to meet customer requirements. The current study entails application of Six Sigma methodology within Traditional Claim Process to reduce the rework percentage. Rework is defined as any claim that resulted in a change to the disposition of the original claim and required reprocessing due to error in original processing or receipt of additional information. In the present case, rework had been suitably segmented keeping in mind the causes of the same. A structured Six Sigma DMAIC methodology had been deployed using statistical techniques on critical controls, rework codes and processors, contributing to high reworked claims. Successful implementation of Six Sigma tools resulted in dramatic reduction in the number of claims being reworked. Significant financial benefit in terms of penalty saved was also realized by the management. Continued monitoring of the process was implemented to ensure sustenance of the benefits.

Keywords: Six Sigma, DMAIC, Rework Claims, Continued Monitoring, Rework Code, Development Action Plan.

1. INTRODUCTION

Total Quality Management is a novel approach in several service industries to handle the challenges posed by the competitive business world. It can be viewed as a process of embedding the quality awareness at every step of production or service. It aims at continuous improvement process by alluring their customers with high quality product at low price.

Six Sigma Philosophy: Voelkel, J.G. contends that Six Sigma blends correct management, financial and methodological elements to make improvement in process and products in ways that surpass other approaches. Mostly led by practitioners, Six Sigma has acquired a strong perspective stance with practices often being advocated as universally applicable. Six Sigma has a major impact on the quality management approach, while still is based on the fundamental methods and tools of traditional quality management (GohandXie, 2004). Six Sigma is a strategic initiative to boost profitability, increase market share and improve customer satisfaction through statistical tools that can lead to breakthrough quantum gains in quality; Mike Harry (2000). Park (1999) believes that Six Sigma is a new paradigm of management innovation for company's survival in this 21st century, which implies three things: Statistical Measurement, Management Strategy and Quality Culture. Six Sigma is a business improvement strategy used to improve profitability, to drive out waste, to improve quality.

2. PROBLEM DEFINITION AND BACKGROUND

Rework is defined as any claim that results in a change to the disposition of the original claim and requires reprocessing due to:

- An error in original processing.
- Receipt of additional information.
- Claim needs to be reprocessed due to an error identified through audit.
- Claim returned by member (Employee of an organization who has health insurance coverage through the employer) or provider (Providers are individuals, corporations, institutions, or facilities who are licensed by the government to provide medical care, services, goods and supplies to patients) e.g. Physicians, Nurses, Hospitals, and Nursing homes for reprocessing.
- Claim needs to be reprocessed due to receipt of additional information requested from claimant.

One of the parameters to gauge the performance of the team is through rework percentage. The SLA set by the client is 0.90%. Rework percentage is calculated based on the rework claims generated for the particular month against the total production achieved for that month. The objective of this project is to reduce the RE rework rate of Allentown team. Currently team has dedicated rework examiner for few bins, however, the entire team is trained on rework handling to ensure any increase in volumes are handled in a timely manner. Reduction of RE% [rework] of Allentown Team from 1.01% [July 2010] to less than 0.90% [SLA – 0.90%].

3. DMAIC METHODOLOGY

3.1 Define Phase

The Allentown National Accounts claims adjudication team adjudicates medical and hospital claims for members and providers. Claims are processed and passed through various stages. Claims incorrectly processed are routed back to processor or various bins for re-processing. Claims are routed with a rework code which signifies the reason for rework.

Examiners have to validate the RE code provided by CSR and use the appropriate code. There are few instances where CSR routes the claim with incorrect RE code and examiner need to validate before finalizing the claim. One of the key focus areas of our customer is to ensure we handle claims in the first instance (FCR – First Claim Resolution) in order to avoid rework.

3.1.1 *Operational Definition*

Claims that are returned as Rework for each individual measured against their production.

$$\text{Rework}\% = \frac{\text{No. of Claims Reworked}}{\text{Total Claims Produced}} \times 100$$

Importance of achieving SLA:

- No penalty levied
- Utilization of resources to process fresh claims
- Enhanced customer satisfaction.

Year 2010 rework data was considered as baseline for the project. A cross-functional team was formed and a project charter was developed. The target for Apr, May and Jun was not met and hence "Cost Benefit Analysis" was done which provided the base for start of the project.

Project scope: There are two categories of rework RE and RA.

RE = Rework resulting due to inefficiency within the client system [Insurance company is responsible].

RA = Rework resulting due to incomplete information provided by member/provider [Insurance company not responsible].

Increase in RE% results in customer dissatisfaction, attracts penalties and retards the possibility of organic expansions.

Project Charter, Process Flow diagram and SIPOC diagram are shown which explain the expected delivery and scope of the project.

Six Sigma Project Charter

Project Name	Reduction of Rework Percentage for Traditional Claims Process	**Project number**	BLR-GRDN-201010-030
Process Name	Polo	**Sponsor**	Soumen Nag Choudhary
Sub Process Name	ACAS	**Project Lead**	Shereena Mody
Date Chartered	**Revision**	**Project Mentor**	Nanda Kumar

Business Case	**Problem Statement**	**Purpose**
The Allentown National Accounts claims adjudication team adjudicates medical and hospital claims for members and providers. Claims are processed and passed through various stages. Claims incorrectly processed are routed back to processor or various bins for re-processing. Claims are routed with a rework code which signifies the reason for rework. Examiners have to validate the RE code provided by CSR and use the appropriate code. There are few instances where CSR routes the claim with incorrect RE code and examiner need to validate before finalizing the claim. One of the Key focus areas of our customer is to ensure we handle claims in the first instance (FCR – First Claim Resolution) in order to avoid rework.	Currently the team is not meeting the Customer defined threshold of 0.9%	**Goal Statement:** Reduction of RE% [rework] of Team from 1.01% [July 2010] to less than 0.90% [SLA – 0.90%] **Potential Benefits:** • Exceeding Client Expectations • Optimum utilization of resource • Potential new business opportunities • Reduction in Rework Penalty

Project scope Includes	**Team Members**
All current scenarios included	Shereena Mody Aravind K Prem Kumar M Somanna
Project scope Excludes	
Any major processing changes.	

Project Milestones			
Start:	End:		
	Target Date	Actual Date	Tollgate Review Date
Define			
Measure			
Analyze			
Improve			
Control			

Project CTB (Critical to Business) Details

Process CTBs/CTQs/CTDs	**Unit**	**Baseline**	**Current**	**Target**
Rework%	%	1.01%	1.01%	0.9%

Project Charter Sign-off with Date

Sponsor	Soumen Nag Choudhary
Project Lead	Shereena Mody
Mentor	Nanda Kumar

Fig. 1: Project Charter

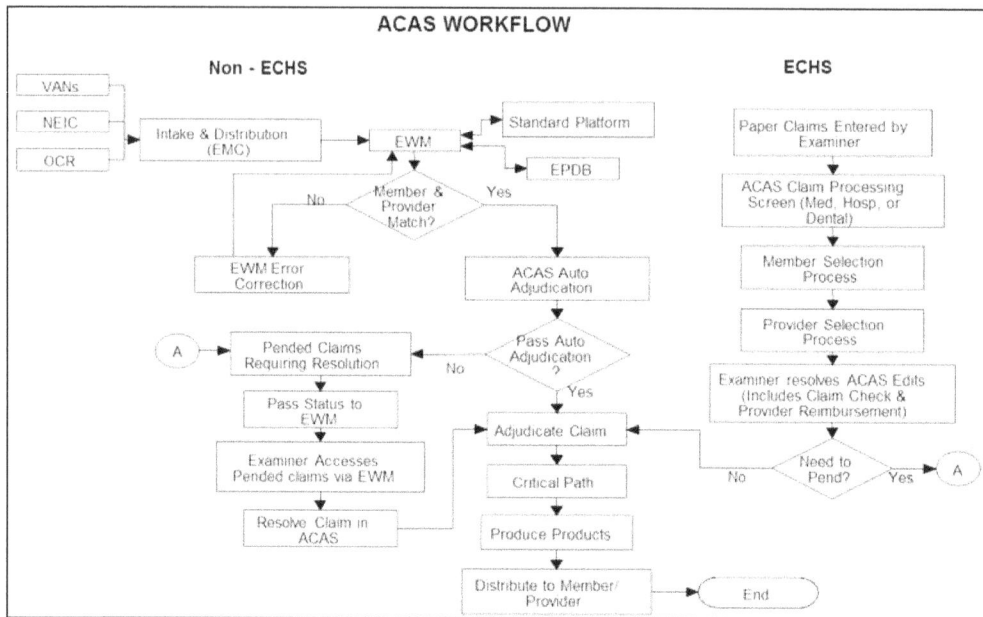

Fig. 2: Process Flow Diagram

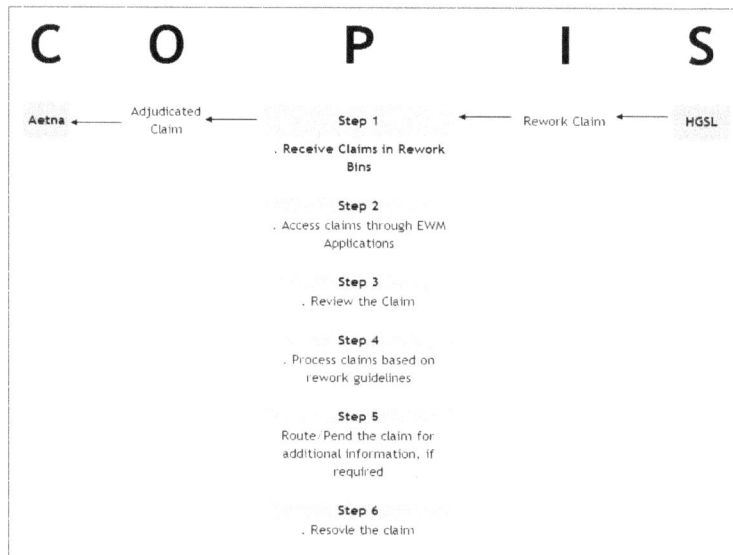

Fig. 3: COPIS Diagram

3.2 Measure Phase

Data collection was done with the help of data received from the customer for the past 6 months. The raw data had the required cuts and dimensions – Rework Reason, Rework Processor ID, Rework month, Owning Key, Control Number.

Table 1: Defects and Opportunities of Error

Sl. No.	Rework Code	Category
1	0	Unk
2	1	Accident details
3	2	Strategic Contract Manager
4	3	Assignment of Benefits
5	4	Bundling/Unbundling
6	5	AST Denied or Allowed Charges Incorrectly
7	6	Pre-Existing Condition
8	7	Date of Service
9	8	Draft/EOB/EBB
10	9	Duplicate Charges
11	10	Pend for other Insurance (COB/ Medicare)
12	11	Eligibility
13	12	Balance Billing
14	13	Financial Maintenance
15	14	Full Time Student Status
16	15	Client General Policy Provisions
17	16	Incorrect Member
18	17	Initial Request Not Resolved
19	18	CCR Instructions Not Followed
20	19	Fee Error: Non-Par
21	20	Dental X-Rays Received
22	21	EDI/Imaging Errors
23	22	Pend Request for Additional Info
24	23	PS Specific Benefit Provisions
25	24	Precert/ Referral
26	25	Incorrect Benefit Level (Pref/non-Pref)
27	26	Provider Info Missing
28	27	NOT USED
29	28	Contract not Automated
30	29	NOT USED
31	30	Unnecessary Pend/ Referral
32	31	AVP/BRF
33	32	Incorrect/ Missed Co-Pay
34	33	Denied or Allowed Charges Incorrectly (Non-AST)
35	34	Missed Medicare
36	35	Incorrect Medicare

Sl. No.	Rework Code	Category
37	36	Missed COB
38	37	Incorrect COB
39	38	NOT USED
40	39	Dental Pre-D
41	40	Orthodontic Error
42	41	Charges Incorrectly submitted, entered or missed
43	42	Contract negotiated free load error
44	43	NYHCRA Missed Payment
45	44	NYHCRA Calculated Incorrectly
46	45	New Information Received to Review Non Pended claim
47	46	Provider PIN/TIN Selection Issues
48	47	Additional Information Received
49	1A	Overpayment Vendor Discovered Rework
50	1B	Overpayment Vendor Discovered Rework
51	1C	Overpayment Vendor Discovered Rework
52	1D	Overpayment Vendor Discovered Rework
53	1E	Overpayment Vendor Discovered Rework
54	1F	Billing Elements
55	1G	Incorrect Payment Calculations
56	1H	Other Insurance
57	1I	Incorrect Payment Calculations
58	1J	Additional Information received
59	1K	Benefit Level
60	1L	Plan of Benefits
61	1M	Policy Provisions
62	1N	Policy Provisions
63	1O	Incorrect Payment Calculations

Overall a claim has 53 opportunities where an error can be charged. Number of opportunities mentioned here are number of defects that could be charged in a claim.

The Sigma level and the level of defects when the project was initiated:

DPU = 0.01009 (July 2010)

DPMO = 10089.41

Baseline Sigma level   =3.8

3.3 Analyze Phase

Cause and Effect diagram:

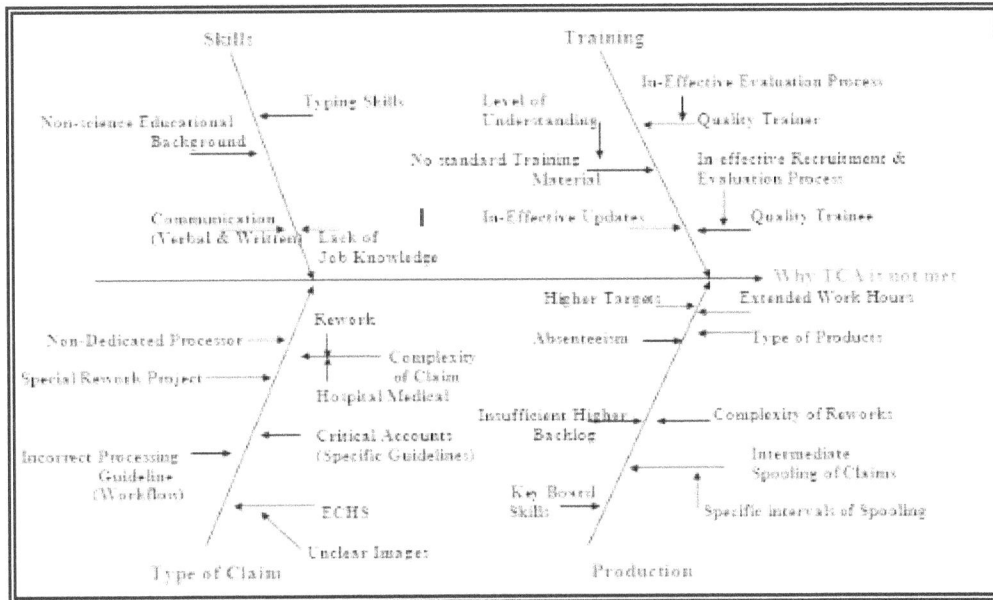

Fig. 4: Cause and Effect Diagram

The causes were identified using Cause and Effect analysis. Following analyses were done to identify the root causes for high rework:

Step 1: Selected Top-10 controls which contributed high number of rework claims.

Step 2: Selected Top-10 Rework Codes which contributed more RE claims.

Step 3: Selected Top-10 Processors who contributed more RE claim.

Step 4:   Analysis was done on the above-mentioned controls and rework codes. Examiners who were on the top of the list of rework generators were placed in a DAP (Development Action Plan) and were provided additional support. The entire team had to review the rework claim IDs generated by themselves as well as the team and come out with action plans to reduce the rework IDs in the future through brainstorming and error-sharing sessions. Team was re-trained on the coding and validation of rework claims.

During this phase, the study also revealed:

(a) High RE error rate from ECHS examiners who manually key in the fields.

(b) Many of the trainees were not aware about the validation of RE codes.

(c) Limited amount of effective monitoring and evaluation by team lead because of large team size.

(d) Increase in unnecessary pends.

(e) Issues primarily because of system enhancements.

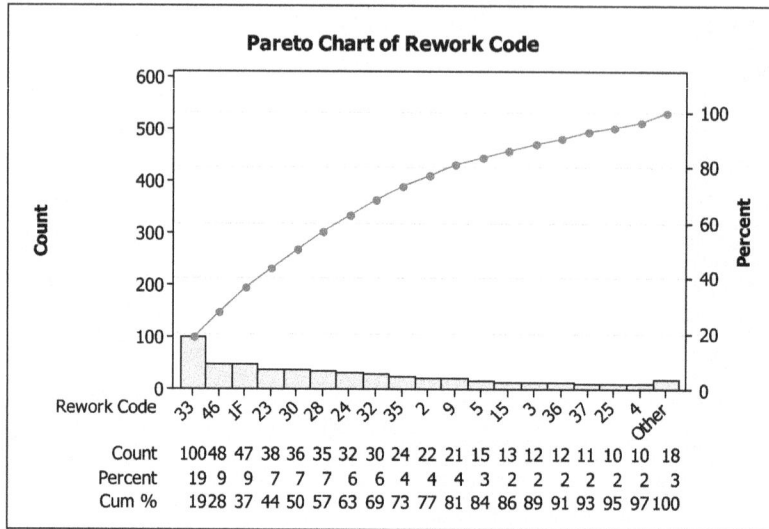

Fig. 5: Pareto Chart for Rework Code

Previous Processor ID	111111	222222	333333	444444	555555	666666	777777	888888	999999	10000	Grand Total	% Contribution
A123456	44		3	4	15	2		1	54	6	129	7.88%
B123456		106					1	8		1	116	7.08%
A123457		13	3		1		1	96		1	115	7.02%
B123457	2		65	29		2			1	3	102	6.23%
A123458		1	1	35		8			1	39	85	5.19%
B123458	1	2	2		55	1		1	13	9	84	5.13%
A123459	1	10	1	3	1	45	1	1	6	15	84	5.13%
B123459		3	1				65	8			77	4.70%
A123460		70	1				1	2	1		75	4.58%
B123460	1		4	3		43		4	4	2	61	3.72%

Fig. 6: Examiner Contribution to Top 10 Controls

Rework Reason	1E+05	222222	333333	444444	555555	666666	777777	888888	999999	10000	Grand Total	% contribution
33	20	59	19	27	26	37	18	55	43	24	328	20.02%
30	24	88	25	11	12	28	11	41	19	10	269	16.42%
46	7	20	1	14	11	11	4	22	11	35	136	8.30%
1F	14	7	7	4	6	12	9	15	9	21	104	6.35%
23	4	21	2	9	3	16	4	5	5	18	87	5.31%
32		12		8	1	7		3	6	45	82	5.01%
28	1	8	1	3	5	8	6	11	8	10	61	3.72%
2	4	16	3	1	8	5	3	9	5	5	59	3.60%
9	4	4	9	2	1	6	10	9	7	7	59	3.60%
35	4	5	12	11		5	7	7	2	1	54	3.30%

Fig. 7: Rework Claims against Control Numbers

3.4 Improve Phase

Each of the important root causes identified was addressed with a specific solution:

(a) Hot spots were created for the top 3 controls which contributed to more numbers of RE claims. Top RE examiners were advised to read the instructions about processing guidelines and update the team on a daily basis. (CCI – Coverage Card Inquiry–gives specific instructions about the plans).

(b) One-on-one counseling was done by the leads and the managers on a weekly basis and updates were provided to the team on where they stood in terms of internal audits on rework claims.

(c) External Pend Audit: Checklist was provided to examiners who were informed to follow the checklist before claims were pended. Claims pended were audited by the gatekeepers and if there were any incorrect pends, gatekeeper used to educate the concerned examiner and claims were corrected on the same day.

(d) GUP – Group Update Process was conducted, wherein the erring examiners were made to update the errors to the team in order to reinforce the correct workflow.

There was a drastic reduction in the number of claims being reworked month on month and the team was able to achieve the glide path targets from August 2010 to date. Absolute numbers were reduced from 1,473 claims (July'10) to 881 (Jan'11). The team's rework stood at 0.70% by the end of Jan 2011.

Table 2: Development Action Plans

Points of Discussion and Action Plans put in place
Reasons for increase in rework
<ul><li>Increase in unnecessary pends - Claims are incorrectly processed by the state-side team</li><li>Focus and rigor went missing due to unavailability of the data</li><li>Rushing through claims</li><li>Transitioned Accounts</li><li>Outsourcing claims to other regions</li><li>Plan setup problems</li><li>Proactive call issues leading to incorrect external pend errors</li><li>Processor skill level is not upto the requirement</li></ul>
Plans in place
<ul><li>Concentration on the top RE generators</li><li>Internal audits are conducted and feedback is been provided</li><li>GUP - Group Updates conducted- examiners with errors provide updates on errors to the group</li><li>Monitoring system in place - CNET (Claims are constantly tracked and audited)</li><li>100% External Pend Audit and developed a change in External Pend Audit approach</li><li>RE data tracked internally discussed with the team</li></ul>
Common discussion/suggestions - KMT and Ops
<ul><li>Zero tolerance on RE codes #17, #45, #46, #30</li><li>Region SPOCs from KMT need to be provided with specific responsibilities</li><li>Instead of auditing AD claims need to Audit PR claims so that examiners are updated on their errors immediately</li><li>A standard on the analysis done by KMT needs to be decided</li><li>Drill down to control level then to processor level</li><li>Once there is a SPOC assigned to each region/Site, a weekly review needs to be conducted on the progress</li><li>DAP reviews need to be conducted - Will bring the rigor back</li><li>The best way is to have KMT on the floor - one-on-one RE counseling to be restarted</li><li>Need more interaction from both ends - Ops and KMT in terms of prepay error sessions and RE counseling</li><li>Training required for rework examiners on validating the RE codes</li></ul>

Table 3: Benefits from Action Plans

Action Plan	Audience	Benefit and Purpose	Workflow
Validating RE vs RA	All Processors	– Helps in reduction of unnecessary RE coding and vice versa – This rework reason should almost always be used with RA – Non-Client Error (RA) indicates that we received information that was not available when the claim was initially handled, therefore it should not be coded as client error – If the rework truly is a client error (RE) use the more appropriate rework reason	– Processor have been educated on the rework codes – Provided the RE validation chart as a ready reference for rework code generation
RE counseling to the entire team	All Processors	– Helps in reduction of unnecessary RE coding and vice versa – This rework reason should almost always be used with RA – Non-Client Error(RA) indicates that we received information that was not available when the claim was initially handled, therefore it should not be coded as client error – If the rework truly is a client error(RE) use the more appropriate rework reason	– All RE claim ids will be traced for the respective team and SGH/Team lead will review the claim for correct coding RE vs. RA and instruct the processor on correct coding of rework – Refresher with rework example will be shared with the team on how to determine RE vs RA
Dedication of Accounts	All Processors	– Helps in avoiding incorrect benefits applied – Examiners will have specific controls to handle – Examiners will be aware of the CCI guidelines	– Weekly profile assignment to be done to the team
Job Aid for external pend, external audit	All Processors and External Pend Auditor	– Avoid unnecessary pend with action codes 067, 125, 068 – Audit all external pend eligible claims by the gate keeper	– Examiners to refer the checklist before routing claims for EP audit – Examiners to route the claims to gatekeeper for audit – Claims will be pended after EP audit only – Unnecessary pend request to be educated to processor and team
5 days TAT for WSF Pends	Plan Sponsordedicated processor	– Avoid unnecessary pend with action codes 750 – Give the time gap for the WSF forms to get uploaded	– Dedicated examiner to maintain 5 days TAT in the account – Check all the member related DCN's to find WSF
CCI Review – Top 5 controls	Dedicated processors	– Helps in avoiding incorrect benefits applied – Examiners will have specific controls to handle – Examiners will be aware of CCI guidelines	– Leads will have review with dedicated examiners on their understanding of CCI benefits

Table 4: Checks and Controls

Documentation	Checks/Controls	Frequency
RE Validation Chart	Random audit of RE validation chart availability and to check the effectiveness of the usage	Training at once
RE Claim tracker	Random audit of RE Claims for rework reason code 45 and 12 to check the effectiveness of team lead and SGH	Daily
Profile Assignment document	Random check in 1Z56 report for unassigned account processing To have backup for all the dedicated accounts Examiners to swap the account only with prior approval from leads	Weekly
External Pend Checklist andTracker	Random audit of job aid availability and to check the effectiveness of the usage Mail the external pend checklist to stateside contact	Daily
Daily TAT Mail	Lead to monitor the TAT of the account	Daily
TOU on the CCI	Monthly review of CCI, Daily RE Claims review	Monthly

Table 5: 5-Why Analysis

Error Description	Impact Contribution	Why?	Why?	Why?	Why?	Why?	Suggestion/Solution	Action Owner
RE33 – Denied or Allowed Charges Incorrectly	23%	Misinterpretation of CCI/Contract Comments	Non-standard comments in CCI Contract	Not Familiar with the comments	Unassigned profile handling	Frequent changes in the profile assignment	Dedicated profile assignment, rework validation	Leads
RE30 – Unnecessary Pend/Referral	15%	Incorrect Rework Coding	Incorrect rework coding validation	Used rework code CSR provided without validation	Not used rework processing tool	Poor knowledge of rework validation	External pend audit, Usage of rework processing tool, Group rework code validation under supervision of lead/SME	Leads
RE46 – PIN/TIN Selection Issues	7%	Incorrect Provider Selection	Missed to select appropriate provider	Details were updated after claim released	RFO not followed		Use of provider selection power tool; Follow FIFO in profile handling	Leads
RE1F – Charges Incorrectly Entered or Missed	6%	IOP Pages not viewed Properly	Contents are not legible	Ignored the content in IOP	Multiple pages submitted		Clarity of the IOP needs to be improved. Processor should spend adequate time for reviewing image	Leads
RE23 – Plan of Benefits	6%	CCI Plan Sponsor Tool not verified	Unassigned profile handling	Frequent changes in profile handling	Examiner will not be aware of the guidelines of the new account		Dedicated profile assignment, Maintaining backup system for all assignment to handle in absence of primary processor	Leads

3.5 Control

The team was able to sustain the results for 7 consecutive months with the help of specific action plans in place. Constant monitoring and updates continue to happen to ensure there are no drastic variations in the results.

The current Sigma level or the current level of defects:

DPU = 0.00701 (January 2011)

DPMO = 7010.48

Baseline Sigma level = 4.0

Table 6: Pre and Post Project Result

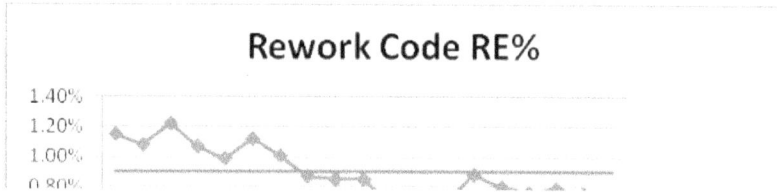

Average RE numbers before the project commenced were 1,593 and after the implementation of project are 970, a difference of 623 claims.

4. FINANCIAL BENEFITS

Penalty paid for the period Jan'10 to July'10 (before the commencement of the project) was USD 881.48 and after the implementation of the project team has had savings of USD 608!

5. CONCLUSION

The same action plans were implemented for one more team in the month of September 2010 and the process has started yielding results for the month of December '10 and January '11. Team is below the glide path (<0.90% - SLA) for 2 consecutive months and continues move with the same trajectory.

REFERENCES

[1] Cheng, Y.H. (2005), The Improvement of Assembly Efficiency of Military Product by Six- Sigma. NCUT Thesis Archive, Taiwan.

[2] Cournoyer, M.E., Renner, C.M., Lee, M.B., Kleinsteuber, J.F., Trujillo, C.M., Krieger, E.W. and Kowalczyk, C.L. (2010), Lean Six Sigma tools, Part III: Input metrics for a Glovebox Glove Integrity Program, Journal of Chemical Health and Safety, Article in press, 412, 1–10.

[3] Edgardo, J., Escalante, V. and Ricardo, A.D.P. (2006), An application of Six Sigma methodology to the manufacture of coal products, World Class Aplications of Six Sigma, 98–124.

[4] Hook, M. and Stehn, L. (2008), Lean Principles in Industrialized Housing Production: the Need for a Cultural Change, Lean Construction Journal, 20–33.

[5] Shrivastava, R.L. and Desai, T.N., "Six Sigma—A New Direction to Quality and Productivity Management", WCECS 2008, Oct. 22–24, 2008, San Francisco, USA.

[6] Shrivastava, R.L., Ahmad, K.I. and Desai, T.N., "Engine Assembly Quality Improvement using Six Sigma", WCE 2008, July 2–4, 2008, London, U.K.

[7] Wang, H., "A Review of Six Sigma Approach: Methodology, Implementation and Future Research", Zhejiang Normal University, Jinhua City, China, pp. 1–4.

Application of DMAIC Approach for Improving the Accuracy of Output from the Current Level of 89% to 98%

Moses Davala

TCS Ltd, Think Campus, Electronic City, Bangalore-560100, India

ABSTRACT: On-time and accuracy are the key metrics, indicates the process health and any deviation from the specified SLA is a defect. The accurate output means that the data has to be delivered, ideally with no correction, thereafter from end user, which is also a customer requirement.

The primary purpose of this project is to reduce the reworks occurring due to human errors and to improve the percentage accuracy of end product delivered using Define, Measure, Analyze, Improve and Control (DMAIC) approach.

1. INTRODUCTION

TCS Latam principally covers both Retail Measurement Services (RMS) and Consumer Panel Services (CPS). The objective of RMS service is the scientific measurement of various dimensions supply side of retail market in terms of sales, and share and could be used to draw inferences on the stance of a particular retailer in the retail market facilitating future business planning. On the other hand CPS service is an art of scientific measurement of the demand side of retail market in terms of consumer preferences and consumption, expenditure pattern.

TCS Latam essentially performs and delivers a total of 1500 tasks in a month. For smooth functioning, a production calendar is prepared and published on a monthly basis which is signed off by Delivery Leaders. Therefore, production time lines are fixed for various tasks.

In this sense, tasks are required to be delivered within the given Expected Completion Date (ECD) and with accuracy. As is noted tasks are time sensitive the tasks needs to be delivered with accuracy else financial risks in terms of penalties exist for delays and inaccurate data. In some cases, inaccuracy may also lead to loss of credibility in the market.

The primary purpose of this project is to apply DMAIC approach to reduce reworks by associates and improve the productivity. Results achieved are reflected in production cost savings and process improvements.

2. PURPOSE OF THE PROJECT

To complete the process, it is necessary to understand the meaning of the task and the objective of the program, but it is also necessary to review the input files. This is a Voice of Customer (VOC)

from one of our clients called Beatriz. Improve the productivity the by reducing the reworks is the agenda for the most of the organizations. By reducing rework organizations can reduce production cost, Turnaround Time (TAT), and increase productivity of the resource.

2.1 Baseline

TCS Latam performs 1500 task across four business streams (Scantrack, Retail Index, Population and CPS). Stratification has been done to find out the stream, process and countries causing maximum deviation from the specified limits i.e., process responsible for maximum reworks. Pareto Charts are constructed to identify the major contributors and given below.

Figure 1 shows that the most of the errors occurred in retail Index Process.

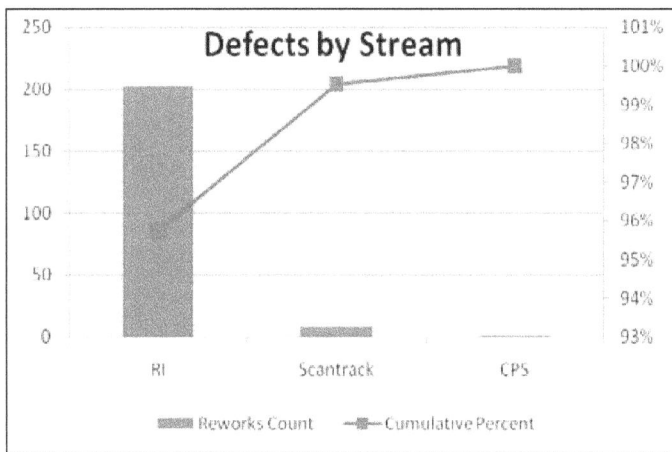

Fig. 1: Defects by Stream of Business

Figure 2 shows that the most of the errors occurred in 3.2 Process, i.e., Sample Inspection.

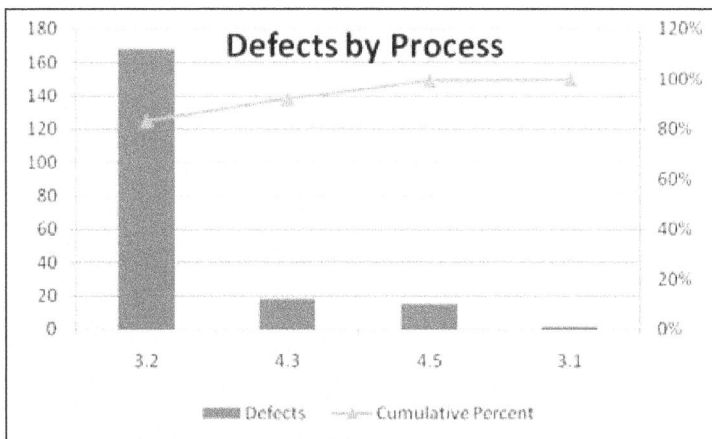

Fig. 2: Defects by Process

Figure 3 shows that the most of the errors occurred for Venezuela, Columbia, PR, Central America and DR.

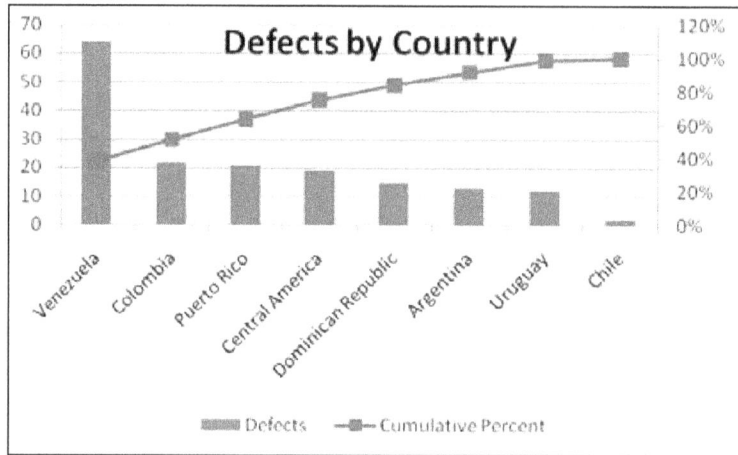

Fig. 3: Defects by Country

Base lining shows that majority of the reworks occurred in RI stream for Execute and Analyze Sample inspection process and for the above mentioned countries, the project was limited to these countries and for the process and rest of the process are out of scope.

3. THE DEFINE PHASE

The define phase of the Six Sigma methodology aims to define the improvement project in terms of customer requirements and identify the underlying process that needs improvement. The first step was to develop a project charter with all necessary details of the project including team composition and schedule for the project.

A basic flowchart of the process was prepared and a COPIS (Customer – Output –Process – Input – Supplier) mapping was carried out to have a clear understanding about the process. The team focused on honing process for improvement, which is defined as the scope of the project.

3.1 Project Details

- *Problem Statement:* Between July'11 and August'11 based on the project activity tracker, we have found that 89% projects met client expected accuracy against the target of 98%. This has been leading to client dissatisfaction as this is a measure of team efficiency.
- *Goal Statement:* To improve the accuracy to 98% by November, 2011.
- *Business Case:* Identify the source and cause of errors and reduce the reworks. This leads to improve the productivity of associate and customer satisfaction.
- *In-scope:* Activities performed by onshore team, L2 level tasks of for Execute and Analyze Sample Inspection.

- *Out-Scope:* Activities performed by offshore team, Input delay from external agencies, tasks performed by GO and DA Team and any adhoc requests.

3.2 Project Team

A cross-functional team was framed to address all needs of project. Team comprised of members from top management and middle level management. This enabled commitment and faster execution of the project.

4. MEASURE PHASE

This phase is concerned with selecting appropriate product characteristics, mapping the respective process, studying the accuracy of measurement system, making necessary measurements, recording the data, and establishing a baseline of the process capability or sigma rating for the process.

4.1 Process Mapping

A high-level process mapping for the Sample Inspection process was made as shown in Figure 4.

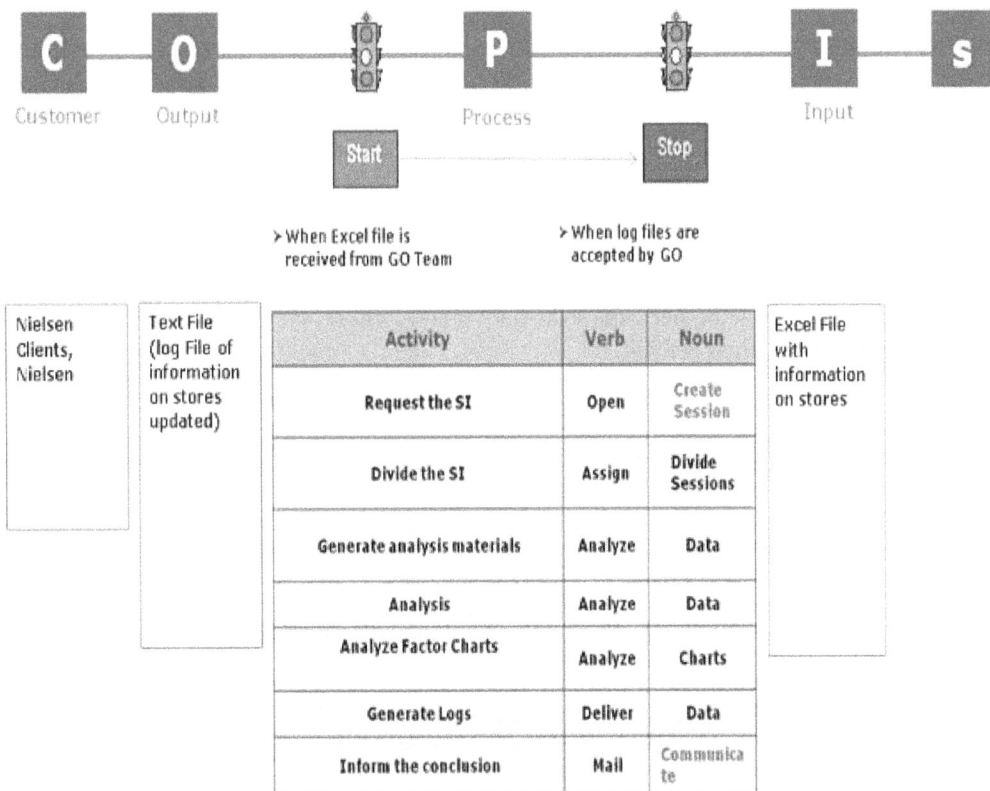

Nielsen Clients, Nielsen	Text File (log File of information on stores updated)	Activity	Verb	Noun	Excel File with information on stores
		Request the SI	Open	Create Session	
		Divide the SI	Assign	Divide Sessions	
		Generate analysis materials	Analyze	Data	
		Analysis	Analyze	Data	
		Analyze Factor Charts	Analyze	Charts	
		Generate Logs	Deliver	Data	
		Inform the conclusion	Mail	Communicate	

Fig. 4: COPIS

VOC-CTQ

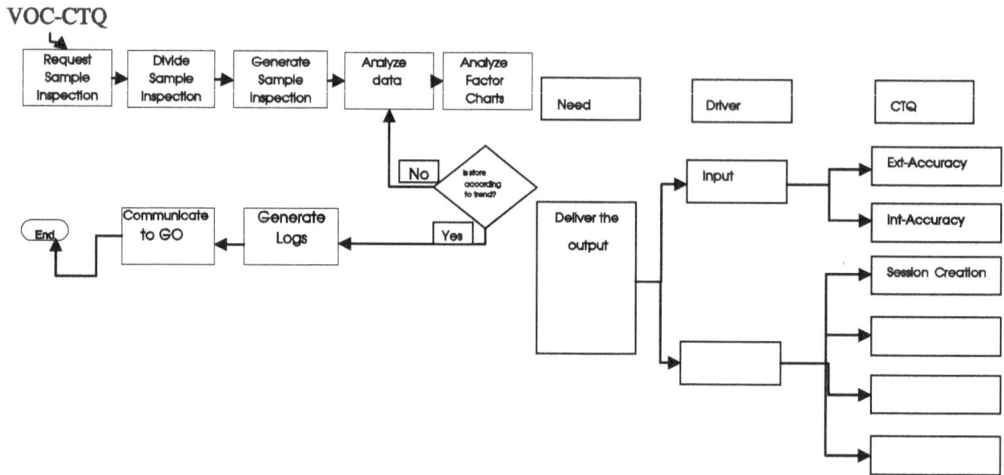

Fig. 5: VOC-CTQ

The process starts with activating Sample Inspection and creating a session for analyzing the data. The most critical part of the process is generating sample inspection material and analyzing the data. Generating the sampling inspection material involves identifying the correct market basket for the index and analyze the data involves adjusting the identified store for 'X' factor and 'Z' factor. Associate needs to adjust the store ACV values accordingly to the previous trend by monitoring the market adjustment factors.

4.2 Data Collection

Two months data (July–August 2011) was collected for process. Data collected was represented in Graph-1 and 2.

The Figure 6 below gives the per cent tasks delivered to client without reworks. The average accuracy for the given period is 89%.

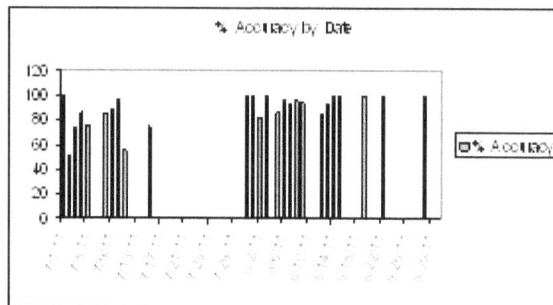

Fig. 6: Percent Accuracy by Date

The Figure 7 below gives the tasks delivered to client with and without reworks. The reworks are proportionate to the number of tasks needs to be delivered for the day.

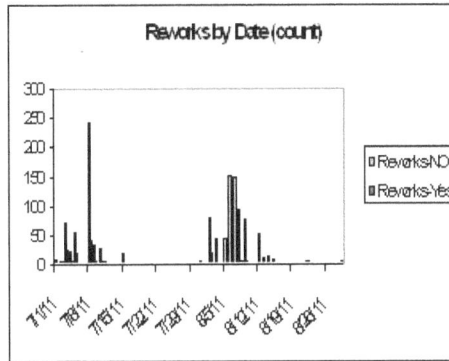

Fig. 7: Count of Reworks by Date

4.3 Cause and Effect Diagram

Figure 8 shows cause and effect diagram representing relation between cause and sub causes with corrections. Causes are collected through brainstorming and suggestions. As depicted in figure substantial number of projects are effected by Man with lack of knowledge in methods.

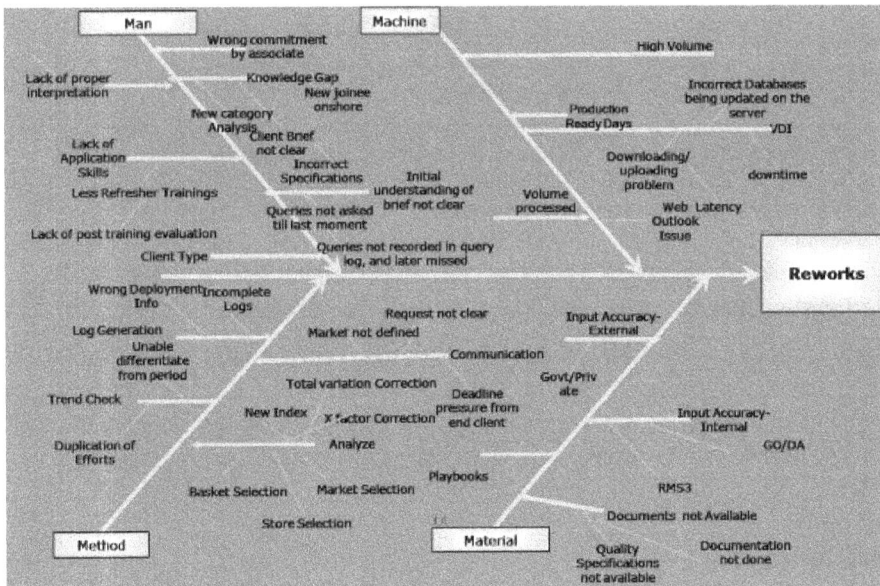

Fig. 8: Fishbone Diagram for the Process

4.4 Process Capability Analysis

Process capability analysis compares the performance of a process against its specifications. Process is capable if virtually all of the possible variable values fall within the specification limits.

The Figure 9 gives the process capability for the current process and it was obeserved that the ppm total is 985467 and corresponding Ppk is- 0.81.

Negative Ppk mean process falls outside the specification limits (because the process is producing a large proportion of defective output. In this case, process producing more number of defective output than the lower specification limit.

The Ppk value shows that the process is substandard for meeting the customer requirement. Process improvement should be given high priority and documented in a corrective action plan. Increased inspection and monitoring is required until ungoing PPk ≥ 1.33.

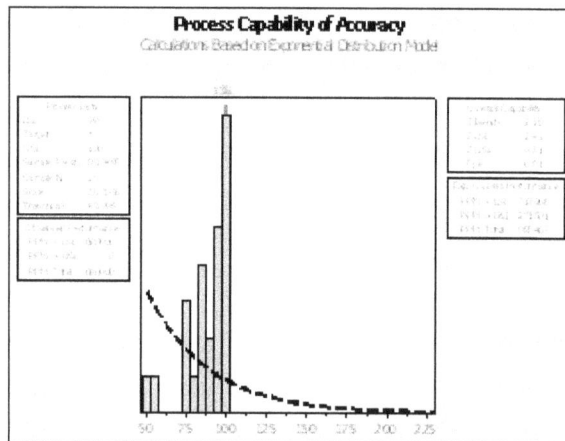

Fig. 9: Process Capability for the Current Process

5. ANALYZE PHASE

Process performance was assessed using Cause-and-Effect diagrams, to isolate key problem areas, to study the causes for the deviation from ideal performance, and to identify if there is a relationship between the variables. The probable causes that can lead to quality nonconformance in a project during different phases of a project life cycle were listed.

The Failure Modes and Effects Analysis (FMEA) was subsequently carried out to arrive at a plan for prevention of causes for failure. This helps in Identifying the potential failure modes in which a process or product may fail to meet specifications, and rating the severity of the effect on the customer. Based on the factors, a Risk Priority Number (RPN) for each failure mode is calculated.

5.1 Stratification

Stratification was conducted to identifying major root causes for reworks (Figure 10). It facilitated to identify the key reasons for reworks. The Pareto charts given below gives the major causes for reworks. Human Error – Subjectivity is the major cause for the reworks. The procedure followed by the associate to correct or update the store is incorrect and not accepted by SME then the error can be stated as subjectivity issue. Correction and updating a store is subjective and depends upon the discretion of associate. Associate should have profound knowledge on market parameters and store structure.

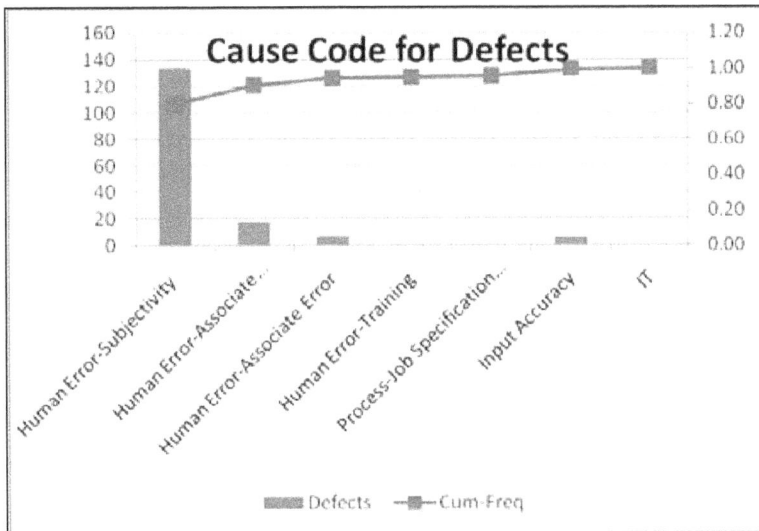

Fig. 10: Causes for Defects

5.2 Identifying the Factors Influencing Accuracy

The effects of the inputs on accuracy per cent was validated by regression analysis. Factors are identified and analyzed using stepwise regression approach. Initially following dependent (Accuracy) and independent variables (Expertise levels in SAS, Expertise levels in MS Excel, Tandem Score, KT Score, SAS Score, Stat Score, Years of Experience Educational Qualifications and Grade (job position)) were considered for the analysis. Partial Correlation Coefficient and VIF was conducted to identify the multiple collinearity.

Results shows that experience SAS (Advance level) high scores in Tandem and good scores in SAS examination (internal) have a positive and significant ($p < 0.005$) affect on delivering the tasks with minimal reworks.

Table 1: Regression Analysis

Predictor	Coeff	SE	T	P	VIF
Constant	−19.27	8.59	−2.24	0.03	
SAS	3.24	1.17	2.77	0.01	1.1
Tandem Score	1.49	0.11	14.13	0	1.3
SAS Score	0.31	0.09	3.51	0	1.3

S = 3.37437 R-Sq = 90.1% R-Sq(adj) = 89.0%

Table 2: ANOVA Analysis

Source	DF	SS	MS	F	P
Regression	3	2804.5	934.83	82.1	0
Error	27	307.43	11.39		
Total	30	3111.94			

5.3 Identifying the Factors Affecting Reworks

An attempt was made to identify the factors influencing the reworks. Delivering index without reworks was considered as dependent variable and First time running the process, subjectivity, years of experience, input accuracy, associate grade, new dex, communication gap (Spanish) and frequency of index (monthly/bimonthly) are considered as independent variables. Among all, variables having significant impact were selected for the final model.

Logistic Regression Table 3 shows the estimated coefficients, standard error of the coefficients, z- values, and p-values.

Table 3: Estimates of the Reworks

Predictor	Coef	SE Coef	Z	P	Odds Ratio
Constant	−2.58	1.52	1.69	0.09	
Subjectivity	1.66	0.65	2.56	0.01	5.27
Grade	−0.76	0.39	1.97	0.05	0.47
First_Time	1.33	0.66	2.03	0.04	3.78

The grade is having negative affect indicating that as grade increases the failure rate increases. This might be due to additional responsibilities has to handled by the associate.

5.4 Validation of Causes and Solutions

The details of validated of all causes is summarized and is given in the following Table 4

Table 4: Causes and Validation Methods

Cause	Validation Method	Conclusion
First time Performer	Logistic Regression	Root Cause
Subjectivity	Logistic Regression	Root Cause
Years of experience	Linear /Multiple Linear Regression	Not a Root Cause
Input accuracy	Logistic Regression	Not a Root Cause
Associate grade	Linear /Multiple Linear Regression	Root Cause
New Index	Logistic Regression	Not a Root Cause
Communication gap (Spanish)	Logistic Regression	Not a Root Cause
Frequency of index	Logistic Regression	Not a Root Cause
SAS Expertise	Multiple Linear Regression	Root Cause
Excel Expertise	Multiple Linear Regression	Not a Root Cause
Tandem Score	Multiple Linear Regression	Root Cause
KT Score	Multiple Linear Regression	Not a Root Cause
SAS Score	Multiple Linear Regression	Root Cause
Statistics Score	Multiple Linear Regression	Not a Root Cause
Educational Qualifications	Multiple Linear Regression	Not a Root Cause

Table 5: Validated Causes and Solutions

Cause	Solution
First time Performer	Stringent QC before delivery to Customer
Subjectivity	QC document in place before production start date duly signed by SME
Associate grade	Reschedule the additional responsibilities during production period
SAS Expertise	Train the Associates
Tandem Score	Retrain the Associates
SAS Score	Train the Associates

6. IMPROVE PHASE

During this phase of a Six Sigma project, solutions were identified for all root causes selected during the analyse phase, implemented them after studying the risk involved in implementation and results were observed. The team had detailed discussions involving all the stakeholders of the process and solutions were identified for all selected root causes. The solutions identified are presented in Table 5. An implementation plan was prepared for all solutions with responsibility and target date for completion for each solution. The solutions were implemented as per the plan and results were observed. The data on number of reworks are were collected from the process after the project. The process capability evaluation was done (Figure 12).

The PPM level of the process was zero and Ppk > 1.33. The corresponding sigma rating was six (Table 6). A dot plot (Figure 12) was made for comparing the process before and after the project, which shows significant reduction reworks after the process improvement this project. A control charts was introduced for monitoring the process along with an out of control action plan. This helps the team lead or delivery manager to take action on the process in case of assignable causes occur. Training was provided for the people working with the process about the improved operational methods so that their confidence level in working with the new process increases.

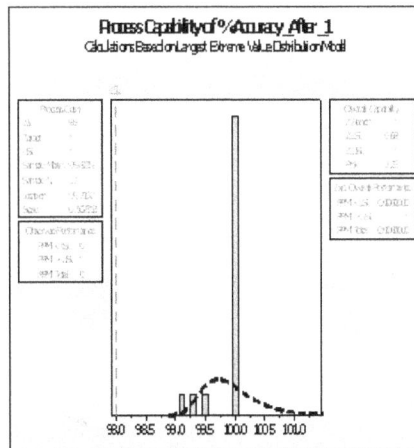

Fig. 11: Process Capability for Accuracy afs.ter Process Improvement

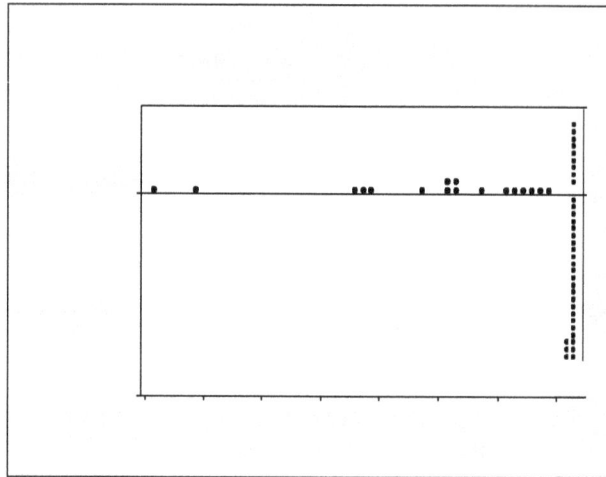

Fig. 12: Dot Plot for Accuracy–Before and after Process Improvement

6.1 Savings

As a result of this study, the first pass yield has improved from 89% to 100%. The team with the help of finance department estimated the tangible savings due to this project. It was found that as a result of this project, the cost associated with reworks has come down drastically. The finacial benefits are.

No. of times process reworked = 31.5

Rework cycle time per process = 8hrs. Total rework time saved = 31.5 man days Average salary per day = $ 96

Saving per month = 96 × 31.5 =$ 3024

Total Saving per annum = $ 36288

$ 36,288 saving due to reduction in reworks.

Table 6: Comparison of Results Before and After

Statistic	Before	After
Sigma Level	2.06	6
DPMO	19607	0
Yield	89%	100%

7. CONTROL PHASE

The real challenge of Six Sigma implementation is not in making improvements in the process but in sustaining the achieved results. The process improvements that were introduced resulted in the reduction of reworks. The process capability for quality of deliverables improved from 2.06 σ to 6 σ. DPMO reduced to zero.

The best practices and lessons learnt in this Six Sigma project for were applied in other project teams and other types of projects Since, the field errors reduced and the process capability for quality deliverables increased to more than five sigma. Since, the time sensitive nature of the process, changes were documented in the Procedures of Quality Management System (IQMS).

The thrust on Six Sigma Quality has helped in creating and sustaining customer focus in the TCS-Latam, leading to improved customer satisfaction as indicated in the feedback from the customer. At the same time, active participation of the team members from all levels in the Six Sigma projects has evolved a culture of effective and creative team work. The goal is to achieve Six Sigma level which is currently at 6σ. The same process will be extended to on-time delivery in near future.

ACKNOWLEDGEMENTS

I would like to thank Indian Statistical Institute, Bangalore, Boby and all the faculties for their in valuable advice and also I would like to thank Subhasis Samantaray, Delivery Head, TCS, Latam Bangalore, India for supporting and guiding throughout the project. My Sincere thanks to all the team members who assisted me by providing valuable inputs.

Author Index

A

Acharya, U.H. 1, 62
Achouri, Ali 52
Akkinapalli, Vidya Sagar 62
Anantharaman, Senthil 142

B

Balasubramanyam, K. 151
Basak, Ishita 328
Bhatia, Satvinder Singh 342
Bhavana, Surapaneni 90

C

Chakraborty, Ashis K. 203, 302, 328
Chatterjee, Moutushi 203
Coelho, Joyson Peter 377
Corera, A.M. Romesh Kumar 394

D

Davala, Moses 430
Deora, Saraswathi 246
Dewal, Sushant Mohan 313
Duraibalan, Arun 80

G

Ghosh, Satadal 302
Gouri, Sudipta 130
Gunasekaran, N. 355
Gupta, V. 174

H

Hudli, Anand V. 222
Hudli, Shrihari A. 222

I

Inoue, Shinji 213

J

John, Boby 1
Joseph, Jones 105

Joseph, Mary 117

K

Kaur, Manwinder 342
Khanuja, Gurupreet Singh 403
Kumar, Lalit 5
Kumar, Prashant 388

L

Lakshmikanthan, P.R. 31
Lal, Arvind Kumar 342

M

Mallya, Yogish 388
Mital, Neena 31
Mody, Shereena 417
Mukhopadhyay, Arup Ranjan 366
Mukhopadhyay, Chiranjit 229

N

NagaRaju, O. 288
Nailwal, Beena 252
Nataraj, M. 355
Neelufur 70

P

Parthiban, P. 90

R

Reddy, Akepati Sivarami 342
Roy, Soumya 229, 302

S

Sabari, Gireesh 105
Samanta, Ranjan 105
Sasun, C. 313
Schoepf, Gerhard 15
Shanmugaraja, M. 355
Singh, S.B. 252
Singh, S.R. 174
Subhani, S.M. 187

Sujir, Sitaram Vikram 161
Suneetha, S.P.V.N.D. 288
Suresh, P.K. 117

T
Taleb, Hassen 52
Tiwari, Richa 90

V
Valsa, B. 5

W
Wade, Vijay 97

Y
Yamada, Shigeru 213

Z
Zubar, H. Abdul 90